Dust Explosion Dynamics

Dust Explosion Dynamics

Russell A. Ogle

Exponent, Inc.
Warrenville, IL, United States

AMSTERDAM • BOSTON • HEIDELBERG • LONDON
NEW YORK • OXFORD • PARIS • SAN DIEGO
SAN FRANCISCO • SINGAPORE • SYDNEY • TOKYO
Butterworth-Heinemann is an imprint of Elsevier

Butterworth-Heinemann is an imprint of Elsevier
The Boulevard, Langford Lane, Kidlington, Oxford OX5 1GB, United Kingdom
50 Hampshire Street, 5th Floor, Cambridge, MA 02139, United States

Notices
Knowledge and best practice in this field are constantly changing. As new research and experience
broaden our understanding, changes in research methods, professional practices, or medical treatment
may become necessary.

Practitioners and researchers must always rely on their own experience and knowledge in evaluating
and using any information, methods, compounds, or experiments described herein. In using such
information or methods they should be mindful of their own safety and the safety of others, including
parties for whom they have a professional responsibility.

To the fullest extent of the law, neither the Publisher nor the authors, contributors, or editors, assume
any liability for any injury and/or damage to persons or property as a matter of products liability,
negligence or otherwise, or from any use or operation of any methods, products, instructions, or ideas
contained in the material herein.

British Library Cataloguing-in-Publication Data
A catalogue record for this book is available from the British Library

Library of Congress Cataloging-in-Publication Data
A catalog record for this book is available from the Library of Congress

ISBN: 978-0-12-803771-3

For Information on all Butterworth-Heinemann publications
visit our website at https://www.elsevier.com

  Working together
to grow libraries in
developing countries

www.elsevier.com • www.bookaid.org

Publisher: Joe Hayton
Acquisition Editor: Fiona Geraghty
Editorial Project Manager: Maria Convey
Production Project Manager: Laura Jackson
Designer: Maria Inês Cruz

Typeset by MPS Limited, Chennai, India

To Donna,

"...I love you just the way you are."

—Billy Joel, The Stranger (1977)

Contents

List of Figures

List of Tables

Nomenclature

ROMAN LETTERS

A	Arrhenius pre-exponential factor
A_v	Specific surface area
a_g	Speed of sound in a gas
a_R	Rosseland mean absorptivity
B	Spalding's transfer number
BM	Bradley–Mitcheson model
b_{FT}	Shvab–Zeldovich variable based on fuel mass fraction and enthalpy values
b_{OxT}	Shvab–Zeldovich variable based on oxidizer mass fraction and enthalpy values
b_{FOx}	Shvab–Zeldovich variable based on fuel and oxidizer mass fraction values
C	Heat capacity of condensed phase material
C_D	Drag coefficient
C_i	Molar concentration of species i
C_L	Local dust concentration in a flammability tube, mass units
C_p	Heat capacity at constant pressure
C_v	Heat capacity at constant volume
c	Speed of light
Da	Damkohler number
d	Diameter or distance
d_n	nth moment of the particle size distribution
d_p	Particle diameter
$\overline{d_p}$	Mean particle diameter
d_{quench}	Flame quenching distance
d_{32}	Sauter mean diameter
E	Expansion ratio, $\mathcal{M}_u T_b / \mathcal{M}_b T_u$
E_a	Activation energy
E_{ign}	Ignition energy
F_{f-0}	Radiation view factor from the flame to the ambient environment
$F(d_p)$	Cumulative distribution function for a particle size distribution
F/A	Fuel–air ratio
$(F/A)_{st}$	Stoichiometric fuel–air ratio
f	Fraction of fuel mass burnt
$f(d_p)$	Frequency distribution function for a particle size distribution
G	Incident radiation on solid surface
H	Height
H_i	Enthalpy of species i

h	Convective heat transfer coefficient, Planck's constant
h_i	Specific enthalpy of species i
HHV	Higher heating value
$I_{\lambda\omega}$	Radiation intensity per unit wavelength per unit solid angle
J	Variable for gravitational settling in the transition regime
k	Thermal conductivity, Boltzmann's constant, kinetic rate constant
k_c	Kinetic rate constant
k_g	Film mass transfer coefficient
k_0	Overall rate coefficient
K_{of}	Burn rate coefficient
K_{st}	Dust explosibility parameter, $K_{st} = \left(dP/dt\right)_{max} \cdot V^{1/3}$
L	Characteristic length
Le	Lewis number
LFL	Lower flammable limit, mass concentration
LFL_{up}	Lower flammable limit, upward flame propagation, mass concentration
LFL_{down}	Lower flammable limit, downward flame propagation, mass concentration
LHV	Lower heating value
LvE	Lewis—von Elbe model
Ma	Mach number
m	Total mass, complex index of refraction
m_i	Mass of species i
$\dot{m}_i$	Mass flow rate of species i
$\mathcal{M}_i$	Molar mass of species i
NCV	Nagy—Conn—Verakis model
Nu	Nusselt number
n_i	Moles of species i
n_p	Number concentration of particles
P	Pressure
P_{max}	Dust explosibility parameter, maximum explosion pressure
Q	Heat interaction; enthalpy of combustion per unit mass of fuel
Q_a	Dimensionless absorption efficiency coefficient
Q_e	Dimensionless extinction efficiency coefficient
Q_s	Dimensionless scattering efficiency coefficient
q_b	Black body hemispherical emission
$q_{b\lambda}$	Black body spectral radiant heat flux
q_x'''	Volumetric heat interaction
Pr	Prandtl number
Re	Reynolds number
R_g	Universal gas constant $= 8314.5$ J/(kmol-K)
r	Radial coordinate in cylindrical or spherical geometry
r_c	Instantaneous radius of particle, radius of dust cloud
r_f, r_{flame}	Radius of flame surrounding a fuel particle
r_s	Particle surface radius

r_w	Radius of spherical vessel
r_0	Initial radius of particle
S_b	Flame speed of a premixed flame
S_u	Laminar burning velocity of a premixed flame
s	Path length for radiative intensity
Sc	Schmidt number
Sh	Sherwood number
T	Temperature
t	Time
t_{wave}	Time scale for a deflagration wave
$t_{1/2}$	Half-life
U_i	Internal energy of species i
u_i	Specific internal energy of species i
V	Volume
V_b	Volume of burnt mixture
v	Velocity
$v_{p,TSV}$	Terminal settling velocity for a single particle
v_{sm}	Smolder wave velocity
v_x	Velocity component in the x-direction
$\bar{v}_x$	Time averaged velocity component in the x-direction
v'_x	Instantaneous velocity fluctuation in the x-direction
W	Work interaction
X_i	Mole fraction of species i
x	Spatial coordinate direction, size parameter for Mie scattering theory
x_B	Fractional conversion of species B
$Y_{F,U}$	Mass fraction of fuel, unburnt condition
Y_i	Mass fraction of species i
Ze	Zeldovich number

GREEK SYMBOLS

α	Absorptivity in thermal radiation
α_{fill}	Volume fraction of enclosure filled with a combustible dust cloud
α_i	Volume fraction of phase i
α_T	Thermal diffusivity
β	Density ratio for gas-solid reaction in Chapter 6
β	Defined constant in Chapter 9, $\beta = (\gamma + 1)/(\gamma - 1)$
β_λ	Extinction coefficient
γ	Ratio of heat capacities
δ	Premixed flame thickness, Frank−Kamenetskii number
δ_d	Thickness of accumulated dust layer
δ_{sm}	Smolder wave thickness

ϵ_0	Turbulent dissipation rate
ε	Emissivity
κ	Flame curvature
λ	Wavelength of electromagnetic radiation
θ	Dimensionless temperature, flame cone angle on Bunsen burner flame
μ	Dynamic viscosity
υ_i	Specific volume of species i
Π_i	Dimensionless ratio of type i
ρ	Mass density
ρ	Reflectivity in thermal radiation
ρ_i	Mass density of chemical species i
ρ_d	Mass concentration of dust−air mixture
ρ_{mix}	Mixture density
ρ_s	Solid phase density
σ	Stefan−Boltzmann constant $= 5.67 \times 10^{-8}$ W/m^2-K^4
σ	Standard deviation
τ	Optical thickness, transmissivity in thermal radiation, burnout time for a single particle
τ_{cloud}	Burnout time for a dust cloud
τ_x	Time scale for process x
τ_0	Shear stress at bounding wall or surface in Chapter 9
Φ	Equivalence ratio
ϕ	Velocity potential in acoustics
φ	Mass ratio of carbon to oxygen for the stoichiometric condition
ω	Solid angle
ω_i'''	Chemical reaction rate per unit volume for species i

SUPERSCRIPTS

0	Standard state for energy value
g	Gas phase
m	Mixture
s	Solid phase
$'$(prime)	Shock-fixed coordinate system
+	Dimensionless quantity for turbulent boundary layer

SUBSCRIPTS

act	Actual value
ad	Adiabatic value at constant pressure
b	Burnt mixture

CJ   Chapman–Jouget detonation
c   Combustion
d   Dust
ex   Explosion, adiabatic value at constant volume
fg   Liquid–vapor phase transition
i Species i
in   Inflow of material
mix   Mixture value
of   one-film combustion model
out   Outflow of material
p   particle
ref   Reference state
s   Solid
stoich   Stoichiometric value
tf   two-film combustion model
th   Threshold value in Chapter 9
u   Unburnt mixture
x Coordinate direction
0   Initial condition
∞ Free stream property in Chapter 9

SPECIAL SYMBOLS

$\hat{A}$ Molar value of quantity A
$\dot{A}$ Flow rate of quantity A
$\hat{A}$ Dimensionless value of A
$\mathcal{A}_{\lambda\omega}$ Radiative absorption per unit wavelength per unit solid angle
$\mathcal{D}_e$ Effective diffusivity in a porous medium
$\mathcal{D}_i$ Diffusivity of species i
$\mathbb{E}$ Oxygen to fuel mass ratio
$\mathcal{F}_{drag}$ Aerodynamic drag force
$\mathcal{E}_{\lambda\omega}$ Radiative emission per unit wavelength per unit solid angle
$\mathcal{L}_{flow}$ Characteristic length scale of flow
$\mathcal{L}_{int}$ Integral microscale
$\mathcal{L}_k$ Kolmogorov microscale
$\mathcal{L}_M$ Markstein length
ℓ_p Interparticle spacing in dust cloud
$\mathcal{M}_i$ Molar mass of species i
s Stoichiometric coefficient, mass of oxidizer per unit mass of fuel

Preface

Combustible dust is both an explosion and a fire hazard. Over the last 300 years, coal mines, grain elevators, and numerous manufacturing facilities have experienced catastrophic accidents resulting in thousands of deaths and injuries and destroying hundreds of structures. The investigation and control of combustible dust hazards has been the subject of intense effort between two primary groups: industrial safety experts and research scientists. The industrial safety experts have concentrated on evaluating the magnitude of the hazard for a given dust and developed various methods for protection. The research scientists have devoted their efforts to understanding the fundamental physical and chemical factors underlying combustible dust behavior. Both groups have produced a large volume of literature that suits the needs of their peers, but there seems to be little interaction or communication between these two groups. I believe that industrial safety experts could benefit greatly from the insights and knowledge produced by combustion scientists. My goal is for this book to help bridge that gap by demonstrating how a deeper understanding of combustible dust hazards can be achieved through combustion science.

Dust explosion dynamics is the investigation of the physical and chemical phenomena that govern dust explosions and related combustible dust hazards. This text explores the use of fundamental principles of combustion science to evaluate the magnitude of combustible dust hazards in a variety of settings. Models are developed to describe dust combustion phenomena using the principles of thermodynamics, fluid dynamics, transport phenomena, and chemical kinetics. Simple, tractable models are described and compared with experimental data when possible. The purpose of these simple models is to give the reader insight into the complex phenomena involved in dust explosions. Later, more sophisticated models are introduced to help the reader gain an appreciation for the challenges yet to be tackled.

Some excellent books have been written on dust explosions. But for the most part, books on dust explosions emphasize the industrial safety perspective. The serious student of combustible dust hazards should consult the following texts:

- Amyotte, P. *An Introduction to Dust Explosions.*
- Bartknecht, W. *Dust Explosions: Course, Prevention, Protection.*
- Barton, K. *Dust Explosion Prevention and Protection.*
- CCPS/AIChE. *Guidelines for Safe Handling of Powders and Bulk Solids.*
- Cheremisinoff, N.P. *Dust Explosion and Fire Prevention Handbook: A Guide to Good Industry Practices.*
- Eckhoff, R. *Dust Explosions in the Process Industries,* Third Edition.
- Frank, W.L. and S.A. Rodgers. *NFPA Guide to Combustible Dusts.*
- Nagy, J. and Verakis, H.C. *Development and Control of Dust Explosions.*

The texts cited above are largely devoted to the prevention of and protection from dust explosions. Three of these books do take a more fundamental approach. The book by Dr. Amyotte offers many conceptual insights into the physical and chemical phenomena involved in combustible dust hazards. The book by Dr. Eckhoff is especially noteworthy and in that he has done a masterful job of summarizing both the pragmatic industrial investigations and the scientific research on combustible dust. The book by Mssrs. Nagy and Verakis summarizes the extensive work performed by the U.S. Bureau of Mines and does a particularly good job of describing and deriving integral models for explosion development and venting. But none of these books provides the foundation in combustion science that can assist the reader interested in gaining a deeper understanding of the scientific research on combustible dusts.

This book is different in that it focuses on exploring the combustion science underlying combustible dust hazards. Using simple models and detailed examples, it demonstrates how the fundamental principles of combustion science can be applied to better understand dust explosion phenomena. The intent of the book is to eliminate some of the empiricism used to explain dust explosions and to give the reader insight into the physical and chemical phenomena involved. To accomplish this goal, I will introduce you to just enough combustion science so that you can read, interpret, and use the scientific literature on combustible dusts.

One of the elements of hazard analysis is the evaluation of hazard severity or consequences. This book is about consequence analysis for combustible dusts. Consequence analysis uses mathematical models to make predictions about the severity of hazardous outcomes. There are four primary hazards of combustible dust: smoldering, flash fires, dust explosions, and flame acceleration effects (shock waves or detonations). The consequences of these hazards are addressed in this book through mathematical models based on combustion science. The models developed herein are simplified versions of more reliable, and strongly empirical, methods that are presented in the safety literature. It is my hope that the insight derived from simple models will give the reader a better appreciation for the strengths and limitations of the more empirical design methods. There are a number of worked examples throughout the text. These are mostly straightforward calculation exercises designed to give the reader an appreciation for the relative magnitude of physical quantities important to dust explosion dynamics.

The plan of this book reflects my philosophy for mathematical model building: start with a simple model, challenge it with empirical data, and improve the model as guided by the data. Thus, we will begin with models based solely on thermodynamics and neglect finite rate processes. Next finite rate processes are introduced through formulations using a burning velocity for the dust flame. Models based on computational fluid dynamics (CFD) will complete the presentation. CFD models offer the opportunity to introduce important concepts from multiphase flow, turbulence, thermal radiation, and chemical kinetics.

There is no question that CFD models are the most powerful and comprehensive formulations, but they are also the most complex to use. CFD models impose a burden on the user to specify a number of model inputs that must either be measured or be estimated. Many CFD codes are proprietary and can be expensive to purchase or lease, and are therefore not always readily available to the analyst. The emphasis on simpler models offers an opportunity to extract physical insight with less investment of resources. Similarly, in addition to the current literature, the reader will notice many citations to the older technical literature. In many of these older citations, one can find clever conceptual approaches and analytical techniques which offer deep insight into dust explosion phenomena. With the onset of robust numerical methods and fast, yet inexpensive, digital computers, one finds that much of the current combustion literature turns immediately to CFD simulations. While CFD is a powerful tool, it requires a great deal more effort to develop results and may not be available to all analysts interested in dust fires and explosions.

This book is written at a level suitable for both students and practicing professionals in the engineering and scientific fields. My objective is to introduce the reader to specific concepts from combustion science that will lead to deeper insight into the behavior of combustible dusts. I hope you will not be deterred by the over 600 equations in this book. There is a large number of equations because I have worked out many derivations in detail so that the you, the reader, will not spend endless hours trying to retrace the footsteps of so many combustion scientists before you. I have purposefully omitted many topics from combustion science because they are not germane to my immediate goals. For the reader desiring a broader introduction to combustion phenomena, I recommend the following books:

- Glassman, I., Yetter, R.A., and Glumac, N.G. *Combustion,* Fifth Edition.
- Kuo, K. *Principles of Combustion,* Second Edition.
- Law, C.K. *Combustion Physics.*
- Linan, A. and Williams, F.A. *Fundamental Aspects of Combustion.*
- McAllister, S., Chen, J-Y., and Fernandez-Pello, A.C. *Fundamentals of Combustion Processes.*
- Turns, S.R. *An Introduction to Combustion,* Third Edition.
- Williams, F.A. *Combustion Theory,* Second Edition.
- Zeldovich, Ya.B., Barenblatt, G.I., Librovich, V.B., and Makhviladze, G.M. *The Mathematical Theory of Combustion and Explosions.*

A brief word about the references cited in this book. I have made no attempt to comprehensively review the scientific literature on combustible dust. I have been selective and have focused mostly on papers, reports, and books which I have found useful in my consulting practice. This is not to say that the omitted works are of an inferior quality, far from it. Eckhoff has presented an excellent survey of the scientific literature in his text on dust explosions. For the interested reader seeking a broader, more balanced introduction to the scientific literature, I strongly recommend his *Dust Explosions in the Process Industries,* Third Edition.

It is my hope that you will find this book useful in your efforts to manage combustible dust hazards. If you have any comments, criticisms, suggestions, or if you discover any typographical errors, I will be most grateful if you would send me an email at dustexplosiondynamics@gmail.com.

LIABILITY DISCLAIMER

Although the information in this book is believed to be factual and accurate, no warranty or representation, expressed or implied, is made with respect to any or all of the content thereof and no legal responsibility is assumed therefore. The equations, methods, and examples used are for illustrative purposes only and may not be suitable for engineering design. The reader assumes all liability in the application of the material in this book.

Acknowledgments

In some ways, this book has been 30 years in the making. My doctoral research project, published in 1986, was an investigation of aluminum dust explosions with a 20-L sphere. I have been interested in the combustion science of dust explosions ever since. Fiona Geraghty, acquisitions editor at Elsevier, opened the door for me to begin this project. Maria Convey and Laura Jackson at Elsevier have guided me through the writing and production process with both grace and patience. I am grateful to Fiona, Maria, and Laura for all of their efforts.

I would like to thank my attorney, Karen Shatzkin (Shatzkin & Mayer, New York), for her professional guidance. I thank my colleague, Timothy Myers, who voiced his support and encouragement on this book project from its inception. I have benefited greatly from the support of my friends and colleagues at Exponent over the years, and I am especially grateful for the guidance and encouragement I have received from my mentor and supervisor Harri Kytomaa. I must also acknowledge the technical assistance of Brenton Cox, Sean Dee, and Stephen Garner, my colleagues at Exponent. They checked my mathematical derivations, corrected errors, and suggested a number of lucid edits to the text. With their assistance, I have made every effort to eliminate errors in the text. Any errors that remain are mine alone.

I am grateful to my parents, Jim and Dorothy, for their love and guidance throughout the years. While writing this book my adult children-Laura (and her husband, Scott), Andrew, and Peter-provided some desperately needed distraction and encouragement.

And finally, I thank my wife, Donna, for her love and support. She makes all good things possible.

Introduction to combustible dust hazards

1

Finely divided solid matter plays an irreplaceable role in modern technology. It has been estimated that three-quarters of all raw materials used in the chemical and process industries and half its products, on a weight basis, are particulate solids. At least 1% of the world's electrical power is consumed in crushing and grinding operations. Particulate solids are an eminently useful form of solid matter. They are especially well suited for storage, handling, transportation, and processing operations. However, simply converting a bulk solid material into a particulate material introduces new hazards. For example, a large block of quartz is relatively inert. But as a finely divided solid (sand) poured into a silo, the quartz becomes an engulfment hazard to personnel who enter the silo.

Any oxidizable material with a sufficiently small particle size is, under the right circumstances, potentially capable of combustion. In general, the finer the particle size, the faster the rate of combustion and thus the greater the hazard. Easily oxidized, finely divided material is called *combustible dust*.[1] Combustible dust presents three types of combustion hazards: smoldering fires, flash fires, and dust explosions. A dust explosion is the most severe of these hazards.

A dust explosion begins when a cloud of combustible dust is ignited. The ignition event can be thought of as a ball of flame. The flame creates a high temperature volume of gas which tends to expand into the unburnt portion of the dust cloud. As the surface of the flame expands it consumes fuel creating a larger volume of combustion products. The continued expansion of the burnt gases accelerates the flame causing it to consume fuel faster and faster. If the dust cloud is confined in a vessel or structure, the high temperature of the combustion products leads to a rise in pressure. If the rising pressure exceeds the strength of the confinement, the vessel or structure will rupture allowing the high pressure gas to expand into the surroundings.

Dust explosion dynamics is the study of the physical and chemical phenomena that govern the fire and explosion hazards of combustible dust. This book is written for the student or the expert who is interested in seeing how simple combustion models can yield insight into the physical and chemical phenomena that govern combustible dust hazards. From the perspective of safety risk assessment, these simple models will help the investigator gain a deeper understanding of the consequences of combustible dust hazards. However, this is not a book devoted

---

[1]This definition is very informal and should not be confused with definitions for combustible dust that are based on government regulations, safety standards, or specific industry guidelines.

to the important subjects of explosion protection, safety standards, or risk assessment. These topics are covered in several books cited in the references.

This book assumes the reader has taken undergraduate level courses in thermodynamics, fluid dynamics, heat and mass transfer, and some chemical kinetics. These are the foundation subjects for combustion. I will introduce certain fundamentals and key results from combustion science as needed to develop the elements of dust fire and explosion dynamics. Citations to combustion textbooks and the literature are given for additional reading and investigation. The combustion behavior of flammable gases and vapors tends to occupy the majority of attention in combustion texts. The unique hazards of combustible dusts garner less attention in these books.

To gain an appreciation for the science behind combustible dust hazards, a brief history of dust explosion research will be presented from the perspective of combustion science. Some basic aspects of combustion behavior are introduced and key terms are defined. I then establish seven important differences between combustible dusts and flammable gases. These differences will become the basis for organizing many of the empirical observations and combustion models to be discussed later in this book.

The goal in studying the combustion behavior of combustible dust is to better manage its hazards. Safety management has three basic components: hazard identification, evaluation, and control. The first two components, identification and evaluation, are based on laboratory testing. In a very brief fashion, I next define the common parameters used to evaluate combustible dust hazards and describe the typical test methods for characterizing combustible dusts. The task of hazard control requires some sense of how the hazards of combustible dust can reveal themselves in the real world. I describe some common hazard scenarios to provide a conceptual framework for thinking about these hazards. Finally, some of the basic dust explosion prevention and control strategies are summarized. These sections are exceptionally brief as there are many good books and review papers available for further study.

1.1 HISTORICAL PERSPECTIVE ON COMBUSTIBLE DUST HAZARDS RESEARCH

When speaking of an awareness of combustible dust hazards, there are two groups to consider: the technical community and the industrial community. Unfortunately, the knowledge of combustible dust hazards has been slow to diffuse beyond specialists in both communities. Even today, combustible dust accidents surprise unsuspecting victims. This may be due in part to the fact that dust explosions are (thankfully) a rare occurrence when compared to the incidence rate of similar hazards like structural fires.

There have been several information campaigns in the 20th century to educate the industrial community about combustible dust hazards, and there have been further efforts in the 21st. These information campaigns have been accompanied by a tremendous growth of technical information regarding the properties and hazards of combustible dust. But while the recognition of combustible dust hazards spans almost three centuries, the understanding of dust explosion phenomena based on the application of fundamental scientific principles is a more recent development. As often happens with the investigation of new phenomena, progress was made on several fronts in a somewhat nonlinear fashion. Occasionally, these independent efforts would intersect, and at other times they diverged. Coal and agricultural grains proved to be the early focal points of investigation. Eckhoff has presented a very comprehensive literature review on all aspects of dust explosion research (Eckhoff, 2003). This brief historical review focuses on the development of the conceptual underpinnings of dust explosion phenomena.

The awareness of combustible dust hazards in the technical community has been traced back to the late 18th century with a publication by the Italian natural philosopher Count Morozzo (Morozzo, 1795). Morozzo's description of a grain dust explosion at a bakery in Turin, Italy, included the identification of the key elements necessary for a dust explosion: finely divided fuel, creation of a dust suspension in air, confinement of the dust cloud, and the presence of an ignition source. In the same publication, Morozzo discussed examples of chemical reactivity hazards (eg, mixing nitric acid with "fat oils") and self-heating and spontaneous ignition involving oil-impregnated sail cloth, linen rags, and hay stacks. However, not all of the article was founded on reliable empirical observation as he also described the "phenomenon" of spontaneous human combustion! Notwithstanding this minor criticism, Morozzo offered perhaps the first description of the five factors for a dust explosion, later identified as the dust explosion pentagon.

The industrial revolution created a new demand for coal. As coal mining activity increased, so did the number of fatal coal mine explosions. These accidents were typically documented in news media accounts, but were rarely investigated by technical experts. An exception to this was the investigation of the Haswell Colliery explosion of 1844 by the chemist Michael Faraday and geologist Charles Lyell (Faraday and Lyell, 1845). The realization that coal dust was explosive was a hotly debated topic (Verakis and Nagy, 1987). While it was generally accepted that fire-damp (methane gas) was an explosion hazard in coal mines, Faraday and Lyell demonstrated the role of coal dust as a fuel for an explosion. The Faraday—Lyell effort was a classic example of the scientific investigation of an accidental explosion. Their inference was based on the observation of coal dust deposited along walls and supports in the direction of flame propagation, and the determination that this deposited coal dust had the appearance of coke (pyrolyzed coal). Their report drew many insights from the stoichiometry of combustion.

The National Fire Protection Association was formed in 1896, reflecting a growing need for fire protection in an increasingly urbanized environment. The fire protection literature of the early 1900s revealed a basic awareness of dust explosion phenomenon. Ingle and Ingle wrote a textbook for fire insurance professionals based largely on descriptive chemistry, stoichiometry, and thermochemistry (Ingle and Ingle, 1900). In their text, they present a discussion of coal combustion and indicate the potential explosion hazard of coal dust and, by analogy, other combustible dusts such as cotton, grain, and "inflammable liquids" (Ingle and Ingle, 1900, pp. 45−57). They even give a cogent description of a secondary dust explosion.

Schwartz wrote a similar textbook and gave a similar description of dust explosion behavior (Schwartz, 1901). Schwartz went a step further, however, and gave recommendations for preventing dust explosions. These recommendations included the use of ventilation, pneumatic dust collection, dust-tight equipment enclosures, and the use of safety lamps (Schwartz, 1901, pp. 57−62). Several chapters in Schwartz's book were devoted to various specific combustible dusts including sugar, grain, paper, wood, coal, and fodder (animal feed). Schwartz repeatedly warned of the danger of creating dust suspensions. These texts revealed an understanding of combustion phenomena based on stoichiometry and thermodynamics.

A detailed account of these early coal dust explosion investigations can be found in the research report by George Rice of the U.S. Geological Survey (Rice, 1910). The report by Rice et al. sets out a systematic presentation of empirical test data generated by both the US and European investigators. The report also discussed data trends in explosion behavior based on particle size, volatile matter, and humidity. One of the more impressive aspects of this research was the performance of full-scale dust explosion tests in mines.

As the USGS increased their commitment to dust explosion research, the U.S. Department of Agriculture also began to assume a role in dust explosion research based on their concerns of combustible grain dust hazards. Price and Brown published two monographs summarizing the state of knowledge of grain dust hazards (Price and Brown, 1920, 1921). There was a great deal of qualitative discussion in this monograph on dust explosion prevention, but the conceptual basis was entirely empirical. In 1922, NFPA formed its committee on dust explosion hazards. The NFPA committee developed safety standards for dust explosion prevention. Also, in 1957 they published a compendium of over a 1000 dust explosion accidents in the United States which spanned the time frame from 1860 to 1956 (NFPA, 1957).

In the period 1930−1950, the state of combustion science took a dramatic leap forward. The scientific needs of the military provided a significant motivation for this growth. Modern developments in fluid mechanics, heat transfer, mass transfer, and chemical kinetics experienced significant growth and integration during this period (Emmons, 1980/81). The growth in sophistication of combustion science was demonstrated in many fields including dust explosion research. The U.S. Bureau of Mines was a significant contributor to numerous aspects of dust explosion phenomena and published many compendia of dust explosion test results (see references in Chapter 1 of Nagy and Verakis, 1983). Many of these studies were of a fundamental nature and ranged from single particle combustion, stationary flames, flame propagation and pressure development in test vessels, dust cloud ignition, and explosion venting. These fundamental studies demonstrated the value and potential benefits of dust explosion research founded on the principles of the new science of combustion.

A series of four fatal grain elevator explosions in a 1-week period in 1977 spurred renewed interest in dust explosion fundamentals (NAS, 1978, pp. 2−3). Researchers embarked on new studies on the role of turbulence and shock wave phenomena in dust explosion dynamics. Combustion science began to leave its imprint on dust explosion research. Much of the literature that has been published since the 1970s has continued to focus on empirical studies regarding dust explosion phenomena and on explosion prevention and protection. A more recent impetus to dust explosion safety came in 2006 with the publication of a combustible dust hazard investigation report by the U.S. Chemical Safety and Hazard Investigation Board (CSB, 2006).

Eckhoff's book (Eckhoff, 2003) is a comprehensive survey of the combustion science-oriented dust explosion literature through 2003, and he has updated this survey in a review paper published in 2009 (Eckhoff, 2009). A great deal of recent activity has focused on better characterization of combustible dust testing equipment at the laboratory scale. Fundamental studies have examined the contributions of thermal radiation, turbulence, and chemical kinetics to dust flame propagation using both asymptotic methods and computational fluid dynamics.

1.2 AN INTRODUCTION TO COMBUSTION PHENOMENA

There is an enormous body of literature dedicated to the study of combustion. In this section, some of the more basic concepts and phenomena are introduced and defined. A good basic introduction to these concepts can be found in Crowl's book (Crowl, 2003). Several combustion textbooks provide a more comprehensive discussion of these concepts (Drysdale, 1997; Glassman, 1997; Kuo, 2005; Law, 2006; Turns, 2012).

Combustion is the rapid, self-sustaining, oxidation of a fuel that generates heat, light, or smoke. The fuel may be a gas, liquid, or solid. Combustion generally occurs in the gas phase. So if the fuel is a condensed phase, then some of the heat liberated by the combustion reaction must be consumed in vaporizing or pyrolyzing the fuel. The term *pyrolysis* refers to the thermal degradation of a chemical. Pyrolysis is an endothermic process and is a common feature in the combustion of most solid carbonaceous fuels. When you overheat bread in a toaster to the point that the bread chars, you are observing the process of pyrolysis.

There are two types of combustion: flaming and smoldering. *Flaming* combustion refers to the formation of a hot column of gaseous combustion products that gives off visible light and sometimes smoke. Flaming combustion can be observed with flammable gases, flammable vapors, and combustible solids. *Smoldering* combustion is flameless combustion. It is often accompanied by glowing char and the emission of smoke. Smoldering only occurs with solid fuels which form porous solid chars.

The combustion reactions occur in a narrow region called the *combustion wave* (the term narrow region is relative to a characteristic length dimension of the fuel body). The combustion wave may be stationary or it may propagate through the unburnt mixture. At steady state, the combustion wave velocity is a characteristic parameter and is a function of chemical reaction rates and transport processes. There are two types of combustion waves in flammable gas systems: deflagration waves and detonation waves. The basic structure of a combustion wave is depicted in Fig. 1.1.

A *flame* is a combustion wave that occurs in a gaseous state with a burning velocity that is less than the speed of sound in the unreacted mixture. The characteristic velocity of a flame is called the *burning velocity*. If the flame is stationary, the burning velocity refers to the velocity of unburnt mixture feeding the flame. Flames that move through the unburnt mixture are called *deflagrations*. In the common usage of the combustible dust literature, a deflagration is envisioned to occur in an enclosure. Thus, the heat release results in a rise in pressure within the enclosure. This phenomenon is called a *dust explosion*. If the enclosure is sufficiently large, the heat release does not cause a rise in pressure; this phenomenon is called a *flash fire*.

Detonations are combustion waves with a speed greater than the sound speed of the unreacted medium. A dust detonation is a particularly hazardous event because, once it occurs, it causes an overpressure that is much greater than the strength of most industrial buildings. A deflagration-to-detonation transition (DDT) is a specific physical situation where a propagating flame accelerates rapidly and becomes a detonation (Crowl, 2003, pp. 203−204). In an accident scenario, a detonation is unlikely to occur spontaneously. It usually requires a DDT event.

There are two types of flame systems: *premixed* and *diffusion* (nonpremixed). In a premixed system, the fuel and oxidizer are mixed prior to combustion. The combustion waves described as deflagrations and detonations are premixed

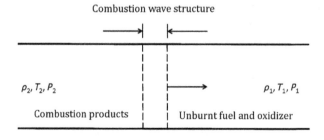

Combustion wave structure

ρ_2, T_2, P_2

Combustion products

ρ_1, T_1, P_1

Unburnt fuel and oxidizer

FIGURE 1.1

Depiction of the structure of a combustion wave.

systems. Flammable gases, flammable vapors, and combustible dust clouds can all form premixed flames. Examples of premixed flames include a Bunsen burner, the burner in a gas-fired furnace, and a dust explosion.

In diffusion flame systems, the fuel and oxidizer are not mixed prior to combustion. The fuel and oxidizer are separated and move toward each other through a combination of flow and diffusion. They meet at a flame surface or flame sheet. Under certain specific circumstances, flammable gases can form diffusion flame systems. Condensed fuels like flammable liquids and combustible solids almost always form diffusion flame systems. When a flammable liquid burns, the fuel must vaporize and diffuse into the flame sheet. When solid fuels burn, fuel vapor is formed by melting and vaporization (eg, candle wax) or the fuel vapor is formed by pyrolysis (eg, wood). Both candle flames and wood-burning campfires are diffusion flames. The combustion of single solid particles is another example of diffusion flames. Thus a burning dust cloud exhibits both types of flames: a diffusion flame at the single particle scale and a premixed flame at the dust cloud scale.

The initiation of combustion is called *ignition*. Ignition is dependent on a number of factors, and as a practical matter the determination of ignition characteristics of a fuel is dependent on the test apparatus and conditions. Ignition can be characterized by an energy input (eg, an electric spark) or by an elevated temperature (eg, a hot surface). Although simple in concept, ignition is difficult to analyze theoretically. It can also be difficult to measure, and there are a large number of standardized laboratory apparatuses and procedures available to the investigator.

A flammable gas can mix with air to form a flammable mixture. Similarly, a combustible dust can mix with air to form a flammable mixture. With gaseous mixtures, there is a range of flammability over which combustion can be initiated. This range is characterized by a *lower and upper limit of flammability* (designated as LFL and UFL, respectively; these limits are sometimes referred to as the lower and upper explosive limits). The determination of flammability limits requires specialized laboratory apparatus and procedures. The specific results obtained are, to a degree, dependent on the magnitude of the ignition source used. Combustible dust exhibits flammability ranges similar to flammable gases. But an important distinction is that combustible dusts do not exhibit a definite upper explosive limit. There are some plausible explanations for this observation and will be discussed later in the book.

Another important concentration that defines combustion behavior is the *stoichiometric concentration*. At the stoichiometric concentration, the concentration of fuel is exactly matched to the oxygen concentration required for complete combustion. Unlike flammability limits, the stoichiometric concentration is theoretical parameter whose value is independent of any test apparatus. It has been found empirically that the maximum burning velocity or maximum explosion pressure for flammable gas mixtures is typically measured at or slightly greater than the stoichiometric concentration. Because of its both theoretical and practical significance, the stoichiometric concentration is an important combustion parameter.

1.3 SEVEN KEY DIFFERENCES BETWEEN COMBUSTIBLE DUSTS AND FLAMMABLE GASES

Whether one considers a combustible dust or a flammable gas, a fire or explosion hazard exists when the fuel is dispersed in air at an ignitable concentration. While there are important similarities between combustible dusts and flammable gases, there are some very important differences as well (Nagy and Verakis, 1983, Chapter 3). Many of these features are related to the simple fact that a flammable gas can form a molecular mixture with air. Combustible dusts are solid particles and so must undergo a physical or chemical transformation to allow the mixing of fuel and air at a molecular level. Alternatively, in some cases, the oxygen in the air must be able to diffuse to and chemically react at the solid particle surface. The solid, discrete nature of the fuel particles imposes an additional constraint on the reacting system, slowing down the combustion process when compared with gaseous fuels. The seven key distinguishing differences between combustible dusts and flammable gases are the following:

- Necessary conditions for a deflagration
- Chemical purity of fuel
- Particle size and shape
- Uniformity of fuel concentration and initial turbulence
- Range of ignitable fuel concentrations
- Heterogeneous and homogeneous chemical reactions
- Incomplete combustion.

Each of these factors is discussed in turn.

1.3.1 NECESSARY CONDITIONS FOR A DUST DEFLAGRATION

In both hazard evaluation and in accident investigation, it is helpful to conceptualize the necessary conditions for a deflagration. With flammable gases and vapors, these conditions are symbolized as the fire *triangle*: fuel, oxidizer, and an ignition source make up the three sides of the triangle. For a combustible dust flash fire, a fourth condition is necessary: the dust (fuel) must be suspended creating a dust concentration within the ignitable range. This is often symbolized as the combustible dust flash fire *square*. A dust explosion requires one more condition beyond the flash fire: containment of the dust cloud such that the combustion process causes an increase in the pressure of the enclosure. Since a dust explosion requires five necessary conditions, it is symbolized as a *pentagon* (Fig. 1.2).

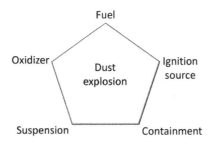

FIGURE 1.2

Combustible dust pentagon.

Another combustible dust hazard is *smoldering*. Smoldering is a combustion wave that travels through a fixed bed of combustible particulate solids. This fixed bed of dust may take the form of a thin layer deposited on horizontal surfaces or it may be a pile of material. Smoldering velocities travel at much slower rates than dust deflagrations. However, if disturbed to create a dust suspension, a smoldering wave can act as an ignition source resulting in a dust explosion. Thus, combustible dusts are a hazard in both the dispersed state and the undisturbed state. The four necessary conditions for smoldering can be symbolized as a square, though these are not the same four conditions required for a combustible dust flash fire. Smoldering requires fuel, oxidizer, an ignition source, and a minimum (or critical) thickness for the deposit. The fuel must satisfy two criteria: it must be porous and form a solid char (Drysdale, 1997, p. 275). The ignition source may be external (independent of the fuel) or internal (self-heating caused by oxidation). The minimum thickness of the deposit arises due to the requirement for thermal insulation, ie, the deposit must be able to trap the chemical energy released by oxidation causing a temperature rise in the deposit mass.

1.3.2 CHEMICAL PURITY OF FUEL

With modern analytical chemistry, it is usually a straightforward task to determine the chemical composition of a flammable gas or vapor. Once the chemical species has been identified, its properties, including its flammability characteristics, can be measured or estimated. The task may not be simple, but even in industrial settings it is usually possible to establish the chemical composition of a hazardous atmosphere.

The same is not true for dusts. Many combustible dusts have the potential to exhibit a surface chemical composition that is different from the composition of their interior. Dust particles are efficient adsorbents and can acquire foreign chemical species (additional chemical compounds that are not of the same composition as the combustible dust) from the atmosphere or their surroundings (Adamson and Gast, 1997, p. 259). These contaminants have the potential to alter the properties of the dust.

Probably, the two most common contaminants come from the ambient environment: air and water. For example, metals and their alloys will frequently form oxide surface coatings due to exposure to air. Many organic dusts have the ability to adsorb—and in some cases absorb—moisture from the atmosphere. The surface composition of a particulate material can have a significant effect on its ignition properties and perhaps, to a lesser extent, its combustion flame speed.

This is illustrated in the work of Benard et al. in which they investigated the effect of both the extent of the oxide coating on aluminum particles and the amount of water trapped under the oxide coating (Benard et al., 2012). They found that while increasing the thickness of the oxide coating reduced the ignition energy and flame velocity of dust clouds, the water content under the coating had the opposite effect. Thus, surface composition of the particles had a nonlinear effect: the intermediate oxide levels were more easily ignited and had higher flame velocities because of the effect of water on aluminum combustion.

Finally, many manufactured particulate materials are coated with additives to improve the storage stability or flow characteristics of the bulk material. The impact of these additives will depend on their chemical composition and their relative thickness compared to the native particle.

1.3.3 PARTICLE SIZE AND SHAPE

Particle size and shape are a primary determinant of explosion severity. The rate of combustion of particulate materials is dependent on the surface area available for mass transfer. The finer the particles, the greater the specific surface area. The specific surface area of a spherical particle can be calculated as shown in Eq. (1.1):

$$A_v = \frac{\pi d_p^2}{\frac{\pi}{6} d_p^3} = \frac{6}{d_p} \tag{1.1}$$

where A_v is the specific surface area of the dust (m^{-1}) and d_p is the mean particle diameter (m). For a given combustible dust, finer particles are easier to ignite and burn more rapidly (Ogle et al., 1988). For fibrous particulate materials, the traditional diameter determined by sieve analysis may be less useful for characterization and instead the thinnest dimension of the fiber will be a better descriptor since that is the geometric feature that will control combustion behavior (Amyotte and Eckhoff, 2010).

Particle shape plays a similar role; as the particle shape deviates from a perfect sphere, the specific surface area increases. If the particle melts during the combustion process, it will revert to a spherical shape. For nonmelting solids, particle shape may play a role due to its incremental increase of the particle's specific surface area. Typically, the impact of particle shape may be viewed as a second-order effect compared to the dominant role of the particle size.

The particle size distribution of a given particulate material can shift toward smaller particle sizes—and higher rates of combustion—due to attrition and particle breakage. An accurate assessment of the particle size distribution for a given material is further complicated by the difficulty of obtaining representative samples of the material. These issues are discussed further in the next chapter.

1.3.4 UNIFORMITY OF FUEL CONCENTRATION AND INITIAL TURBULENCE

Flammable gases and vapors can mix with air and achieve a uniform concentration through forced convection, natural convection, and diffusion. These processes are time dependent but, in general, the degree of mixing of fuel with air increases with time. Once the fuel concentration has entered the flammable range, the combustion hazard persists even in a quiescent environment.

The opposite is true for combustible dusts. A pile of dust will remain a pile of dust until it is disturbed. The formation of an ignitable dust cloud requires either a mechanical disturbance of the dust pile or entrainment by fluid convection. The resulting dust cloud will usually be spatially nonuniform in dust concentration. If the disturbance or source of entrainment is removed, the dust will settle out of suspension. Thus, the dust cloud concentration is transient; the longer the observation time, the lower the dust concentration until eventually it becomes a dust deposit.

In the laboratory, it is very difficult to create a dust cloud with a spatially uniform dust concentration. It seems reasonable to infer then that in an accident setting a spatially uniform dust cloud would be an exceptional event. And yet in most theoretical investigations of dust explosion behavior, it is generally assumed that the dust dispersion exhibits a uniform dust concentration. This observation is not a criticism; it is a caution. Throughout this text, when discussing dust flames I will be assuming the cloud has a uniform dust concentration. This ideal case is a very important benchmark for theoretical study. But it is also an approximation.

1.3.5 RANGE OF IGNITABLE FUEL CONCENTRATIONS

Flammability limits have been determined for a wide range of flammable gases and vapors. The limits are determined with standardized laboratory instruments and procedures (Britton et al., 2005). The results have good comparability between different laboratories. A useful empirical observation is that the LFL is approximately one-half of the stoichiometric concentration and the UFL is three times the stoichiometric concentration (Glassman, 1997, p. 163).

The situation for combustible dusts is more complicated. While the LFL (for dusts this is called the minimum explosible concentration or MEC) can be determined with some confidence, the UFL (maximum explosible concentration) is a more difficult matter, and investigators generally agree that a well-defined

UFL does not exist for dusts (Cashdollar, 2000; Myers and Ibarreta, 2013). The determination of ignitable dust concentrations is strongly dependent on the dust dispersion process.

1.3.6 HETEROGENEOUS AND HOMOGENEOUS CHEMICAL REACTIONS

Over the past 80 years, a great deal has been learned about the chemical reactions involved in the combustion of flammable gases and vapors (Lewis and von Elbe, 1987, Chapters II, III, and IV; Glassman, 1997, Chapter 3; Law, 2006, Chapter 3; Kuo, 2005, Chapter 2). These chemical reactions all occur in the gas phase and are termed homogeneous reactions. Consider flame propagation through a cloud of flammable liquid droplets: the liquid must first vaporize and then the fuel vapor must mix with the surrounding air to form a flammable vapor concentration that can support combustion. The combustion reactions are homogeneous reactions. Surface chemical reactions (chemical reactions influenced by collisions with vessel walls) tend to play a minor role.[2]

The burning of combustible dusts will typically involve a network of homogeneous and heterogeneous reactions. Consider a cloud of carbonaceous particles. The carbonaceous solids undergo pyrolysis creating a flow of fuel vapor from the solid particle into the surroundings. Flame propagation can occur when the fuel vapor, called pyrolysate, mixes with the surrounding air forming a flammable concentration. The pyrolysis reactions leave behind a solid char particle. Depending on the oxygen concentration, the char particle can undergo oxidation. This is a heterogeneous reaction. Heterogeneous reactions involve reactions at a fluid–solid interface. Evidence of surface oxidation reactions has been documented in single particle combustion experiments with both char-forming materials (eg, coal and wood) and metals (Eckhoff, 2003, pp. 251–265).

1.3.7 INCOMPLETE COMBUSTION

The final aspect distinguishing combustible dust from flammable vapors is the dominant occurrence of incomplete combustion. With flammable gases and vapors in the flammable range from the LFL to the stoichiometric concentration, combustion tends to go to completion. The fuel is consumed and oxygen is depleted in accordance with the stoichiometric requirements. With a flammable gas or vapor in a premixed system, the mixture forms a gaseous "solution" at the molecular level. There are no finite rates of interphase mass transfer slowing down the combustion process. Confined deflagration tests with flammable gases typically find the maximum explosion pressure to occur at or just beyond the stoichiometric concentration.

---

[2]I am neglecting the occurrence of elementary reactions that involve collisions of radicals with the wall of the containing vessel. These interactions can be especially important in the determination of flammability limits.

Combustible dust, on the other hand, tends to burn incompletely during flame propagation (Nagy and Verakis, 1983, pp. 39−41). This is due to the slower rates of transport and kinetics. In combustible dusts, the combustion process at the particle scale requires several steps in series: heating of the particle, pyrolysis of the particle forming fuel vapor, diffusion of the fuel vapor into the surroundings, and finally the chemical reactions between the fuel molecules and oxygen. Each of these steps occurs at its own finite rate. Thus, for the same combustion test configuration, the mass burning rates of flammable gases or vapors will be faster than the mass burning rate for a combustible dust.

The slower mass burning rate of combustible dusts means as a deflagration wave passes through a dust cloud more fuel is left behind. With combustible dusts, the maximum explosion pressure often occurs at concentrations that are at least two to four times greater than the stoichiometric concentration (Nagy and Verakis, 1983, p. 39; Eckhoff, 2003, p. 35).

1.4 COMBUSTIBLE DUST HAZARD PARAMETERS

The key parameters used to describe combustible dust hazards have two distinct and related characteristics. They are rooted in fundamental combustion behavior but are measured using standardized equipment operated to satisfy consensus industry standards. The key parameters are identified in this section and are related to fundamental concepts in combustion science. The measurement of these parameters is discussed in the following section. The key parameters fall into three categories: explosion parameters, concentration limits, and ignition parameters.

The explosion parameters are the explosion pressure and the maximum rate of pressure rise. These parameters are typically measured in a vessel with the intent of achieving ignition in the center of a uniform dust cloud. The explosion pressure is the final pressure attained in a test vessel following a dust deflagration. The explosion pressure is one measure of the potential severity of an explosion. The higher the explosion pressure, the greater the damage imparted to the surroundings. The explosion pressure is directly related to the average flame temperature in the test vessel.[3] Since a real test vessel will normally experience some heat loss, the average flame temperature inferred from an explosion pressure is related to, but not the same as the constant volume adiabatic flame temperature. As an ideal, however, the constant volume adiabatic flame temperature is a thermodynamic parameter and is a useful point of reference. In theory, the explosion pressure recorded in a test vessel should be independent of the vessel size or shape. In practice, it has been found that the test vessel must satisfy certain

---

[3]By average flame temperature, I mean the mass-averaged temperature of the combustion products. The procedure for calculating this average is explained in Chapter 3, Thermodynamics of Dust Combustion.

volume and shape constraints in order to obtain empirically measured explosion pressures that match approximately the constant volume explosion pressure calculated by thermodynamics.

The maximum rate of pressure rise is directly related to the mass burning rate of fuel. The maximum rate of pressure rise is very sensitive to the size and shape of the test vessel. At a practical level, the size and shape of the vessel have an important effect on both the dispersion of the dust in the vessel and heat losses to the vessel wall. The methods of analysis of the pressure rise data vary depending on the model chosen to describe the combustion process. An important feature of the maximum rate of pressure rise is that, to a first approximation, the test data obtained from spherical vessels with central ignition can be scaled with the simple relationship

$$K_{st} = \left(\frac{dP}{dt}\right)_{max} * V^{1/3} \qquad (1.2)$$

where K_{st} is the combustible dust parameter, $\left(\frac{dP}{dt}\right)_{max}$ is the maximum rate of pressure rise measured in the test vessel (bar/s), and V is the volume of the test vessel (m^3). This relationship holds reasonably well for spheres with a volume of at least 20 L. A great deal of explosion protection design methods has been based on K_{st} data (Frank and Rodgers, 2012, pp. 201−220). However, to be useful it is important that K_{st} data are obtained using appropriate equipment and procedures (as discussed in the following section) (Eckhoff, 2003, pp. 340−341).

The chemical energy content of a dust cloud is directly proportional to the dust (fuel) concentration. There are three important dust concentration parameters: the stoichiometric concentration, the minimum explosible concentration, and the maximum explosible concentration. The stoichiometric concentration is a theoretical parameter determined by a mass balance assuming complete combustion of the fuel. It is the concentration of fuel that will (at least in theory) consume all of the oxygen in the test chamber. It is an important theoretical parameter because it serves as a reliable point of reference when describing the fuel load for any given combustible dust hazard scenario. In the absence of empirical test data, it can be predicted to a first approximation that a combustible dust cloud will achieve its highest explosion pressure and its maximum burning rate above its stoichiometric concentration.

The minimum and maximum explosible concentrations have conceptual merit, but their values depend on how they are obtained. They can be calculated using a theoretical model or the measured using a specific test method and laboratory apparatus. The minimum and maximum explosible concentrations represent the extreme values of dust concentration that will support combustion. The minimum explosible concentration (MEC) is the dust cloud concentration that can just support the propagation of a flame. A smaller concentration results in the failure of the dust flame to propagate. The minimum explosible concentration is a measure of the minimum fuel chemical energy density needed to support combustion.

If measured in an appropriate test apparatus using a standardized procedure, the minimum explosible concentration is sufficiently reproducible that it can be used for the engineering design of explosion protection safeguards.

The maximum explosible concentration is the maximum concentration of dust that will support flame propagation. It can be thought of as the concentration at which the thermal capacity of the dust is sufficiently large that it quenches the flame by absorbing the flame's thermal energy faster than the combustion process can generate it. The maximum explosible concentration has been found to be quite sensitive to the design of the test apparatus and the particulars of the test procedure. Measurement data for the maximum explosible concentration tend to have poor reproducibility and are generally considered to not be appropriate for engineering design purposes.

A final concentration limit of interest is the limiting oxygen concentration (MOC). This is the minimum oxygen concentration that will support combustion. The measured value is both apparatus and procedure dependent. It is useful as an engineering parameter to guide the design of inert atmospheres for explosion prevention (Crowl, 2003, pp. 117−120).

There are two categories of useful ignition tests for combustible dusts, dust cloud measurements and dust layer measurements. There are two ignition tests for dust clouds, the minimum autoignition temperature (AIT) and the minimum ignition energy (MIE). In each case, the dust concentration is varied so as to achieve the lowest value of the ignition parameter. The minimum AIT is measured in a temperature-controlled furnace. An example application of the AIT is to use it as a reference for comparison with hot surfaces present in a given industrial setting. The MIE is performed in a chamber with the cloud subjected to an electric discharge of known energy. The MIE is typically interpreted as a measure of the ignition energy from an electric arc or an electrostatic discharge (static electricity).

For combustible dust layers, the standard ignition test is based on thermal ignition theory. A dust layer of prescribed depth is placed on the temperature-controlled test surface. Ignition is defined by an observed temperature rise or by the appearance of smoldering (glowing) or flaming combustion. The hot surface ignition temperature can be used to determine the maximum accumulation of dust deposits that can be tolerated on a heated surface.

1.5 COMBUSTIBLE DUST TESTING

While the hazards of combustible dust are best examined from the viewpoint of combustion science, it is equally important to gain an appreciation for some of the key empirical tests which have evolved over time to become the accepted basis for explosion protection. The combustible dust hazard parameters described in the previous section are operational definitions, that is, they are the outputs of tests performed in a specified apparatus with a specified procedure. It is important to recognize that operational measurements are not universal physical constants. Measurements performed for a hazard parameter in two different test apparatuses or using different methods may yield different results. Over the last 100 years of dust explosion research, certain standards for the measurement of hazard parameters have emerged.

At the beginning of the 20th century in the United States, dust explosion testing was carried out in combustion bombs capable only of measuring the maximum pressure attained in the test. It was recognized that the violence of a deflagration was characterized by not just the maximum pressure but also by the rate of pressure rise. Although many investigators were active in this area, the Bureau of Mines in particular developed several test chambers and instrumentation for performing such measurements. The cylindrical Hartmann bomb became the Bureau's standard dust explosibility chamber for laboratory testing (typically with a volume of 1.2 L). By 1968 investigators at the Bureau of Mines had published dust explosibility data on thousands of dust tests (Verakis and Nagy, 1987). The Hartmann bomb suffers from two problems: the small volume and its cylindrical shape cause it to experience significant heat losses and distortion of the initially spherical flame surface. This makes the scaling of dust explosibility data to larger vessels unreliable.

In the 1980s, the 20-L sphere became the de facto standard apparatus for dust explosibility parameters. This is because the spherical shape is the sensible choice for spherical flame propagation. The 20-L volume has been determined to be the minimum volume that correlates well with larger instruments due to the smaller heat losses at the wall (Bartknecht, 1989, pp. 56−65). The basis for this judgment is the relative degree to which the K_{st} values derived from a 20-L sphere apparatus seem to follow the cube root scaling law. Thus, explosibility data obtained from the 20-L sphere can be used for scaling to larger vessels. A sketch of the test setup is shown in Fig. 1.3.

The standardized procedures used with the 20-L sphere result in a tendency to "overdrive" the deflagration. This is a consequence of the strong igniters used to initiate the test. These overdriven deflagration results have sometimes resulted in false positives (Myers and Ibarreta, 2013). Alternative procedures using a cubic meter sphere have been developed to compensate for this effect. Dust explosion testing has become standardized making interlaboratory comparisons of data more reliable. Table 1.1 is a list of test methods for dust explosion testing using the 20-L sphere.

Dust explosibility data obtained using these methods have both a theoretical foundation and have been correlated with explosion protection safeguards like explosion venting.

Ignition testing of combustible dusts has required the development of specific laboratory equipment and procedures for dust clouds and dust layers. A list of accepted test methods for combustible dust ignition characteristics is shown in Table 1.2.

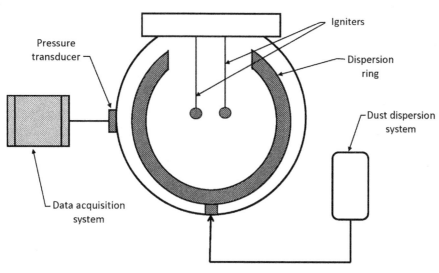

FIGURE 1.3

Schematic of the 20-L sphere explosion testing device.

Table 1.1 Test Methods for Dust Explosibility Parameters
(Myers and Ibarreta, 2013)

Test Methods	Description
ASTM E1226, *Standard Test Method for Explosibility of Dust Clouds*	This test method addresses two issues: (1) Is the dust explosible? (2) If so, at what concentration does the maximum pressure and maximum rate of pressure rise?
	The first test is called the go/no-go test. The dust is tested at two or more concentrations. If the test generates a pressure ratio greater than 2, the dust is declared combustible. Allowances must be made for the pressure contribution of the igniter system used in the test
	The second test consists of measuring the maximum pressure and maximum rate of pressure rise at a range of dust concentrations. The maximum value of the rate of pressure rise is used to calculate the K_{st} value
ASTM E1515, *Standard Test Method for Minimum Explosible Concentration of Combustible Dusts*	The dust is tested at a trial concentration. If the pressure ratio is greater than or equal to 2, the test concentration is greater than the MEC and a lower concentration must be tested. This procedure is continued until the MEC is bracketed by the test results
ASTM E2931, *Limiting Oxygen Concentration in Dust Clouds*	This procedure is similar to the MEC procedure. Air is blended with an inert gas like nitrogen to create the desired oxygen test concentration. The dust concentration is varied at a given oxygen concentration to determine if there is a failure to propagate a deflagration

Table 1.2 Test Methods for Ignition Characteristics of Combustible Dust (Myers and Ibarreta, 2013)

Test Methods	Description
ASTM E1491, *Standard Test Method for Minimum Autoignition Temperature of Dust Clouds*	Dust is dispersed at various concentrations within a heated furnace. The autoignition temperature is determined by the minimum temperature at which a flame propagates from the furnace
ASTM E2019, *Standard Test Method for Minimum Ignition Energy of a Dust Cloud in Air*	The minimum ignition energy is based on an electrostatic discharge. Several variables must be varied to determine the minimum energy value: dust concentration, spark energy, spark timing, and the electrode spacing
ASTM E2021, *Standard Test Method for Hot-Surface Ignition Temperature of Dust Layers*	Varying dust layer thicknesses are tested on a temperature-controlled surface and observations of ignition or exothermic behavior are recorded

While these test methods have theoretical foundation, it is more difficult to apply the results to a larger scale or to a physical environment different from the test conditions.

1.6 COMBUSTIBLE DUST HAZARD SCENARIOS

There are four basic hazard scenarios for combustible dust: smoldering, flash fires, dust explosions, and flame acceleration effects. This is not to suggest that in performing a risk assessment that there are only four accident scenarios to consider. Instead these four basic hazard scenarios identify four basic types of combustion behavior with combustible dusts. There are an almost infinite number of ways that a given accident sequence could result in one of these four basic hazard scenarios. Guidance on performing combustible dust hazard analyses can be found in the book by Frank and Rodgers (Frank and Rodgers, 2012, pp. 101−124) and the CCPS/AIChE book (CCPS/AIChE, 2005, pp. 117−147). There are a variety of hazard analysis techniques available to the analyst, but whichever technique is chosen, it is important that the analyst and all participants have an acquaintance with accident case histories. Thoughtful discussions of combustible dust accidents can be found in several books and technical papers (Eckhoff, 2003; Amyotte, 2013; Frank and Rodgers, 2012).

The discussion of the four basic hazard scenarios that follows is organized in order of increasing mass burning rate and, in general, increasing order of severity.

1.6.1 SMOLDERING

Smoldering was previously defined as flameless combustion. If the physical conditions are conducive to self-heating and spontaneous ignition, smoldering can initiate in an accumulation or pile of combustible dust. A typical situation where piles of bulk solids can accumulate is in storage silos and bins. One potential ignition source in piles of stagnant material is self-heating and spontaneous ignition. Agricultural grains, coal, and many other carbonaceous materials are susceptible to this ignition scenario. The self-heating is due to the slow oxidation of the material. If the pile is sufficiently large, the heat released from the exothermic oxidation reaction leads to a rise in temperature and, eventually, ignition. A comprehensive review of spontaneous ignition can be found in the monograph by Bowes (Bowes, 1984). The ignition event tends to occur in the center of the material and results in a smoldering region of fuel. Fig. 1.4 illustrates the smoldering fire hazard scenario.

The hazard of smoldering in an unconfined setting depends on the size of the pile. One possibility is the potential for a transition to flaming combustion. However, a more insidious hazard is the potential for the creation of a fireball during the suppression of the fire. The high-velocity stream from a fire extinguisher can disperse the smoldering material and result in a flash fire. Even a "small" pile of smoldering material is capable of generating a dangerous fireball if dispersed.

Smoldering in a confined space presents two hazards: the generation of flammable gas/smoke and toxic gas emission. Smoldering is a less efficient form of combustion compared to flaming in the sense that smoldering tends to generate a higher yield of incomplete combustion products, including carbon monoxide and smoke (Ogle et al., 2014; Quintiere et al., 1982). Carbon monoxide and smoke

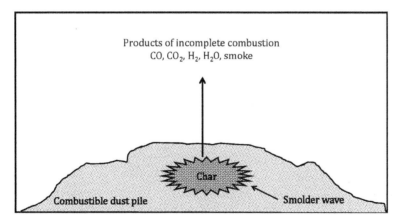

FIGURE 1.4

The smoldering fire hazard of combustible dust.

present two hazards: they are both flammable and toxic. The flammable gas/smoke generated by smoldering can become an explosion or flash fire hazard to personnel and property. It is also a toxic exposure hazard to personnel.

1.6.2 FLASH FIRE

Consider the following situation. A mass of combustible dust is suddenly dispersed into the air forming an ignitable concentration. The ignition of the dust cloud results in a transient flame called a flash fire. These are the requirements for the flash fire square. A flash fire is a dust deflagration that does not result in a rise in pressure. This requires that the confinement volume must be much larger than the volume of the unburnt dust cloud. A quantitative estimate of this volume ratio will be developed in Chapter 3.

To become a flash fire hazard, a deposit of dust must become dispersed. The severity of the hazard will depend on a number of practical factors such as the size of the dust deposit, its location, the potential for dispersal, the location and frequency of occupation of an area by personnel, and by the nature of the work activities in the area. One factor to be especially aware of is the elevation of the dust deposit. If it is located at the elevation of the floor, it represents one level of hazard. But if the dust deposit is located above the occupants of the work space, the hazard level is greater because of the ease of dispersal should the dust deposit be disturbed resulting in a suspended dust cloud. Fig. 1.5 illustrates the flash fire hazard scenario.

The hazards of a combustible dust flash fire are flame engulfment (or contact), radiant heat, and direct contact with burning particles (Grimard and Potter, 2011).

1.6.3 DUST DEFLAGRATION (EXPLOSION)

The next hazard scenario begins with the same physical setting as the flash fire. The difference is an additional factor: confinement. These are the five factors that comprise the dust explosion pentagon. The distinction between a flash fire

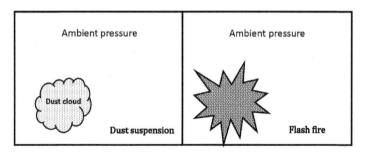

FIGURE 1.5

The flash fire hazard of combustible dust.

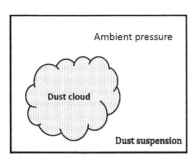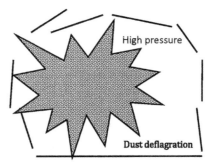

FIGURE 1.6

The dust explosion hazard scenario.

and a dust deflagration is due to the volume of the confinement. If the ratio of the combustible dust cloud to the confinement volume is sufficiently large, the heat release results in a pressure rise. If the maximum pressure attained exceeds the strength of the structure, the structure will rupture resulting in the projection and displacement of fragments. Fig. 1.6 illustrates the dust explosion hazard scenario.

The hazards of a dust explosion include the overpressure, missiles, flames, and radiant heat. Generally, the missiles are the greatest hazard to personnel in the far field. Overpressure, flames, and radiant heat are all hazards to personnel in the near field. Overpressure is an especially problematic hazard of dust explosions because it can lead to structural collapse, a hazard to personnel in both the near and far field (CCPS/AIChE, 1996, pp. 58–65).

1.6.4 FLAME ACCELERATION EFFECTS

Flame acceleration effects arise when the mass burning rate of the deflagration increases due to the influence of turbulence. The combustion process releases heat causing a localized expansion of the burnt gas. The burnt gas expansion results in an acceleration of the flame speed, but more importantly, increases the turbulent characteristics of the flame resulting in an increase in the mass burning rate. The increasing mass burning rate results in an increasing heat release rate which, in turn, accelerates the flame motion. Flame acceleration results in the magnification of dust explosion hazards. I separate flame acceleration effects into two distinct categories: unconfined and confined.

Unconfined flame acceleration is commonly called a secondary dust explosion. It is due to two factors: fugitive dust deposits and the presence of obstacles. The primary dust explosion occurs in a confinement and causes its rupture. The sudden release of high pressure from the ruptured confinement causes the formation of a shock wave which disperses the fugitive dust deposits in its adjacent surroundings. The shock wave propagates outwards and entrains a fraction of

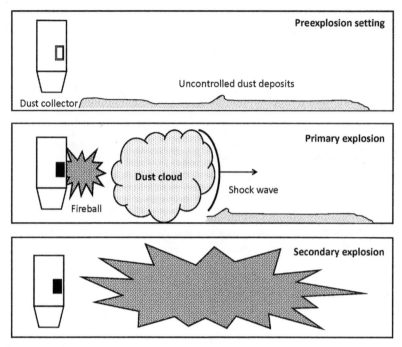

FIGURE 1.7

The secondary explosion hazard scenario.

the dust deposits creating a dust cloud. The slower flame exiting the ruptured confinement then ignites the secondary cloud and chases the shock wave.

If the secondary dust explosion occurs inside an enclosure, there is the potential for the secondary explosion to cause the enclosure to rupture. I still consider this to be an unconfined system because the enclosure does not affect the turbulent flame acceleration: the enclosure simply experiences the consequences. Fig. 1.7 illustrates the typical secondary explosion hazard scenario.

The ingestion of new fuel into the dust deflagration results in an increase in the mass burning rate thus accelerating the flame. In general, the deflagration wave will become increasingly turbulent during the acceleration process. The presence of obstacles in the flame path can generate additional turbulence which further increases the mass burning rate. An unconfined accelerating dust flame will rarely generate overpressure and blast wind magnitudes sufficient to cause human injury and structural damage. The primary hazard will be the thermal effect of flame impingement and hot gases.

Flame acceleration is significantly more hazardous when it occurs in confinement. Confined flame acceleration can lead to overpressures and blast wind at levels capable of inflicting human injury and structural damage. Three variations of confined flame acceleration must be considered: flame acceleration

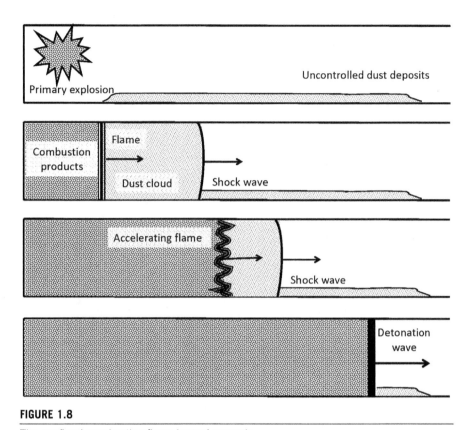

FIGURE 1.8

The confined accelerating flame hazard scenario.

in a channel, pressure piling, and detonation. Flame acceleration in a channel is more rapid and severe than unconfined flame acceleration. This is because the hot, compressed gaseous combustion products are constrained by the channel to expand in only one direction. Confined flame propagation will result in the formation of a shock wave and an intensification of the burning rate due to turbulence. This leads to much higher levels of overpressure and blast wind than can be attained in unconfined settings. In rare circumstances, confined flame acceleration has been observed to transition to a detonation. The confined accelerating flame hazard is illustrated in Fig. 1.8.

Many process plants have a physical layout that includes a series of interconnected vessels. In the handling of combustible dust, the physical connection of different enclosures creates a special type of flame acceleration hazard called pressure piling (Eckhoff, 2003, p. 74). Pressure piling occurs when the primary dust explosion is transmitted to another vessel or enclosure. The explosion pressure in the second confinement is magnified due to the increase in initial pressure in the second enclosure. Fig. 1.9 illustrates the pressure piling hazard scenario.

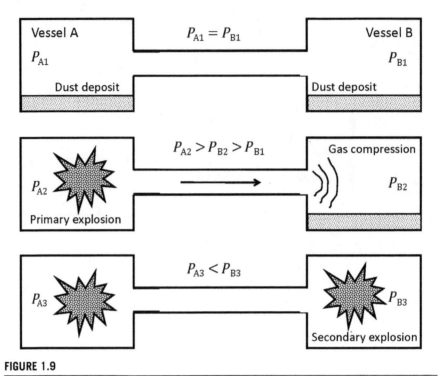

FIGURE 1.9

The pressure piling hazard scenario.

The process safety literature speaks of a special type of hazard called the domino effect. Domino effects are defined as the triggering of secondary events by a primary event such that the result is an increase in consequences or area of an effect zone (Amyotte, 2013, pp. 128–137). The destructive forces of dust explosions and flame acceleration—secondary explosions and pressure piling—are examples of domino effects.

1.7 PREVENTION AND CONTROL OF DUST FIRES AND EXPLOSIONS

The ultimate reason for studying dust fire and explosion dynamics is to prevent their occurrence or to control their consequences. How does one turn combustion theory into combustion safety? The answer lies in the discipline of system safety. System safety provides the strategy and tools for systematically identifying, evaluating, and controlling hazards. System safety evolved as a discipline during World War II (1939–1945). Prior to World War II, the practice of safety was largely an empirical and intuitive art (Roland and Moriarty, 1990, pp. 8–12).

Stated another way, safe practices and technology were largely developed by trial and error. The situation changed dramatically during the war with the introduction of many revolutionary, science-based technologies—radar, sonar, the atomic bomb—intuition was no longer an acceptable basis for safety management.

System safety grew out of the need to prevent costly accidents instead of just simply managing the consequences. Initially, it found its strongest proponents in the aviation field, as both pilots and their aircraft were precious resources that could not be squandered on accidents. In the 1960s, system safety was adopted in the nuclear and aerospace industries. By the 1970s, the chemical process industries were enthusiastically adapting system safety methods to aid in the prevention of explosions, fires, and toxic releases. This adaptation became known as chemical process safety, more simply as process safety. Through the efforts of European and American engineering organizations, the discipline of process safety management has spread throughout the world, and its practitioners have developed an impressive body of knowledge. For example, in the last 30 years the Center for Chemical Process Safety (CCPS), a technology alliance of the American Institute of Chemical Engineers (AIChE), has published over 100 books on various aspects of process safety technology and management.

Many of the strategies, methods, and concepts of process safety have direct applicability with combustible dust hazard management. Over the last 80 years, the combustible dust safety standards published by the National Fire Protection Association (NFPA) have evolved from a strict safety rule format into a set of performance-based standards that have incorporated many aspects of process safety (Frank and Rodgers, 2012, NFPA 654, 2013; NFPA 652, 2016). The proper application of process safety to combustible dust hazards can be greatly improved by acquiring a solid background in combustion science. In applying process safety to combustible dust hazards, the objective is to defeat the combustible dust pentagon.

The system safety hierarchy is an important concept in the control of hazards. The basic idea behind the hierarchy is to rank in decreasing order of effectiveness the strategies one may choose to implement for hazard control. Table 1.3 summarizes the hierarchy.

At the top of the list is the most reliable means of hazard control: prevention. The next most reliable method of hazard control relies on physical safeguards designed to mitigate the consequences of the hazard. These are often called engineering controls. Next in line are procedural controls, often called administrative controls, which require deliberate human action to be successfully employed. The final element of the hierarchy is no action or risk acceptance.

Table 1.3 System Safety Hierarchy for Hazard Control

Eliminate the hazard
Control the hazard with physical safeguards
Control the hazard with procedural safeguards
Accept the hazard

Table 1.4 Application of Inherently Safer Design Principles to Combustible Dust Hazard Management

Inherently Safer Design Principles
Avoid hazards
Reduce severity
Reduce likelihood
Segregate
Apply passive safeguards
Apply active safeguards
Apply procedural safeguards
Apply residual risk reduction measures

The design philosophy of inherently safer design offers a more comprehensive approach to combustible dust hazard management than the simple system safety hierarchy. Inherently safer design was originally based on the four principles: minimization, substitution, moderation, and simplification (CCPS/AIChE, 2008). Amyotte et al. have expanded these principles and specialized them to combustible dust hazards (Amyotte et al., 2003). As with the system safety hierarchy, the emphasis is first on prevention. A summary of the inherently safer approach is presented in Table 1.4.

There is a large number of hazard control measures that one can implement at an industrial facility. The decision on which safeguards to implement can be challenging. The combustible dust pentagon is a convenient tool for organizing and evaluating hazard control strategies as illustrated in Table 1.5.

Fuel control could refer to any one of several strategies: eliminating or reducing the quantity of combustible dust inside the equipment or enclosure housing the dust, adding an inert chemical agent to suppress ignition, or shifting the particle size distribution of the dust to make it more resistant to ignition (Amyotte, 2013, pp. 168–180). Probably, the most common method for implementing this control strategy is housekeeping. Repairing equipment leaks and removing deposits of fugitive combustible dust on a scheduled basis is a proven technique for reducing the potential for secondary explosions.

Oxidizer control refers to either reducing or eliminating oxidizer from the atmosphere inside the equipment or enclosure housing the combustible dust. This is typically done with an inert gas such as nitrogen or carbon dioxide. To maximize its effectiveness, the oxygen concentration should be monitored and controlled. Depending on the dynamics of the process, control may be achieved manually or automatically (with programmable logic controllers).

Ignition control means eliminating or reducing the number or intensity of ignition sources. Engineering safeguards such as explosion-proof electrical service and bonding/grounding to control static electricity, a hot work safety program are all examples of ignition control. Explosion suppression can also fall under this category.

Table 1.5 Combustible Dust Hazard Control Strategies Organized Around the Combustible Dust Pentagon

Fuel	Oxidizer	Ignition Source	Suspension	Confinement
Substitute noncombustible material Add inert material Agglomerate to larger particle size	Reduce oxygen concentration in equipment with inert gas Install oxygen sensors and interlocks to shut down equipment if oxygen level is exceeded	Eliminate open flames Install explosion-proof electrical service Control static electricity Administer hot work safety program Explosion suppression	Avoid excessive air current velocities Control fugitive dust accumulations Apply inert liquid to dust deposits to prevent entrainment by convection	Design equipment to contain explosion pressures Install explosion vents panels Install explosion isolation

Dust suspension control means changing the material handling processes to eliminate or reduce the formation of dust clouds. As mentioned earlier, housekeeping is an important safeguard when working with combustible dusts. In some work environments, it may be possible to apply an inert liquid to fugitive dust deposits to prevent entrainment by convection currents.

Control of dust containment means to reduce the effect of containment in the event of a dust deflagration. This can be done by creating a more open and unconfined work area (a strategy followed in the US grain industry) and by deflagration venting.

It should be obvious that the successful implementation of these hazard control strategies will require estimation or measurement of the dust explosibility parameters.

1.8 CHALLENGES IN MODELING DUST FIRES AND EXPLOSIONS

The goal of this book is to survey the available tools for modeling the effects of dust fires and explosions. Combustion science offers a sophisticated array of tools for this purpose. The primary challenge to this effort is the complexity of describing the particulate nature of combustible dusts. Unlike flammable gases, a given combustible dust may vary in terms of its chemical composition and its particle size distribution. Depending on the chemical composition of the dust, the combustion mechanism may depend on both homogeneous and heterogeneous chemical kinetics. Obtaining representative samples of combustible dust from an industrial process can present significant difficulties. Describing the dust dispersion process requires the description of the turbulent flow of a multiphase mixture. The mathematical tools for describing turbulent, multiphase reactive flows have been under development since the 1960s and are still under development (Smoot and Smith, 1985; Oran and Boris, 1987; Kuo and Acharya, 2012a,b; Crowe et al., 2012). But the use of models based on these formal methods is computationally taxing, time-consuming, and dependent on an accurate specification of the dust's particulate properties (particle size distribution, density, thermodynamic properties, transport properties, and optical properties) and the chemical kinetics of the combustion process. This discussion will be elaborated upon in Chapter 10.

A further difficulty in analyzing the development of a dust fire or explosion is the sensitivity of the dynamics to the quality of the dust dispersion, the nature of the ignition source, and the boundary conditions imposed by the surroundings (the geometry of the structure, the strength of the enclosure, and the presence of obstacles in the flame path to name just a few). The development of a combustible dust event—whether smoldering, a flash fire, a deflagration, or an accelerated flame—is sensitive to changes in the boundary conditions. Whether modeling is performed for hazard analysis (prediction) or for accident investigation (analysis), the potential significance of parameter uncertainties must be borne in mind.

As the physical realism of the combustion process increases, the number of input parameters also increases. This is the essential tension in mathematical modeling. Simple models require few input parameters and are thus easier to use. But simple models fail to provide a detailed description of the physical and chemical processes of combustion. The lack of realism in simple models introduces uncertainty in the predictions of the model. Sophisticated models offer a more comprehensive, realistic description of combustion. But sophisticated models require the specification of a larger number of input parameters, many of which are difficult to measure or estimate. Thus, the imprecise specification of input parameters introduces uncertainty into the model predictions. Fig. 1.10 illustrates how the uncertainty of model predictions depends on input data uncertainty and combustion model uncertainty. This relationship suggests that for a given modeling effort there is an optimal modeling level of effort that corresponds to a minimum in total uncertainty. A model can always be made more sophisticated, but if the solution of the more sophisticated model requires specification of input parameters that are indeterminate or unmeasurable, then the model predictions will grow more uncertain.

Mathematical model building is an exercise in the scientific method. The model can be considered a hypothesis. To justify the model (hypothesis), it must be tested. I broadly define two measures of a model's utility: accuracy and insight. The test of the accuracy of a model's accuracy is to challenge it with empirical data. The degree to which the model predictions comport with the empirical data is a direct measure of the model's accuracy. Further refinement of the model is justified only if it can improve the agreement of model predictions with empirical data.

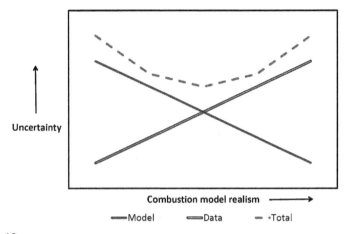

FIGURE 1.10

Total uncertainty of model predictions as a function of combustion model uncertainty and input data uncertainty.

On the other hand, the value of some models is that they provide insight into a complex, otherwise intractable problem. As a matter of preference, for any given modeling problem I like to begin with the simplest description of the problem that still retains the basic physical and chemical events of interest. If this first approximation fails to offer the desired level of insight, then I gradually add more realism into the model until I am satisfied with the results. Whether you wish to achieve accuracy or insight, model building is an iterative process of discovery.

1.9 OVERVIEW OF THIS BOOK

In this chapter, the three fundamental combustible dust hazards—smoldering, flash fires, and explosions—were introduced. An historical review of dust explosion research indicated that although combustible dust hazards have been well known for over 300 years, the application of combustion science to dust explosion phenomena is a relatively recent initiative. Some basic concepts from combustion science were introduced next, and the key parameters used to characterize combustible dust hazards were presented. The elements of combustible dust testing were described as they relate to the determination of the key combustible dust hazard parameters. Four basic combustible dust hazard scenarios were introduced, and the strategies for controlling the hazards were described in the context of the dust explosion pentagon. Some of the challenges in the mathematical modeling of dust fires and explosions were presented.

Chapter 2, The Key Physical Properties of Combustible Dust, will introduce briefly several important concepts regarding the physical properties of combustible dust including the properties of single particles, populations of particles, and bulk properties of dusts. Of these properties, probably the most important is the particle size distribution of the dust, as this property plays a central role in the combustion rate of particles.

The use of thermodynamic principles to analyze dust combustion behavior is discussed in Chapter 3, Thermodynamics of Dust Combustion. Several important hazard parameters will be discussed including the adiabatic flame temperature, the maximum explosion pressure, and the expansion ratio of the combustion products. The variation of these properties with dust concentration will be explored and the important role of the stoichiometric condition will be emphasized. It will be shown that simple thermodynamic models can be used to predict secondary explosion pressures and to help discriminate between a flash fire and an explosion.

Chapter 4, Transport Phenomena for Dust Combustion, presents a summary of simplified conservation equations for combustion analysis. This chapter is not intended to replace course work in fluid dynamics, transport phenomena, or combustion science. Rather, it will introduce some basic models and equations that will be used in subsequent chapters. The Rankine—Hugoniot analysis of combustion waves will be presented as a vehicle for distinguishing between deflagrations and detonations. After reviewing the equations of change for laminar reacting flows, the analysis of planar premixed flames for gaseous mixtures will be discussed. This is followed by the spherical diffusion flame model for liquid fuel droplets. Simple concepts for multiphase flow analysis are explored for both porous media (relevant to smoldering) and dispersed flows. This chapter closes with a review of some of the key concepts from thermal radiation and turbulence.

The perspective of Chapter 5, Smoldering Phenomena, is on the combustion behavior of a fixed bed of particles (porous medium). This chapter explores the combustion process of smoldering. A brief presentation of ignition behavior caused by self-heating and thermal runaway is presented. The physical description of a smoldering wave is developed and the key properties governing its behavior are identified. A simple models for one-dimensional and two-dimensional smolder waves are presented. The hazards of smoldering in a confined space are discussed. This chapter concludes with a brief discussion of strategies to prevent or control smoldering hazards.

Chapter 6, Dust Particle Combustion Models, presents the single particle combustion model with consideration for both diffusion and kinetic processes. Analysis of single particle combustion is the first step toward understanding dust cloud combustion. The physical description of single particle burning phenomena is discussed first followed by a scaling analysis of the heat and mass transfer processes versus the chemical kinetics. The combustion of single fuel particles are an example of noncatalytic gas–solid reactions. Two particle models are introduced in Chapter 6: the shrinking particle model and the shrinking unreacted core model. Both models can exhibit two extremes of combustion behavior: diffusion-controlled and kinetic-controlled. The most common single particle combustion model, the d-squared model for diffusion-controlled kinetics of shrinking particle combustion is introduced. This is followed by a discussion of combustion studies of both organic solids and metallic solids. This survey affords an opportunity to learn how particle size and the physical microstructure of solid particles can influence their combustion behavior. One of the key conclusions of Chapter 6 is that the relative combustion rate of a particle increases with increasing volatility.

With Chapter 7, Unconfined Dust Flame Propagation, the focus is on the propagation of a flame through a cloud of particles without confinement. The chapter begins with the physical description of the one-dimensional planar dust flame. The relative importance of physical processes is demonstrated with scaling analysis. The thermal theory of flame propagation is presented and illustrated with models of increasing complexity. Next, Ballal's unified theory of dust, mist, and gaseous combustion is reviewed. Limit phenomena—ignition and extinction—are discussed next. This is followed by a survey of the literature on the two main divisions of combustible dust: organic solids and metallic solids. The chapter concludes with a survey of flash fire phenomena and their control.

Chapter 8, Confined Unsteady Dust Flame Propagation, examines the confined deflagration: flame propagation through a cloud of particles confined by an enclosure. The confined deflagration is an important model because it serves as the primary basis for the measurement of dust explosibility parameters and as the paradigm for the dust explosion. The physical description of spherical flame propagation in a closed vessel is the first topic. The relative importance of physical processes is explored through scaling analysis. Two integral models of confined flame propagation are introduced, one based on the adiabatic compression of the fluid mixture and one based on an isothermal compression. The origin and limitations of the cubic pressure rise and cube root scaling law are explained. The role of the igniter strength and initial turbulent flow field in dust explosion testing is discussed. A survey of illustrative results for both organic and metallic solids are presented and a discussion of dust explosion hazard controls completes the chapter.

The role of flame acceleration is investigated in Chapter 9, Dust Flame Acceleration Effects. Flame acceleration magnifies the violence of dust explosions. The consequences of flame acceleration depend on whether the flame propagation occurs in an unconfined or confined setting. In unconfined settings, the flame acceleration hazard can lead to secondary dust explosions. In a confined setting, such as long channels or interconnected vessels, the flame acceleration hazard is shock wave formation, detonation, and pressure piling. The modeling of secondary dust explosions is presented in the context of dust lofting and suspension. The discussion of confined flame acceleration includes a discussion of shock wave formation and the transition to detonation. Next, the phenomenon of pressure piling in interconnected vessels is analyzed. The chapter ends with a summary of methods for controlling flame acceleration effects.

Dust explosions are multiphase, turbulent, chemically reacting flows. Chapter 10, Comprehensive Dust Explosion Modeling, explores the challenges of comprehensive dust flame modeling using computational fluid dynamics (CFD).

The challenges presented by disparate time and length scales are explored and a strategy for CFD modeling is suggested. Examples are presented to show how more sophisticated models have been developed using CFD resulting in improvements in the description of how multiphase flow, turbulence, thermal radiation, and chemical kinetics influence dust explosion dynamics.

REFERENCES

Adamson, A.W., Gast, A.P., 1997. Physical Chemistry of Surfaces, sixth ed. John Wiley & Sons, New York.

Amyotte, P., 2013. An Introduction to Dust Explosions. Butterworth Heinemann, Elsevier, New York.

Amyotte, P., Eckhoff, R.K., 2010. Dust explosion causation, prevention and mitigation: an overview. J. Chem. Health Saf. 5, 15–28.

ASTM E1226, 2012. Standard Test Method for Explosibility of Dust Clouds. ASTM International, West Conshohocken, PA.

ASTM E1491, 2012. Standard Test Method for Minimum Autoignition Temperature of Dust Clouds. ASTM International, West Conshohocken, PA.

ASTM E1515, 2007. Standard Test Method for Minimum Explosible Concentration of Combustible Dusts. ASTM International, West Conshohocken, PA.

ASTM E2019, 2007. Standard Test Method for Minimum Ignition Energy of a Dust Cloud in Air. ASTM International, West Conshohocken, PA.

ASTM E2021, 2009. Standard Test Method for Hot-Surface Ignition Temperature of Dust Layers. ASTM International, West Conshohocken, PA.

ASTM E2931, 2013. Limiting Oxygen Concentration in Dust Clouds. ASTM International, West Conshohocken, PA.

Bartknecht, W., 1989. Dust Explosions: Course, Prevention, Protection. Springer-Verlag, Berlin.

Benard, S., Gillard, P., Foucher, F., Mounaim-Rousselle, C., 2012. MIE and flame velocity of partially oxidised aluminium dust. J. Loss Prevent. Process Ind. 25, 460–466.

Bowes, P.C., 1984. Self-heating: Evaluating and Controlling the Hazards. Elsevier, New York.

Britton, L.G., Cashdollar, K.L., Fenlon, W., Frurip, D., Going, J., Harrison, B.K., et al., 2005. The role of ASTM E27 methods in hazard assessment. Part II: Flammability and ignitibility. Process Saf. Prog. 24, 12–28.

Cashdollar, K.L., 2000. Overview of dust explosibility characteristics. J. Loss Prevent. Process Ind. 13, 189–199.

CCPS/AIChE, 1996. Guidelines for Evaluating Process Plant Buildings for External Explosions and Fires. American Institute of Chemical Engineers, New York.

CCPS/AIChE, 2005. Guidelines for Safe Handling of Powders and Bulk Solids. American Institute of Chemical Engineers, New York.

CCPS/AIChE, 2008. Inherently Safer Chemical Processes: A Life Cycle Approach, second ed. American Institute of Chemical Engineers, New York.

Crowe, C.T., Schwarzkopf, J.D., Sommerfield, M., Tsuji, Y., 2012. Multiphase Flows with Droplets and Particles, second ed. CRC Press, Boca Raton, FL.

Crowl, D., 2003. Understanding Explosions. Center for Chemical Process Safety, American Institute of Chemical Engineers.

CSB, 2006. Investigation Report: Combustible Dust Hazard Study. U.S. Chemical Safety and Hazard Investigation Board, Washington, DC.

Drysdale, D., 1997. An Introduction to Fire Dynamics, second ed. John Wiley & Sons, New York.

Eckhoff, R., 2003. Dust Explosions in the Process Industries, third ed. Gulf Professional Publishing, Elsevier, New York.

Eckhoff, R., 2009. Review article: dust explosion prevention and mitigation, status and developments in basic knowledge and in practical application. Int. J. Chem. Eng.1−12, Volume 2009, article ID 569825.

Emmons, H.W., 1980. The growth of fire science. Fire Saf. J. 3, 95−106.

Faraday, M., Lyell, C., 1845. On the subject of the explosion in the Haswell Colleries and on the means of preventing similar accidents. Philos. Mag. 26, 16−35.

Frank, W.L., Rodgers, S.A., 2012. NFPA Guide to Combustible Dusts. National Fire Protection Association, Quincy, MA.

Glassman, I., 1997. Combustion, third ed. Academic Press, New York.

Grimard, J.K., Potter, K., 2011. Effect of Dust Deflagrations on Human Skin. Major qualifying project submitted in partial fulfillment of B.S. degree at Worcester Polytechnic Institute., Worcester, MA.

Ingle, H., Ingle, H., 1900. The Chemistry of Fire and Fire Prevention. Spon & Chamberlain, London.

Kuo, K., 2005. Principles of Combustion, second ed. John Wiley & Sons, New York.

Kuo, K., Acharya, R., 2012a. Applications of Turbulent and Multiphase Combustion. John Wiley & Sons, New York.

Kuo, K., 2012b. Fundamentals of Turbulent and Multiphase Combustion. John Wiley & Sons, New York.

Law, C.K., 2006. Combustion Physics. Cambridge University Press, Cambridge.

Lewis, B., von Elbe, G., 1987. Combustion, Flames, and Explosions of Gases, third ed. Academic Press, New York.

Morozzo, Count, 1795. Account of a violent explosion which happened in a flour warehouse at Turin, December the 14th, 1785, to which are added some observation on spontaneous inflammations. The Repertory of Arts and Manufactures, Volume II, London, pp. 416−432.

Myers, T.J., Ibarreta, A., 2013. Tutorial on combustible dust. Process Saf. Prog. 32, 298−306.

Nagy, J., Verakis, H.C., 1983. Development and Control of Dust Explosions. CRC Press, Boca Raton, FL.

NAS, 1978. International Symposium on Grain Elevator Explosions. National Academy of Sciences.

NFPA, 1957. Report of Important Dust Explosions—A Record of Dust Explosions in the United States and Canada Since 1860. National Fire Protection Association, Boston, MA.

NFPA 652, 2016. Standard on the Fundamentals of Combustible Dust. National Fire Protection Association, Quincy, MA.

NFPA 654, 2013. Standard for the Prevention of Fire and Dust Explosions from the Manufacturing, Processing, and Handling of Combustible Particulate Solids. National Fire Protection Association, Quincy, MA.

Ogle, R.A., Beddow, J.K., Chen, L.-D., Butler, P.B., 1988. An investigation of aluminum dust explosions. Combust. Sci. Technol. 61, 75−99.

Ogle, R.A., Dillon, S.E., Fecke, M., 2014. Explosion from a smoldering silo fire. Process Saf. Prog. 33, 94−103.

Oran, E.S., Boris, J.P., 1987. Numerical Simulation of Reactive Flow. Elsevier, New York.

Price, D.J., Brown, H.H. (assisted by H.R. Brown and H.F. Roethe), 1921. Dust Explosions—Theory and Nature of, Phenomena, Causes and Methods of Prevention. National Fire Protection Association, Boston, MA.

Quintiere, J., Birky, M., McDonald, F., Smith, G., 1982. An analysis of smoldering fires in closed compartments and their hazard due to carbon monoxide. Fire Mater. 6, 99−110.

Rice, G.S., 1910. The Explosibility of Coal Dust. U.S. Geological Survey Bulletin 425, Washington, DC.

Roland, H.E., Moriarty, B., 1990. System Safety Engineering and Management, second ed. Wiley-Interscience, New York.

Schwartz, V, translated by Salter, C.T.C., 1901. Fire and Explosion Risks—A Handbook Dealing with the Detection, Investigation and Prevention of the Dangers Arising from Fires and Explosions of Chemico-technical Substances and Establishments. Charles Griffin & Company.

Smoot, L.D., Smith, P.J., 1985. Coal Combustion and Gasification. Plenum Press, New York.

Turns, S.R., 2012. An Introduction to Combustion, third ed. McGraw-Hill, New York.

Verakis, H.C., Nagy, J., 1987. A brief history of dust explosions. In: Cashdollar, K.L., Hertzberg, M. (Eds.), Industrial Dust Explosions, ASTM STP 958. American Society for Testing and Materials, West Conshohocken, PA, pp. 342−350.

The key physical properties of combustible dust

This chapter is a brief departure from the theme of applying combustion theory to combustible dust fire and explosion phenomena. In this chapter, I address the geometric approximations that are typically applied to combustible dust investigations. This chapter is intended for someone who has little or no experience in particle characterization. My goal is to introduce such a person to a few of the key concepts from particle measurement and characterization. My objective is not to make you an expert in this field. For those interested in digging deeper into this fascinating topic, I will suggest references for further study.

A significant portion of the world economy can be envisioned as a flow network of materials. Particulate materials—finely divided solids—represent a significant fraction of this flow. The storage and handling of particulate materials is a dominant aspect of materials management in many industries: agricultural grains and food ingredients, primary metal ore processing, commodity chemicals, plastics and elastomers, and pharmaceuticals. A great number of the particulate materials handled in the stream of commerce are combustible and thus present a combustion hazard. The first determinant of combustibility is the chemical composition of the material. That will determine if it is oxidizable. The next determinant is the characteristic size of the particle.

One can turn to a number of different references to find a definition for combustible dust. For example, *combustible dust* is defined in NFPA 654 as a finely divided combustible particulate solid that, if suspended in air or some other oxidizing medium, is capable of presenting a flash fire or explosion hazard over a range of solids concentrations (NFPA 654, 2013, p. 8). In the annex material of the standard, dust is defined as particle size less than 500 microns based on sieve analysis. As a point of reference, the period at the end of this sentence is roughly 400 microns in diameter.

In thinking about fine particulate matter like dust particles, it is quite natural to envision them as spheres of uniform size. This mental model is very useful, but not entirely accurate. Whether naturally occurring or manmade, granular materials and powders rarely consist of particles of a single size with a perfectly spherical shape. There is usually a distribution of particle sizes and shapes. The distribution of particle sizes gives rise to a distribution of particle masses and surface areas. We will see later that the particle diameter plays a central role in determining the relative contributions of chemical and transport processes to the overall rate of particle combustion.

The distribution of particle shape confounds the determination of particle size. The mathematical measurement and description of particle shape has been a fertile area for research (Beddow, 1980; Rodriguez et al., 2013). Even if a definitive method for shape measurement was available, it is difficult to apply this information to the mathematical modeling of chemical kinetics and transport processes. The key point about nonspherical particle shapes is that the lack of perfect sphericity introduces measurement and description error into particle size characterization. The combustion models presented in this text are predicated on the assumption that the combustible dust particles are perfect spheres of uniform size. Whenever evaluating the goodness of fit of a dust combustion model to empirical data, you must always bear in mind that the idealization of dust particles as perfect spheres has introduced a definite but unquantified error.

To evaluate the combustible dust hazard of any given particulate material, it is also important to remember that the particle size distribution of the material may change over time as the material is transported and handled. Many combustible particulate solids are friable. NFPA 654 cautions that during the handling of combustible particulate solids, it is possible through attrition to create combustible dust (NFPA 654, 2013, pp. 27−28). Thus, it should be anticipated that the particle size distribution for a given material may evolve with successive handling or movement events causing a longer tail in the particle size distribution toward smaller particle diameters.

Particle size, shape, and surface area measurement of irregularly shaped particles are briefly surveyed in the following section. This is followed by a discussion of particle size distributions and their statistics. The physical properties of single particles and then particle assemblies (dust deposits) are described in the following two sections. A short presentation on the challenges of representative sampling of particulate materials is next. The final section is a summary of how combustible dust hazard parameters vary with particle size.

2.1 PARTICLE SIZE, SHAPE, AND SURFACE AREA MEASUREMENT

The ideal representation of particulate materials is the perfect sphere of a single diameter. The reality of particulate materials is that they typically come in a variety of sizes and shapes. The goal of particle size and shape characterization is to derive a fundamentally useful description of the population of particles in terms of some distribution. From this distribution, mean values or other useful statistics can be obtained.

There are a number of different particle size measurement technologies available to the investigator (Beddow, 1980). The majority of them are based on a physical principle involving the interaction of a particle with a medium. An equivalent spherical diameter for the particle is calculated from the particle's

Table 2.1 Selected Values of Sieve Sizes

Tyler Equivalent Designation (mesh)	U.S. Sieve Series (microns)
40	420
65	210
100	149
200	74
400	37

behavior. The most common methods for particle size analysis are sieving, sedimentation, light scattering, and microscopy/image analysis.

Sieving is the process of running a powder sample of known mass through a series of sieve plates with successively smaller screen sizes. The sieve stack is agitated to cause the particles to slide across the screen surface. Particles smaller than the screen size will fall through the screen onto the next sieve plate. When the sieving process is complete, a portion of the sample will remain on each sieve plate. The mass fraction of the sample residing on each sieve becomes the basis for formulating a frequency distribution for the powder.

The sieves use screens with square-shaped holes. The progression of screen sizes is governed by standards organizations. Table 2.1 shows the sieve screen sizes for some commonly used sieves.

Sieving is a simple, relatively easy means of determining particle size. However, there is little theoretical basis for its performance. Sieving works better with particles whose length, width, and thickness are approximately equal ("chunky"). Sieving may be less useful for sizing particles that are very thin in one dimension ("plate-like") or very thin in two dimensions ("fibers").

Sedimentation and light scattering are indirect measurement techniques that depend on the physical behavior of particles. Particle sizing instruments based on sedimentation determine particle sizes from the settling rates of the particles. The particle diameter is derived from the settling rate using Stokes' law for a sphere (Bird et al., 2002, pp. 58–61). Light scattering determines particle size by measuring the diffraction angle of a stream of particles as they pass through a beam of laser light. The angle of diffraction from a particle depends on its diameter. In both of these technologies, the particles are assumed to be spherical in shape.

To a certain degree, microscopy/image analysis permits a direct measurement of particle size and shape. There are two limitations of this technique. First, the image of the particle profile is a two-dimensional projection of a three-dimensional object. Second, one must choose how to measure the size and shape of an irregularly shaped particle. There are commercially available systems which perform these measurements, but caution is required when applying these results to physical or chemical behavior. This is very much an active area of research.

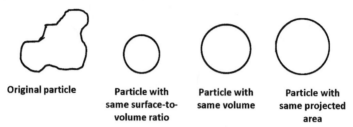

Original particle | Particle with same surface-to-volume ratio | Particle with same volume | Particle with same projected area

FIGURE 2.1

Equivalent spherical diameters.

The choice of which particle sizing technique to use will depend on many factors including access to equipment, cost, sample size requirements, and time constraints. It is important to recognize that the particle size measurements from one technique will be different from those obtained using a different technique. An irregularly shaped particle can be described as an equivalent spherical particle based on any particular constraint. Fig. 2.1 illustrates the range of equivalent spherical diameters which could be used to describe an irregularly shaped particle.

In general, one definition for an equivalent sphere is no better than another. In any given situation, the best definition to use will depend on the application.

The specific surface area has a first-order effect on the rate of combustion. This is because both transport processes and chemical kinetics depend on the magnitude of the surface area. The specific surface area (m²/kg) increases dramatically as the particle size decreases. The surface-to-volume ratio is easy to compute for spheres:

$$A_v = \frac{\pi d_p^2}{\frac{\pi}{6} d_p^3} = \frac{6}{d_p} \tag{2.1}$$

For irregularly shaped particles, the surface area is more easily measured than calculated. The surface area of a powder can be measured by gas adsorption (Adamson and Gast, 1997, Chapters XVI and XVII).

2.2 PARTICLE SIZE STATISTICS

If combustible dusts were composed of perfectly spherical particles of a single size, then the population could be described by a single particle diameter. But the reality is that real combustible dusts are almost always composed of irregularly shaped particles with a distribution of particle diameters. Once an appropriate size measurement technique has been selected, the next step is to analyze the distribution of size measurements.

The particle size distribution can be formulated as either a frequency distribution or a cumulative distribution. The frequency distribution is a plot of the

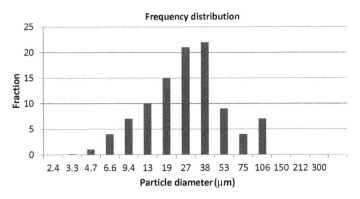

FIGURE 2.2

Discrete frequency distribution of atomized aluminum particle size measurements by light scattering.

From Ogle, R.A., 1986. A New Strategy for Dust Explosion Research: a Synthesis of Combustion Theory, Experimental Design and Particle Characterization. Ph.D. thesis. University of Iowa, Iowa City, Iowa.

fraction of the sample population in each size interval versus the size intervals. An example of a discrete frequency distribution is shown in Fig. 2.2.

Detailed instructions on how to develop a frequency distribution from raw particle size analysis data can be found in the books by Hinds and Beddow (Hinds, 1999, pp. 75−84; Beddow, 1980, pp. 328−338). The frequency distribution can be used to create a weighted average of some characteristic of the particle population such as an average particle velocity or average reaction rate. The frequency distribution in particle size can also be used to calculate a distribution in other geometric properties such as surface area, volume, or mass. If there is a sufficient degree of resolution in the data, the discrete distribution can be converted into a continuous distribution. The advantage of continuous frequency distributions is that it enables the analyst to apply the full range of tools from calculus. The mean value is a measure of the central tendency of the size distribution. The mean particle diameter of the sample population can be calculated directly from the frequency distribution using the mean value theorem from calculus.

$$\overline{d_p} = \int_0^\infty d_p f(d_p) \, dd_p \tag{2.2}$$

The variance is a measure of the spread of the distribution. The variance is defined by the following equation:

$$\sigma^2 = \int_0^\infty \left(d_p - \overline{d_p}\right)^2 f(d_p) \, dd_p = \int_0^\infty d_p^2 f(d_p) \, dd_p - \overline{d_p^2} \tag{2.3}$$

The coefficient of variation, COV, is defined as $\sigma/\overline{d_p}$. If the COV is less than 0.1, the size distribution can be considered to be monodisperse (Crowe et al., 2012, p. 43). The continuous form of the discrete frequency distribution is shown in Fig. 2.3.

The cumulative distribution is the area under the curve of the frequency distribution. The cumulative distribution is a convenient way to determine the fraction of the particle population that has a particle size less than or equal to a specific value. An example of a discrete cumulative distribution is shown in Fig. 2.4.

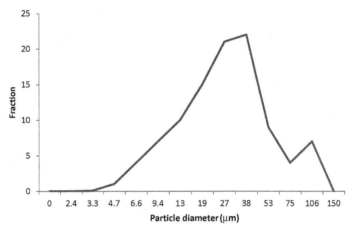

FIGURE 2.3

Continuous frequency distribution of atomized aluminum particle size measurements by light scattering.

From Ogle, R.A., 1986. A New Strategy for Dust Explosion Research: a Synthesis of Combustion Theory, Experimental Design and Particle Characterization. Ph.D. thesis. University of Iowa, Iowa City, Iowa.

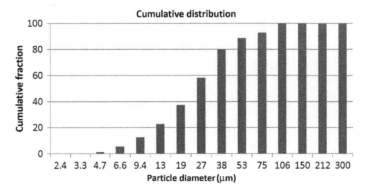

FIGURE 2.4

Cumulative distribution of atomized aluminum particle size measurements by light scattering.

From Ogle, R.A., 1986. A New Strategy for Dust Explosion Research: a Synthesis of Combustion Theory, Experimental Design and Particle Characterization. Ph.D. thesis. University of Iowa, Iowa City, Iowa.

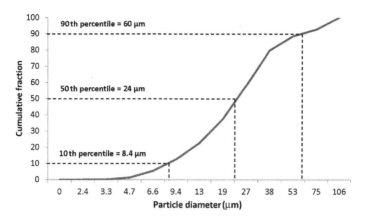

FIGURE 2.5

Continuous cumulative distribution of atomized aluminum particle size measurements by light scattering.

From Ogle, R.A., 1986. A New Strategy for Dust Explosion Research: a Synthesis of Combustion Theory, Experimental Design and Particle Characterization. Ph.D. thesis. University of Iowa, Iowa City, Iowa.

There are advantages to converting a discrete cumulative distribution into a continuous one. The cumulative distribution is the integral of the frequency distribution over the range of particle size. To calculate the fraction of the particle population that has a particle size less than or equal to A, you solve the following integral:

$$F(A) = \int_0^A f(d_p)dd_p \qquad (2.4)$$

It is not unusual for particle size distributions that span 2 orders of magnitude or more. In such cases, the mean value may not be the best representation of the central tendency of the population. A better description of the population can be obtained by considering percentiles such as the 10 percentile, 50 percentile (the median), and 90 percentile. Fig. 2.5 shows the continuous form of the cumulative distribution.

The frequency distribution can be used to calculate weighted averages of particle properties such as surface area or volume. The n-th moment of a particle size distribution is defined by

$$d_n = \int_0^\infty d_p^n f(d_p)dd_p \qquad (2.5)$$

Moment averaged particle diameters can be formed by taking the ratio of moments. For example, a volume-to-surface area mean diameter, d_{32} is defined by the ratio of the third moment to the second moment.

$$d_{32} = \frac{\int d_p^3 f(d_p)dd_p}{\int d_p^2 dd_p} \qquad (2.6)$$

This moment average is called the Sauter mean diameter. It has a special significance because it represents a volume-to-surface area ratio, a physical quantity that is very important in both transport phenomena and chemical kinetics.

Finally, a word of caution regarding particle size statistics. Most size measurement technologies report size distributions in terms of numbers of particles. Sieve size analysis is an exception. Sieve analysis data is reported in mass per size class. A mean particle diameter derived from sieve analysis is a mass-averaged quantity. Number averaging a distribution with a large number of very fine particles will produce a small mean particle diameter. Mass averaging the same population of particles will result in a much larger mean particle diameter, because the mass average is essentially a third moment average. The best mean particle diameter is a choice dictated by the particular application.

2.3 PHYSICAL PROPERTIES OF SINGLE PARTICLES

Solid particles can be created by a number of different physical and chemical processes including gas phase chemical reactions, condensation spray drying, and vapor—solid deposition. Perhaps the most commonly encountered physical process for particle formation is by attrition from larger pieces of solid parent material. Thus, the physical properties of the particle will generally be the same as the parent material (excluding possibly nanoscale effects). This section considers the properties of single particles. There are five categories of physical properties of single particles that can be important for an investigation into the behavior of a combustible dust. These properties are summarized in Table 2.2.

The first question to consider is the chemical composition of the particles. For pure chemicals, the chemical composition of a particle is simply its chemical formula. One useful chemical distinction for combustible dusts can be the determination if the particle is a metal or nonmetal. Typically, nonmetallic combustible dusts are organic materials. Naturally occurring organic materials like coal, wood, and agricultural grains are very complex materials that defy simple chemical

Table 2.2 Single Particle Properties for Combustible Dust

Category	Properties
Chemical composition	Metal or nonmetal, element or compound, volatile or pyrolyzing organic solid, proximate analysis, ultimate analysis
Physical structure	Particle size, particle shape, density, porosity
Thermal properties	Heat capacity, enthalpy of combustion, melting point, boiling point, thermal conductivity
Electrical properties	Electrical conductivity: conductor, semiconductor, or insulator
Optical properties	Refractive index, light scattering behavior

identification. Two useful means for characterizing such materials are proximate and ultimate analysis.

Proximate analysis is the characterization of a material in terms of four classifications: moisture, volatiles, carbon, and ash (Speight, 2005, pp. 41–66). Although originally developed for the analysis of coal, it is frequently used for other complex organic materials to characterize their combustion behavior. Proximate analysis is a gravimetric technique and is based on the mass loss of the sample when subjected to successively higher temperatures. The test is most easily performed with a thermogravimetric analysis instrument. Moisture is driven off by exposing the sample to a temperature of 110°C and the quantity of water is determined by the difference in sample mass. The volatiles content is the loss of mass when the dried sample is exposed to an oven temperature of approximately 900°C. The ash content is the mass of the sample that remains after complete combustion, and the carbon content is determined by difference from the mass loss after volatilization and the mass of ash.

Ultimate analysis is a form of elemental analysis (Speight, 2005, pp. 67–91). It is an analytical technique that determines the mass fraction of carbon, hydrogen, nitrogen, and oxygen in the sample. There are automated chemical analysis instruments that perform this determination. Depending on the user's needs, additional elements can be measured as well.

The next question to consider about a combustible dust is the physical structure of the particles. The geometry of nonporous solid particles can be described in terms of a particle size and shape. Porosity introduces another variable to consider. Porous particles have additional surface area beyond the size and shape of their exterior surface. The rates of heterogeneous chemical reactions and interphase heat and mass transfer depend on the available surface area of the particle. The added surface area due to porosity may lead to larger particle combustion rates. The porous surface area can also increase the potential for adsorption of contaminants from the particle's environment. The density of the particle, accounting for its porosity, governs the inertial behavior of the particle in a fluid flow field, while the particle size and shape will influence viscous interactions.

The thermal properties of particles determine a variety of phenomena of importance to combustion. The enthalpy of combustion is the energy content that creates the high temperature environment of combustion. The heat capacity and thermal conductivity govern the rate of temperature rise within the particle mass. The melting point and boiling point are useful reference temperatures for metals and pure chemical materials. For most naturally occurring materials, however, the chemical constitution of the particle undergoes thermal degradation (pyrolysis) before the melting transition is attained.

The electrical properties of greatest interest are the electrical conductivity and dielectric constant. These properties are usually not relevant to combustion processes. They are very important in determining the electrostatic behavior and electrical behavior of combustible dusts; these are important issues in ignition behavior. Ignition by static electricity is beyond the scope of this book. Useful discussions can be found in many of the combustible dust books identified in the preface.

The optical properties of particles are important because they govern the thermal radiation behavior of dust clouds. The key optical property of interest is the index of refraction. The optical behavior of particles depends on the index of refraction, the particle size and shape, and the wavelength of electro-magnetic radiation. This will be developed further in the chapter on transport phenomena.

2.4 PHYSICAL PROPERTIES OF PARTICULATE MATERIALS IN BULK

Particulate materials have physical properties that individual particles do not possess. These physical properties arise from particle—particle and particle—wall interactions. These interactions and their consequences are not apparent from examination of a single particle. These physical properties can be important to dust fire and explosion phenomena (Eckhoff, 2009b).

Particle assemblies are collections of particles in a static pile or deposit. In this book, these assemblies will be called particulate materials. Particulate materials exhibit physical behavior due to particle—particle interactions and particle—wall (solid surface) interactions (Seville et al., 1997, Chapters 3—5; Rhodes, 2008, Chapters 10 and 11; Schulze, 2008, Chapters 2 and 3). In some circumstances, the material behaves like a solid and in other circumstances, it behaves like a fluid. For example, when a particulate material is poured into a container, it will exhibit an apparent or bulk density. If the container is vibrated, the particulate material will settle and exhibit a decrease in its volume and an increase in its apparent density. If this same particulate material possesses a broad particle size distribution, then as the container is vibrated the smaller particles will migrate toward the bottom of the container. This phenomenon is called particle segregation.

In a static pile, particulate materials can exhibit shear strength giving rise to solid-like behavior. For example, if a storage silo is not designed taking the shear strength into account, the particulate material can form a stable arch in the silo preventing the outflow of material. If the silo has been designed taking into account the shear strength, the particulate material will flow out of the silo like a fluid. The adherence of particulate material to the walls of a silo, or even the formation of a stable arch, is a consequence of both the particle—particle interactions (internal friction) and particle—wall interactions (wall friction).

The shear strength of a particulate material can be an important factor in the development of a secondary dust explosion. The ability of a precursor shock wave to loft and fluidize a dust deposit will depend on the magnitude of the shock loading compared to the shear strength of the dust. In general, the shear strength of a dust deposit increases with decreasing particle size. For many materials, the moisture content of the material can also have a significant impact.

A dust deposit consists of solid particles and the air that fills the void space within the deposit. The static dust–air mixture is a porous medium. The thermal properties of a porous medium—the heat capacity and the thermal conductivity— are a function of the properties of the individual phases (solid particle and air) and of the structure of the porous medium (Kaviany, 1995, pp. 119–153). These thermal properties influence the potential for a dust deposit to undergo spontaneous ignition and smoldering. The mixture heat capacity of a porous medium is a straightforward calculation which can be performed with confidence. Although sophisticated models have been proposed for calculating the effective thermal conductivity of model porous media structures, there is considerable uncertainty in applying these models to real dust deposits. The investigator must often resort to estimation or measurement.

2.5 SAMPLING OF PARTICULATE MATERIALS

To characterize the properties of a particulate material, a sample of the material must be collected. Manufactured particulate materials are usually produced with a specification for chemical composition and in many cases a particle size distribution. Samples of the particulate material can be obtained while the material is in motion going from one unit operation to another. Sampling the material while it is in motion enhances the potential for random—and therefore representative— sampling.

Obtaining a representative sample of a static particulate material can be difficult, especially if the material is a fugitive dust deposit created by random, uncontrolled events. Instead of collecting a representative sample, it may be more practical to collect a judgment sample. In judgment sampling, the investigator uses their knowledge about the process operations to collect samples that will likely reflect worse case conditions. For example, one could choose to obtain samples where the dust deposits likely contained the largest fraction of fine dust. Such a sample will most likely exhibit a higher explosibility hazard than a representative sample. The wisdom of such a choice will depend on numerous case-specific factors and should be considered carefully. The drawback of judgment sampling to reveal worse case explosibility characteristics is that it could exaggerate the magnitude of the combustible dust hazard leading, in turn, to an inefficient allocation of resources for hazard control.

Another source of difficulty in obtaining representative samples of combustible dust is the cascade of diminishing sample sizes. The mass of combustible dust in a manufacturing facility may literally amount to several metric tonnes. The

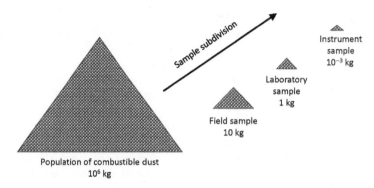

FIGURE 2.6

Subdivision of sample sizes.

sample size to be submitted to an image analysis instrument may be less than one gram. Thus the mass ratio of the population of combustible dust material to the mass of material submitted to size analysis may range over a factor of 10^9. Fig. 2.6 illustrates this broad range of sample masses as one progresses from the field to the laboratory.

Selecting a field sample of combustible dust and reducing it to a representative quantity of material measured in grams can be a daunting challenge. There are a variety of sampling techniques and tools available to the investigator (Jillavenkatesa et al., 2001, pp. 7—25). These tools are especially helpful for procuring field samples and subdividing them for the laboratory in a nonbiased manner.

Perhaps the most important aspect of sampling is the preparation of a sampling plan. There is no accepted guidance for the content and format of sampling plans for combustible dust characterization. The following suggestions in Table 2.3 should assist the investigator in developing a sampling plan that is relevant and useful for their particular application.

Following these suggestions will assist the investigator in maximizing the amount of useful data obtained from a sampling effort.

2.6 THE SIGNIFICANCE OF PARTICLE SIZE ON COMBUSTIBLE DUST HAZARDS

The particle size distribution of a particulate material has a significant if not dominant impact on dust explosion parameters. Smaller particles have faster combustion rates than larger particles. For example, the maximum rate of pressure rise increases as the mean particle diameter decreases. The maximum explosion pressure exhibits a slower but definite increase with decreasing particle diameter. These trends are demonstrated by the data in Fig. 2.7 (Dufaud et al., 2010).

Table 2.3 Suggestions for Developing a Combustible Dust Sampling Plan

No.	Suggestion	Rationale
1.	Define the objectives of the sampling program	The end purpose of the sample data will determine the best method of sampling
2.	Define the boundaries of the sampling program	The extent of the sampling activity should be specified in terms of time and space
3.	Select a sampling strategy	The sampling strategy should follow naturally from the objectives and boundaries of the program
4.	Select the particle size measurement technique and other laboratory measurements to be performed on the samples	The measurement techniques should be selected to best reflect the physical attributes of interest
5.	Determine the number, size, and location of samples to be procured	These specifications relate to the quality of the sampling program
6.	Specify a protocol for sample labeling and documentation of the sampling process	If the data are important, then it is best to document the sampling process sufficiently to ensure reproducibility of the program
7.	Consider what steps if any must be taken to preserve the integrity of the samples	Both organic and metallic materials are potentially subject to degradation over time due to exposure to air and humidity

In an attempt to capture information about the breadth of a particle size distribution, Castellanos et al. recommended the use of a polydispersity index, σ_D (Castellanos et al., 2014). The polydispersity index is defined as

$$\sigma_D = (D_{90} - D_{10})/D_{50} \qquad (2.7)$$

The symbol D_i designates the ith percentile in the cumulative particle size distribution. The polydispersity index can be thought of as the (approximate) range of the distribution divided by its median. The breadth of a given particle size distribution increases with the polydispersity index. These researchers created five blends of aluminum powder with a similar median particle size (D_{50}) and a range of polydispersity index values. They found that both the rate of pressure rise and the explosion pressure increased with increasing polydispersity (Fig. 2.8).

Both the minimum explosible concentration and the minimum ignition energy decrease as the particle diameter decreases. To achieve ignition, particles must be heated to high temperature. Given the same thermal environment, smaller particles will achieve the requisite temperature more quickly than larger particles. Fig. 2.9 shows ignition data as a function of particle size for Pittsburgh bituminous coal (Hertzberg et al., 1982).

In summary, for a fixed chemical composition and dust concentration, smaller particle sizes are easier to ignite and will result in greater explosion pressures and rates of pressure rise.

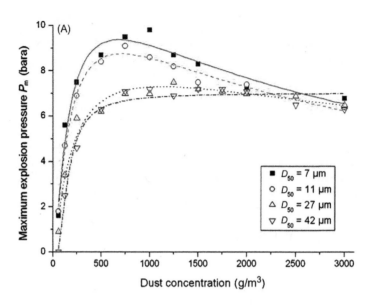

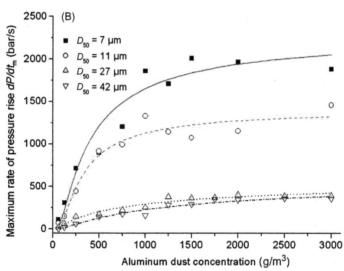

FIGURE 2.7

The effect of dust concentration and particle size on the dust explosbility parameters:
(A) explosion pressure and (B) maximum rate of pressure rise.

From Dufaud, O., Traore, M., Perrin, L., Chazelet, S., Thomas, D., 2010. Experimental investigation and modelling of aluminium dusts explosions in the 20L sphere. J. Loss Prevent. Process Ind. 23, 226–236.

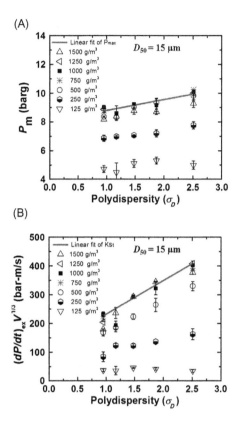

(A)

(B)

FIGURE 2.8

Explosion parameters as a function of polydispersity for aluminum dust: (A) explosion pressure and (B) the deflagration index, Kst.

From Castellanos, D., Carreto-Vazquez, V.H., Mashuga, C.V., Trottier, R., Mejia, A.F., Mannan, M.S., 2014.
The effect of particle size polydispersity on explosibility characteristics of aluminum dust. Powder Technol.
254, 331–337.

2.7 SUMMARY

This chapter briefly introduced several important concepts regarding the physical properties of combustible dust. The first concept discussed was particle size and shape. Visualizing a collection of combustible dust particles as perfect sphere of uniform shape is an important and useful conceptual model for exploring the hazards of combustible dust. However, it is also an approximation that introduces some definite, but perhaps unknown, error into the analysis. Real particles often have nonspherical shapes and come in a range of particle sizes. In addition to particle size and shape, I discussed briefly the properties of single particles,

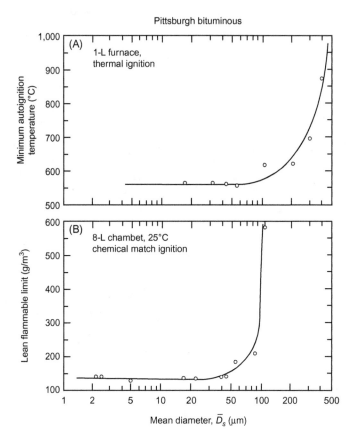

FIGURE 2.9

Ignition properties for Pittsburgh bituminous coal as a function of particle size:
(A) minimum ignition temperature and (B) minimum explosible concentration.

From Hertzberg, M., Cashdollar, D.L., Ng, D.L., Conti, R.S., 1982. Domains of flammability and thermal ignitability for pulverized coals and other dusts: particle size dependences and microscopic residue analyses. In: Nineteenth Symposium (International) on Combustion. The Combustion Institute, Elsevier, pp. 1169–1180.

populations of particles, and bulk properties of dusts. Of these properties, probably the single most important issue raised in this chapter is the description of particle size distribution of the dust, as this property plays a central role in the
combustion rate of particles.

Realistic characterization of a specific combustible dust requires a sampling
protocol that provides meaningful data. Some of the major challenges in sampling
dust deposits were reviewed and guidelines for developing a sampling protocol
were presented.

In terms of the combustible dust hazard parameters (explosion pressure and the maximum rate of pressure rise), smaller particles present greater hazards than larger particles. For a given ignition stimulus, smaller particles are more easily ignited with lower minimum explosible concentrations and lower ignition energies or temperatures. In comparing the behavior of different combustible dusts, it is important to keep the dominant role of particle size in mind.

REFERENCES

Adamson, A.W., Gast, A.P., 1997. Physical Chemistry of Surfaces, sixth ed. John Wiley & Sons, New York.

Beddow, J.K., 1980. Particulate Science and Technology. Chemical Publishing Company, Los Angeles.

Bird, R.B., Stewart, W.E., Lightfoot, E.N., 2002. Transport Phenomena, second ed. John Wiley & Sons, New York.

Castellanos, D., Carreto-Vazquez, V.H., Mashuga, C.V., Trottier, R., Mejia, A.F., Mannan, M.S., 2014. The effect of particle size polydispersity on explosibility characteristics of aluminum dust. Powder Technol. 254, 331–337.

Crowe, C.T., Schwarzkopf, J.D., Sommerfield, M., Tsuji, Y., 2012. Multiphase Flows with Droplets and Particles, second ed. CRC Press, Boca Raton, FL.

Dufaud, O., Traore, M., Perrin, L., Chazelet, S., Thomas, D., 2010. Experimental investigation and modelling of aluminium dusts explosions in the 20 L sphere. J. Loss Prevent. Process Ind. 23, 226–236.

Eckhoff, R., 2009b. Understanding dust explosions—the role of powder science and technology. J.Loss Prevent. Process Ind. 22, 105–116.

Hertzberg, M., Cashdollar, D.L., Ng, D.L., Conti, R.S., 1982. Domains of flammability and thermal ignitability for pulverized coals and other dusts: particle size dependences and microscopic residue analyses. Nineteenth Symposium (International) on Combustion. The Combustion Institute, Elsevier.

Hinds, W.C., 1999. Aerosol Technology: Properties, Behavior, and Measurement of Airborne Particles, second ed. Wiley-Interscience, New York.

Jillavenkatesa, A., Dapkunas, S.J., and Lum, L.-S.H., 2001. Particle size characterization. NIST Recommended Practice Guide, Special Publication 960-1, National Institute of Standards and Technology, Washington, DC.

Kaviany, M., 1995. Principles of Heat Transfer in Porous Media, second ed. Springer-Verlag, Berlin.

NFPA 654, 2013. Standard for the Prevention of Fire and Dust Explosions from the Manufacturing, Processing, and Handling of Combustible Particulate Solids. National Fire Protection Association, Quincy, MA.

Ogle, R.A., 1986. A New Strategy for Dust Explosion Research: a Synthesis of Combustion Theory, Experimental Design and Particle Characterization. Ph.D. thesis. University of Iowa, Iowa City, Iowa.

Rhodes, M., 2008. Introduction to Particle Technology, second ed. John Wiley & Sons, New York.

Rodriguez, J.M., Edeskar, T., Knutsson, S., 2013. Particle shape quantities and measurement techniques—a review. Electron. J. Geotech. Eng. 18, 169–198.

Schulze, D., 2008. Powders and Bulk Solids: Behavior, Characterization, Storage and Flow. Springer, Berlin.

Seville, J.P.K., Tüzün, U., Clift, R., 1997. Processing of Particulate Solids. Blackie Academic and Professional, London.

Speight, J.G., 2005. Handbook of Coal Analysis. Wiley-Interscience, New York.

Thermodynamics of dust combustion

3

Many of the theoretical limits of combustion behavior can be established through thermodynamic analysis. This chapter begins with a brief overview of chemical mixture relations, control volumes, and mass and energy balances. The next section shows how the application of stoichiometry enables the calculation of the stoichiometric dust concentration, the reference point for distinguishing between fuel-lean and fuel-rich conditions. The equivalence ratio, a traditional measure of fuel and oxygen concentrations is introduced. The section on thermochemistry describes how the chemical energy content of a combustible dust is calculated. The next two sections introduce two important limiting cases: constant pressure and constant volume combustion. The key performance parameters for constant pressure combustion are the adiabatic isobaric flame temperature and the expansion ratio of combustion products. The key performance parameters for constant volume combustion are the isochoric flame temperature and the explosion pressure. These are important metrics which characterize the maximum theoretical output of a fire or explosion. The basic means for calculating these metrics are presented and some useful approximations relating these two limiting cases are presented.

Following that, three specific applications are investigated with thermodynamic models. These three applications each represent a combustible dust hazard scenario. The first application derives the simple linear relation between the fuel burn fraction and the fractional pressure rise. This relation is invoked as an assumption in most investigations of constant volume combustion. The next application considers the question, "when is a confined deflagration a flash fire?" The third application considers the important problem of evaluating the magnitude of a secondary explosion.

Thermodynamic calculations are not typically performed when evaluating combustible dust hazards. Experts rely instead on the laboratory measurement of dust explosibility. That is because design methods for explosion protection rely on these measurements. The calculation methods presented in this chapter are not intended to replace the empirical methods. Instead, the objective here is to gain insight into combustible dust phenomena. Much of the uncertainty in describing the combustion behavior of dust is caused by the uncertainty of transport and kinetic rate processes. Thermodynamics provides predictions of limiting behavior that are independent of rate processes. Hence, the thermodynamic analysis provides a base case against which to compare real combustion behavior.

3.1 CHEMICAL MIXTURE RELATIONS

The composition of a chemical mixture can be described in terms of mass or mole units. Two convenient measures of composition are the mass fraction Y_i and the mole fraction X_i which are defined as

$$Y_i = \frac{m_i}{m} \quad \text{and} \quad X_i = \frac{n_i}{n} \tag{3.1}$$

where m_i is the mass of species i, m is the total mass of the mixture, n_i is the moles of species i, and n is the total moles of the mixture. The mass and mole fractions sum to unity for N chemical species in the mixture.

$$\sum_{i=1}^{N} Y_i = 1 \quad \text{and} \quad \sum_{i=1}^{N} X_i = 1 \tag{3.2}$$

The mass of a chemical species is related to its mole number by the relation $m_i = n_i \mathcal{M}_i$ where $\mathcal{M}_i$ is the molar mass of species i. The mixture molar mass, $\mathcal{M}_{\text{mix}}$, can be calculated using either a mole or mass fraction basis.

$$\mathcal{M}_{\text{mix}} = \sum_{i=1}^{N} X_i \mathcal{M}_i \quad \text{and} \quad \frac{1}{\mathcal{M}_{\text{mix}}} = \sum_{i=1}^{N} \frac{Y_i}{\mathcal{M}_i} \tag{3.3}$$

The following relations can be used to convert from mole to mass fractions or vice versa.

$$Y_i = \frac{X_i \mathcal{M}_i}{\sum_{i=1}^{N} X_i \mathcal{M}_i} \quad \text{and} \quad X_i = \frac{Y_i / \mathcal{M}_i}{\sum_{i=1}^{N} Y_i / \mathcal{M}_i} \tag{3.4}$$

Throughout this text, it is assumed that the ideal gas equation of state can be used to describe the volumetric properties of both unburnt and burnt gas mixtures. The ideal gas equation of state takes the following form for a gas consisting of a single chemical species.

$$P \mathcal{M} = \rho R_g T \tag{3.5}$$

where P is the absolute pressure, $\mathcal{M}$ is the molar mass of the single chemical species, ρ is the gas density, R_g is the universal gas constant (8314 J/kmol-K), and T is the absolute temperature.

It will be assumed throughout this text that gas mixtures form ideal gas mixtures. In an ideal gas mixture, it is assumed that the molecules do not interact with each other. Each chemical species exerts its own pressure contribution independent of the other species. Thus the total pressure of an ideal gas mixture is equal to the sum of the partial pressures of each species. The same is true for the gas volume.

$$P = \sum_{i=1}^{N} P_i \quad \text{and} \quad V = \sum_{i=1}^{N} V_i \tag{3.6}$$

Table 3.1 Composition of Air

Chemical Species	Mole Fraction	Mass Fraction	Molar Mass (kg/kmol)
N_2	0.79	0.77	28.0
O_2	0.21	0.23	32.0
Mixture	1.00	1.00	28.85

The thermodynamic properties of ideal gas mixtures take a particularly simple form: the property of the mixture is equal to the sum of that property for each individual component weighted by its mole fraction (O'Connell and Haile, 2005, Chapter 4). For example, the mixture heat capacity at constant pressure $C_{p,mix}$ is calculated in mole units by the relation:

$$\hat{C}_{p,mix} = \sum_{i=1}^{N} X_i \, \hat{C}_{p,i} \tag{3.7}$$

For the ideal gas mixture, the equation of state takes the same form as for the pure component form, but it is written for the molar mass and density of the mixture.

$$P \, \mathcal{M}_{mix} = \rho_{mix} \, R_g \, T \tag{3.8}$$

The mixture density can be calculated using mass fractions

$$\frac{1}{\rho_{mix}} = \sum_{i=1}^{N} \frac{Y_i}{\rho_i} \tag{3.9}$$

The N chemical species are all assumed to be in the gas phase. The adjustment of the ρ_{mix} term for solid components requires some discussion about multiphase systems and will be addressed in Chapter 4.

To simplify mass balance calculations, the molar composition of air will be assumed to be fixed at 79.0% nitrogen (N_2) and 21% oxygen (O_2). Expressed as a mole ratio, there are 3.76 moles of N_2 for every mole of O_2 (0.79/0.21=3.76). For this nominal composition, the molar mass of air is equal to 28.85 kg/kmol. For reference, the composition of air in both mass and mole units is given in Table 3.1.

The pressure for standard atmosphere equals 1.01325×10^5 Pa or 1.01325 bars (absolute). Absolute pressures are reported in this text as bar and gauge pressures are reported as barg.

EXAMPLE 3.1

This example illustrates the calculation of ideal gas mixture properties. Calculate the mixture molar mass for a dust cloud consisting of 500 g/m^3 of iron dust in air. Assume the cloud is at atmospheric pressure and a temperature of 300 K.

Solution

Mole units are selected for the calculation (mass units could be used equally well). Select a basis of 1 m³ of the dust cloud for the calculation. The dust cloud consists of 500 g iron and 1 m³ air. The moles of iron is calculated directly from its mass:

$$\text{Moles Fe} = \frac{500 \text{ g Fe}}{55.845 \text{ g/mol}} = 8.953 \text{ mol Fe}$$

The moles of air are calculated from the ideal gas equation of state:

$$n_{air} = \frac{P\,V}{R_g\,T} = \frac{(1.01325 \times 10^5 \text{ Pa}) (1 \text{ m}^3)}{(8314 \text{ J/(kmol} \cdot \text{K)})(300 \text{ K})} = 40.6 \text{ mol air}$$

$$X_{air} = \frac{n_{air}}{n_{total}} = \frac{40.6}{8.953 + 40.6} = 0.819$$

$$X_{Fe} = 1 - X_{air} = 0.181$$

The mixture molar mass is calculated as

$$\mathcal{M}_{mix} = \sum_{i=1}^{N} X_i \mathcal{M}_i = (0.819)(28.85) + (0.181)(55.845) = 33.7 \text{ g/mol}$$

This is the desired result.

3.2 MASS AND ENERGY BALANCES

A first step toward understanding the physical and chemical behaviors of a system is to understand its mass and energy balance relationships. Thermodynamics supplies the tools needed to investigate these relationships. A thermodynamic analysis begins with the definition of the system of interest. There are two types of systems in thermodynamics: closed systems which do not exchange mass with the environment and open systems which do exchange mass with the environment. The definition of a system requires a specification of its boundary. The boundary is the interface that separates the system from its environment. Fig. 3.1 illustrates these two types of systems.

A closed system is called a control mass. The 20-L spherical test chamber is an example of a control mass. A finite quantity of combustible dust dispersed inside the test chamber is ignited. The dust reacts to form combustion products. The chamber walls are impermeable so no material leaves the chamber. There may or may not be significant heat loss through the chamber walls. This is a closed system and the dust−air mixture can be described as a control mass.

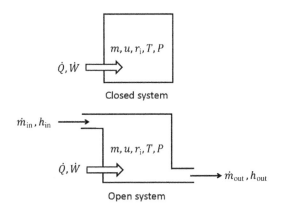

FIGURE 3.1

Illustration of closed and open thermodynamic systems.

An open system is called a control volume. A dust collector venting flame and hot gases through its explosion vent is an example of a control volume. Dusty air enters the dust collector and flows through the bag filters. Clean air is exhausted from the dust collector leaving the dust to accumulate inside. Since the inflow and outflow cross the boundary of the dust collector, it is an open system. Similarly, the venting of flame and hot gases through the explosion vent and into the surroundings is an outflow from an open system.

In control volume analysis, gradients within the system are generally neglected, ie, the system is assumed to be spatially uniform. Balance equations are used to describe the changes that occur within the system. The balance equations take the following form:

$$\text{Accumulation} = \text{Inflow} - \text{Outflow} + \text{Generation} \qquad (3.10)$$

The accumulation term describes the changes that occur within the system, the inflow and outflow terms are the flows that cross the system boundary, and the generation term accounts for chemical reactions that occur within the system. Three balance equations are required for analyzing combustible dust phenomena: total mass, species mass, and energy. The total mass balance is

$$\frac{dm}{dt} = \dot{m}_{\text{in}} - \dot{m}_{\text{out}} \qquad (3.11)$$

where m is the mass of the system and $\dot{m}$ is the mass flow into or out of the system. The species mass balance has an extra term for chemical reactions.

$$\frac{dm_i}{dt} = \dot{m}_{i,\text{in}} - \dot{m}_{i,\text{out}} + r_i \qquad (3.12)$$

where m_i is the mass of chemical species i, $\dot{m}_i$ is the mass flow of species i, and r_i is the rate of chemical reaction of species i. When writing species mass balances, there are only $n-1$ independent species balances. The total mass balance provides the additional n-th mass balance.

The energy balance written in terms of energies takes this form (neglecting changes in kinetic or potential energy):

$$\frac{d}{dt}(mu) = \dot{m}_{\text{in}} h_{\text{in}} - \dot{m}_{\text{out}} h_{\text{out}} + \dot{Q} + \dot{W} \tag{3.13}$$

where u is the specific internal energy (energy per unit mass), h is the specific enthalpy, $\dot{Q}$ is the heat flow rate into the system and $\dot{W}$ is the rate of work (or power) added to the system.[1] Each of these equations can be generalized to account for multiple inflow or outflow streams. The specific energy quantities, u and h, are intensive quantities because they are normalized by the mass of the system or the mass flow rate of the stream. Capital letters are used to designate extensive quantities. Extensive quantities are formed by multiplying the specific energy quantity by its associated mass. For example, the internal energy U is calculated as $U = mu$.

For each chemical species, there are three potential contributions to its enthalpy: the sensible enthalpy, the enthalpy of formation, and the enthalpy of a phase transition. The sensible enthalpy for a chemical species undergoing a temperature change can be written in this form:

$$h_i(T) = \int_{T_{\text{ref}}}^{T} C_{\text{p}} \, dT \tag{3.14}$$

The sensible enthalpy is the thermal contribution and for ideal gases is a function of temperature only. The enthalpy of formation $\Delta h_{f,i}^0$ is the chemical energy required to form a particular chemical species from its constituent elements. The enthalpy of phase transition is the energy change required to transform the material from one phase to another. In this book, the only phase transitions of interest will be melting Δh_{m} and vaporization Δh_{fg}.

3.3 STOICHIOMETRY OF COMBUSTION

The combustible dust hazard parameters are a strong function of the fuel concentration. The relative importance of the physical and chemical processes responsible for dust flame propagation can vary depending on whether the fuel concentration is fuel lean or fuel rich. The equivalence ratio, one of the concentration measures introduced in this section, is a convenient way to express fuel-lean or fuel-rich conditions with a dimensionless number. Stoichiometric measures will be discussed in both mole and mass units. It is usually easier to work out the stoichiometry of a chemical reaction in mole units, but mass units are more convenient for calculations involving balance equations.

---

[1]The convention followed here is that work is a positive quantity when work is performed on the system by the surroundings.

The first step toward understanding the thermodynamics of combustible dust is to write out the stoichiometry of the combustion reaction. The reaction stoichiometry establishes the quantity of oxidizer needed to react with a given fuel. Combustible dust comes in two broad chemical categories: organic and inorganic. Most of the inorganic combustible dusts are metals, but there are two notable exceptions: silicon and sulfur (Cashdollar and Zlochower, 2007). This discussion on stoichiometry will concentrate on organic and metallic fuels in atmospheric air, and it is assumed that the combustible dust under consideration can be described by a single molecular formula.

The stoichiometry for a combustion reaction is based on the premise that the reaction proceeds to the most stable oxidation states of the combustion products. The combustion reaction for an elemental metal $\mathbb{M}$ can be written in a generalized format:

$$x\,\mathbb{M}(\text{solid}) + y(O_2 + 3.76\ N_2) \rightarrow \mathbb{M}_xO_{2y}\ (\text{solid}) + 3.76\ yN_2 \qquad (3.15)$$

In the case of metal combustion, at stoichiometric conditions there is a net decrease in the number of gas moles. In an isothermal system held at constant volume, the decrease in the number of gas moles favors a decrease in system pressure.

EXAMPLE 3.2

Calculating initial and final quantities for complete combustion of aluminum dust in air. Assume a basis of 1 mole of aluminum.

Solution

This example illustrates the use of species mass balances. Begin by writing a balanced chemical reaction for aluminum combustion in air.

$$Al(s) + \frac{3}{4}(O_2 + 3.76\ N_2) \rightarrow \frac{1}{2}Al_2O_3(l) + \frac{3}{4}(3.76)N_2$$

The table below summarizes the initial and final quantities of each chemical species.

Species	Molar Mass (g/mol)	Initial Moles (mol)	Initial Mass (g)	Final Moles (mol)	Final Mass (g)
Al	26.98	1.00	26.98	0	0.00
O_2	32.00	0.75	24.00	0	0.00
N_2	28.00	2.82	78.96	2.82	78.96
Al_2O_3	101.96	0	0.00	0.50	50.98
Total		4.57	129.94	3.32	129.94

As dictated by the conservation of mass, the final mass equals the initial mass.

With organic fuels with an elemental composition of carbon, hydrogen, and oxygen (CHO fuels), the completely oxidized combustion products are carbon dioxide, CO_2, and water, H_2O. With the combustion reaction so written, the relative quantities of fuel and oxidizer are specified. The stoichiometry for the complete combustion of a hydrocarbon fuel, represented generally as $C_xH_yO_z$, can be written as a balanced chemical reaction (mole units):

$$C_xH_yO_z + \left(x + \frac{y}{4} - \frac{z}{2}\right)(O_2 + 3.76N_2) \rightarrow xCO_2 + \frac{y}{2} H_2O + 3.76\left(x + \frac{y}{4} - \frac{z}{2}\right)N_2 \quad (3.16)$$

This equation gives the stoichiometric mole quantity of air required for complete combustion of a single mole of fuel. For combustible dusts the fuel is a condensed phase, so with combustion there is a net increase in the number of gas moles. In an isothermal system held at constant volume, the increase in the number of gas moles favors an increase in system pressure.

EXAMPLE 3.3

Calculating initial and final quantities for complete combustion of corn dust. Assume corn dust has the empirical formula $CH_{2.01}O_{0.80}$ (Medina et al., 2013).

Solution

Assume a basis of 1 mole of corn dust. The balanced chemical reaction can be developed from Eq. (3.15) with $x = 1$, $y = 2.01$, and $z = 0.8$.

$$CH_{2.01}O_{0.80} + 1.1025(O_2 + 3.76 N_2) \rightarrow CO_2 + 1.005 H_2O + 7.1534 N_2$$

The table below summarizes the initial and final quantities of chemical species:

Species	Molar Mass (g/mol)	Initial Moles (mol)	Initial Mass (g)	Final Moles (mol)	Final Mass (g)
$CH_{2.01}O_{0.80}$	26.84	1.00	26.84	0.00	0.00
O_2	32.00	1.1025	35.28	0.00	0.00
N_2	28.00	4.1454	116.07	4.1454	116.07
CO_2	44.011	0.00	0.00	1.00	44.011
H_2O	18.016	0.00	0.00	1.005	18.106
Total	–	6.2479	178.19	6.1604	178.19

As dictated by the conservation of mass, the mass of products equals the mass of reactants.

The fuel/air ratio F/A is the next step toward defining the equivalence ratio Φ. The F/A ratio is defined as the ratio of fuel mass to air mass. The stoichiometric fuel/air ratio $(F/A)_{st}$ can be calculated for the general hydrocarbon fuel $C_xH_yO_z$ by the relation:

$$(F/A)_{st} = \frac{\mathcal{M}_F}{4.76\left(x + \dfrac{y}{4} - \dfrac{z}{2}\right)\mathcal{M}_A} \tag{3.17}$$

The subscripts F and A denote fuel and air, respectively. The equivalence ratio Φ is defined as the actual fuel/air ratio divided by the stoichiometric fuel/air ratio:

$$\Phi = \frac{(F/A)_{actual}}{(F/A)_{st}} = \frac{m_{F,actual}}{m_{F,st}} \cdot \frac{m_{A,st}}{m_{A,actual}} = \frac{n_{F,actual}\mathcal{M}_F}{n_{F,st}\mathcal{M}_F} \cdot \frac{n_{A,st}\mathcal{M}_A}{n_{A,actual}\mathcal{M}_A} = \frac{n_{F,actual}}{n_{F,st}} \cdot \frac{n_{A,st}}{n_{A,actual}} \tag{3.18}$$

Thus the equivalence ratio can be expressed in units of mass or moles. The value of Φ can range from 0 to ∞. The stoichiometric condition corresponds to $\Phi = 1$. The mixture is fuel lean when $\Phi < 1$ and is fuel rich when $\Phi > 1$. Conversely, one can refer to a fuel-lean mixture as having an excess air content, and a fuel-rich mixture as having an air deficient content. In fuel-lean conditions, the fuel is the limiting reactant for combustion. In fuel-rich conditions, the oxidizer is the limiting reactant for combustion.

With combustible dusts, one often finds the fuel quantity expressed as a dust concentration in air (g/m^3). The equivalence ratio can be expressed in terms of the dust concentration. Consider first an organic fuel of the form $C_xH_yO_z$. Assuming the dust is dispersed in air, the ideal gas equation of state, and the initial conditions are 1 atmosphere and the temperature is 298 K, the stoichiometric concentration of dust $\rho_{d,st}$ can be calculated as (remember dust = fuel)

$$\rho_{d,st} = \frac{m_{d,st}}{V} = \frac{n_{d,st}\mathcal{M}_d}{V} = \frac{n_{O_2,st}\mathcal{M}_d}{\left(x + \dfrac{y}{4} - \dfrac{z}{2}\right)V} = \frac{0.21\mathcal{M}_d}{\left(x + \dfrac{y}{4} - \dfrac{z}{2}\right)}\left(\frac{P}{R_gT}\right)$$

$$= \frac{0.21\mathcal{M}_d}{\left(x + \dfrac{y}{4} - \dfrac{z}{2}\right)}\left[\frac{1.01325 \times 10^5 \text{ Pa}}{(8.314 \text{ J/mol} \cdot \text{k})(298 \text{ K})}\right] \tag{3.19}$$

This gives a simple result for calculating the stoichiometric dust concentration in g/m^3 for any organic fuel of the form $C_xH_yO_z$ with a molar mass $\mathcal{M}_d$ expressed in g/mol:

$$\rho_{d,st} = \frac{8.59\mathcal{M}_d}{\left(x + \dfrac{y}{4} - \dfrac{z}{2}\right)} \tag{3.20}$$

With this calculation established, the equivalence ratio for hydrocarbon fuels can now be expressed in terms of the actual dust concentration:

$$\Phi = \frac{(m_d/m_{O_2})_{actual}}{(m_d/m_{O_2})_{st}} = \frac{[(\rho_d V)/m_{O_2}]_{actual}}{[(\rho_d V)/m_{O_2}]_{st}} = \frac{\rho_{d,actual}}{\rho_{d,st}} \tag{3.21}$$

Substituting the expression above for the stoichiometric dust concentration gives the following expression for the equivalence ratio in terms of measurable quantities:

$$\Phi = \frac{\rho_{d,actual}}{8.59 \mathcal{M}_d} \left(x + \frac{y}{4} - \frac{z}{2} \right) \tag{3.22}$$

This simple derivation has brought us to a very important point: for a given combustible dust, if the molecular formula or ultimate analysis is available, it is a simple task to calculate the stoichiometric dust concentration and, more importantly, the equivalence ratio for any given dust concentration.

Similarly, for metal combustion with the metal oxide product $M_x O_{2y}$, the stoichiometric dust concentration and equivalence ratio are given by the following expressions:

$$\rho_{d,st} = \frac{8.59 \mathcal{M}_d}{y} \tag{3.23}$$

$$\Phi = \frac{(\rho_{d,actual})y}{8.59 \mathcal{M}_d} \tag{3.24}$$

Even at high dust concentrations, the volume fraction of solid particles occupies a very small portion of the dust cloud. This means that although the total number of moles of oxygen in the dust cloud is lower compared to the same volume of pure air, the concentration of oxygen in the gas phase is unchanged. This will become an important observation in Chapter 8 when we discuss the lack of a true maximum explosion concentration for combustible dusts. To demonstrate the effect of particle volume first requires the introduction of the concept of a multiphase mixture. Consider a dust cloud dispersed in air. The cloud volume consists of two components: the solid phase volume and the gas volume. These two components can be expressed as a dust (solid) volume fraction α and a gas volume fraction $1 - \alpha$.

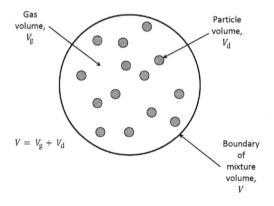

FIGURE 3.2

Definition sketch for a multiphase mixture.

$$\alpha = V_d/V \quad \text{and} \quad 1-\alpha = V_g/V \tag{3.25}$$

where V_d is the volume of dust particles, V_g is the volume of gas excluding the dust particles, and V is the combined volume of dust particles and gas (Fig. 3.2).

In combustible dust investigations, the fuel concentration is often expressed as the dust concentration ρ_d in grams of dust per cubic meter of dust cloud. The dust concentration is the mixture density ρ_m and can be calculated from the phase densities by the expression $\rho_d = Y_{F,0}\,\rho_m = Y_{F,0}\left[\alpha\rho_s + (1-\alpha)\,\rho_g\right]$.

If the dust can be idealized as an assembly of monosized spheres with diameter d_p, the dust volume fraction can be calculated by the following expression:

$$á = \frac{\text{volume of particles}}{\text{volume of mixture}}$$

$$= (\text{volume of single particle})\left(\frac{\text{number of particles}}{\text{volume of mixture}}\right) \tag{3.26}$$

$$= \left(\frac{\pi}{6}d_p^3\right)\left[\frac{\rho_d}{\rho_s\left(\frac{\pi}{6}d_p^3\right)}\right]$$

Which gives the simple result:

$$\alpha = \frac{\rho_d}{\rho_s} \tag{3.27}$$

For most practical combustible dust hazard scenarios, the volume fraction of dust particles will be much less than 1 ($\alpha \ll 1$). To get a feeling for the volume concentration of dust particles in a dense cloud, consider a cloud of graphite dust with a dust concentration of $1000\ \text{g/m}^3$ (equivalence ratio of approximately 10) and a solid density of $2300\ \text{kg/m}^3$. The volume fraction is calculated as $\alpha = \rho_d/\rho_s = (10^3\ \text{g/m}^3)/(2.3 \times 10^6\ \text{g/m}^3) = 4.3 \times 10^{-4}$. Even though this dust cloud is extremely fuel-rich, the dust particles occupy only 0.043% of the cloud volume. Thus, the volume fraction of solid particles occupies a very small portion of the dust cloud.

The following example illustrates how to calculate the stoichiometric dust concentration using the traditional mass balance approach.

EXAMPLE 3.4

(A) Calculate the stoichiometric concentration of dust for the following organic materials: cellulose, corn flour, lycopodium, and carbon. (B) Calculate the stoichiometric concentration for the following inorganic materials: aluminum, sulfur, and iron assuming that the products of combustion are Al_2O_3, SO_3, and Fe_2O_3, respectively. The ambient conditions are atmospheric pressure at 1.01325×10^5 Pa and a temperature of 300 K. Assume that air obeys the ideal gas equation of state.

Solution

Select a basis of 1 m³ of dust–air mixture. To determine the stoichiometric concentration of dust, the solution strategies are (1) determine the number of moles of O_2 in the basis volume of dust–air mixture; (2) calculate the number of moles of fuel that can react with the moles of O_2; (3) convert the mole quantity of fuel into a mass quantity; (4) divide the stoichiometric mass of fuel by the volume of the basis to determine the stoichiometric dust concentration.

$$n_{O_2} = X_{O_2} \, n_{air} = \frac{0.21 \, PV}{R_g \, T} = \frac{0.21(1.01325 \times 10^5 \text{ Pa}) \, (1 \text{ m}^3)}{(8314 \text{ J/kmol} \cdot \text{K}) \, (300 \text{ K})}$$

$$n_{O_2} = 8.531 \times 10^{-3} \text{ kmol } O_2 = 8.531 \text{ mol } O_2$$

For the organic fuels, the number of moles of fuel is calculated from the relation:

$$n_{dust} = \frac{n_{O_2}}{\left(x + \dfrac{y}{4} - \dfrac{z}{2}\right)}$$

The mass of fuel is then obtained by multiplying the moles of fuel by its molar mass. The table below summarizes the calculations for the organic materials. The sources for the empirical formulas are cellulose (Sami et al., 2001, p. 178); corn flour and lycopodium (Medina et al., 2013, p. 103).

Organic Fuel	Empirical Formula	Molar Mass (g/mol)	$x + \dfrac{y}{4} - \dfrac{z}{2}$	Mol. Fuel	Stoich. Dust Concentration (g/m³)
Cellulose	CH_2O	30.03	1	8.531	256
Corn flour	$CH_{2.01}O_{0.80}$	26.84	1.103	7.737	208
Lycopodium	$CH_{1.65}O_{0.22}$	17.19	1.303	6.550	113
Carbon	C	12.011	1	8.531	103

The results indicate that as the fuel molar mass increases, the stoichiometric dust concentration increases.

For the inorganic fuels, the number of moles of fuel is calculated from the relation:

$$n_{dust} = \frac{n_{O_2}}{y}$$

The mass of fuel is then obtained by multiplying the moles of fuel by its molar mass. The table below summarizes the calculations for the inorganic materials.

Inorganic Fuel	Molar Mass (g/mol)	$\frac{1}{y}$	Mol. Fuel	Stoich. Dust Concentration (g/m^3)
Aluminum	26.98	1.333	11.37	307
Sulfur	32.06	0.667	5.687	182
Iron	55.847	1.333	11.37	635

The results indicate that the relationship between the stoichiometric dust concentration and the molar mass is not strictly related to the molar mass because of the influence of the stoichiometric coefficients.

3.4 THERMOCHEMISTRY OF COMBUSTION

Combustion transforms the chemical energy content of a combustible dust into the destructive potential of a fire or explosion. This destructive potential can be gauged by the maximum potential temperature or pressure rise that can be achieved by combustion. The heat released by the combustion of a particular fuel with air is reported in one of three ways: the enthalpy of combustion, the higher heating value (HHV), or the lower heating value (LHV). While the enthalpy of combustion can be reported in either mole or mass units, HHV and LHV are usually reported in mass units. The enthalpy of combustion is negative in value signifying that it is an exothermic reaction. The HHV and LHV values are reported as positive quantities. The enthalpy of combustion is calculated from the enthalpies of formation for the reactants and product with the assumption that any water formed by the combustion reaction remains in the vapor phase. The enthalpy of formation for chemical species is usually reported on a mole basis. The calculation of the enthalpy of combustion is most easily performed on a basis of a unit mole of fuel.

$$\Delta \hat{H}_c^0 = \sum_{i=1}^{p} n_i(\text{products}) \, \Delta \hat{h}_{f,i}^0 - \sum_{i=1}^{r} n_i(\text{reactants}) \, \Delta \hat{h}_{f,i}^0 \tag{3.28}$$

The summation indices p and r represent the number of product and reactant species, respectively. The enthalpy of combustion for a large number of elements and chemical compounds is tabulated in standard references for thermodynamic data (Lide, 2006; Dean, 1999).

LHV is equal to the negative value of the enthalpy of combustion, ie, LHV is reported as a positive value. LHV assumes that the water formed by combustion remains in the vapor phase. The HHV value is the enthalpy of combustion assuming water is condensed to the liquid phase. The HHV value is greater than LHV and is calculated by adding the enthalpy of vaporization for water to the LHV value.

$$HHV = LHV + \Delta h_{fg} \tag{3.29}$$

The enthalpy of vaporization h_{fg} for water is 2.44 MJ/kg or 43.92 MJ/kmol (McAllister et al., 2011, pp. 23−24).

The situation is more complex for natural materials like wood, coal, and agricultural grains due to the inherent compositional variability of these materials. For natural materials, the enthalpy of combustion must either be measured using calorimetry or estimated from published correlations. An example of a correlation for natural materials based on the ultimate analysis of the fuel is shown here (Channiwala and Parikh, 2002)

$$HHV \, (MJ/kg) = 34.91C + 117.83H + 10.05S - 10.34O - 1.51N - 2.11A \tag{3.30}$$

In the equation, the variables C, H, S, O, N, A are the mass fractions for carbon, hydrogen, sulfur, oxygen, nitrogen, and ash, respectively, on a dry basis. The applicability and limitations of the correlation are explained in the original reference. It is interesting to note that the above correlation was derived from data for gases and liquids as well as solids.

EXAMPLE 3.5

Calculate the HHV and LHV of wheat dust and brown coal with the proximate analysis and ultimate analysis data provided by Wolinski and Wolanski (1993).

Solution

The proximate analysis data are presented in the table below:

Proximate Analysis (%)	Wheat	Brown Coal
Fixed carbon	5	39.1
Volatile matter	78	54.5
Ash	6.7	3.7
Moisture	9.5	2.7

The ultimate analysis data are presented in the table below:

Ultimate Analysis (%)	Wheat	Brown Coal
C	50.95	77.76
H	5.93	6.42
N	1.56	1.13
S	0.3	2.96
O	41.26	11.73

The HHV values are calculated directly using the empirical correlation developed by Channiwala and Parikh (2002). LHV values are obtained by subtracting the enthalpy of vaporization of water.

$$HHV(wheat) = 34.91(0.5095) + 117.83(0.0593) + 10.05(0.003)$$
$$- 10.34(0.4126) - 1.51(0.0156) - 2.11(0.067)$$

$$HHV(wheat) = 20.37 \text{ MJ/kg}$$

$$LHV(wheat) = HHV - \Delta h_{fg}(water) = 20.37 - 2.44$$

$$LHV(wheat) = 17.93 \text{ MJ/kg}$$

Similarly for brown coal

$$HHV(brown\ coal) = 33.70 \text{ MJ/kg}$$
$$LHV(brown\ coal) = 31.26 \text{ MJ/kg}$$

The calculated LHV values are compared with the measured values in the table below:

LHV (MJ/kg)	Wheat	Brown Coal
Calculated	17.9	31.3
Measured	16.1	30.2
Error %	11.2	3.6

The agreement of the calculated values with the measured values is reasonably good.

3.5 FLAMES AND CONSTANT PRESSURE COMBUSTION

Constant pressure combustion can be achieved in both laboratory settings and in accident scenarios. In the laboratory, investigators have devised a number of test fixtures that allow them to create both stationary and propagating dust flames (Cassel, 1964; Horton et al., 1977; Goroshin et al., 1996; Shoshin and Dreizin, 2002). On an industrial scale, coal-fired power plants use large burner systems to burn pulverized coal in stationary, but very turbulent, flames (Borman and Ragland, 1998, pp. 506–530). In the context of an accident scenario, a flash fire is a form of constant pressure combustion with the flame propagating through a dust cloud.

In constant pressure combustion, the maximum dust flame temperature that can be achieved is the adiabatic flame temperature because there are no heat losses. The adiabatic condition also results in the maximum expansion ratio or density ratio for the gaseous combustion products. The adiabatic flame temperature for a combustible dust is a function of the equivalence ratio. For fuel-lean conditions ($\Phi < 1$), the flame temperature increases with increasing fuel concentration. The flame temperature for most combustible dusts reaches a maximum in fuel-rich conditions ($\Phi > 1$), and then gradually declines with increasing fuel loading.

In some portions of this discussion, I will refer to adiabatic isobaric flame temperatures to emphasize the constant pressure requirement in contrast to the adiabatic isochoric (constant volume) flame temperature.

3.5.1 PHYSICAL DESCRIPTION

A dust flame is a hot, luminous zone of gas and burning particles. A thermodynamic model of a constant pressure flame must, by necessity, neglect the finite rate effects of transport phenomena and chemical kinetics because the model is based on the assumption of thermodynamic equilibrium. The model assumes that the combustible dust cloud is instantaneously converted from reactants into combustion products. Although this chemical conversion occurs in a finite volume, the flame is allowed to expand freely to maintain a constant pressure equal to the pressure of its surroundings. The flame temperature calculated for the constant pressure scenario is an ideal limit of thermal behavior. Real flame behavior may not match these assumptions.

Departures of real dust flame behavior from the predictions of the thermodynamic model will be caused by the finite rate processes. To evaluate the impact of these finite rate processes, one should consider physical and chemical phenomena at two length scales: the macroscale and the microscale. The macroscale dimension corresponds to the length scale of the flame (the burner diameter if a stationary flame or the fireball diameter if a propagating flame). The microscale dimension corresponds to a characteristic particle diameter (Fig. 3.3). At the macroscale, heat loss due to convection and thermal radiation can contribute to departures from equilibrium. At the microscale, the temperature of the burning particles may differ from the gas phase temperature. Thus there will be heat transfer between the particles and the gas phase. The microscale environment is strongly influenced by the volatility of the fuel. For combustible dusts of low volatility, each particle may be surrounded by its own flame. High volatility combustible dusts may form a flame volume that surrounds the boundary of the dust cloud.

The power of the thermodynamic model for the constant pressure flame is that it stands as a reliable reference point from which to judge the impact of finite rate processes. Real flame temperatures will be lower than the adiabatic flame temperature due primarily to heat losses to the surroundings and incomplete combustion of the dust.

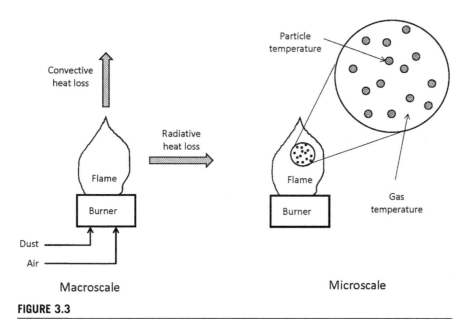

FIGURE 3.3

Illustration of the macroscale and microscale dimensions for a dust flame burning at constant pressure.

3.5.2 CALCULATION OF THE ADIABATIC ISOBARIC FLAME TEMPERATURE

Two methods for calculating the adiabatic isobaric flame temperature are now described. The first method is very approximate as it relies on two major simplifying assumptions: (1) constant thermodynamic properties and (2) complete combustion. Complete combustion here refers to the complete consumption of the fuel in fuel-lean and stoichiometric conditions and it refers to the complete consumption of oxygen in fuel-rich conditions. In fuel-lean conditions, the assumption of complete combustion describes well the distribution of combustion products. In fuel-rich conditions, the assumption of complete combustion begins to break down. As the reactant mixture becomes more fuel rich, products of incomplete combustion increase in both variety and concentration. The chemical equilibrium analysis of combustion is discussed at length in the combustion textbooks by Turns and Glassman and coworkers (Turns, 2012, pp. 38–53; Glassman et al., 2015, pp. 1–34 and 478–501).

It is assumed that the initial state of the combustible dust mixture is at ambient temperature T_0 and atmospheric pressure P_0. The final condition is the adiabatic isobaric flame temperature T_{ad} and atmospheric pressure P_0 (Fig. 3.4).

First consider the simple method for calculating the isobaric flame temperature. This approach is especially useful if the only thermodynamic data available to the investigator is the enthalpy of combustion in the form of LHV or HHV data. The calculation is based

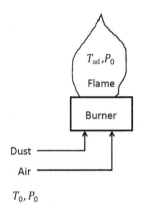

FIGURE 3.4

Definition sketch for the adiabatic isobaric flame temperature.

on the following assumptions: constant heat capacities for each of the gaseous combustion products and negligible dissociation effects. The adiabatic flame temperature is determined from an energy balance written for the combustion process. Neglecting heat losses, work interactions, and changes in the kinetic or potential energy of the flow streams, the energy balance for isobaric combustion in mole units is

$$\hat{H}_{\text{reactants}} = \hat{H}_{\text{products}} \qquad (3.31)$$

Substituting the enthalpy expressions into the energy balance yields

$$\hat{H}_{\text{reactants}} = \sum_{i=1}^{r} n_i \, \hat{h}_i = \sum_{i=1}^{r} n_i \left[\hat{h}_{\text{f},i}^0 + \hat{C}_{\text{p},i}(T_0 - T_{\text{ref}}) \right] \qquad (3.32)$$

$$\hat{H}_{\text{products}} = \sum_{i=1}^{p} n_i \, \hat{h}_i = \sum_{i=1}^{p} n_i \left[\hat{h}_{\text{f},i}^0 + \hat{C}_{\text{p},i}(T_{\text{ad}} - T_{\text{ref}}) \right] \qquad (3.33)$$

For hydrocarbon−air mixtures, it is usually a good assumption to assign equal heat capacities for both reactants and products. We assume throughout this derivation that the mixture forms an ideal solution. The temperature of the reactant is the initial temperature T_0. The usual convention for the reference temperature T_{ref} for the enthalpy of formation is 298 K and the reference pressure is atmospheric pressure or 101.325 kPa. Recalling the definition of the enthalpy of combustion

$$\Delta \hat{H}_c = \sum_{i=1}^{p} n_i(\text{products}) \, \Delta \hat{h}_{\text{f},i}^0 - \sum_{i=1}^{r} n_i(\text{reactants}) \, \Delta \hat{h}_{\text{f},i}^0 \qquad (3.34)$$

Solving for the flame temperature yields the following expression in mole units (assuming the initial temperature is equal to the reference temperature):

$$T_{\text{ad}} = T_{\text{ref}} + \frac{n_F \left(\Delta \hat{h}_c \right)}{\sum_{i=1}^{p} n_i \, \hat{C}_{\text{p},i}} \qquad (3.35)$$

In this equation, n_F is the moles of fuel and $\Delta \ddot{h}_c$ is the molar enthalpy of combustion. In fuel-lean conditions, the adiabatic flame temperature is limited by the mass of fuel available for combustion. Following McAllister et al. (McAllister et al., 2011, pp. 31−34), the adiabatic flame temperature relation can be written in a convenient form using mass units for fuel-lean conditions and LHV

$$T_{ad} = T_0 + \frac{\Phi (F/A)_{st} LHV}{\left[1 + (F/A)_{st} \Phi\right] C_{p,mean}} \tag{3.36}$$

where $C_{p,mean}$ designates the mean specific heat capacity for the gaseous product mixture calculated at some mean temperature (where it is assumed that the mean heat capacity of the reactants equals the mean heat capacity of products). It is defined by the relation:

$$C_{p,mean} = \frac{1}{n_{p,total}} \sum_{i=1}^{p} n_i \, C_{p,i} \tag{3.37}$$

The mean temperature generally recommended is the mean temperature between the reactant temperature and the flame temperature. For hydrocarbon fuels, a good first estimate of the mean temperature is 1200 K (Turns, 2012, pp. 31−35).

In fuel-rich conditions, the adiabatic flame temperature is limited by the mass of oxygen available for combustion. For fuel-rich conditions, the adiabatic flame temperature takes the following form:

$$T_{ad} = T_0 + \frac{(F/A)_{st} LHV}{\left[1 + (F/A)_{st} \Phi\right] C_{p,mean}} \tag{3.38}$$

The procedure for calculating the adiabatic isobaric flame temperature T_{ad} is shown in Table 3.2.

Detailed examples on adiabatic temperature calculations can be found in combustion textbooks (McAllister et al., 2011, Chapter 2; Turns, 2012, Chapter 2; Kuo, 2005, Chapter 1). The flame temperature depends not only on the fuel concentration in air but also on the composition of combustion products formed. Thus far the discussion has assumed that combustion is complete. For organic fuels, this means that only carbon dioxide and water are formed.

Table 3.2 Calculation Procedure for the Adiabatic Flame Temperature

Calculation Steps

1. Write the complete combustion reaction equation
2. Determine the equivalence ratio
3. Compute the combustion mass balance and determine the composition of the combustion products
4. Gather the thermochemical data: LHV, C_p
5. Compute the adiabatic flame temperature from the energy balance

This assumption is reasonably met with gaseous fuels when the fuel concentration ranges from fuel lean to stoichiometric. In this concentration range, the adiabatic flame temperature is approximately linear with increasing dust concentration. The situation becomes more complicated with condensed fuels like combustible dust, especially at fuel-rich concentrations. At fuel-rich conditions, the combustion process tends to create products of incomplete combustion. For organic fuels, this means the production of carbon monoxide and solid carbon in the form of soot or char. In the fuel-rich range, the adiabatic flame temperature is nonlinear with respect to the dust concentration: the flame temperature reaches a peak value and then diminishes with increasing dust concentration. The investigation of fuel-rich conditions is best suited for formal chemical equilibrium analysis.

The second method for calculating adiabatic flame temperatures is based on the use of a chemical equilibrium code. The prediction of the composition and thermodynamic state of combustion products requires an analysis of the chemical equilibrium of the reacting system. Algorithms for equilibrium combustion analysis have been developed that can be implemented with manual calculations or spreadsheets (Turns, 2012, pp. 38−52). However, there are powerful computer software products such as the NASA Lewis Chemical Equilibrium with Applications (CEA) program that are available free of charge which liberate the investigator from tedious numerical calculations. These algorithms are based on the minimization of the free energy of the reacting system (Gordon and McBride, 1994). A comprehensive discussion of chemical equilibrium analysis can be found in the monograph by Smith and Missen (Smith and Missen, 1982).

Chemical equilibrium analysis has one serious drawback: it requires enthalpy and entropy of formation data for the fuel compound. For fuels that are elements or pure chemical compounds, there is a reasonable chance that this information is available in the chemical literature. For natural materials, and with many combustible dusts, this is a hurdle that is difficult to overcome. In such cases, one can invoke the assumption of complete combustion with oxygen as the limiting reactant. While this will not give insight into the relative amounts of products of incomplete combustion, it will at least ensure that the requirement for conservation of mass is obeyed.

EXAMPLE 3.6

Calculate the adiabatic isobaric flame temperature for aluminum dust at its stoichiometric concentration in air. Do the calculation (1) manually and (2) using the NASA Lewis CEA chemical equilibrium code or its equivalent. For the manual calculation, assume complete combustion with no dissociation. Estimate the product mixture enthalpy using constant heat capacity values evaluated at the approximate melting point temperature Al_2O_3, 2400 K. Assume that the initial temperature is 298 K. (data source: Glassman et al., 2015, Appendix A).

Solution

From Example 3.2, we know that the stoichiometric reaction for aluminum combustion is written as

$$Al + \frac{3}{4}(O_2 + 3.76\,N_2) \rightarrow \frac{1}{2}Al_2O_3(l) + \frac{3}{4}(3.76)N_2$$

A basis of 1 mole of aluminum is chosen. The mass balance and thermodynamic properties are presented in the table below (Glassman et al., 2015, Appendix A):

Species	Moles	$\Delta\hat{h}_{f,i}^{0}$ (kJ/mol at 298 K)	$\hat{C}_p$ (J/mol at 2400 K)
Al	1	0	–
O_2	0.75	0	–
N_2	2.82	0	36.511
Al_2O_3	0.5	−1557.989	192.464

From Eq. (3.31), the change in enthalpy across the flame is constant: $\hat{H}_{reactants} = \hat{H}_{products}$. Writing the enthalpy in expanded terms

$$\hat{H}_{reactants} = \sum_{i=1}^{3} n_i\,\hat{h}_i = n_{Al}\,\hat{h}_{f,Al}^{0} + n_{O_2}\,\hat{h}_{f,O_2}^{0} + n_{N_2}\,\hat{h}_{f,N_2}^{0} = 0$$

$$\hat{H}_{products} = \sum_{i=1}^{2} n_i\,\hat{h}_{f,i}^{0} = n_{Al_2O_3}\left[h_{f,Al_2O_3}^{0} + \hat{C}_{p,Al_2O_3}(T_{ad} - 298)\right] + n_{N_2}\,C_{p,N_2}(T_{ad} - 298)$$

Substituting the numerical values and solving for T_{ad}, the result is $T_{ad} = 3660$ K.

Using the CEA code, the isobaric flame temperature is $T_{ad} = 3540$ K. The close agreement between the simple manual calculation and the chemical equilibrium code is somewhat fortuitous.

EXAMPLE 3.7

Using Eq. (3.38), calculate the adiabatic isobaric flame temperature for a sawdust material with an ultimate analysis corresponding to CH_2O (cellulose) and HHV $= 18,149$ kJ/kg. Assume constant heat capacities evaluated at 1200 K. Evaluate the stoichiometric condition at an initial temperature $T_0 = 298$ K.

Solution

Eq. (3.38) is suitable for the calculation of the adiabatic flame temperature at fuel-lean conditions.

$$T_{ad} = T_0 + \frac{n_F \mathcal{M}_F LHV}{\sum_{i=1}^{p} n_i \hat{C}_{p,i}}$$

Begin by writing the stoichiometric equation for combustion.

$$C_x H_y O_z + \left(x + \frac{y}{4} - \frac{z}{2}\right)(O_2 + 3.76 N_2) \rightarrow x CO_2 + \frac{y}{2} H_2O + 3.76 \left(x + \frac{y}{4} - \frac{z}{2}\right) N_2$$

From the ultimate analysis of the sawdust, $x = 1$, $y = 2$, and $z = 1$. Substituting these values give the following mole numbers: $n_{CO_2} = 1$ mol, $n_{H_2O} = 1$ mol, and $n_{N_2} = 3.76$ mol. The molar mass of the fuel is $\mathcal{M}_F = 30.03$ g/mol. Thus,

$$LHV = HHV - \Delta h_{fg} = 18.149 - 2.44 = 15.709 \text{ MJ/kg}$$

$$n_F \mathcal{M}_F LHV = 1 \text{ mol} \cdot 30.03 \frac{g}{mol} \cdot 15.709 \frac{MJ}{kg} \cdot \frac{kg}{10^3 g} = 471.27 \text{ kJ}$$

$$\sum_{i=1}^{p} n_i C_{p,i} = n_{CO_2} C_{p,CO_2} + n_{H_2O} C_{p,H_2O} + n_{N_2} C_{p,N_2}$$

$$= 1(56.342) + 2(43.768) + 3.76(33.723) = 270.677 \text{ J/K}$$

The heat capacity data have been evaluated at 1200 K using the thermochemical data in the Appendix. Substituting these values into Eq. (3.38) gives the desired result:

$$T_{ad} = 2040 \text{ K}$$

A final observation about the adiabatic isobaric flame temperature is that it is an indication of the change in gas density in going from the unburnt state to the burnt state (Nagy, 1985, pp. 57−58). For typical combustible dusts, the flame temperature is in the range of 2100−2400 K, so the ratio of final to initial mixture temperature (300 K) is approximately

$$\frac{T_{ad}}{T_0} \cong 7-8 \tag{3.39}$$

From the ideal gas equation of state, the volume ratio of expansion of the burnt mixture at atmospheric pressure is

$$\frac{v_b}{v_u} = \frac{\rho_u}{\rho_b} \cong 7-8 \tag{3.40}$$

Recall the specific volume is related to the density by the relation $v = 1/\rho$.

The specific volume ratio is an indication of the growth in mixture volume due to combustion. A flash fire is an isobaric flame, so the specific volume ratio due to combustion is an indication of the size of the flash fire. In real flash fire events, heat losses and the effect of finite rate processes will reduce the magnitude of this ratio to perhaps 5 or 6.

EXAMPLE 3.8

Calculate the fireball volume from the flash fire of a dust cloud with an expansion ratio $v_b/v_u = 8$.

Solution

For this example, assume that the dust cloud is a cylinder with an aspect ratio of 1 (the height equals its diameter). Additionally, assume that the volume of burnt gas preserves the cylindrical shape and aspect ratio. Writing the formula for the volume of a cylinder for initial and final state gives

$$V_1 = \frac{\pi}{4} D_1^2 H_1 = \frac{\pi}{4} D_1^3 \quad \text{and} \quad V_2 = \frac{\pi}{4} D_2^2 H_2 = \frac{\pi}{4} D_2^3$$

where V is the cylinder volume, D is its diameter, and H is its height. The subscripts 1 and 2 refer to the initial and final conditions, respectively. An expansion ratio of 8 implies the following:

$$\frac{V_2}{V_1} = \left(\frac{D_2}{D_1}\right)^3 = 8 \rightarrow D_2 = 2 D_1$$

Thus, for this geometric model of a flash fire, the diameter of the fireball is twice its initial dust cloud volume. Likewise, the height doubles in size. The interested reader can repeat this calculation assuming a spherical geometry and verify that the result is the same.

This simple geometric model is one way to evaluate the consequences of a flash fire. Two caveats must be remembered. First, real dust clouds are rarely perfect cylinders, so the impact of geometric imperfection must be considered. Second, gravity exerts a significant effect on fireball development: fireballs have a natural tendency to rise due to the effect of buoyancy. The buoyant motion tends to stretch the fireball upwards and reduce its radial expansion. Flash fires will be discussed further in Chapter 7.

3.5.3 DUST FLAME TEMPERATURE MEASUREMENT

In this section, some laboratory studies are reviewed in which dust flame temperatures were measured and compare these results with chemical equilibrium

Table 3.3 Maximum Measured Flame Temperatures for Elemental Dust Flames in Isobaric Combustion. Dust Concentration Corresponds to the Maximum Measured Flame Temperature

Elements	Dust Concentration (g/m³)	Measured Temperature (K)	Calculated Temperature (K)	Source
Aluminum	500	3250	3400	Goroshin et al. (2007)
Iron	650	1350	2250	Sun et al. (2000)
Sulfur	250	1400	1900	Proust (2006b)

calculations. In addition to the variation with dust concentration, the measurement results are dependent on the size of the laboratory burner or test fixture. Some measured dust flame temperatures are within 20% or so of chemical equilibrium calculations, but many of the measurements are 30−50% below the equilibrium values.

Combustion studies with pure elements are especially attractive because the full suite of thermodynamic data is available to the investigator making it possible to accurately compute adiabatic isobaric flame temperatures. Flame temperature measurements for aluminum, iron, and sulfur dust flames are summarized in Table 3.3 and compared with chemical equilibrium calculations. The aluminum flame experiments were conducted with a dust flame burner. The aluminum flame temperatures were inferred from spectrometric measurements taken from a stationary dust flame. The instrumentation allowed separate measurements of the particle and gas phase temperatures, and the difference between them was small. The iron and sulfur tests were conducted by propagating dust flames in vertical tubes. The iron and sulfur flame temperatures were measured with thermocouples taken from freely propagating dust flames. Thermocouple measurements can measure gas phase temperatures only, so no direct comparison of particle and gas phase temperatures was possible. The experimental methods sections of each of these three studies are a testament to the difficulty of performing such measurements in the laboratory.

Two observations are apparent. First, as expected, due to the presence of heat losses and finite rate processes, the measured flame temperatures are lower than the calculated (adiabatic isobaric) flame temperatures. Second, the discrepancy between calculated and measured temperatures is not consistent for the three elements: the discrepancy is smallest for aluminum and greatest for sulfur. The discrepancy may be due in part to the particle size distribution of the powder tested.

Isobaric flame temperature measurements and their corresponding chemical equilibrium values for organic combustible dusts are shown in Table 3.4. The starch, lycopodium, and 1-eicosanol tests were all conducted by propagating dust flames in vertical tubes. The coal dust tests were conducted with a dust flame burner.

Table 3.4 Maximum Measured Flame Temperatures for Organic Dust Flames in Isobaric Combustion

Dusts	Dust Concentration (g/m³)	Measured Temperature (K)	Calculated Temperature (K)	Source
Starch	250	1500	2300	Proust (2006a)
Lycopodium	100	1500	2200	Proust (2006a)
1-Eicosanol (C_{20} alcohol)	250	1500	N/A	Gao et al. (2012)
Pittsburgh coal	200	1550	N/A	Horton et al. (1977)

When compared at a level of precision of two significant figures and similar dust concentrations, the measured isobaric flame temperatures are remarkably similar.

3.6 EXPLOSIONS AND CONSTANT VOLUME COMBUSTION

Constant volume combustion is important as the basis for laboratory test methods and as the conceptual model for a dust explosion accident scenario. Dust explosibility tests are conducted in constant volume chambers (see Chapter 1). The combustion reaction causes a rise in pressure which is a direct consequence of the heat release and temperature rise. Similarly, a dust deflagration within a piece of equipment or in a confined space is well represented as constant volume combustion.

In most analyses of isochoric combustion, the walls of the vessel are assumed to be rigid, adiabatic, and impermeable. As with constant pressure combustion, the adiabatic condition for constant volume (isochoric) combustion results in the greatest flame temperature and the maximum increase in pressure. The isochoric flame temperature is greater than the corresponding isobaric flame temperature because the work done in pushing back the atmosphere in isobaric combustion is absent in isochoric combustion. The increment of chemical energy that went into isobaric work is instead transformed into an incremental increase in thermal energy for isochoric combustion. The internal energy for an ideal gas is related to the enthalpy by the relation (O'Connell and Haile, 2005, p. 76)

$$\Delta \hat{u}_i = \Delta \hat{h}_i - \Delta n_i \, R_g \, T \tag{3.41}$$

It is often observed that in combustion problems with air as the oxidizer the change in total moles Δn_i is small and can be neglected (McAllister et al., 2011, p. 28).

$$\Delta \hat{h}_c \cong \Delta \hat{u}_c \tag{3.42}$$

While this is true, it does not mean that the isochoric and isobaric flame temperatures will also be nearly equal. In fact, with hydrocarbon fuels the isochoric flame temperature is often on the order of 10–20% larger than the isobaric flame temperature due to the difference in specific heats at constant volume versus constant pressure.

3.6.1 PHYSICAL DESCRIPTION

The thermodynamic model for isochoric combustion assumes that in the initial (unburnt) condition the dust mixture is dispersed uniformly within the closed vessel. In the final (burnt) condition the combustion products, their temperature, and pressure are uniformly distributed within the vessel. The assumption of thermodynamic equilibrium means that spatial gradients within the closed vessel are neglected. Thus, the ignition and combustion events of the dust cloud are assumed to be spatially uniform processes. Fig. 3.5 is a definition sketch for the thermodynamic equilibrium analysis of isochoric combustion.

As with the isobaric flame, the thermodynamic model of the isochoric flame must neglect finite rate processes. Finite rate processes in isochoric combustion occur at both the macroscale (the diameter of the test chamber) and the microscale (characteristic particle diameter). At the macroscale, heat losses may occur at the vessel wall or the dust mixture may be nonuniform within the vessel. At the microscale, there may be a temperature jump between the burning particles

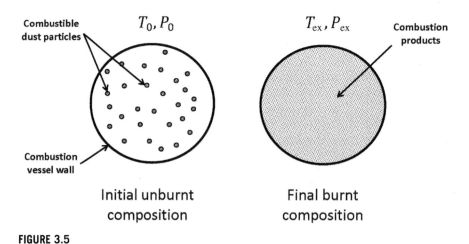

Combustible dust particles

T_0, P_0

Combustion products

Combustion vessel wall

T_{ex}, P_{ex}

Initial unburnt composition

Final burnt composition

FIGURE 3.5

Thermodynamic model for constant volume combustion.

and the gas phase. Real dust flames may exhibit significant departures from thermodynamic equilibrium. The value of the thermodynamic equilibrium model is that it establishes a reference case from which the magnitude of finite rate effects can be determined.

An additional complication is that isochoric flames are, generally speaking, propagating flames. The propagation of a flame in a constant volume vessel leads to additional thermodynamic processes that are not a concern in isobaric combustion. The standard model for isochoric combustion is that the dust mixture is ignited at the center of the vessel. The dust flame then propagates outwards as an expanding spherical flame moving toward the vessel wall. The motion of the flame in the closed vessel causes an adiabatic compression of the unburnt mixture near the wall. The propagating flame is quenched and extinguished when it impinges on the vessel wall. The compression effect causes the burnt mixture to exhibit a radial temperature profile with the center temperature being a maximum. This phenomenon, called the Flamm–Mache effect, will be described in greater detail in Chapter 8 (Lewis and von Elbe, 1987, pp. 381–395; Zeldovich et al., 1985, pp. 470–485).

3.6.2 CALCULATION OF THE ADIABATIC ISOCHORIC FLAME TEMPERATURE

As with the adiabatic isobaric flame temperature, two methods are described for calculating the adiabatic isochoric flame temperature: (1) the method based on constant thermodynamic properties and (2) chemical equilibrium. The natural energy quantity to use for evaluating constant volume processes is the internal energy. Compilations of thermodynamic properties of chemical species rarely include the internal energy; instead the property usually tabulated is the enthalpy. The method for calculating the isochoric flame temperatures with constant properties presented here is based on the presentation by Turns (Turns, 2012, pp. 36–37). Turns's method uses enthalpy data directly and then adjusts the calculation to convert into internal energies.

Neglecting heat losses, work interactions, and changes in the kinetic or potential energy of the control mass, the energy balance for isochoric combustion in mole units is

$$\hat{U}_{\text{reactants}} = \hat{U}_{\text{products}} \tag{3.43}$$

The internal energy of combustion is calculated from enthalpy data using the standard thermodynamic definition in mole units

$$\hat{H}_i = \hat{U}_i + P\hat{V}_i \tag{3.44}$$

The symbol $\hat{V}_i$ represents the molar specific volume of species i. We assume throughout this derivation that the mixture forms an ideal solution. Substituting Eq. (3.44) into (3.43) gives the following expression:

$$\hat{H}_{\text{reactants}} - \hat{H}_{\text{products}} - \hat{V}(P_0 - P_{\text{ex}}) = 0 \tag{3.45}$$

The initial pressure is P_0, the final pressure is the isochoric combustion (explosion) pressure P_{ex}, and V is the fixed volume. With the ideal gas equation of state, the PV term in Eq. (3.44) can be written in terms of temperature

$$P_0 V = \sum_{i=1}^{r} n_i \, R_u \, T_0 = n_{\text{reactants}} \, R_g \, T_0 \tag{3.46}$$

$$P_{\text{ex}} V = \sum_{i=1}^{p} n_i \, R_u \, T_{\text{ex}} = n_{\text{products}} \, R_g \, T_{\text{ex}} \tag{3.47}$$

Eqs (3.46) and (3.47) are now substituted into Eq. (3.45) to give the following result

$$\hat{H}_{\text{reactants}} - \hat{H}_{\text{products}} - R_g \left(n_{\text{reactants}} T_0 - n_{\text{products}} T_{\text{ex}} \right) = 0 \tag{3.48}$$

The isochoric (explosion) temperature can be calculated from Eq. (3.48). The explosion pressure is calculated from the ideal gas equation of state

$$P_{\text{ex}} = \left(\frac{\rho_{\text{mix}} R_g}{\mathcal{M}_{\text{mix}}} \right) T_{\text{ex}} \tag{3.49}$$

The manual calculation of the isochoric flame temperature requires only thermochemical data. The disadvantage of the manual calculation is that it requires the specification of the combustion products (complete combustion assumption). More accurate calculation can be made with chemical equilibrium software if the necessary thermodynamic data are available.

As with the isobaric flame temperature, the isochoric flame temperature is a function of dust (fuel) concentration. In fuel-lean conditions, the isochoric flame temperature increases almost linearly with dust concentration. The isochoric flame temperature achieves a maximum value somewhat beyond the stoichiometric concentration. In chemical equilibrium calculations, the flame temperature decreases from the maximum value, but in experimental measurements the flame temperature tends to reach a plateau value that is less than the maximum value but greater than the value calculated by chemical equilibrium methods. Fig. 3.6 shows a comparison of maximum pressure measurements with chemical equilibrium calculations for cornstarch and aluminum powders. As indicated in Eq. (3.49), the explosion pressure is a measure of the average combustion temperature.

Another important insight from thermodynamics is that for a given dust at any given dust concentration, the isochoric flame temperature is always greater than the isobaric flame temperature due to the absence of work interactions in a constant volume system.

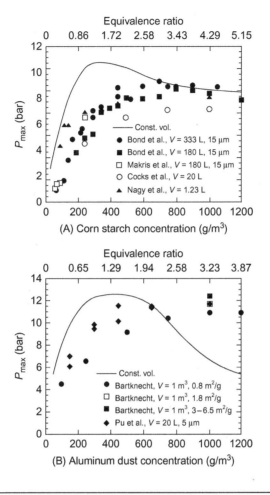

FIGURE 3.6

Maximum pressure measurements as a function of dust concentration: (A) cornstarch and (B) aluminum. The solid lines indicate the chemical equilibrium calculation.

From Lee, J.H.S., Zhang, F., Knystautas, R., 1992. Propagation mechanisms of combustion waves in dust-air mixtures. Powder Technol. 71, 153–162.

EXAMPLE 3.9

Calculate the adiabatic isochoric flame temperature for aluminum dust at its stoichiometric concentration in air. Do the calculation (1) manually and (2) using the NASA Lewis CEA chemical equilibrium code or its equivalent. For the manual calculation, assume complete combustion with no dissociation. Estimate the product mixture enthalpy using constant heat capacity values evaluated at the approximate melting point temperature Al_2O_3, 2400 K. Assume that the initial temperature is 298K (data source: Glassman et al., 2015, Appendix A).

Solution

From Example 3.2, we know that the stoichiometric reaction for aluminum combustion is written as

$$Al + \frac{3}{4}(O_2 + 3.76\,N_2) \rightarrow \frac{1}{2}Al_2O_3(l) + \frac{3}{4}(3.76)N_2$$

A basis of 1 mole of aluminum is chosen. The mass balance and thermodynamic properties are presented in the table below:

Species	Moles	$\Delta \hat{h}_{f,i}^0$ (kJ/mol at 298 K)	$\hat{C}_p$ (J/mol at 2400 K)
Al	1	0	—
O_2	0.75	0	—
N_2	2.82	0	36.511
Al_2O_3	0.5	−1557.989	192.464

From Eq. (3.31), the change in enthalpy across the flame is constant: $\hat{H}_{reactants} = \hat{H}_{products}$. Writing the enthalpy in expanded terms

$$\hat{H}_{reactants} = \sum_{i=1}^{3} n_i \hat{h}_i = n_{Al}\hat{h}_{f,Al}^0 + n_{O_2}\hat{h}_{f,O_2}^0 + n_{N_2}\hat{h}_{f,N_2}^0 = 0$$

$$\hat{H}_{products} = \sum_{i=1}^{2} n_i \hat{h}_{f,i}^0 = n_{Al_2O_3}\left[\hat{h}_{f,Al_2O_3}^0 + \hat{C}_{p,Al_2O_3}(T_{ad}-298)\right] + n_{N_2}\hat{C}_{p,N_2}(T_{ex}-298)$$

$$\hat{H}_{reactants} - \hat{H}_{products} - R_g\left[(n_{Al}+n_{O_2}+n_{N_2})T_0 - n_{N_2}T_{ex}\right] = 0$$

Substituting the numerical values and solving for T_{ex}, the result is $T_{ex} = 3816$ K.

Using the CEA code, the isochoric flame temperature is $T_{ex} = 4030$ K. The agreement with the chemical equilibrium calculation in the isochoric case is reasonably good.

3.6.3 MEASUREMENTS OF CONSTANT VOLUME FLAME TEMPERATURES, PRESSURES, AND COMBUSTION PRODUCT COMPOSITION

There are fewer measurements of adiabatic isochoric flame temperatures than have been published on isobaric flame studies. It is easier to establish a (mostly)

Table 3.5 Temperature and Pressure Measurements in Isochoric Combustion Experiments for Elemental Combustible Dusts

Element	Stoich. Dust Concentration (g/m³)	Calculated Equilibrium Temperature (K)	Measured Temperature (K)	Calculated Equilibrium Pressure (bar)	Measured Pressure (bar)	Pressure Calculated from Ideal Gas EOS (bar)
Al-3	310	4060	2800	12.4	7.5	9.3
Al-4	310	4060	2400	12.4	6.6	8
Mg	420	3610	2800	15.7	8.5	9.3
Ti	420	3990	2850	11.6	5.7	9.5
Fe-1	650	2490	1800	6.8	4.5	7.2
Fe-2	650	2490	1600	6.8	3.1	5.3
S	280	2200	1300	6.5	3.6	4.3

All data from Cashdollar and Zlochower (2007), except sulfur data taken from Kalman et al. (2015).

uniform dust cloud inside a closed vessel compared to experiments in tubes, but temperature measurements are hindered by the optical thickness of the dust mixture. Temperature measurements have been performed with optical pyrometers (Cashdollar and Zlochower, 2007) or, less commonly, with thermocouples (Kalman et al., 2015).

Combustion studies with elements are considered first. Table 3.5 summarizes a small portion of data from research programs using a 20-L test vessel at the Pittsburgh Research Laboratory (formerly the U.S. Bureau of Mines Laboratory) of the National Institute of Occupational Safety and Health (NIOSH) (Hertzberg et al., 1992; Cashdollar, 1994; Cashdollar and Zlochower, 2007). Additional data on sulfur dust is taken from the work of Kalman et al. using a 31-L cubical test chamber (Kalman et al., 2015). The measured temperatures and pressures come from experiments conducted with the NIOSH 20-L sphere. The NIOSH temperatures were obtained with an optical pyrometer. The sulfur temperatures were measured with a thermocouple. The calculated temperatures and pressures were obtained from the NASA Lewis CEA program. The measured temperatures and pressures in Table 3.5 are significantly less than the chemical equilibrium calculations, ranging from 50% to 70% of the predicted values. The discrepancy between measured and theoretical values indicates that finite rate processes have an important contribution to the combustion behavior of dust clouds. In particular, factors which will cause a departure from chemical equilibrium include the Flamm–Mache temperature gradient, the local temperature difference between the particles and the gas phase, finite rates of chemical kinetics, and heat losses at the vessel wall.

The final column shows the pressure calculated from the ideal gas equation of state using the measured temperatures.

$$P_{ex} = P_0 \left(\frac{T_{ex}}{T_0} \right) \qquad (3.50)$$

Table 3.6 Temperature and Pressure Measurements in Isochoric Combustion Experiments for Organic Combustible Dusts (Cashdollar and Hertzberg, 1983)

Dusts	Volatiles (%)	Sauter Mean Diameter (μm)	Measured Temperature (K)
Pittsburgh coal	34	5	2020
Pittsburgh coal	36	27	1900
Anthracite coal	7	5	1700
Gilsonite	85	12	1950

All measurements at a dust concentration of 250 g/m³.

The pressures calculated from the ideal gas relation are consistently larger than the measured pressures but less than the pressures calculated by chemical equilibrium. Generally speaking, the measured pressures correlate positively with the measured temperatures.

Similar results are seen with organic dusts. The measurement data in Table 3.6 also comes from a NIOSH publication (Cashdollar and Hertzberg, 1983). Temperatures were obtained with an optical pyrometer and reflect particle temperatures. Data suggest that higher temperatures were achieved with smaller particle diameters or higher volatiles content.

In addition to measuring the explosion pressure and temperature, conducting combustion experiments in a closed vessel permits the measurement of the composition of the combustion products. A small number of studies have investigated the chemical composition of the combustion products following the isochoric combustion of organic fuels (Lemos and Bouriannes, 1991; van der Wel et al., 1991; Sattar et al., 2012). Under fuel-lean conditions, the primary gas measured on a dry basis was carbon dioxide. Under fuel-rich conditions (typically $\Phi > 2$), the concentration of carbon dioxide diminished while the concentrations of carbon monoxide and hydrogen increased (to be expected based on the water−gas shift equilibrium reaction (Turns, 2012, pp. 46−52).

3.6.4 RELATIONS BETWEEN FLAME TEMPERATURES AND EXPLOSION PRESSURE

There is an interesting approximate relation between the constant pressure flame temperature, T_{ad}, and the constant volume flame temperature, T_{ex} (Ogle, 1999). This relation is useful for quick estimates of T_{ex} based on knowing the T_{ad} value. The energy balance for isochoric combustion is written in mole units

$$\Delta \hat{u}_c = n \, \hat{C}_v (T_{ex} - T_0) \tag{3.51}$$

An energy balance for isobaric combustion of the same mixture can be written in mole units as follows:

$$\Delta \hat{h}_c = n \, \hat{C}_p (T_{ad} - T_0) \tag{3.52}$$

If the change in the number of moles during combustion can be neglected, then there is an approximate equality between the internal energy and enthalpy of combustion.

$$\Delta \hat{u}_c \cong \Delta \hat{h}_c \qquad (3.53)$$

This assumption works reasonably well when air is the oxidizer. Equating Eqs. (3.50) and (3.51) and solving for T_{ex} yields the following relation (Zeldovich et al., 1985, p. 471):

$$T_{ex} = T_0 + \gamma \, (T_{ad} - T_0) \qquad (3.54)$$

where γ, the specific heat ratio, is defined as $\gamma = C_p/C_v$. Since the ratio $T_0/T_{ad} \ll 1$, Eq. (3.54) can be simplified to the approximate expression

$$T_{ex} = \gamma \, T_{ad} \qquad (3.55)$$

The explosion pressure is related to the explosion temperature by the equation of state

$$P_{ex} = P_0 \left(\frac{T_{ex}}{T_0} \right) \qquad (3.56)$$

Substituting Eq. (3.54) into (3.56) to eliminate T_{ex} gives an expression for the explosion pressure in terms of the isobaric flame temperature.

$$P_{ex} = P_0 \left[1 + \gamma \left(\frac{T_{ad}}{T_0} - 1 \right) \right] \qquad (3.57)$$

Because $T_{ad}/T_0 \gg 1$, this relation can be simplified to

$$P_{ex} = \gamma P_0 \left(\frac{T_{ad}}{T_0} \right) \qquad (3.58)$$

As a practical matter, Eq. (3.57) or (3.58) can be used to estimate the explosion pressure P_{ex}, an important measure of the hazard potential, for a confined dust explosion at any dust concentration. For most hydrocarbon fuel mixtures in air, the specific heat ratio of the burnt gas composition is approximately 1.1–1.2. Thus, the explosion pressure ratio is approximately 10–20% greater than the isobaric flame temperature ratio. As is evident from inspection of Table 3.5, predictions of the isochoric combustion temperature and pressure are approximate at best.

EXAMPLE 3.10

Using the simple thermodynamic relations derived in this section, calculate the isochoric flame temperature and explosion pressure for the stoichiometric condition of three inorganic powders: aluminum, sulfur, and iron. Assume that the adiabatic isobaric flame temperatures are known, assume the initial temperature is 298 K, the initial pressure was 1.01 bars, and use a value of 1.2 for the specific heat ratio. Compare the results with a chemical equilibrium calculation.

Solution

The calculations are summarized in the table below:

Parameters	Al	C	Fe
$T_{ad}(K)$	3540	1970	3100
$T_{ex} = T_0 + \gamma \, (T_{ad} - T_0)(K)$	4190	2300	3660
$T_{ex} = \gamma \, T_{ad}(K)$	4250	2360	3720
T_{ex} (CEA)(K)	4030	2670	
$P_{ex} = P_0\left[1 + \gamma\left(\dfrac{T_{ad}}{T_0} - 1\right)\right]$(bar)	14.2	7.81	12.4
$P_{ex} = \gamma P_0\left(\dfrac{T_{ad}}{T_0}\right)$(bar)	14.4	8.01	12.6
P_{ex} (CEA)(bar)	12.2	9.17	

Although convenient to use, the simple thermodynamic relations yield very approximate results.

An alternative use of these relations is to use them to estimate the explosion temperature T_{ex} from a measured explosion pressure P_{ex}. Since explosion pressure data are almost always obtained by central ignition followed by progressive combustion, the temperature profile of the combustion products will be nonuniform (Flamm—Mache temperature gradient). A calculated explosion temperature based on the measured explosion pressure is equivalent to a mass-averaged temperature. It should be anticipated that the calculated explosion temperature will be less than the isochoric flame temperature due to heat losses at the vessel wall and other finite rate processes.

3.7 RELATION BETWEEN BURN FRACTION AND FRACTIONAL PRESSURE RISE

Most investigations of flame propagation in constant volume combustion assume a linear relation between the fraction of fuel burnt and the rise in pressure. This relation follows directly from an energy balance on the dust mixture in the closed vessel. It can be derived with the following assumptions: (1) the temperature and concentration fields within the vessel are well mixed; (2) constant specific heats; (3) constant number of moles in reactants and products; and (4) the gas obeys the ideal gas equation of state. The derivation begins with writing an energy balance on the combustion vessel in mole units (Zeldovich et al., 1985, pp. 470–472).

$$f\Delta\hat{u}_c = \hat{C}_v \, (T - T_0) \text{ with } \Delta\hat{u}_c = \hat{C}_v \, (T_{ex} - T_0) \tag{3.59}$$

The average temperature within the vessel depends on the burn fraction (extent of reaction) f, which is defined as the fraction of the moles (or mass) of fuel consumed in the combustion reaction at any given point in the course of the reaction. A similar relation is often derived in chemical engineering kinetics books when describing the thermal behavior of adiabatic chemical reactors (Schmidt, 1998, pp. 218−219).

$$f = \frac{n_F}{n_{F,0}} = \frac{n_F \, \mathcal{M}_F}{n_{F,0} \, \mathcal{M}_F} = \frac{m_F}{m_{F,0}} \tag{3.60}$$

where $f=0$ at the start of combustion (zero quantity of fuel consumed) and $f=1$ when combustion is complete (all fuel consumed in fuel-lean systems or all oxygen consumed in fuel-rich systems). Eq. (3.59) can be solved for f

$$f = \frac{T - T_0}{T_{ex} - T_0} \tag{3.61}$$

Solving for the temperature as a function of the extent of reaction f

$$T = T(f) = T_0 + f(T_{ex} - T_0) \tag{3.62}$$

The vessel temperature is a linear function of the extent of reaction. The pressure in the vessel can be calculated by substituting the temperature function into the equation of state

$$P = P(f) = \left(\frac{n_0 R_g}{V_0}\right)[T_0 + f(T_{ex} - T_0)] \tag{3.63}$$

where V_0 is the volume of the combustion vessel. Thus, the vessel pressure is a linear function of the extent of reaction (the burn fraction).

$$f = \frac{P - P_0}{P_{ex} - P_0} \tag{3.64}$$

The results derived in this section are valid for a spatially uniform combustion process that occurs in a closed vessel. As mentioned earlier in this chapter, most constant volume combustion experiments are conducted with central ignition followed by radial flame propagation. This progressive flame phenomenon or progressive combustion is also observed in accidental dust explosions. In a progressive combustion event, the assumption of well-mixed temperature and concentration profiles is violated. However, subject to certain conditions, the linear pressure rise relation is approximately valid during progressive combustion. The derivation of this relation for a propagating flame model is considered in Chapter 8.

3.8 WHEN IS A DEFLAGRATION A FLASH FIRE?

A flash fire is a confined dust deflagration that does not generate significant overpressure. There is no rigorous distinction between a flash fire and a confined deflagration. The distinction is usually based on whether the overpressure generated by the deflagration was sufficiently large to cause structural or mechanical damage. In a closed space, any deflagration releases heat into the space raising its average temperature, and hence, the room pressure. A flash fire is created in one of two circumstances: either the space is not truly closed or only a fraction of the space has been filled with an ignitable dust concentration. If the space is not truly closed, then there are openings in the room that can vent the rising pressure and hot gases formed by the deflagration. If the space is indeed closed, then a simple thought experiment shows why the dust cloud must have a volume smaller than the closed space.

Consider a volume completely filled with suspended dust at its stoichiometric concentration. If ignited, the resulting deflagration will achieve an explosion pressure on the order of 8−10 barg. Most industrial structures will suffer catastrophic collapse at pressures in excess of 0.3 barg (CCPS, 1996, p. 40). Stoichiometric conditions result in pressures on the order of 30 times greater than needed to cause complete failure of the structure. If the dust concentration is reduced to the MEC, the resulting pressure will range anywhere from 1 barg (the minimum pressure response required for dust explosibility testing) to 4 barg, which is on the order of 10 times the catastrophic failure pressure. Therefore, the MEC condition still produces an explosion pressure which is far too great, ie, the pressure will still cause catastrophic failure of a typical industrial structure.

A confined deflagration overpressure that is sufficiently small to ignore will depend on the structural strength of the enclosure. A reasonable value for a nonhazardous overpressure is 0.03 barg (CCPS, 1996, p. 40). The only way to achieve such a low pressure is to fill only a portion of the enclosure volume with a combustible dust cloud. The resulting deflagration of the dust cloud creates a high-pressure "bubble" of combustion products which then expand and fill the enclosure. The expansion leads to a final pressure that is much less than the constant volume explosion pressure, and the mixing of combustion products with the ambient air reduces their temperature far below the isochoric flame temperature.[2] The equalization of the pressure throughout the enclosure volume is a reasonable approximation (this will be evaluated in Chapter 4). The equilibration of the gas temperature throughout the enclosure volume is not likely to occur. In reality, the mixing effect on the temperature will be more localized. Nevertheless, in the model the difference between a confined deflagration and a flash fire is the size (volume) of the dust cloud in relation to the size (volume) of the enclosure. This phenomenon is called a fractional volume deflagration.

The development that follows was originally developed for hazard analysis with flammable gases (Ogle, 1999), but since this is a thermodynamic argument, it is readily adapted to combustible dusts. The same cautions offered earlier

---

[2]The expansion of the high-pressure "bubble" also performs work as it compresses the ambient gas, and the expansion adiabatically cools the high-pressure combustion products. This aspect of the problem is set aside for the moment.

about thermodynamic models hold here as well. The thermodynamic analysis is useful in establishing a theoretical limit on combustion behavior. Real dust deflagration behavior will probably exhibit significant departures from the thermodynamic limit. When predicting the consequences of a dust deflagration event, the investigator can take some comfort in the knowledge that the thermodynamic model places a maximum or minimum limit on the hazard scenario under consideration.

There are other important limitations to a thermodynamic analysis of a complex phenomenon like this. Fluid dynamic effects like shock wave interactions, flame acceleration due to the presence of obstacles, or the initial turbulence with dust suspension are not considered. This analysis also neglects fluid—structure interactions, venting of high-pressure gas from the structure, the strength of the structure, and the consequences of structural failure. The investigator must consider whether these approximations are acceptable in their particular situation.

Fig. 3.7 is a definition sketch for the thermodynamic analysis of a dust deflagration in which the dust cloud does not fill the entire volume of the enclosure. The subscript C is for the dust cloud and subscript A is the ambient environment; the subscripts 1 and 2 represent the initial and final conditions. For the pre-explosion condition, the dust cloud is assumed to be at its stoichiometric

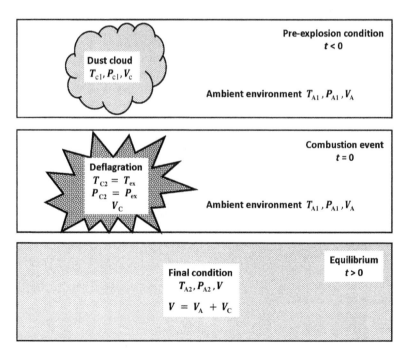

FIGURE 3.7

Definition sketch for the analysis of a fractional volume deflagration.

concentration. Selecting an alternative concentration, such as the MEC, is less satisfactory because the MEC value for a given dust is dependent on the particle size distribution of the dust, the test vessel used, and the test protocol followed. The stoichiometric condition is selected for this analysis because it is an unambiguous thermodynamic state. For flammable gases, the stoichiometric condition will result in an explosion pressure that is very nearly the maximum value as a function of concentration. For combustible dust, the stoichiometric condition does not correspond to the maximum explosion pressure. So this analysis may not be as conservative for combustible dusts as it is for flammable gases.

The enclosure is assumed to be rigid, impermeable, and adiabatic. In the pre-explosion state, the temperature and pressure of the dust cloud are equal to the ambient environment: $T_{A1} = T_0$ and $P_{A1} = P_0$. The volume of the dust cloud, V_C, is distinct from the volume of the ambient environment, V_A, such that the total volume of the enclosure, V, is equal to the sum of the volumes: $V = V_C + V_A$. At the instant of the combustion event, the dust cloud burns at constant volume such that $T_{C2} = T_{ex}$ and $P_{C2} = P_{ex}$, where the explosion temperature and pressure correspond to the stoichiometric dust concentration. The volume of high pressure, burnt gas products expands and mixes with the ambient air. The final pressure of the ambient environment within the enclosure, P_{A2}, is much lower than the explosion pressure, and the final ambient temperature, T_{A2}, is much lower than the explosion temperature. Since the contents of the two volumes are assumed to be well mixed at the conclusion of the event, we can set $T_{A2} = T_2$. Qualitatively, this highly simplified model seems to capture the basic physical features that one would expect for this situation.

The analysis begins with the specification of the system. I select the contents of the enclosure as the control mass. Heat and work interactions are neglected, as are changes in kinetic or potential energy. An energy balance written for the cloud and the ambient environment takes the following form in mole units

$$\Delta \hat{U}_A + \Delta \hat{U}_C = 0 \qquad (3.65)$$

Internal energy within the enclosure is conserved. Writing the internal energy in terms of temperature, the energy balance becomes

$$n_{A1} \hat{C}_v (T_2 - T_0) = n_{ex} \hat{C}_v (T_{ex} - T_2) \qquad (3.66)$$

The specific heats are assumed to be constant and equal. The number of moles in the ambient space and in the dust cloud can be calculated using the ideal gas equation of state:

$$n_{A1} = P_0 V_A / R_g T_0, \quad n_{ex} = P_{ex} V_C / R_g T_{ex} \qquad (3.67)$$

Substituting these expressions into the energy balance and solving for the final temperature, T_2

$$T_2 = T_0 T_{ex} \left[\frac{P_0 V_A + P_{ex} V_C}{P_0 V_A T_{ex} + P_{ex} V_C T_0} \right] \qquad (3.68)$$

The final pressure, P_2, can be written with the aid of the ideal gas equation of state

$$P_2 = \frac{(n_{A1} + n_{ex}) R_g T_2}{(V_A + V_C)} \tag{3.69}$$

The equations for T_2, n_{A1}, and n_{ex} can be substituted into the equation for P_2 and solved for the final pressure in the enclosure

$$P_2 = \frac{1}{V}(P_{ex} V_C + P_0 V_A) \tag{3.70}$$

This result can be used to calculate the final pressure in the enclosure if the two volumes are known or specified.

Alternatively, Eq. (3.70) can be solved for the fractional fill, α_{fill}, which is defined as the ratio of the unburnt dust cloud volume to the enclosure volume. This equation can be used to determine the fractional volume of the deflagration that yields a final enclosure pressure P_2.

$$\alpha_{fill} = \frac{V_C}{V} = \left[\frac{P_2 - P_0}{P_{ex} - P_0}\right] \tag{3.71}$$

This result has the same functional form as the result for the fractional pressure rise that was derived in Section 3.7 (Eq. 3.65). Zalosh has described the use of the fractional pressure rise equation to evaluate the pressure rise for a fractional volume deflagration (Zalosh, 2002). Eq. (3.70) can be solved for the final compartment pressure P_2

$$P_2 = P_0 + \alpha_{fill}(P_{ex} - P_0) \tag{3.72}$$

Fig. 3.8 is a plot of the fractional deflagration model for a combustible dust with an isochoric combustion (explosion) pressure of 8 bars at the stoichiometric condition. When the fractional fill is zero, the compartment pressure is equal to 1 bar (approximately 1 atmosphere). As the volume of the dust cloud increases (the fractional fill increases), the final compartment pressure increases. In the limit that the dust cloud fills the entire compartment, the final pressure is equal to the isochoric combustion pressure of 8 bars.

Recall that the model is based on the assumption that the dust cloud is at its stoichiometric concentration, $\rho_{d,stoich}$. The mass quantity of dust suspended in the cloud is then given, by definition, as

$$m_d = \rho_{d,stoich} V_C \tag{3.73}$$

This last equation is the quantity of dust that is suspended in the cloud and participates in combustion. It is not the total quantity of dust in the enclosure. In real accident scenarios, only a fraction of the dust layer is actually suspended to form a cloud and participates in combustion. The complex fluid dynamics associated with dust suspension from stagnant layers is explored in Chapter 9.

For an enclosure of volume V, the criterion for a confined deflagration to behave as a flash fire is that the final pressure in the enclosure must be less than

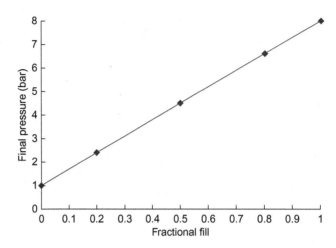

FIGURE 3.8

Fractional deflagration model for a combustible dust with an isochoric combustion (explosion) pressure of 8 bars at the stoichiometric condition.

0.03 barg (using the suggested CCPS damage criterion). To behave as a flash fire, the fractional fill must not exceed the value

$$\alpha_{fill}(\text{flash fire}) \leq \frac{0.03}{(P_{ex} - 1.013)} \tag{3.74}$$

where P_{ex} is expressed in bar. Eq. (3.74) should be used as a test for distinguishing between a flash fire and a confined deflagration only if the overpressure damage criterion is sensible for the specific application under study. There will be situations where this criterion is not appropriate. For example, in enclosures with venting capacity, a larger dust cloud than that indicated by Eq. (3.74) may be tolerable. Example 3.11 explores some typical results with this relation.

EXAMPLE 3.11

Consider a fractional volume deflagration in an enclosure with no venting capability. The stoichiometric explosion pressure for the dust is 7 bars. Calculate the fractional fill corresponding to 0.03 barg.

Solution

Eq. (3.71) gives the fractional fill criterion for a flash fire. Substituting the value for $P_{ex} = 7$ bars, the result is

$$\alpha_{fill} \leq \frac{0.03}{(P_{ex} - 1.013)} \leq \frac{0.03}{(7 - 1.013)}; \rightarrow \alpha_{fill} \leq 0.0050 \text{ or } 0.5\% \text{ of the enclosure volume}$$

To avoid pressurization of the enclosure during a dust deflagration, the volume of the dust cloud at the stoichiometric concentration must not exceed 0.5% of the enclosure volume. This is a very small volume that is likely overestimated due to neglecting the effect of finite rate processes. In a real dust deflagration, finite rate processes will prevent complete combustion of the dust and so the pressure rise due to the calculated fractional fill will be less than 0.03 barg.

To place the fractional fill criterion in perspective, consider an industrial workspace of 1000 m^2 with a ceiling height of 5 m. Applying the fractional fill criterion of 0.5%,

$$V_{cloud} = 0.005 \, V_{building} = 0.005 \cdot (1000 \text{ m}^2) \cdot (5 \text{ m}) = 25 \text{ m}^3$$

Assume the cloud is a cylinder with an aspect ratio $D/H = 1$. The dimensions of the cylinder corresponding to the fractional fill criterion are

$$D = H = (4 \cdot V_{cloud}/\pi)^{1/3} = (100 \text{ m}^3/\pi)^{1/3} = 3.2 \text{ m}$$

Thus, a dust cloud shaped like a cylinder with a height and diameter of 3.2 m or less would yield, upon ignition, a flash fire rather than a dust explosion. For a material that approximates pure carbon (eg, fresh carbon black), the stoichiometric dust concentration is 103 g/m^3. The quantity of carbon dust in the dust cloud required by the model is 2.58 kg.

Other models for the fractional volume deflagration for flammable gas mixtures have been presented by other investigators along with experimental data (Sibulkin, 1980; Vykhristyuk et al., 1988; Stamps et al., 2009; Thomas and Oakley, 2010). The combustion tests used flammable gases like hydrogen−air or methane−air mixtures. These studies all demonstrate the approximate linear relationship between the fractional pressure rise and the fractional fill of the enclosure volume. Similar experiments have yet to be performed for combustible dusts, so a direct comparison of the model with experimental data cannot be performed. Again I remind the reader that a thermodynamic model like the fractional volume deflagration model is an approximation to real behavior and does not account for the influence of finite rate processes.

3.9 THERMODYNAMIC MODEL FOR SECONDARY DUST EXPLOSION PRESSURES

Secondary dust explosions occur when a first dust explosion, called the primary explosion, disperses fugitive dust and initiates a second dust explosion. A secondary explosion is a consequence of poor housekeeping in which combustible dust is leaked or spilled into a building enclosure and allowed to accumulate. This physical situation is similar to the fractional volume deflagration model, but now the surrounding environment adds additional fuel to the expanding fireball created by the primary dust explosion.

In the original fractional volume deflagration model, the high-pressure combustion products from the primary dust explosion were diluted as they expanded into the surroundings. The final pressure was less than the isochoric deflagration pressure. In a secondary explosion, the additional fuel ingested into the fireball counteracts the reduction in pressure due to expansion. Therefore, the secondary explosion gains strength and yields a higher explosion pressure than the fractional volume deflagration pressure. Depending on the ratio of the primary volume to the secondary volume, and depending on the dust loading, the final pressure can approach the value of the isochoric deflagration pressure. Fig. 3.9 summarizes the basic events in a secondary dust explosion.

The thermodynamic model is founded on the assumption that a secondary dust explosion has occurred. The intent of the thermodynamic model is to evaluate the minimum consequences of a secondary explosion. The model cannot determine whether or not a secondary explosion will occur. The answer to that question requires more than just the tools of thermodynamics.

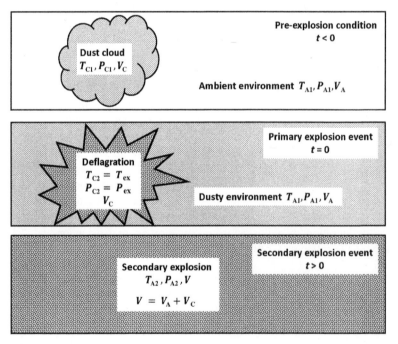

FIGURE 3.9

Thermodynamic model for the secondary dust explosion pressure.

To evaluate the potential severity of a secondary dust explosion, it is desired to calculate the final pressure P_2 of the event. The same energy balance used to analyze the fractional volume deflagration applies to the secondary explosion. The difference is that more fuel is consumed in the secondary explosion. The additional fuel is consumed in the ambient compartment. For the secondary explosion model, the initial ambient conditions correspond to the isochoric combustion of the dust loading in compartment A evaluated for the volume of the compartment, V_A. This means that the dust loading in compartment A must be known or specified. The MEC is selected as the smallest secondary explosion event that can be predicted. The corresponding accumulated dust layer in the second compartment is calculated by a mass balance on the dust (Eckhoff, 2003, pp. 9–10).

$$\rho_b\, \delta_{d,\min} = (MEC)\, H \tag{3.75}$$

This equation states that the mass of dust in the dust layer, calculated by the bulk density of the dust times the minimum dust layer thickness, is equal to the dust mass in a dust cloud with height H and suspended dust concentration equal to the MEC. Solving for the minimum dust thickness,

$$\delta_{d,\min} = H\left(\frac{MEC}{\rho_b}\right) \tag{3.76}$$

Caution must be exercised in using this equation to calculate a minimum dust thickness as it assumes that all of the dust layer can be suspended, and that the dust cloud will be uniform in dust concentration. This is an unlikely occurrence and will tend to exaggerate the accumulated dust hazard. Whether this overestimation of the hazard is tolerable or not is a decision that the safety analyst must carefully consider.

Returning to the fractional volume deflagration, Eq. (3.72) gave the linear pressure rise as a function of the volume of the two volumes of the compartment

$$P_2 = \frac{1}{V}(P_{ex}V_C + P_{A1}V_A) \tag{3.77}$$

In the fractional volume deflagration, at the moment of combustion within the dust cloud, the initial pressure of the ambient compartment was the atmospheric pressure, $P_{A1} = P_0$. In the secondary explosion, the initial pressure of the ambient compartment is assigned a constant volume explosion pressure equal to an assigned initial condition based on the dust loading. Thus, in this model for a secondary explosion, the final pressure is obtained by the adiabatic mixing of the internal energies of the two compartments. This equation can be cast into a more convenient form by recognizing that $\alpha_{fill} = V_C/V$ and $1 - \alpha_{fill} = V_A/V$

$$P_2 = P_0 + \alpha_{fill}\,(P_{ex} - P_0) \tag{3.78}$$

This equation indicates that the final enclosure pressure consists of two contributions: the dust loading of the ambient compartment plus a contribution from the primary explosion. The contribution from the primary explosion is weighted by the fractional volume of the compartment occupied by the primary explosion.

Consider two limiting cases. In the first case, assume that the dust loading in the ambient compartment is sufficiently large to yield an explosion pressure equal to the isochoric combustion pressure of the dust. Then $P_{A1} = P_{ex}$, and $P_2 = P_{ex}$ for all values of the fractional volume α_{fill}.

In the second case, assume that the dust loading in the ambient compartment results in a dust concentration equal to the MEC. The consensus dust explosibility testing standard uses the criterion that the MEC corresponds to an explosion pressure of 1 barg (2 bars corrected for the ignitor contribution). So the MEC condition can be represented as $P_{A1} = 1$ barg. For large values of the fractional volume α_{fill}, the final compartment pressure approaches the isochoric combustion pressure. For small values of the fractional volume, the final compartment pressure approaches 1 barg. Recall that the suggested criterion for negligible structural damage is $P_2 < 0.03$ barg, and catastrophic structural damage is predicted for $P_2 > 0.21$ barg. Thus, even at the MEC, a secondary dust explosion results in pressures on the order of 10 times greater than the catastrophic structural damage criterion.

Fig. 3.10 illustrates the linear relationship between the fractional fill and final explosion pressure for the secondary explosion model. The combustible dust is assumed to have an isochoric combustion (explosion) pressure of 8 bars at the stoichiometric condition. In addition, the performance line for the fractional deflagration model is plotted on the same figure for comparison. In the limit of a fractional fill of zero, the secondary explosion pressure is

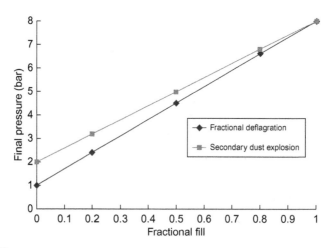

FIGURE 3.10

Thermodynamic model for a secondary dust explosion compared with the fractional deflagration model.

2 bars, the value corresponding to the entire volume of the compartment filled to the MEC. In comparison, at a fractional fill value of 0 the compartment pressure remains 1 bar (approximately atmospheric pressure). At small values of the fractional fill $\alpha_{fill} < 0.2$, the secondary explosion model predicts a final pressure that is much greater than that for the fractional deflagration. This indicates the importance of the fugitive dust accumulation in the second compartment. At large values of the fractional fill, the final pressures for the two models converge to the isochoric combustion (explosion) pressure of 8 bars.

There are some important caveats for the thermodynamic model for a secondary dust explosion. First, it is a thermodynamic model so finite rate processes have been neglected. Second, it is difficult to estimate what fraction of the dust in the enclosure will be entrained and burnt in the secondary explosion. Assuming that all of the dust participates in the secondary explosion may result in a significant overestimate of the final explosion pressure. Third, it is reasonable to infer that at some small fractional fill value, $\alpha_{fill} \ll 1$, the primary deflagration will be unable to initiate a secondary explosion. The thermodynamic model is not capable of determining this de minimis quantity. Finally, it is important to remember that fluid dynamic processes like turbulence, flame acceleration, and shock wave interactions have been neglected. This model can be used to develop intuition and insight into the problem, but it is unlikely that it will provide accurate predictions for combustible dust behavior.

3.10 SUMMARY

This chapter introduced the use of thermodynamics to establish key parameters for evaluating combustible dust behavior. The first parameter was the stoichiometric dust concentration. This establishes the criterion for determining if a particular scenario is fuel lean or fuel rich, a primary determinant for establishing if the combustion products will be fully oxidized (fuel lean) or if there will be products of incomplete combustion. The stoichiometric condition is also a useful benchmark that indicates whether an observed dust loading is approaching its maximum hazard level. Expressed in terms of the equivalence ratio, the maximum hazard level tends to occur in the fuel-rich regime.

Combustible dust hazards are well represented by two limiting types of combustion: constant pressure (isobaric combustion) and constant volume (isochoric combustion). Constant pressure combustion is an analog for flash fire behavior. The key performance parameters for constant pressure combustion are the adiabatic isobaric flame temperature and the volume expansion ratio. Higher isobaric flame temperatures indicate potentially greater volume expansion ratios, both of which reflect greater thermal hazards. Experimental measurements of isobaric flame temperatures tend to be lower than those predicted by chemical equilibrium calculations, reflecting the importance of finite rate processes.

Constant volume combustion is an analog for confined deflagration (explosion) behavior. The key performance parameters for constant volume combustion are the adiabatic isochoric flame temperature and the explosion pressure. Higher isochoric flame temperatures cause higher explosion pressures, both of which indicate greater explosion hazards. Experimental measurements of flame temperatures and explosion pressures tend to be lower than the chemical equilibrium predictions. As a general trend, in the fuel lean condition the explosion pressure increases with dust concentration. As the dust concentration crosses the stoichiometric value and becomes fuel-rich, the explosion pressure asymptotically reaches a maximum value.

Three examples were presented to illustrate how simple thermodynamic analysis can give insight into the complex behavior of combustible dusts. The first example was the relationship between the burn fraction and the fractional pressure rise. The resulting analysis demonstrated the linear relationship between the burn fraction and the fractional pressure rise. The second example examined the flash fire scenario and developed one possible criterion to evaluate when a dust deflagration is a flash fire and when it is an explosion. The third example presented a simple model for secondary dust explosions and demonstrated the hazard presented by fugitive dust accumulations within enclosures.

The performance parameters developed in this chapter define the minimum or maximum limits of combustible dust behavior. In real combustible dust hazard scenarios, the actual deflagration behavior will likely be significantly different from that which is predicted by thermodynamics due to irreversible finite rate processes. But the general trends of combustion behavior with increasing dust concentrations are correctly predicted by thermodynamics.

REFERENCES

Borman, G.L., Ragland, K.W., 1998. Combustion Engineering. McGraw-Hill, New York.

Cashdollar, K.L., 1994. Flammability of metals and other elemental dust clouds. Process Saf. Prog. 13, 139−145.

Cashdollar, K.L., Hertzberg, M., 1983. Infrared temperatures of coal dusts explosions. Combust. Flame 51, 23−35.

Cashdollar, K.L., Zlochower, I.A., 2007. Explosion temperatures and pressures of metals and other elemental dust clouds. J. Loss Prev. Process Ind. 20, 337−348.

Cassel, H.M., 1964. Some Fundamental Aspects of Dust Flames. Report of Investigations 6551. U.S. Department of the Interior, Bureau of Mines.

CCPS/AIChE, 1996. Guidelines for Evaluating Process Plant Buildings for External Explosions and Fires. American Institute of Chemical Engineers, New York.

Channiwala, S.A., Parikh, P.P., 2002. A unified correlation for estimating HHV of solid, liquid, and gaseous fuels. Fuel 81, 1051−1063.

Dean, J.A., 1999. Lange's Handbook of Chemistry, fifteenth ed. McGraw-Hill, New York.

Eckhoff, R., 2003. Dust Explosions in the Process Industries, third ed. Gulf Professional Publishing, Elsevier, New York.

Gao, W., Dobashi, R., Mogi, T., Sun, J., Shen, X., 2012. Effects of particle characteristics on flame propagation behavior during organic dust explosions in a half-closed chamber. J. Loss Prev. Process Ind. 25, 993−999.

Glassman, I., Yetter, R.A., Glumac, N.G., 2015. Combustion, fifth ed. Academic Press, New York.

Gordon, S., McBride, B.J., 1994. NASA Report NASA RP-1311. Computer Program for Calculation of Complex Chemical Equilibrium Compositions and Analysis. NASA Lewis Research Center.

Goroshin, S., Fomenko, I., Lee, J.H.S., 1996. Burning velocities in fuel-rich aluminum dust clouds. Twenty-Sixth Symposium (International) on Combustion. The Combustion Institute, Elsevier.

Goroshin, S., Mamen, J., Higgins, A., Bazyn, T., Glumac, N., Krier, H., 2007. Emission spectroscopy of flame fronts in aluminum suspensions. Proc. Combust. Inst. 31, 2011−2019.

Hertzberg, M., Zlochower, I.A., Cashdollar, K.L., 1992. Metal dust combustion: explosion limits, pressures, and temperatures. Twenty-Fourth Symposium (International) on Combustion. The Combustion Institute, Elsevier.

Horton, M.D., Goodson, F.P., Smoot, L.D., 1977. Characteristics of flat, laminar coal-dust flames. Combust. Flame 28, 187−195.

Kalman, J., Glumac, N.G., Krier, H., 2015. Experimental study of constant volume sulfur dust explosions. J. Combust. article ID 817259, 11 pages.

Kuo, K., 2005. Principles of Combustion, second ed. John Wiley & Sons, New York.

Lee, J.H.S., Zhang, F., Knystautas, R., 1992. Propagation mechanisms of combustion waves in dust-air mixtures. Powder Technol. 71, 153−162.

Lemos, L., Bouriannes, R., 1991. Starch dust combustion characteristics in a closed spherical vessel. Progress Astronaut. Aeronaut. 132, 93−106.

Lewis, B., von Elbe, G., 1987. Combustion, Flames, and Explosions of Gases, third ed. Academic Press, New York.

Lide, D.R., 2006. CRC Handbook of Chemistry and Physics, 86th ed. Taylor and Francis, Boca Raton, FL: CRC Press.

McAllister, S., Chen, J.-Y., Fernandez-Pello, A.C., 2011. Fundamentals of Combustion Processes. Springer, Berlin.

Medina, C.H., Phylaktou, H.N., Sattar, H., Andrews, G.E., Gibbs, B.M., 2013. The development of an experimental method for the determination of the minimum explosible concentration of biomass powders. Biomass Bioenergy 53, 95−104.

Nagy, J., 1985. Informational Report 1119. The Explosion Hazard in Mining. U.S. Department of Labor, Mine Safety and Health Administration.

O'Connell, J.P., Haile, J.M., 2005. Thermodynamics: Fundamentals for Applications. Cambridge University Press, Cambridge.

Ogle, R.A., 1999. Explosion hazard analysis for an enclosure partially filled with a flammable gas. Process Saf. Prog. 18, 170−177.

Proust, C., 2006a. A few fundamental aspects about ignition and flame propagation in dust clouds. J. Loss Prev. 19, 104−120.

Proust, C., 2006b. Flame propagation and combustion in some dust−air mixtures. J. Loss Prev. Process Ind. 19, 89−100.

Sami, M., Annamalai, K., Wooldridge, M., 2001. Co-firing of coal and biomass fuel blends. Prog. Energy Combust. Sci. 27, 171−214.

Sattar, H., Slatter, D., Andrews, G.E., Givvs, B.M., and Phylaktou, H.N. ,2012. Pulverised biomass explosions: investigation of the ultra rich mixtures that give peak reactivity. In: Proceedings of the IX International Seminar on Hazardous Process Materials and Industrial Explosions (IX ISHPMIE), Krakow.

Schmidt, L.D., 1998. The Engineering of Chemical Reactions. Oxford University Press, Oxford.

Shoshin, Y., Dreizin, E., 2002. Production of well-controlled laminar aerosol jets and their application for studying aerosol combustion processes. Aerosol Sci. Technol. 36, 953−962.

Sibulkin, M., 1980. Pressure rise generated by combustion of a gas pocket. Combust. Flame 38, 329−334.

Smith, W.R., Missen, R.W., 1982. Chemical Reaction Equilibrium Analysis: Theory and Algorithms. Wiley-Interscience, New York.

Stamps, D., Cooper III, E., Egbert, R., Heerdink, S., Stringer, V., 2009. Pressure rise generated by the expansion of a local gas volume in a closed vessel. Proc. Royal Soc. A 465, 3627−3646.

Sun, J.-H., Dobashi, R., Hirano, T., 2000. Combustion behavior of iron particles suspended in air. Combust. Sci. Technol. 150, 99−114.

Thomas, G.O., Oakley, G.L., 2010. Overpressure development during the combustion of a hydrogen−air mixture partial filling a confined space. Process Saf. Environ. Protect. 88, 24−27.

Turns, S.R., 2012. An Introduction to Combustion, third ed. McGraw-Hill, New York.

van der Wel, P.G.J., Lemkowitz, S.M., Scarlett, B., van Wingerden, C.J.M., 1991. A study of particle factors affecting dust explosions. Particle Particle Syst. Character. 8, 90−94.

Vykhristyuk, A.Ya, Babkin, V.S., Kudryavtsev, E.A., Kudryavstev, N.D., Krivulin, N., 1988. Pressure rise in a closed container during the combustion of a localized gas volume. Combust. Explos. Shock Waves 24, 15−20.

Wolinski, M., Wolanski, P., 1993. Shock-wave induced combustion of dust layers. Prog. Astronaut. Aeronaut. 152, 105−118.

Zalosh, R., 2002. Explosion protection. In: P.J. Di Nenno et al., eds., The SFPE Handbook of Fire Protection Engineering, third ed., Section 3, Chapter 16. National Fire Protection Association, Quincy, MA, pp. 3-408 to 3-410.

Zeldovich, Ya.B., Barenblatt, G.I., Librovich, V.B., Makhviladze, G.M., 1985. The Mathematical Theory of Combustion and Explosions, translated from Russian by McNeill, D., Consultants Bureau, Plenum Publishing Corporation, New York.

Transport phenomena for dust combustion

4

It is assumed that the reader has had a previous introduction to the equations of change for chemically reacting flows. The objective of this chapter is to summarize the fundamental forms of the equations of change, also called the conservation equations that will become useful in investigating combustible dust phenomena. Combustion science has contributed enormously to the body of knowledge comprising the dynamics of chemically reacting fluids. But much of this literature has been focused on the dynamics of chemically reacting gases. Moreover, the investigation of combustion phenomena has resulted in a number of concepts and models that are specific to the field. This chapter will introduce some of these concepts and models developed for gaseous systems as a reference point for comparison as later chapters apply these concepts to combustible dust.

Regardless of the type of combustion phenomenon—smoldering, flames, or detonations—they all share one remarkable characteristic in common: they are all combustion waves. Each of these phenomena can be envisioned as a narrow region where the combustion process occurs with the unburnt fuel mixture entering the wave and the combustion products exiting. This similarity means that there is also a similarity in the methods of analysis used to explore their behavior.

This chapter follows a progression from the simplest form of the equations of change and gradually introduces the complexity of multiphase mixtures, thermal radiation, and turbulence. After a brief précis on the equations of change for chemically reacting fluids, I begin with the simplest description of a combustion wave based on the jump conditions of fluid dynamics. This analysis, based on the work originally developed by Chapman and Jouguet, neglects the effects of transport processes and assumes infinite rates of chemical reaction. This analysis based on jump conditions leads to relations that can describe the essential features of deflagrations and detonations. These relations, originally developed by Chapman and Jouguet for the analysis of premixed gaseous systems, will be applied to combustible dusts in Chapter 9.

The next section examines the fundamental behavior of premixed gas flames (deflagrations). From the macroscale perspective, there is a direct analogy between premixed gas flames and premixed dust-air flames. The dynamics of premixed flames are governed by the effects of the finite rates of transport processes and chemical kinetics. After introducing the governing equations for laminar premixed flames, the topics of flame propagation, ignition, and quenching are discussed and illustrated with some typical results.

Dust Explosion Dynamics.
© 2017 Elsevier Inc. All rights reserved.

The section that follows, the diffusion flame analysis of a liquid fuel droplet, is directly related to the burning of a single dust particle. If a dust flame is envisioned as a cloud of burning particles, single particle combustion is the micro-scale model of a dust flame. The behavior of burning droplets is characterized by the temperature and concentration profiles surrounding the particle, and these fields govern the burning time (the time required to burn the droplet to completion) and the mass combustion rate.

The next two sections consider transport phenomena in multiphase systems. First I discuss the transport phenomena in porous media, a subject that is relevant to the dynamics of smoldering. The structure of the porous medium determines the transport behavior of the fluid as it migrates through the pore space. A mathematical concept called local volume averaging is invoked to derive the field equations for transport phenomena in porous media. Because we will restrict ourselves to thermal theories of ignition and wave propagation, we will only need the thermal energy equation. Many organic combustible dusts are capable of undergoing self-heating and spontaneous ignition leading to the formation of a smoldering wave. This is caused by an interaction between heat generation by the oxidation reaction and heat losses to the surroundings. Under certain conditions, the smoldering wave can lead to a flaming fire or an explosion and thus, smoldering is an important combustible dust phenomenon to understand.

The second topic in multiphase systems is the flow of a dispersed cloud of particles; this topic is relevant to dust deflagrations. The presence of particles influences the dynamics of fluid in motion. The magnitude of this influence depends on both characteristic length scales and characteristic time scales. The diameter of the particles must be small enough compared to the length scale of the flow field so that the motion of the particles can adapt to changes in the flow field. Here too the formulation of the transport equations for dispersed multiphase flow can be formalized through the concept of local volume averaging. The motion of the fluid and the particles can be weakly or strongly coupled depending on the magnitude of interphase momentum, heat, or mass transfer. Criteria for this coupling are derived in terms of dimensionless numbers. The degree of flow coupling is also dependent on the particle concentration; guidelines are suggested for evaluating this phenomenon. Finally, the important concept of equilibrium flows is described.

Thermal radiation is an important factor in many aspects of dust flame behavior. While many aspects of gaseous flame behavior can be adequately described without accounting for thermal radiation, dust clouds have the potential to create a strongly participating medium for thermal radiation. The dynamics of thermal radiation are accounted for through an energy transport equation for photons called the radiative transfer equation. While difficult to solve in general, two limiting cases of behavior called the optically thin and optically thick regimes simplify the solution to the transfer equation. A discussion of the radiative properties of particle clouds is the final topic of this section.

The fluid flow phenomenon of turbulence is an important factor in most practical combustible dust hazard scenarios. This section begins with a brief description of turbulence and its effect on transport processes. The level of sophistication for this discussion is kept at the most elementary level. Time and length scales useful for describing turbulent transport are introduced and examples are presented to illustrate their estimation. The specific case of how turbulence influences the combustion behavior of premixed flames is discussed. Again, specific time and length scales provide an important set of criteria for predicting turbulent premixed flame behavior.

4.1 EQUATIONS OF CHANGE FOR LAMINAR REACTING FLOWS

In this section, we review the equations of change for laminar reacting flows. There is an extensive body of knowledge on this topic. Our focus will be on the form of these equations that are most useful for the analysis of dust combustion. Specifically, to investigate flame propagation behavior we will rely primarily on the one-dimensional forms of the equations of change in rectangular or spherical coordinates. The specification of the equations of change in a one-dimensional format implies that variations in the other two coordinate directions are negligible. Fig. 4.1 illustrates the use of a rectangular coordinate system for the analysis of a planar flame and the use of the radial coordinate for spherical flame propagation.

   For the investigation of confined flame propagation, we will first rely on transient control volume methods (so-called integral models or lumped parameter models). A comprehensive presentation of the equations of change can be found in any of several excellent textbooks on transport phenomena (Bird et al., 2002; Slattery, 1981) or combustion theory (Turns, 2012; Law, 2006; Williams, 1985; Kuo, 2005).

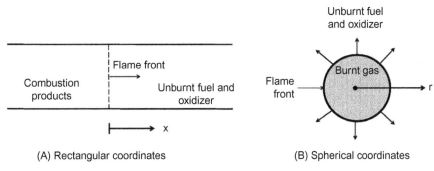

(A) Rectangular coordinates   (B) Spherical coordinates

FIGURE 4.1

Illustration of one-dimensional rectangular and spherical coordinate systems.

4.1.1 PRELIMINARIES

The equations of change are derived from the fundamental assumption that the body (or system) under consideration, whether it is a fluid or solid, can be treated as a continuum. This means that the medium has physical properties that are defined at each point within the body. The equations of change apply everywhere within the body, but not at the interface between the body and its surroundings. At the interface, a special form of the equations of change is required; these are called the jump conditions. For the types of fluids and solids of interest in combustible dust investigations, the equations of change required express the balance of mass, chemical species, linear momentum, and energy. This holds similarly for the jump conditions.

In most combustible dust hazard scenarios, we are interested in the flow of particles dispersed in a moving carrier fluid. The carrier fluid is a gas mixture and, in most cases, the carrier fluid is air. Our focus at the beginning of this chapter is on the carrier fluid. The dispersed particles will be introduced later. Likewise, the equations of change will be introduced first in a form suitable for the analysis of laminar flow. Turbulence, a major feature of many combustible dust hazard scenarios, will be introduced later in this chapter.

4.1.2 CONTINUITY

The continuity equation is a partial differential equation that describes the balance of mass in a continuous body (the fluid or region of interest). It is valid at any arbitrary point within the continuous medium. The two most common forms useful for the analysis of dust combustion are the one-dimensional forms of the continuity equation expressed in rectangular or spherical coordinates. The one-dimensional flow of a fluid that follows a rectilinear path is called a planar flow. In rectangular coordinates, the one-dimensional form of the continuity equation is

$$\frac{\partial \rho}{\partial t} + \frac{\partial}{\partial x}(\rho v_x) = 0 \tag{4.1}$$

where the equation has been written for change in the x-direction, ρ is the density of the fluid, and v_x denotes the x-component of the fluid velocity.

Stationary flames are a common laboratory system for combustion studies. In steady flow, the time derivative disappears and the one-dimensional continuity equation takes the form

$$\frac{\partial}{\partial x}(\rho v_x) = 0 \quad \text{or} \quad \dot{m}'' = \rho v_x = \text{constant} \tag{4.2}$$

Thus in steady, one-dimensional planar flow, the mass flux ρv_x is a constant that arises naturally in the description of the flow field.

A one-dimensional flow with spherical symmetry that begins at the center and expands outwards is called a radial flow. The one-dimensional form of the continuity equation in spherical coordinates is

$$\frac{\partial \rho}{\partial t} + \frac{1}{r^2}\frac{\partial}{\partial r}\left(r^2 \rho v_r\right) = 0 \tag{4.3}$$

where the equation has been written with the r-direction as the direction of change and v_r denotes the radial component of the fluid velocity. If the flow is steady, Eq. (4.3) is readily integrated to yield

$$\frac{1}{r^2}\frac{\partial}{\partial r}\left(r^2 \rho v_r\right) = 0 \Rightarrow r^2 \rho v_r = \text{constant} \tag{4.4}$$

Note that in radial flow within a spherical geometry, the mass flow is the natural constant of the flow field and not the mass flux.

The continuity equation governs the balance of the total mass of the continuous medium. If the medium consists of a mixture of several chemical species, then the continuity equation governs the behavior of the mixture, not the behavior of the individual chemical species. To describe the behavior of the individual chemical species in a mixture, we need an additional continuity relation.

4.1.3 CHEMICAL SPECIES CONTINUITY

The chemical species continuity equation is a partial differential equation that describes the balance of mass for a given chemical species in a mixture. The species continuity equation has two additional terms not found in the ordinary continuity equation, a diffusion term and a source term for chemical reactions. The species continuity equation can be written in many different forms using molar or mass units. The equations presented below are in mass units and assume that the mixture is an ideal gas. It is also implicitly assumed that the diffusion of chemical species is adequately represented by Fick's law for binary diffusion, and that the binary diffusivity $\mathcal{D}$ is a constant. There are many nuances to the formulation of the species continuity equation and the representation of diffusion fluxes. For more discussion on these points, the reader is referred to books by Bird et al. (2002, Chapters 17 and 19), Slattery (1981, Chapter 8), Turns (2012, Chapters 3 and 7), or Law (2006, Chapters 4 and 5).

For one-dimensional planar flow, the species continuity equation takes the form:

$$\rho\left(\frac{\partial Y_i}{\partial t} + v_x\frac{\partial Y_i}{\partial x}\right) = \rho\mathcal{D}\frac{\partial^2 Y_i}{\partial x^2} + \omega_i''' \tag{4.5}$$

where Y_i is the mass fraction of species i in the mixture, $\mathcal{D}$ is the binary diffusivity, and ω_i''' is the production of chemical species i per unit volume due to chemical reaction.

In spherical coordinates with radial flow, the species continuity equation is

$$\rho\left(\frac{\partial Y_i}{\partial t} + v_r \frac{\partial Y_i}{\partial r}\right) = \rho D\left[\frac{1}{r^2}\frac{\partial}{\partial r}\left(r^2 \frac{\partial Y_i}{\partial r}\right)\right] + \dot{w}_i''' \tag{4.6}$$

For a chemically reacting mixture with N chemical species, there are only $N-1$ independent species continuity equations. The value of the nth species can be determined from the mass fraction relation $Y_N = 1 - \sum_{i=1}^{N-1} Y_i$. An additional important relation involving species continuity can be inferred by realizing that the sum of the species continuity equations over N species should result in the overall continuity equation. This then implies a relation for the chemical reaction rates

$$\sum_{i=1}^{N} \dot{w}_i''' = 0 \tag{4.7}$$

This relation will be useful later in our discussion of the energy equation.

4.1.4 MOMENTUM

The momentum equation is a partial differential equation that governs the balance of linear momentum in a continuous medium. In most flame propagation studies, the viscous forces can be neglected. This is the inviscid flow assumption. The exception to this will be a brief discussion of boundary layer transport in Chapter 9. The momentum equation for one-dimensional planar flow takes the following form:

$$\rho\left(\frac{\partial v_x}{\partial t} + v_x \frac{\partial v_x}{\partial x}\right) = -\frac{\partial P}{\partial x} \tag{4.8}$$

For radial flow in a spherical geometry, the momentum equation is

$$\rho\left(\frac{\partial v_r}{\partial t} + v_r \frac{\partial v_r}{\partial r}\right) = -\frac{\partial P}{\partial r} \tag{4.9}$$

These equations are essential for the study of unsteady flame propagation, especially as the flame accelerates. In steady laminar flame propagation, the momentum equation implies that the flame is isobaric, that is, the pressure does not change appreciably across the flame. To demonstrate the validity of the isobaric approximation, consider the following scaling analysis of the steady momentum equation for planar flow (Law, 2006, pp. 168–169).

We begin by casting the steady momentum equation into nondimensional form by introducing reference values for the density, velocity, and pressure, $\rho_0, x_0, v_{x,0}$, and P_0, respectively. Define the following nondimensional variables: $\hat{\rho} = \rho/\rho_0, \hat{x} = x/x_0, \hat{v}_x = v_x/v_{x,0}$, and $\hat{P} = P/P_0$. Note that in this chapter the "hat" above the variable denotes the nondimensional form of the variable.

Substituting these definitions into the momentum equation gives the following result:

$$\frac{\rho_0 v_{x,0}^2}{P_0}\left(\hat{\rho}\hat{v}_x \frac{d\hat{v}_x}{d\hat{x}}\right) = -\frac{d\hat{P}}{d\hat{x}}$$

(4.10)

For an ideal gas, the sound speed squared, a_0^2, is given as $\gamma P_0/\rho_0$. The ratio of the local velocity to the sound speed is the Mach number, Ma. The momentum equation now becomes

$$\gamma\,\mathrm{Ma}^2\left(\hat{\rho}\hat{v}_x \frac{d\hat{v}_x}{d\hat{x}}\right) = -\frac{d\hat{P}}{d\hat{x}}$$

(4.11)

Typical laminar burning velocities for premixed gas flames are on the order of $0.3-1.0$ m/s. The sound speed for the unburnt state of such flames is on the order of $300-400$ m/s. Therefore, $\mathrm{Ma}^2 \ll 1$, making the left-hand side of Eq. (4.11) approach the limit of zero. If the pressure gradient is zero, the pressure is constant. Thus, the pressure drop across a laminar premixed flame is negligible.

4.1.5 THERMAL ENERGY

The thermal energy equation describes the balance of thermal energy in a continuous medium. The thermal energy equation can be written in many different formats. For one-dimensional planar flow of an ideal gas, a convenient form written for constant physical properties is

$$\rho C_p\left(\frac{\partial T}{\partial t} + v_x \frac{\partial T}{\partial x}\right) = k\frac{\partial^2 T}{\partial x^2} + \sum_{i=1}^{N} h_{f,i}^0\,\omega_i'''$$

(4.12)

This form is for systems with constant pressure systems like flames. The constant k is the thermal conductivity of the medium. Thermal radiation and viscous dissipation are neglected. The source term represents the heat release due to chemical reaction; it is the product of the enthalpy of formation $h_{f,i}^0$ and The mass rate of production per unit volume ω_i''' for each chemical species. For radial flow in a spherical geometry, the equation takes the form:

$$\rho C_p\left(\frac{\partial T}{\partial t} + v_r \frac{\partial T}{\partial r}\right) = k\left[\frac{1}{r^2}\frac{\partial}{\partial r}\left(r^2 \frac{\partial T}{\partial r}\right)\right] + \sum_{i=1}^{N} h_{f,i}^0\,\omega_i'''$$

(4.13)

The source term for the chemical reaction can be transformed into a source term for combustion; it is derived from the following assumptions. It is assumed that the combustion reaction can be represented by a single-step reaction. Written in mass units, the combustion reaction is

$$1\text{kg fuel} + s\,\text{kg oxidizer} \rightarrow (s+1)\,\text{product}$$

(4.14)

The symbol s is a stoichiometric coefficient with the units s kg oxidizer per kg fuel. A species continuity equation is required for each of the species: fuel, oxidizer, and product. The reaction rates for the oxidizer and product can be expressed in terms of the fuel reaction rate with the aid of the stoichiometric coefficients

$$\omega_F''' = \frac{1}{s}\omega_{Ox}''' = -\frac{1}{(s+1)}\omega_{Pr}'''$$
(4.15)

The source term in the thermal energy equation must account for the enthalpy contributions of each species. With the aid of Eq. (4.7), the enthalpy source term can be written as

$$-\sum_{i=1}^{3} h_{f,i}^0\, \omega_i''' = -\left[h_F^0\,\omega_F''' + h_{Ox}^0 s\,\omega_F''' - h_{Pr}^0\,(s+1)\,\omega_{Pr}'''\right] = -\Delta h_c\,\omega_F''' = Q\,\omega_F'''$$
(4.16)

Eq. (4.16) tells us that the enthalpy source term due to chemical reaction is simply the reaction rate of fuel times the heat of combustion.

4.1.6 SOLUTION OF THE EQUATIONS OF CHANGE

In addition to the equations of change, auxiliary relations are required for closure of the system of equations. The first of these is the volumetric equation of state which relates the density of the fluid to the pressure and temperature, $\rho = \rho(P, T)$. The second relation is the caloric equation of state which relates the heat capacity to the temperature of the fluid, $C_p = C_p(T)$, (Bird et al., 2002, pp. 339–342).

The equations of change are the basis for formulating mathematical models of combustion behavior. For any given combustion model, the solution of the model begins with a specification of the independent variables for the velocity, temperature, and species mass fraction fields. The specification will look something like this:

$$v_x = v_x(x, t),\ T = T(x, t),\ Y_i = Y_i(x, t)$$
(4.17)

Initial and boundary conditions are then determined from the physical setting of the problem. The equations of change are then written to reflect the velocity, temperature, and mass fraction specifications. The model is then solved using the initial and boundary conditions. While it is intuitively appealing to solve the model in its native form using the primitive (dimensional) variables, it is often helpful to cast the problem in dimensionless form. Scaling analysis of the dimensionless form of a transport model is a powerful way to gain insight into the properties of the model without actually solving it. The book by Krantz is an especially good reference for scaling analysis of transport models (Krantz, 2007).

4.2 CHAPMAN–JOUGUET ANALYSIS OF COMBUSTION WAVES

Consider a long thin tube filled with a premixed flammable gas mixture. If the mixture is ignited on one end of the tube, a combustion wave will travel through the tube to the opposite end. In the course of conducting such experiments, early investigators discovered that the two types of combustion waves could occur: a slow wave, called a deflagration or flame, and a very fast wave called a detonation. The deflagration resulted if the tube was open at the point of ignition, and a detonation resulted if the tube was closed at the point of ignition.

Several investigators in the late 19th century contributed to our understanding of the underlying physics of deflagrations and detonation (Lee, 2008, pp. 4–7). Their work built on the newly discovered phenomena of shock waves. The properties of a shock wave can be derived by idealizing the shock wave as a surface of discontinuity and applying the equations of change (or conservation equations) across the surface. The form of the equations of change which apply across a surface of discontinuity is called jump conditions. The jump conditions are algebraic equations that describe the conservation of mass, linear momentum, and energy across the surface of discontinuity.

The simplest description of a combustion wave is to visualize it as a surface of discontinuity. We will see that this approach works better for detonations than it does for deflagrations. The analysis of combustion waves based on jump conditions, due to Chapman and Jouguet, results in a computationally convenient mathematical model based on algebraic equations. Depending on the equation of state and physical property relations chosen, the speed of the combustion wave and the changes in temperature and pressure across the wave can be computed in a straightforward way. In premixed flammable gas systems, the calculations of detonation properties are reasonably accurate. This is because there is a strong relationship between shock waves and detonation waves. The calculations are less satisfactory for deflagrations because the finite rate effects of transport processes and chemical kinetics are neglected in the Chapman–Jouguet analysis.

In the next section, I will describe in more detail the behavior of combustion waves traveling in long tubes. Next I introduce the jump conditions for a surface of discontinuity and present a method due to Rankine and Hugoniot for evaluating the range of admissible solutions. Then I show how to generalize this analysis for combustion waves, the so-called Chapman–Jouguet analysis, and illustrate its use in the calculation of detonation wave properties.

4.2.1 PHYSICAL DESCRIPTION OF COMBUSTION WAVES

Consider again a long pipe filled with a premixed flammable gas mixture. It has been stated that the combustion wave behavior is dependent on the boundary conditions, specifically, it depends on whether ignition occurs at an open or closed end. To see why this occurs, we will need to learn a little about pressure waves in a compressible medium (sound waves) and their transformation into shock waves. A very good discussion, a bit dated but relevant nevertheless, can be found in G. H. Markstein's classic book, *Nonsteady Flame Propagation* (Markstein, 1964, Chapter E).

Assume a long tube open at both ends is filled with a premixed flammable gas. The gas is ignited at one end of the tube. What happens after ignition? The combustion wave (deflagration) travels into the unburnt mixture and consumes the fuel and oxidizer. The combustion of the gas mixture raises the temperature of the combustion products to the adiabatic flame temperature. The hot combustion products begin to expand and cool by venting out of the open end of the tube. The pressure of the combustion products is just slightly greater than the ambient pressure; this creates the driving force for flow out of the open end of the tube. This will continue until the deflagration wave reaches the other end of the tube. The speed of the deflagration can only be as fast as it can consume the fuel and oxidizer. Fig. 4.2 is a depiction of this process.

Now consider a similar combustion experiment only this time one end of the tube is closed. Ignition occurs at the closed end. Once again, the combustion wave (initially a deflagration) travels into the unburnt mixture and consumes the fuel and oxidizer. The combustion process raises the temperature of the combustion products to its adiabatic flame temperature. The hot combustion products cannot expand out of the end of the tube because it is closed. The pressure of the hot combustion products begins to increase. A sound wave is generated by each

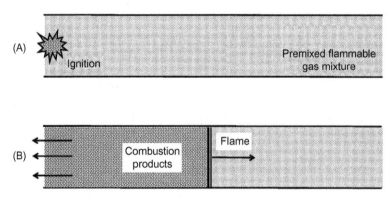

FIGURE 4.2

Open tube ignition of premixed flammable gas mixture: (A) the ignition event and (B) the propagation of the flame into the unburnt mixture with the combustion products flowing out of the tube in the opposite direction.

infinitesimal increase in pressure. The sound wave is a signal that the pressure is increasing. The sound wave reflects back from the closed end of the tube and chases the deflagration wave.

The speed of sound in a gas mixture increases with temperature. This means that each successive sound wave travels faster than the previous sound wave. The sounds waves travel faster than the combustion wave and propagate through the length of the unburnt mixture. Since a sound wave is a compression wave, the sound waves compress and preheat the unburnt mixture. The combustion wave accelerates into the unburnt mixture due to the expansion of the burnt gas column behind it and due to the precompression effect of the sound waves. The sound waves coalesce into a shock wave of increasing strength. If the strength of the shock wave can continue to grow, it eventually ignites the gas ahead of the combustion wave. At that point, the shock wave and combustion wave form a steady wave system called a detonation wave. This fluid dynamic sequence of events is called a deflagration-to-detonation transition or DDT. Fig. 4.3 depicts the DDT process in a premixed flammable gas mixture.

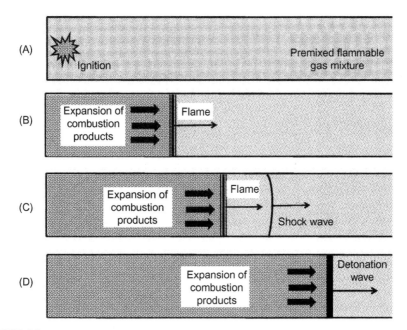

FIGURE 4.3

Closed tube ignition of premixed flammable gas mixture: (A) the ignition event, (B) the propagation of the flame into the unburnt mixture, (C) the formation of a shock wave ahead of the combustion wave, and (D) the merging of the shock and combustion waves into a detonation wave.

4.2.2 JUMP CONDITIONS AND THE RANKINE–HUGONIOT EQUATIONS

A shock wave is a surface of discontinuity in a high speed gas flow.[1] The term "high speed" is relative to the speed of sound in the fluid: shock waves always travel faster than the speed of sound. The strength of a shock wave is defined as its pressure ratio. A shock wave of infinitesimal strength is a sound wave; a strong shock wave has a shocked state pressure much greater than atmospheric pressure. The properties of a shock wave are related to a coordinate frame of reference fixed on the shock wave (Fig. 4.4).

The change in properties across a shock wave is described by the jump conditions of fluid dynamics. With reference to Fig. 4.4, the jump conditions for a shock wave with chemical reaction take the following form:

$$\text{mass: } \dot{m}''_x = \rho_1 v'_{x,1} = \rho_2 v'_{x,2} \tag{4.18}$$

$$\text{momentum: } P_1 + \rho_1 v'_{x,1} = P_2 + \rho_2 v'_{x,2} \tag{4.19}$$

$$\text{energy: } h_1 + \frac{1}{2}\left(v'_{x,1}\right)^2 + Q = h_2 + \frac{1}{2}\left(v'_{x,2}\right)^2 \tag{4.20}$$

The enthalpy of the chemical reaction is denoted by Q.

(A) Laboratory coordinates

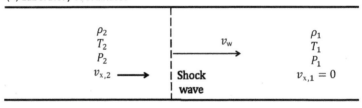

(B) Shock-fixed coordinates

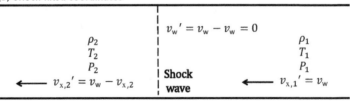

FIGURE 4.4

Shock wave properties relative to laboratory and shock-fixed coordinate frames of reference.

---

[1] Shock waves can be induced in liquids and solids, but we shall limit our introduction to shock waves in the gas phase only.

Combustion scientists have devised a convenient graphical method for analyzing combustion wave behavior. This method requires two relations, a linear equation called the *Rayleigh relation* and a hyperbolic equation called the *Rankine–Hugoniot relation*. These two equations will be presented in a graphical format that will help us establish feasible solutions to the jump conditions for a combustion wave.

We will consider first the derivation of the Rayleigh relation. We begin by first solving the mass jump condition for $v'_{x,2}$:

$$v'_{x,2} = \left(\frac{\rho_1}{\rho_2}\right) v'_{x,1} \tag{4.21}$$

Insert the velocity expression into the momentum jump condition and collect like terms:

$$\rho_1 (v'_{x,1})^2 - \left(\frac{\rho_1^2}{\rho_2}\right)(v'_{x,1})^2 = (P_2 - P_1) \tag{4.22}$$

Dividing through by ρ_1^2 and solving for $(v'_{x,1})^2$, Eq. (4.22) becomes

$$(v'_{x,1})^2 = \frac{1}{\rho_1^2}\left[\frac{P_2 - P_1}{\left(\frac{1}{\rho_1} - \frac{1}{\rho_2}\right)}\right] \tag{4.23}$$

A simpler form is obtained by replacing the density ρ with its reciprocal, the specific volume v. The resulting equation is the Rayleigh relation:

$$-(\dot{m}''_x)^2 = -\left(\rho_1 v'_{x,1}\right)^2 = -\left(\rho_2 v'_{x,2}\right)^2 = \frac{(P_2 - P_1)}{(v_2 - v_1)} \tag{4.24}$$

Notice that the Rayleigh equation was derived without invoking a specific equation of state.

The equation for the Rayleigh line specifies a family of pressure and specific volume values for a fixed mass flux $\dot{m}''_x$. This relation is called the Rayleigh line because the pressure is a linear function of the specific volume for a fixed value of the mass flux (Turns, 2012, pp. 620–621). Solving Eq. (4.24) for the pressure yields the following linear equation:

$$P = a\, v_2 + b \tag{4.25}$$

The slope $a = -\dot{m}''^2_x$ and the intercept $b = P_1 + \dot{m}''^2_x v_1$.

The Rankine–Hugoniot equation is obtained by combining the mass, momentum, and energy equations, applying the ideal gas equation of state, and using the heat capacity relations for an ideal gas. We follow here the procedure outlined in Glassman et al. (2015, pp. 260–262). We begin with a heat capacity relation for an ideal gas of constant properties:

$$C_{\mathrm{p}} = \frac{R_g}{M}\left(\frac{\gamma}{\gamma - 1}\right) \tag{4.26}$$

As before, R_g is the universal gas constant and $\mathcal{M}$ is the molar mass of the mixture. Substituting this into the energy equation gives

$$\frac{R_g}{\mathcal{M}}\left(\frac{\gamma}{\gamma-1}\right)T_1 + \frac{1}{2}(v'_{x,1})^2 + Q = \frac{R_g}{\mathcal{M}}\left(\frac{\gamma}{\gamma-1}\right)T_2 + \frac{1}{2}(v'_{x,2})^2 \qquad (4.27)$$

Substituting the ideal gas equation of state, $R_g T/\mathcal{M} = P/\rho$, and collecting like terms yields

$$\left(\frac{\gamma}{\gamma-1}\right)\left(\frac{P_2}{\rho_2} - \frac{P_1}{\rho_1}\right) - \frac{1}{2}\left[(v'_{x,1})^2 - (v'_{x,2})^2\right] = Q \qquad (4.28)$$

If we can eliminate the velocities from Eq. (4.28), we can express the combustion wave properties in terms of two variables, pressure and density, and two parameters, the heat capacity ratio and the enthalpy of combustion. With reference to Eq. (4.23) and the mass jump condition, we can write

$$(v'_{x,1})^2 - (v'_{x,2})^2 = \left(\frac{1}{\rho_1^2} - \frac{1}{\rho_2^2}\right)\left[\frac{P_2 - P_1}{\left(\frac{1}{\rho_1} - \frac{1}{\rho_2}\right)}\right] = \frac{\rho_2^2 - \rho_1^2}{\rho_1^2 \rho_2^2}\left[\frac{P_2 - P_1}{\left(\frac{1}{\rho_1} - \frac{1}{\rho_2}\right)}\right] = \left(\frac{1}{\rho_1} + \frac{1}{\rho_2}\right)(P_2 - P_1)$$

$$(4.29)$$

Substituting Eq. (4.29) into (4.28) and replacing the density with the specific volume gives us the desired form of the Rankine−Hugoniot relation:

$$\frac{\gamma}{\gamma-1}(P_2 v_2 - P_1 v_1) - \frac{1}{2}(P_2 - P_1)(v_1 + v_2) = Q \qquad (4.30)$$

The relationship between the Rayleigh line and the Rankine−Hugoniot curve is often shown in a $P - v$ diagram as shown in Fig. 4.5.

The effect of chemical energy addition is to form a new Rankine−Hugoniot curve located above the curve with zero energy addition. This is shown in Fig. 4.6.

The use of the Rayleigh equation allows you to divide the Rankine−Hugoniot curve into different regions. This interpretation is due to the scientists Chapman and Jouguet.

4.2.3 CHAPMAN−JOUGUET INTERPRETATION OF THE RANKINE−HUGONIOT DIAGRAM

The Chapman−Jouguet (CJ) interpretation is based on a consideration of the change in entropy across the shock wave. It can be shown that equating the slope of the Rayleigh equation with the slope of the Rankine−Hugoniot equation at fixed entropy yields the following result:

$$\left(\frac{\partial P}{\partial \rho}\right)_S = \frac{(P_2 - P_1)}{(v_1 - v_2)} = a^2 = v_1^2 \qquad (4.31)$$

At the point of tangency, the wave velocity equals the speed of sound.

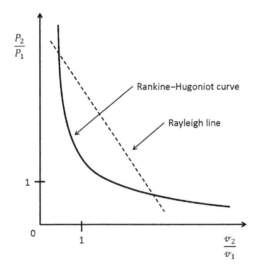

FIGURE 4.5

$P - v$ diagram showing the Rayleigh line and the Rankine–Hugoniot curve.

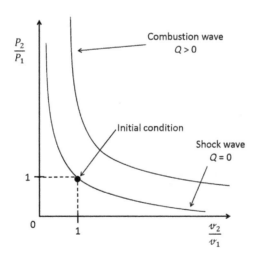

FIGURE 4.6

Rankine–Hugoniot curve with chemical energy addition ($Q > 0$).

It turns out that there are restrictions on the property values that a chemically reacting shock wave can assume. The CJ points on the curve provide a convenient reference point for identifying the different regimes of combustion wave behavior. Fig. 4.7 is a nondimensional plot of the Rankine–Hugoniot curve with the curve divided into different regimes.

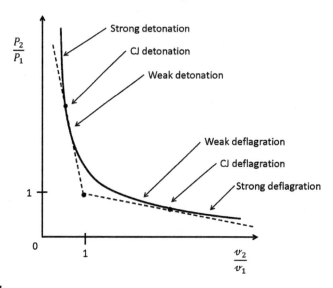

FIGURE 4.7

Rankine–Hugoniot diagram divided into different regimes.

There are two CJ points in the figure corresponding to a detonation wave and a deflagration wave. Compared to its initial pressure and specific volume, the CJ detonation wave corresponds to an increase in pressure and a decrease in specific volume. So the CJ detonation is a compression wave. The CJ deflagration exhibits the opposite trend: compared to its initial conditions, the pressure is decreased and the specific volume is increased. These are the characteristics of an expansion wave.

There is a region on the RH curve that is denoted as an unrealistic solution. Values of pressure and specific volume in this region result in imaginary values for the wave speed and are therefore not realistic solutions. It has been observed experimentally that steady premixed gas detonations tend to propagate at the CJ conditions. Neither the weak nor strong detonation occurs as a steady phenomenon. Deflagration waves tend to behave as weak deflagrations. The CJ deflagration velocity is not a reliable indicator of the deflagration wave speed because of the omission of finite transport and reaction rates from the analysis.

4.2.4 CALCULATION OF DETONATION PARAMETERS

The CJ analysis for combustion waves works quite well for premixed gas detonations. I present a simple and approximate derivation for calculating the properties of a detonation wave. In this derivation, it is assumed that the heat capacities and molar masses of the unburnt and burnt mixture are constant and that the gas mixture obeys the ideal gas equation of state. It is also assumed that the heat capacity ratio is the same for both mixtures. These assumptions can be relaxed if greater accuracy is desired, but the calculations are then more easily performed with a chemical equilibrium code like the NASA Lewis CEA code.

This derivation follows the presentation of Churchill (1980, pp. 102–107).

We begin by writing the jump conditions for mass, momentum, and energy in terms of the Mach number $Ma = v/a$.

$$\text{mass}\quad Ma_1(P_1\rho_1)^{1/2} = Ma_2(P_2\rho_2)^{1/2} \tag{4.32}$$

$$\text{momentum}\quad P_1(1 + \gamma\, Ma_1^2) = P_2(1 + \gamma\, Ma_2^2) \tag{4.33}$$

$$\text{energy}\quad C_p T_1 + \left(\frac{\gamma\, P_1\, Ma_1}{2\rho_1}\right) + Q = C_p T_2 + \left(\frac{\gamma\, P_2\, Ma_2}{2\rho_2}\right) \tag{4.34}$$

Dividing the mass equation by the momentum equation and squaring both sides yields

$$\frac{\rho_1}{P_1}\left(\frac{Ma_1}{1 + \gamma\, Ma_1^2}\right)^2 = \frac{\rho_2}{P_2}\left(\frac{Ma_2}{1 + \gamma\, Ma_2^2}\right)^2 \tag{4.35}$$

Multiplying Eq. (4.35) times the energy equation and simplifying yields

$$\left(\frac{Ma_1}{1 + \gamma Ma_1^2}\right)^2\left[1 + \left(\frac{\gamma - 1}{2}\right)Ma_1^2 - \frac{Q}{C_p T_1}\right] = \left(\frac{Ma_2}{1 + \gamma Ma_2^2}\right)^2\left[1 + \left(\frac{\gamma - 1}{2}\right)Ma_2^2\right] \tag{4.36}$$

For a CJ detonation, $Ma_2 = 1$, so Eq. (4.36) reduces to

$$\left(\frac{Ma_1}{1 + \gamma\, Ma_1^2}\right)^2\left[1 + \frac{\gamma - 1}{2}Ma_1^2 - \frac{Q}{C_p T_1}\right] = \frac{1}{2(\gamma + 1)} \tag{4.37}$$

Multiplying both sides of Eq. (4.37) by $2(\gamma + 1)(1 + \gamma Ma_1^2)^2/Ma_1^2$, simplifying, and rearranging gives the following expression for Ma_1

$$Ma_1^4 - 2\left[1 + \frac{(\gamma + 1)Q}{C_p T_1}\right]Ma_1^2 - 1 = 0 \tag{4.38}$$

Solving this quadratic equation and taking the positive root gives the desired expression for Ma_1

$$Ma_1 = \left[\left(\frac{\gamma + 1}{2}\right)\frac{Q}{C_p T_1}\right]^{1/2} + \left[\left(\frac{\gamma + 1}{2}\right)\frac{Q}{C_p T_1} + 1\right]^{1/2} \tag{4.39}$$

Recall that the detonation velocity can be computed from $v'_{x,1} = Ma_1 a_1$. The other properties are computed by the following expressions:

$$\frac{P_2}{P_1} = \frac{1 + \gamma Ma_1^2}{\gamma + 1} \tag{4.40}$$

$$\frac{T_2}{T_1} = \left[\frac{1 + \gamma Ma_1^2}{(\gamma + 1)Ma_1}\right]^2 \tag{4.41}$$

$$\frac{\rho_2}{\rho_1} = \frac{(1 + \gamma)Ma_1^2}{1 + \gamma Ma_1^2} \tag{4.42}$$

Eqs. (4.39)–(4.42) comprise the complete set of relations needed to specify the properties of a premixed gas detonation.

Churchill suggests a rule of thumb that one should use an effective heat of combustion in this model: $Q_{eff} = 2/3Q_{actual}$ (Churchill, 1980, p. 107). Thompson has derived the corresponding equations for the case where the heat capacity ratios are not equal, ie, $\gamma_1 \neq \gamma_2$ (Thompson, 1972, pp. 352–354).

EXAMPLE 4.1

Assume an isolated pipe section contains a stoichiometric mixture of methane and oxygen at a temperature of 300 K and a pressure of 1.01 bars. If a CJ detonation is initiated in this pipe, what is the detonation wave velocity, the final pressure, and the final temperature that the pipe section experiences? Assume the sound speed in the unreacted mixture is $c = 350$ m/s, a mean heat capacity $C_p = 41.9$ J/mol-K, negligible change in the mixture molar mass, a specific heat ratio $\gamma = 1.4$, and the heat of combustion $Q = 197.7$ kJ/mol (after Churchill, 1980, p. 109). Use Churchill's rule of thumb and use $Q_{eff} = 2/3Q_{actual}$.

Solution

The first step is to calculate the detonation wave Mach number M_1 of the resulting detonation. The detonation pressure and temperature can then be calculated directly. Substitute the appropriate values into the following equation:

$$M_1 = \left[\left(\frac{\gamma+1}{2} \right) \frac{Q}{C_p T_1} \right]^{1/2} + \left[\left(\frac{\gamma+1}{2} \right) \frac{Q}{C_p T_1} + 1 \right]^{1/2}$$

The Mach number of the detonation wave is $M_1 = 7.23$, which give a detonation wave velocity of 2530 m/s. The detonation pressure and temperature are calculated from

$$\frac{P_2}{P_1} = \frac{1+\gamma M_1^2}{1+\gamma} = 30.9, \qquad \frac{T_2}{T_1} = \left[\frac{1+\gamma M_1^2}{(\gamma+1)M_1} \right]^2 = 18.3$$

The numerical results are $P_2 = 31.2$ bars and $T_2 = 5480$ K. Data cited in Lewis and von Elbe (1987, p. 551, Table 9) give the experimentally measured detonation velocity of 2146 m/s, which is in reasonable agreement with these simple calculations.

A comparison of the simple results with NASA CEA calculations is shown in the following table:

Parameters	Simple Model	NASA CEA
Ma	7.23	6.71
v_1 (m/s)	2170	2390
P_2/P_1	30.9	29.1
T_2/T_1	18.3	12.4

> The greatest discrepancy is the temperature result. The error in final temperature in the simple calculation is due to the omission of chemical equilibrium considerations.

It is instructive to compare the CJ detonation model with some experimental data. Detonation velocities are easier to measure than detonation pressures or temperatures. The data summarized in Table 4.1 from Strehlow compare experimental measurements of detonation velocity with chemical equilibrium calculations using the NASA Lewis CEA program (Strehlow, 1984, pp. 303–304).

The agreement between experimental and theoretical detonation velocities is within a few percent.

As a final note, there is a simple relationship between constant volume combustion and the CJ detonation properties (Nettleton, 1987, pp. 25–28). We begin by simplifying the expressions for the detonation properties of an ideal gas mixture with constant properties. For large values of the dimensionless heat of combustion, $Q/C_p T_1 \gg 1$, Eq. (4.39) can be approximated as

$$Ma_1 = \left[\left(\frac{\gamma+1}{2}\right)\frac{Q}{C_p T_1}\right]^{1/2} + \left[\left(\frac{\gamma+1}{2}\right)\frac{Q}{C_p T_1}+1\right]^{1/2} \cong 2\left[\left(\frac{\gamma+1}{2}\right)\frac{Q}{C_p T_1}\right]^{1/2} = \left[2(\gamma+1)\frac{Q}{C_p T_1}\right]^{1/2}$$

$$(4.43)$$

With the aid of the ideal gas heat capacity relations, we can write the following relations:

$$C_p = C_v + \frac{R_g}{\mathcal{M}}, \gamma = \frac{C_p}{C_v} = 1 + \frac{R_g}{\mathcal{M} C_v}, C_v = \frac{R_g}{\mathcal{M}(\gamma+1)} \qquad (4.44)$$

$$C_p = \gamma C_v = \frac{\gamma R_g}{\mathcal{M}(\gamma-1)}, \frac{1}{C_p} = \frac{\mathcal{M}(\gamma-1)}{\gamma R_g} \qquad (4.45)$$

Table 4.1 Detonation Properties for Various Fuel–Oxidizer Mixtures ($P_1 = 1.01$ bars; $T = 298$ K)

System	Measured	Calculated		
	Velocity (m/s)	Velocity (m/s)	Pressure (bars)	Temperature (K)
80% H_2 + 20% O_2	3390	3408	18.0	3439
66.7% H_2 + 33.3% O_2	2825	2841	18.8	3679
25% H_2 + 75% O_2	1163	1737	14.2	2667
CH_4 + O_2	2528	2639	31.6	3332
CH_4 + 3/2 O_2	2470	2535	31.6	3725
41.2% C_2N_2 + 58.8% O_2	2540	2525	46.2	5120

Data from Strehlow, R.A., 1984. Combustion Fundamentals. McGraw-Hill, pp. 303–304.

We now insert these ideal gas heat capacity relations into the simplified expression for the incoming Mach number of the detonation wave

$$Ma_1 = \left[\frac{2Q}{T_1} \left(\frac{M(\gamma+1)(\gamma-1)}{\gamma R_g} \right) \right]^{1/2} \tag{4.46}$$

This result is now substituted into the relation for the detonation pressure, and using the sound speed relation for an ideal gas, $a_1^2 = \gamma R_g T_1/M$, yields

$$\frac{P_2}{P_1} = \frac{1 + \gamma\, Ma_1^2}{\gamma + 1} = \frac{1}{(\gamma + 1)} \left[1 + 2\gamma\,(\gamma+1)(\gamma-1)\frac{Q}{a_1^2} \right] \tag{4.47}$$

For strong detonations (large Mach numbers, $Ma_1 \gg 1$), this equation simplifies to the following:

$$\frac{P_2}{P_1} = \frac{1 + \gamma\, Ma_1^2}{\gamma + 1} \cong \frac{\gamma\, Ma_1^2}{\gamma + 1} = \frac{1}{(\gamma + 1)} \left[2\gamma\,(\gamma+1)(\gamma-1)\frac{Q}{a_1^2} \right] = \frac{2\gamma\,(\gamma-1)\,Q}{a_1^2} \tag{4.48}$$

$$\frac{P_{CJ}}{P_1} = \frac{2\gamma\,(\gamma-1)Q}{a_1^2} \tag{4.49}$$

This expression is a well-known approximation for strong ideal gas detonations. The subscript CJ reminds us that this is the detonation pressure for a strong Chapman–Jouguet detonation.

We need a way to compare this relation for the detonation pressure, P_{CJ}, with the constant volume explosion pressure, P_{ex}. We start by writing the Rankine–Hugoniot relation and impose the constant volume condition, $v_2 = v_1$.

$$\frac{\gamma}{\gamma - 1}(P_2 v_2 - P_1 v_1) - \frac{1}{2}(P_2 - P_1)(v_1 + v_2) = Q$$

$$\rightarrow \frac{\gamma}{\gamma - 1}(P_2 - P_1)v_1 - \frac{1}{2}(P_2 - P_1)(2v_1) = Q \tag{4.50}$$

Multiplying this out and collecting like terms gives us

$$\left(\frac{\gamma}{\gamma - 1} - 1 \right)(P_2 - P_1)(2v_1) = Q \tag{4.51}$$

This equation can be further simplified by invoking the sound speed relation for an ideal gas, $a_1^2 = \gamma P_1 v$:

$$\frac{P_2}{P_1} - 1 = \frac{(\gamma - 1)\,Q}{P_1 v_1} \rightarrow \frac{P_{ex}}{P_1} = 1 + \frac{\gamma\,(\gamma - 1)Q}{a_1^2} \cong \frac{\gamma\,(\gamma - 1)Q}{a_1^2} \tag{4.52}$$

The approximation is based on the scaling argument that $Q/a_1^2 \gg 1$ and the subscript ex reminds us that this is an approximate relation for the constant volume explosion pressure.

We now can compare the detonation pressure expression with the explosion pressure:

$$\frac{P_{CJ}}{P_1} = \frac{2\gamma\,(\gamma - 1)\,Q}{a_1^2} = 2\frac{P_{ex}}{P_1} \tag{4.53}$$

Which gives us our final, somewhat remarkably simple result:

$$P_{CJ} = 2P_{ex} \tag{4.54}$$

Thus the detonation pressure is twice the value of the isochoric combustion (deflagration) pressure. In Chapter 3 it was indicated that the constant volume deflagration pressure for typical combustible dusts presented a hazard level much greater than the strength of most industrial enclosures. The detonation pressure for the same combustible dust presents a hazard level twice that of the constant volume explosion. However, the hazard level of a detonation is so much more than just twice the constant volume explosion. Both pressure levels are dangerously large in comparison with the strength of most built structures. But the speed of a detonation can be as much as 3 orders of magnitude faster than a confined deflagration. The pressure rise of a deflagration can be safely vented to prevent the rupture of a building. The pressure rise of a detonation is too fast and cannot be vented.

Similar reasoning allows one to conclude that

$$T_{CJ} = \left(\frac{2\gamma}{\gamma+1}\right) T_{ex} \tag{4.55}$$

This simplified relation predicts that the temperature of the gas in a CJ detonation is hotter than the constant volume explosion temperature. This additional incremental temperature rise is due to the shock compression process in a detonation. For a typical burnt gas heat capacity ratio of 1.2, the detonation temperature is approximately 9% hotter than the constant volume explosion temperature.

EXAMPLE 4.2

Consider stoichiometric mixtures for the following fuels in air: methane, propane, ethanol, and octane. The first two chemicals are common fuel gases and the second two are common liquid fuels (with octane being an analog for gasoline). Use published adiabatic flame temperature data for the fuel mixtures (Glassman et al., 2015, Appendix B). Compare the final temperatures and pressures for constant volume combustion and CJ detonation using the approximate models of this section. Assume initial conditions of 300 K and 1.01 bars and let the heat capacity ratio equal 1.2.

Solution

The calculated results are summarized in the following table:

Parameters	Methane	Propane	Ethanol	Octane
T_{ad} (K)	2226	2257	2195	2266
$T_{ex} = T_0 + \gamma(T_{ad} - T_0)$ (K)	2610	2650	2570	2660
$T_{CJ} = \left(\frac{2\gamma}{\gamma+1}\right) T_{ex}$ (K)	2850	2890	2800	2900
$P_{ex} = P_0\left[1 + \gamma\left(\frac{T_{ad}}{T_0} - 1\right)\right]$ (bar)	8.79	8.92	8.67	8.95
$P_{CJ} = 2P_{ex}$ (bar)	17.6	17.8	17.3	17.9

It is commonly observed that under similar fuel–air ratios, hydrocarbon fuels are very similar in their combustion characteristics. This comparison bears that out and supports a similar conclusion for detonation properties.

4.2.5 SUMMARY

This section presented the analysis of the properties of combustion waves using the jump conditions of fluid dynamics. The Rankine–Hugoniot diagram played a central role in the analysis of combustion waves and demonstrated the existence of both supersonic combustion waves (detonations) and subsonic combustion waves (deflagrations). The CJ analysis works well for predicting the detonation properties of ideal gases. It does not work well for deflagrations due to the omission of the finite rate processes of transport phenomena and chemical kinetics. Simplified equations were presented for the calculation of detonation properties of ideal gases with constant heat capacity ratios and molar masses. The relationship between the CJ detonation properties and the constant volume combustion parameters was also demonstrated.

4.3 PREMIXED FLAME ANALYSIS OF GASEOUS FUELS

In Chapter 3 the macroscale dimension of a dust flame was described as the characteristic length scale of the flame such as the burner diameter or the fireball diameter. At the macroscale dimension, the premixed gaseous flame is an important analog for dust flames. As a flame sweeps through a cloud of combustible dust particles, the speed of the flame movement is governed by the rate of burning of the particles and the transport of heat and mass between the unburnt region and the flame. The primary difference between a premixed gas flame and a dust flame is the nature and distribution of the fuel. Since the fuel of the dust flame consists of solid particles, we know that an analysis of the dust flame must account for the discrete nature of the fuel particles. The premixed gas flame affords the opportunity to examine flame propagation processes without the complexity of discrete particle processes. In Chapter 6 we will investigate the combustion behavior of individual particles and consider how to incorporate this behavior into dust flame analyses.

This section begins with a description of the physical picture for a premixed flame. Key assumptions are introduced to simplify the analysis and the governing equations are presented. A particularly useful approximation called the thermal theory of flame propagation is presented and the basic combustion characteristics of premixed flames are derived: the laminar burning velocity and the flame thickness. The laminar burning velocity is the characteristic speed of flame travel with respect to the unburnt mixture. The flame thickness is the narrow zone over which the temperature and concentration profiles change from the unburnt to the burnt condition. The ratio of the flame thickness to the burning velocity gives a characteristic combustion time for the flame. Two theories of flame propagation will be presented. The first, due to Annamalai and Puri (2007) is based on a simple energy balance argument. The second method, due to Spalding (1979), uses an integral formulation of the energy equation with an assumed linear temperature profile within the flame.

The important topics of ignition and quenching are presented next using the unifying concept of the flame energy balance to formulate criteria for the initiation or failure of combustion. The discussion of ignition and quenching will conclude this section. There is an extensive body of literature on premixed flame combustion. Only a very small selection of topics is covered here with the goal of providing insight into the macroscale analysis of premixed dust flames. For a deeper dive into the premixed flame literature, I suggest Kuo (2005, Chapter 5), Glassman et al. (2015, Chapter 4), Law (2006, Chapter 7), Zeldovich et al. (1985, Chapter 4), or Williams (1985, Chapter 5).

4.3.1 PHYSICAL DESCRIPTION

In a premixed gas flame, the fuel and oxidizer are mixed together prior to ignition. The flame consumes the fuel and oxidizer and releases heat in a thin zone of chemical reaction. As the unburnt mixture approaches the flame, it is preheated before it enters the reaction zone. In the analysis of premixed flames, it is convenient to divide the flame into two spatial zones, a preheat zone and a reaction zone. The combined distance of these two zones is called the flame thickness. For premixed hydrocarbon–air gas flames, the flame thickness is typically on the order of 1 mm (Law, 2006, p. 246).

Fig. 4.8 shows typical temperature and chemical species mass fraction profiles through a premixed flame.

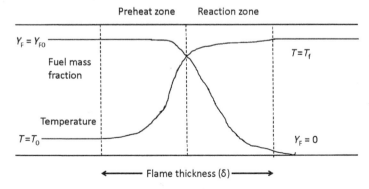

FIGURE 4.8

Temperature and fuel mass fraction profiles in a planar premixed gas flame.

The rate at which fresh unburnt mixture enters the flame is called the burning velocity. The magnitude of the burning velocity is dependent on the chemical reaction rate, the rate of heat conduction, and the rate of chemical species diffusion. Typical values for the burning velocity of hydrocarbon—air flames range from 1 to 100 cm/s (Law, 2006, p. 246). The ratio of the flame thickness to the burning velocity is a characteristic time scale for the flame called the combustion time.

Combustion reactions tend to manifest themselves in very narrow spatial or temporal zones. In other words, combustion reactions tend to occupy a very thin dimension compared to the length scale of the flow or they occur in a very short time frame compared to the time scale of observation. This peculiar feature of combustion reactions is caused by the large relative magnitude of two energies: the enthalpy of the combustion reaction and the activation energy of the chemical reactions. Both features are conveniently translated into by characteristic temperatures. The relative magnitude of the combustion reaction energy can be characterized by the adiabatic isobaric flame temperature. The relative magnitude of the Arrhenius temperature is defined below. The relative magnitude of both quantities is naturally defined by casting them in nondimensional form.

In Chapter 3 it was explained that the adiabatic flame temperature was the maximum temperature that could be attained in an isobaric system. For combustion reactions, the adiabatic reaction temperatures tend to be very large compared to many exothermic noncombustion reactions. The natural definition for a dimensionless flame temperature would be to divide the adiabatic flame temperature by the initial unburnt mixture temperature, $\hat{T}_{ad} = T_{ad}/T_0$. For hydrocarbon—air gas mixtures, this dimensionless number has a typical range from 7 to 8.

The temperature dependence of chemical reactions is usually incorporated into a rate coefficient and modeled by the Arrhenius function (or some modification of it).

The Arrhenius function is given by the equation

$$k = A\exp(-E_a/R_gT) \tag{4.56}$$

where k is the rate coefficient, A is the Arrhenius pre-exponential factor, E_a is the activation energy, and R_g is the universal gas constant. In mechanistic theories of chemical kinetics, the quantity $\exp(-E_a/R_gT)$ is identified as the fractional quantity of reactant that has an energy in excess of the threshold energy (i.e., the activation barrier) necessary to undergo chemical reaction. It is convenient to introduce the concept of the activation temperature $T_a = E_a/R_g$.

The activation energies of combustion reactions tend to be large, often ranging from 20 to 60 kcal/mol (Law, 2006, p. 60). The Arrhenius equation with large activation energies exhibits interesting behavior. The maximum rate is attained at the maximum system temperature T_{ad}. But the larger the activation energy, the higher the value of the temperature must be before the rate begins to increase.

Table 4.2 Fractional Arrhenius Rate Coefficient Versus Fractional Temperature Rise for Different Activation Energies

Activation energy	20 kcal/mol	40 kcal/mol	60 kcal/mol
Activation temperature	10,065 K	20,130 K	30,196 K
T_a/T_{ad}	4.19	8.39	12.6
$T/T_{ad} = 0$	0	0	0
$T/T_{ad} = 0.2$	5.2×10^{-8}	2.6×10^{-15}	1.3×10^{-22}
$T/T_{ad} = 0.4$	1.9×10^{-3}	3.4×10^{-6}	6.2×10^{-9}
$T/T_{ad} = 0.6$	6.1×10^{-2}	3.7×10^{-3}	2.2×10^{-4}
$T/T_{ad} = 0.8$	3.5×10^{-1}	1.2×10^{-1}	4.3×10^{-2}
$T/T_{ad} = 1.0$	1	1	1

The temperature dependence can be explored by taking the ratio of the rate coefficient evaluated at the current temperature and the adiabatic temperature, $k(T)/k(T_{ad})$

$$\frac{k(T)}{k(T_{ad})} = \frac{\exp(-T_a/T)}{\exp(-T_a/T_{ad})} = \exp\left[\frac{T_a}{T_{ad}}\left(1 - \frac{T_{ad}}{T}\right)\right] \qquad (4.57)$$

Here is a table of values to illustrate the influence of larger activation energy on the rate coefficient ratio. The value of T_{ad} is taken as 2400 K (Table 4.2).

These results indicate that at large values of the activation energy, the combustion reaction is suppressed until the system approaches the maximum temperature, the adiabatic flame temperature. The natural consequence of this mathematical behavior is that stationary flames tend to be very thin or transient combustion events will be very short.

EXAMPLE 4.3

Another way to gauge the importance of the activation energy in chemical reactions is to consider the half-life of a first-order, irreversible chemical reaction. The half-life is defined for a batch reactor (constant mass system) as the time required for one-half of the reactants to be consumed by the chemical reaction. Derive the expression for the half-life. Compute the fractional change in half-life from the ambient temperature of 300 K to the assumed reaction temperature of 400 K. Do this for three activation energies: 20, 40, and 60 kcal/mol.

Solution

First we need to derive the expression for the half-life for the first-order, irreversible reaction, $A \rightarrow B$. The rate equation and its solution with the initial condition $C_A(t = 0) = C_{A0}$:

$$\frac{dC_A}{dt} = -kt \rightarrow \ln\left(\frac{C_A}{C_{A0}}\right) = -kt$$

The half-life is determined by setting the condition $C_A(t_{1/2}) = 1/2 C_{A0}$. This gives the desired expression for the half-life, $t_{1/2} = \ln 2 / k$. To compare the fractional change in half-lives at two different temperatures, we take the ratio of half-lives

$$\frac{t_{1/2}(T = 400 \text{ K})}{t_{1/2}(T = 300 \text{ K})} = \frac{\exp[-E/R_g(300)]}{\exp[-E/R_g(400)]} = \exp\left(\frac{-E}{1200 R_g}\right)$$

The numerical results are summarized in the following table:

Activation Energy (kcal/mol)	Half-Life Fraction
20	2.4×10^{-4}
40	5.8×10^{-8}
60	1.4×10^{-11}

Inspection of the results indicates that a 100 K temperature rise causes very short half-lives compared to room temperature. This is another example of the sensitivity of the reaction rate to temperature. Furthermore, as the activation energy increases, the temperature effect becomes more extreme.

4.3.2 KEY ASSUMPTIONS

In this chapter, we confine ourselves to a simple physical model of a premixed flame. This section presents the basic assumptions upon which the physical model is based. Two mathematical models will be derived from this physical model, one based on a scaling argument and the other based on an assumed temperature profile in the flame. What these two models share in common is that they are based on an energy balance that equates the heat release of the reaction zone with the heat conducted through the preheat zone. The assumptions commonly invoked in the analysis of premixed flames concern four main areas: thermodynamics, fluid dynamics, heat and mass transport, and chemical kinetics.

There are two thermodynamic assumptions. First, that the unburnt mixture is fuel-lean or at most at its stoichiometric concentration. Second, that the mixture heat capacities are assumed to be constant with regard to both temperature and composition.

The fluid dynamic assumptions posit that the flow of unburned gas into the flame is a steady, one-dimensional laminar flow. An additional constraint, discussed in Section 4.1, is that the velocity of the unburnt gas flow is very slow compared to its sound speed or Ma ≪ 1. The small Mach number constraint means that the pressure drop across the flame is negligible. These fluid dynamic assumptions mean that the momentum equation is not needed in the analysis of laminar flame propagation.

The heat and mass transfer considerations are greatly simplified by three assumptions: (1) the multicomponent diffusion process can be modeled as binary diffusion; (2) the Lewis number is equal to unity (Le $= \alpha/\mathcal{D} = 1$); and (3) constant physical properties. The assumption of binary diffusion works reasonably well in dilute mixtures such as hydrocarbon fuels in air. This allows the use of Fick's diffusion model in place of the more rigorous Stefan−Maxwell equations (Turns, 2012, pp. 226−233). The unit Lewis number assumption causes the species continuity equation to have the same form as the energy equation.

Finally, one assumption regarding chemical kinetics is invoked. The assumption is that the combustion reaction can be adequately represented by a single-step reaction. This permits the investigator to conceptualize the chemical system consisting of fuel, oxidizer, inert, and products.

4.3.3 ENERGY BALANCE MODEL FOR FLAME PROPAGATION DUE TO ANNAMALAI AND PURI

Annamalai and Puri have presented energy balance model for premixed flame propagation that captures the essential features of the burning velocity (Annamalai and Puri, 2007, pp. 651−654). Their model is similar to, but somewhat different from, one of the first models for premixed flame propagation published 1889 by Mikhel'son (as reported in Linan and Williams, 1993, pp. 22−23).

Consider an energy balance on the flame shown in Fig. 4.9.

Unburnt mixture (fuel, oxidizer, and inert) flow from the left-hand side, react inside the flame, and then the products flow out of the flame toward the right-hand side of the figure. The heat released within the flame thickness δ heats the

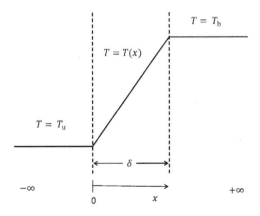

FIGURE 4.9

Definition sketch for Annamalai and Puri's energy balance for a premixed flame.

incoming unburnt mixture via heat conduction. The burnt gas temperature is given by the equation:

$$Y_{F,u} \, \Delta h_c = C_p(T_b - T_u) \tag{4.58}$$

where $Y_{F,u}$ is the fuel mass fraction in the unburnt mixture, Δh_c is the enthalpy of combustion, C_p is the mixture heat capacity on a mass basis, T_b is the burnt gas temperature (approximated by the adiabatic flame temperature), and T_u is the temperature of the fresh, unburnt gas mixture.

The energy balance on the flame can then be written on a unit area basis

$$k(T_b - T_u)/\delta = w'''_{avg} \, \Delta h_c \, \delta \tag{4.59}$$

where w'''_{avg} is the reaction rate for fuel consumption evaluated at some average temperature within the flame and k is the mixture thermal conductivity. Solving for the flame thickness δ,

$$\delta = \left[\frac{k(T_b - T_u)}{w'''_{avg} \, \Delta h_c}\right]^{1/2} = \left(\frac{kY_{F,u}}{C_p \, w'''_{avg}}\right)^{1/2} = \left(\frac{\alpha_T \rho_u Y_{F,u}}{w'''_{avg}}\right)^{1/2} \tag{4.60}$$

Writing the fuel mass balance for the flame gives

$$\rho_u S_u Y_{F,u} = w'''_{avg} \, \delta \tag{4.61}$$

Substituting Eq. (4.60) into Eq. (4.61) to eliminate δ

$$S_u = \left(\frac{\alpha_T \, w'''_{avg}}{\rho_u Y_{F,u}}\right)^{1/2} \tag{4.62}$$

This is the desired result. It indicates that the laminar burning velocity S_u is equal to the square root of the thermal diffusivity and the mean reaction rate. Recognizing that $\rho_u Y_{F,u}$ is the mass concentration of fuel entering the flame, then dividing the fuel mass concentration by the average reaction rate gives the characteristic chemical time t_{chem}.

$$S_u = \left(\frac{\alpha_T}{t_{chem}}\right)^{1/2} \quad \text{and} \quad \delta = (\alpha_T \, t_{chem})^{1/2} \tag{4.63}$$

The expression of the burning velocity and flame thickness in terms of the chemical time will be an especially convenient format when analyzing dust flames. For dust flames, the chemical time can be interpreted as burnout time, the time to completely burn a particle. In Chapter 6 we will see that the burnout time is a natural characteristic for describing the combustion of single particles. In Chapter 7 we will use the burnout time to describe dust flame propagation.

Despite the simplicity of this approach, it has given us physical insight into the key parameters controlling premixed flame propagation. The burning velocity has a square root dependence on both the rate of heat conduction and on the rate of chemical reaction (the inverse of the chemical time). Perhaps the greatest weakness of this approach is that it does not indicate how to compute the average reaction rate ω'''_{avg}. Spalding's thermal theory, presented in the next section, will resolve that question.

4.3.4 THERMAL THEORY OF FLAME PROPAGATION DUE TO SPALDING

Turns has summarized a simplified thermal theory of premixed flame propagation due to Spalding (Turns, 2012, pp. 266–271; Spalding, 1979, pp. 338–343). Consider a one-dimensional planar premixed flame of thickness δ. The unburnt mixture is upstream of the flame with a temperature of T_u and extends to $x \rightarrow -\infty$; the burnt mixture is downstream of the flame with a temperature T_b and extends to $x \rightarrow +\infty$. Since the energy equation is second order in temperature, its solution requires the specification of two boundary values. Additionally, it will become clear that there are two additional unknown parameters in the premixed flame model, the burning velocity S_u and the flame thickness δ. We will need two additional boundary conditions. The natural choice is the specification of the first derivative of temperature $dT/dx = 0$ indicating that there is no heat loss at either end of the flame's spatial domain $(x \rightarrow -\infty, x \rightarrow +\infty)$.

The physical properties of the mixture are assumed to be constant. The speed of the flame is assumed to be slow $(Ma \ll 1)$ so the momentum equation is not needed. Fig. 4.10 is a definition sketch for the problem.

This analysis of premixed flame propagation falls into the category of thermal theories for flame propagation. A thermal theory derives the expressions for the burning velocity and flame thickness based on considerations of the energy equation. If the Lewis number of the mixture is equal to 1, the species equations take the same mathematical form as the energy equation. Thus, solving the energy equation is equivalent to solving the species equations. Turns discusses this

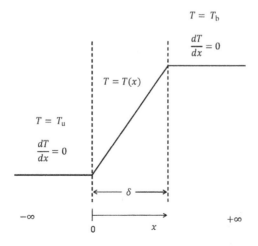

FIGURE 4.10

Definition sketch for Spalding analysis of premixed flame propagation.

simplification at length in terms of a mathematical transformation called the Schvab–Zeldovich transformation. For more details, see Turns's book (2012, pp. 241–245 and 266–273).

The governing equations for the premixed flame are the continuity equation and the thermal energy equation.

$$\frac{\partial}{\partial x}(\rho v_x) = 0 \tag{4.64}$$

$$\dot{m}'' C_{\mathrm{p}} \frac{dT}{dx} = k \frac{d}{dx}\left(\frac{dT}{dx}\right) + \Delta h_{\mathrm{c}}\, \omega_{\mathrm{F}}''' \tag{4.65}$$

The continuity equation is readily integrated using the boundary condition at the unburnt condition

$$\dot{m}'' = \rho v_x = \rho_{\mathrm{u}} S_{\mathrm{u}} = \text{constant} \tag{4.66}$$

The solution of the energy equation requires four boundary conditions: two for the fact that it is a second-order differential equation and two more to specify the values of the two unknown quantities, the burning velocity and the flame thickness. The energy equation is integrated over the domain $(-\infty, +\infty)$

$$\dot{m}''[T]_{T=T_{\mathrm{u}}}^{T=T_{\mathrm{b}}} - \frac{k}{C_{\mathrm{p}}}\left[\frac{dT}{dx}\right]_{dT/dx=0}^{dT/dx=0} = \frac{\Delta h_{\mathrm{c}}}{C_{\mathrm{p}}} \int_{-\infty}^{+\infty} \omega_{\mathrm{F}}''' \, dx \tag{4.67}$$

The second term on the left-hand side is identically zero at $-\infty$ and $+\infty$ and so can be simplified to

$$\dot{m}''(T_{\mathrm{b}} - T_{\mathrm{u}}) = \frac{\Delta h_{\mathrm{c}}}{C_{\mathrm{p}}} \int_{-\infty}^{+\infty} \omega_{\mathrm{F}}''' \, dx \tag{4.68}$$

The evaluation of the integral on the right-hand side is facilitated by invoking a change in the integration variable. If the temperature profile within the flame is assumed to be linear, a natural integration variable is defined by the slope of the profile

$$\frac{dT}{dx} = \frac{T_{\mathrm{b}} - T_{\mathrm{u}}}{\delta} \rightarrow dx = \left(\frac{\delta}{T_{\mathrm{b}} - T_{\mathrm{u}}}\right) dT \tag{4.69}$$

Substituting for the change in the integration variable,

$$\dot{m}''(T_{\mathrm{b}} - T_{\mathrm{u}}) = \frac{\Delta h_{\mathrm{c}} \,\delta}{C_{\mathrm{p}}(T_{\mathrm{b}} - T_{\mathrm{u}})} \int_{T_{\mathrm{u}}}^{T_{\mathrm{b}}} \omega_{\mathrm{F}}''' \, dT = \frac{\Delta h_{\mathrm{c}} \,\delta \omega_{\mathrm{avg}}'''}{C_{\mathrm{p}}},$$
$$\omega_{\mathrm{avg}}''' = \frac{1}{(T_{\mathrm{b}} - T_{\mathrm{u}})} \int_{T_{\mathrm{u}}}^{T_{\mathrm{b}}} \omega_{\mathrm{F}}''' \, dT \tag{4.70}$$

where the average of the reaction rate over the temperature interval $T_{\mathrm{u}} - T_{\mathrm{b}}$ has been defined as ω_{avg}'''. There are two unknowns in Eq. (4.56), S_{u} and δ, so we need another relation to specify the system. Thus, the integrated form of the thermal energy equation reduces to

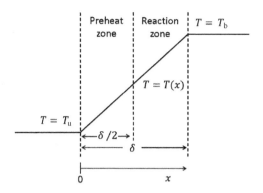

FIGURE 4.11

Premixed flame thickness divided into a preheat zone and a reaction zone.

$$\dot{m}''(T_b - T_u) = \frac{\Delta h_c \delta \dot{\omega}'''_{avg}}{C_p} \tag{4.71}$$

There are two unknowns in Eq. (4.71), S_u (through the mass flux on the left hand side of the equation, $\dot{m}'' = \rho_u S_u$) and δ on the right hand side of the equation, so we need another relation to specify the system.

This relation can be developed from the following conceptual model using the energy equation, Eq. (4.65). Assume that the flame thickness can be divided into two regions: a preheat zone ranging from $x = 0$ to $x = \delta/2$ and a reaction zone ranging from $x = \delta/2$ to $x = \delta$. Fig. 4.11 depicts the geometry of the premixed flame. It is assumed that the chemical reaction rate is negligible in the preheat zone. This assumption removes the source term from the energy equation.

Integrating Eq. (4.67) over the limits $x \to -\infty$ to $x = \delta/2$ and recognizing that $T(x = \delta/2) = (T_b - T_u)/2$,

$$\dot{m}''(T_b - T_u) = \frac{2k(T_b - T_u)}{C_p \delta} \to \dot{m}'' = \rho_u S_u = \frac{2k}{C_p \delta} \tag{4.72}$$

Eqs. (4.70) and (4.72) form a set of two equations and two unknowns. We can substitute for the enthalpy of combustion the following relation: $\Delta h_c = C_p(T_b - T_u)/Y_{F,u}$. Solving for S_u and δ gives the following relations:

$$S_u = \left(\frac{2\alpha_T \dot{\omega}'''_{avg}}{\rho_u Y_{F,u}} \right)^{1/2} \tag{4.73}$$

$$\delta = \left(\frac{2\rho_u \alpha_T}{\dot{\omega}'''_{avg}} \right)^{1/2} = \frac{2\alpha}{S_u} \tag{4.74}$$

These results have the same functional form as those that were derived using the simple energy balance analysis. The difference, however, is that this analysis provided a definition for the calculation of the average reaction rate.

EXAMPLE 4.4

Calculate the burning velocity and flame thickness for a stoichiometric mixture of propane in air using the Spalding premixed flame model. This example follows the one set forth in Turns's book, although Turns presents it in much greater detail (Turns, 2012, pp. 271–273). The fuel mass fraction for a stoichiometric propane–air mixture is $Y_{F,u} = 0.0602$. Assume the unburnt gas condition is at 300 K and 1.01 bars; the adiabatic flame temperature is 2260 K, and the average propane reaction rate is -100 kg/s $\times m^3$ (defined over the temperature range $T_{\omega''_{avg}}$ (mean) $= (T_b + T_{preheat})/2$). The thermal diffusivity is 6.0×10^{-5} m²/s (over the temperature range $T_{preheat} = (T_b + T_u)/2$. Assume the gas mixture obeys the ideal gas equation of state.

Solution

The solution requires estimating both the thermal diffusivity and the averaged reaction rate at the appropriate mean temperatures. In this example, these values have been given. We begin by calculating the mean temperatures:

$$T_{preheat} = \frac{1}{2}(T_b + T_u) = \frac{1}{2}(2260 + 300 \text{ K}) = 1280 \text{ K}$$

$$T_{\omega'''_{avg}} \text{ (mean)} = \frac{1}{2}(T_b + T_{preheat}) = \frac{1}{2}(2260 + 1280 \text{ K}) = 1770 \text{ K}$$

Next we calculate the unburnt gas density using the ideal gas equation of state.

$$\rho_u = \frac{P\mathcal{M}}{R_g T} = \frac{(1.01 \times 10^5 \text{ Pa})(29 \text{ kg/kmol})}{(8314 \text{ j/kmol} \times K)(300 \text{ K})} = 1.17 \text{ kg/}m^3$$

The laminar burning velocity is calculated from the Spalding model:

$$S_u = \left(\frac{2\alpha_T \omega'''_{avg}}{\rho_u Y_{F,u}}\right)^{1/2} = \left[\frac{2(6.0 \times 10^{-5} \text{ m}^2/s)(100 \text{ kg/s} \times m^3)}{(1.17 \text{ kg/}m^3)(0.0602)}\right]^{1/2} = 0.413 \text{ m/s}$$

This is reasonably close to experimental values.

As a final note, there are a number of measurement techniques available for the measurement of the laminar burning velocity. Good overviews of the subject can be found in Kuo (2005, Chapter 5), Law (2006, Chapter 7), and the review paper by Ibarreta et al. (2015).

4.3.5 IGNITION AND QUENCHING OF FLAMES

Ignition is the initiation of combustion in a fuel–oxidizer system. Quenching is a form of extinguishment; this leads to the cessation of the combustion reaction.

In this section, I present a simple analysis of ignition and quenching of flames based on a global energy balance on the flame. While the analysis is approximate at best, the resulting models yield acceptable order-of-magnitude results without intensive computation. A good introduction to this subject can be found in Turns (2012, pp. 287–300). More sophisticated models based on activation energy asymptotics can be found in Williams (1985, Chapter 8) and Law (2006, Chapter 8).

Ignition occurs when the rate of heat release within the reactive gas volume exceeds the heat losses to the surroundings. This initial deposit of thermal energy triggers a thermal runaway with the flame volume growing larger and sweeping through the unburnt mixture. Williams formulated the following criterion for ignition: *Ignition will only occur if enough energy is added to the gas to heat a slab about as thick as a steadily propagating laminar flame to the adiabatic flame temperature* (Williams, 1985, p. 268).

Quenching occurs when the heat losses to the surroundings exceed the rate of heat release within the reacting gas volume. Williams formulated the following criterion for quenching due to heat losses: *The rate of liberation of heat by chemical reaction inside the slab must approximately balance the rate of heat loss from the slab by thermal conduction* (Williams, 1985, p. 268). If the rate of heat loss exceeds the rate of heat generation, the flame temperature rapidly decreases, the combustion reaction rate decreases, and heat losses dominate the system leading to the quenching of the flame.

4.3.5.1 Ignition Energy of a Premixed Flame

Let us consider the problem of ignition first. We will follow the presentation by Turns (2012, pp. 287–300). Assume a spherical volume of unburnt gas is ignited instantaneously by an ignition stimulus (eg, an electric spark). Using Williams' second criterion, it is possible to define a *critical radius* of gas volume that is the minimum volume of burnt gas that can propagate into the remaining unburnt mixture. If the gas volume is less than this critical radius, the flame is quenched. The critical radius r_{crit} is determined by an energy balance on the initial burnt gas volume where the heat generation rate is balanced by the heat loss rate due to thermal conduction.

$$q'''_{generation} = q'''_{loss} \tag{4.75}$$

Substituting expressions for a spherical geometry,

$$\frac{4\pi}{3} r_{crit}^3 (\overline{\omega}''_F Q) = 4\pi r_{crit}^2 k \frac{dT}{dr}\Big|_{r=r_{crit}} \tag{4.76}$$

where Q is the enthalpy of combustion and the other symbols retain their previously defined meanings. The temperature gradient must be evaluated at $r = r_{crit}$. This means we need to first calculate the temperature profile outside the critical radius. The heat loss from the spherical volume of gas is by heat conduction. The temperature inside the sphere is assumed to be spatially uniform and equal to the adiabatic flame temperature. Fig. 4.12 is a definition sketch for the problem.

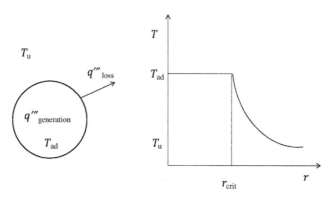

FIGURE 4.12

Definition sketch for critical radius of ignition.

The temperature field surrounding the sphere is given by the energy equation written in spherical coordinates for steady heat conduction

$$\frac{1}{r^2}\left[\frac{d}{dr}\left(r^2\frac{dT}{dr}\right)\right] = 0 \tag{4.77}$$

This differential equation is solved with the boundary conditions $T(r = r_{\text{crit}}) = T_{\text{ad}}$ and $T(r \to \infty) = T_{\text{u}}$. The resulting temperature profile outside the volume of flame is

$$T(r) = T_{\text{u}} + (T_{\text{ad}} - T_{\text{u}})\frac{r_{\text{crit}}}{r}, \quad r_{\text{crit}} < r < \infty \tag{4.78}$$

The gradient is evaluated at $r = r_{\text{crit}}$ to give the desired intermediate result

$$k\frac{dT}{dr}\bigg|_{r=r_{\text{crit}}} = -\frac{k(T_b - T_u)}{r_{\text{crit}}}$$

Substituting this expression for the temperature gradient into the energy balance and solving for the critical radius gives the result:

$$r_{\text{crit}} = \left[\frac{3k(T_b - T_u)}{\overline{\omega}_F''' Q}\right]^{1/2} \tag{4.79}$$

The heat of combustion can be expressed in terms of temperature as $Q = C_p(T_{\text{ad}} - T_{\text{u}})$. The above relation can be expressed in terms of the flame thickness δ. Solving the Spalding flame thickness model for $\overline{\omega}_F'''$ and substituting into the r_{crit} relation gives

$$r_{\text{crit}} = \left(\frac{\sqrt{6}}{2}\right)\delta \cong 1.22\delta \tag{4.80}$$

Thus, the critical radius is slightly larger than the laminar flame thickness. If a thermal stimulus like an electric spark creates a volume with a radius that equals or exceeds the critical radius, flame propagation will continue radially outwards into the unburnt mixture and ignition is said to have occurred.

Alternatively, the critical radius can be expressed in terms of the burning velocity. Substituting the Spalding burning velocity model into the critical radius equation gives the result:

$$r_{crit} = \sqrt{6}\left(\frac{\alpha_T}{S_u}\right) \tag{4.81}$$

The critical radius defines the volume of burnt gas created by the ignition stimulus. Assuming the ignition volume is adiabatic, an energy balance on the ignition volume gives the following expression for the ignition energy E_{ign}:

$$E_{ign} = m_b C_p (T_b - T_u) \tag{4.82}$$

The burnt gas mass for a spherical volume is

$$m_b = \rho_b V_b = \rho_b \left(\frac{4\pi}{3} r_{crit}^3\right) \tag{4.83}$$

Substituting the expression for the critical radius in terms of the burning velocity, and applying the ideal gas equation of state to the burnt gas, the final result for the ignition energy is

$$E_{ign} = \frac{4\pi}{3}(6)^{3/2}\left[P\left(\frac{C_p \mathcal{M}}{R_g}\right)\left(\frac{T_b - T_u}{T_b}\right)\left(\frac{\alpha_T}{S_u}\right)^3\right] \quad \text{where} \frac{4\pi}{3}(6)^{3/2} \cong 61.6 \tag{4.84}$$

Hence, if the adiabatic flame temperature and the laminar burning velocity are known, the ignition energy can be calculated.

4.3.5.2 Quenching of Premixed Flames in a Cylindrical Tube

Let us now take up the problem of quenching. We limit ourselves to the problem of flame quenching while passing through a narrow passage. It has been empirically observed that a flame can pass through larger diameter tubes filled with a fuel−oxidizer mixture with a concentration in the flammable range. If one repeats the experiment with smaller and smaller tubes, eventually one finds a tube diameter that does not permit the passage of the flame. This is called the *quenching diameter*. The heat losses to the tube exceed the heat generation by the combustion reaction and the flame is extinguished or quenched. This phenomenon is the basis for flame arrester technology. Sir Humphry Davy was the first scientist to put this into practice with his invention of his miner's lamp (Turns, 2012, p. 287).

Consider a cylindrical tube filled with a flammable gas mixture. The system of interest is the flame as it enters the tube. Referring to Williams criteria, we begin with an energy balance on the flame and equate the heat generation from the flame with the heat loss by conduction to the tube wall.

$$\dot{q}''_{generation} = \dot{q}''_{loss} \tag{4.85}$$

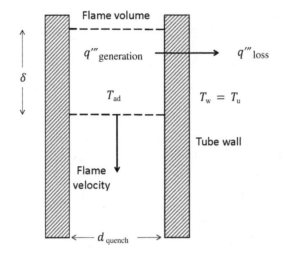

Flame volume

q''' generation q''' loss

T_{ad} $T_w = T_u$

Tube wall

Flame velocity

d quench

FIGURE 4.13

Definition sketch for flame quenching in a tube.

The volume of the flame at the quenching point is given by the flame thickness (along the axial coordinate) and the quench diameter. Fig. 4.13 is the definition sketch for the problem.

Substituting the expression for the heat generation rate and the geometric relation for the volume of a cylinder gives

$$q'''_{generation} = (\overline{\omega}''_F Q) V_{cylinder} = \frac{\pi}{4} d^2_{quench} \delta (\overline{\omega}''_F Q) \qquad (4.86)$$

The temperature of the flame volume is assumed to be spatially uniform and equal to the adiabatic flame temperature. Following the presentation given by Annamalai and Puri (2007, pp. 677−678), we model the heat loss from the flame to the tube wall using the convective heat transfer relation. The area for heat loss is the flame contact area with the cylinder wall. This gives

$$q'''_{loss} = h_{HT} \left(\pi d_{quench} \delta \right) \left(T_{avg} - T_w \right) \qquad (4.87)$$

The temperature of the flame volume is designated as an average temperature T_{avg} and the wall temperature is T_w. Let us estimate the average flame temperature as $(T_{ad} + T_u)/2$ and assign the wall temperature equal to the unburnt mixture temperature. Substituting this into the heat loss expression and now equating the heat generation with the heat loss we get

$$\frac{1}{2} (\overline{\omega}''_F Q) d_{quench} = h_{HT}(T_{ad} - T_u) \qquad (4.88)$$

We again substitute the relation $Q = C_p(T_{ad} - T_u)$. Solving the Spalding flame thickness relation for the mean combustion rate $\overline{\dot{w}}_F'''$ and substituting that into the above expression, we can now solve for the quench diameter:

$$d_{quench} = 2\left(\frac{h_{HT}}{C_p}\right)\frac{\delta^2}{2\rho_u\alpha_T} = \frac{h_{HT}\delta^2}{k} \tag{4.89}$$

The definition of the Nusselt number is $Nu = h_{HT}d_{quench}/k$. Substituting this into the above expression yields

$$d_{quench} = \delta\sqrt{Nu} \tag{4.90}$$

For fully developed laminar flow in a tube with heat transfer at a constant wall temperature, $Nu = 3.66$ (Kays and Crawford, 1980, pp. 96−97). Our final result is

$$d_{quench} \cong 1.91\delta \tag{4.91}$$

Thus, the quench diameter is roughly twice the flame thickness. If a flame propagates into a tube with a diameter smaller than d_{quench}, it will extinguish. However, the reader is cautioned that the foregoing analysis required a number of assumptions that will likely be met only in very rare circumstances. The results for both the critical radius and the quenching diameter should be used for physical insight and should not be used when accuracy is required.

EXAMPLE 4.5

Consider a stoichiometric mixture of propane in air at 300 K and 1.01 bars. Assume the burnt gas (adiabatic flame) temperature is 2260 K, the burning velocity is 46 cm/s, the heat capacity is 1.195 kJ/kg-K, the mixture molar mass is 30 kg/kmol, and the mean thermal diffusivity (over the temperature range of 300−2260 K) is 2.58×10^{-4} m^2/s (estimated using air properties).
(A) What is its critical radius and ignition energy? (B) What is its quenching parameter?

Solution

(A) The model for the critical radius of ignition is

$$r_{crit} = \sqrt{6}\left(\frac{\alpha_T}{S_u}\right) = \sqrt{6}\left(\frac{2.58 \times 10^{-4}}{0.46}\right) = 1.37 \times 10^{-3}\,m$$

The ignition energy is

$$E_{ign} = 61.6\left[P\left(\frac{C_p\mathcal{M}}{R_g}\right)\left(\frac{T_b - T_u}{T_b}\right)\left(\frac{\alpha_T}{S_u}\right)^3\right] = 4.12 \times 10^{-3}\,J$$

The calculated ignition energy is within 25% of the experimental value (Turns, 2012, p. 298, Table 8.5).

(B) A model for the quenching diameter can be obtained by equating the relations for the quench diameter and the critical radius

$$d_{\text{quench}} = 1.91\delta = 1.91\left(\frac{r_{\text{crit}}}{1.22}\right) = 1.57 r_{\text{crit}} = 1.57(1.37 \times 10^{-3}) = 2.14 \times 10^{-3} m$$

This is in agreement with the experimentally determined value of 2.0 mm.

4.3.6 SUMMARY

It must be noted here that combustion engineers have had great success in the investigation of premixed flames. There are literally thousands of research papers published on a variety of chemical systems exploring the effects of transport processes, chemical kinetics, and thermodynamic properties using experimental, theoretical, and numerical methods. The simple models presented thus far are crude and simplistic; their value lies in their ability to capture the essence of the physics and chemistry of premixed flames. The premixed gas flame model is a useful analog for dust flame propagation studies. Further information about gaseous premixed flame studies are summarized in the various combustion textbooks cited earlier in this chapter.

4.4 DIFFUSION FLAME ANALYSIS FOR A LIQUID DROPLET

This is an important analog for dust flame behavior at a microscale (particle) level; since the fuel and oxidizer are initially separated, the rate of combustion of diffusion flames is governed by the finite rate of transport processes. With some basic modification, the diffusion flame analysis of liquid droplets can be transformed into the diffusion flame analysis of solid particles.

This discussion opens with a description of the physical situation and key assumptions that simplify the analysis. The governing equations are presented and solved for the temperature and concentration profiles. The solution is facilitated by the introduction of a mathematical transformation called the Shvab–Zeldovich transformation. Combustion characteristics such as the flame radius, mass burning rate, and burnout time are derived from the temperature and concentration profiles. Illustrative results are presented to convey a sense of the order of magnitude of the key combustion characteristics.

There is an extensive body of literature on the subject of diffusion flames (see Kanury, 1975, Chapter 5; Turns, 2012, Chapter 10; Kuo, 2005, Chapter 6; Strehlow, 1984, Chapter 7). What follows is a very brief survey to convey only the essential ideas necessary for the microscale analysis of dust particle combustion in Chapter 6. Many combustion books, such as those recommended earlier, begin with a discussion of liquid droplet evaporation as a means of introducing some of the mathematical apparatus needed to describe droplet combustion. For the sake of brevity, I will not follow that approach and instead will address only droplet combustion.

4.4.1 PHYSICAL DESCRIPTION AND KEY ASSUMPTIONS

Let us consider a single liquid fuel droplet suspended in air. In the absence of external forces, the droplet assumes a spherical shape. Upon ignition, a flame envelops the fuel droplet at some distance from the droplet surface. The combustion of the droplet can be divided into three stages: a preheat stage in which the hot environment raises the temperature of the droplet from its initial temperature to its boiling point; a shrinking droplet stage where the flame evaporates and burns fuel from the droplet surface; and a burnout stage where the liquid has completely evaporated and the residual fuel vapor is consumed in the flame. The second stage, the shrinking droplet stage, accounts for approximately 90% of the droplet lifetime (Strehlow, 1984, p. 229). The flame that surrounds the fuel droplet is called a flame sheet. Fig. 4.14 is a depiction of the physical system under consideration where r_s denotes the radius of the droplet surface and r_f denotes the flame radius. The thickness of the flame sheet δ is much smaller than the diameter of the fuel droplet $\delta/r_s \ll 1$.

The heat from the flame provides the energy for vaporizing the fuel liquid. A portion of the heat released by the flame is lost to the ambient environment. The heat fluxes for a single burning droplet are shown in Fig. 4.15.

The flame sheet is a sink for fuel vapor and oxygen. The vaporizing droplet provides a mass flux of fuel vapor that diffuses toward the flame sheet, and oxygen diffuses to the flame sheet from the ambient environment. The flame sheet acts as a source for combustion products which diffuse outwards into the ambient environment. The mass fluxes are shown in Fig. 4.16.

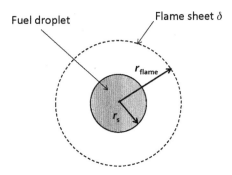

FIGURE 4.14

Isolated liquid fuel droplet surrounded by a flame sheet.

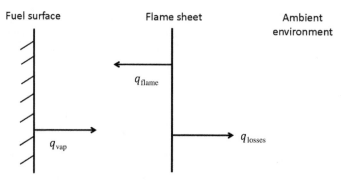

FIGURE 4.15

Heat fluxes for a single burning particle.

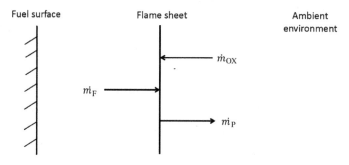

FIGURE 4.16

Sketch showing relationship of species flux directions with respect to the flame sheet (F is fuel, OX is oxidizer, and P is products).

The analysis of a single burning fuel droplet is based on several assumptions. The primary assumptions are

- Infinite rate kinetics
- Finite rates of heat and mass transport
- 1D spherical geometry
- Constant properties
- Single component fuel
- Phase equilibrium: liquid droplet temperature is uniform and equal to its boiling point
- Binary mass diffusion
- $Le = \alpha_T / \mathcal{D} = k / \rho C_p \mathcal{D} = 1$.
- Boundary conditions at the fuel droplet surface:
  $Y_F(r = r_s) = 1$, $Y_{Ox}(r = r_s) = 0$, $Y_P(r = r_s) = $ finite, $T(r = r_s) = T_{vap}$
- Boundary conditions at the flame sheet surface:
  $Y_F(r = r_f) = 0$, $Y_{Ox}(r = r_f) = 0$, $Y_P(r = r_f) = $ finite, $T(r = r_f) = T_{ad}$

- Boundary conditions ambient environment:

$$Y_F(r \to \infty) = 0, \quad Y_{Ox}(r \to \infty) = Y_{Ox,\infty}, \quad Y_P(r \to \infty) = 0, \quad T(r \to \infty) = T_\infty.$$

While these assumptions may seem overly restrictive, they lead to results that are reasonably borne out through empirical investigations. The extension of this basic model to less restricted forms is summarized well in Sirignano's book (Sirignano, 2010).

4.4.2 GOVERNING EQUATIONS

The ultimate objective of this analysis is to derive a model that predicts the burnout time, the time required for complete combustion of a single fuel droplet. We adopt a quasi-steady state analysis strategy. The transient nature of the process will be modeled with a mass balance on the fuel droplet. The mass rate of combustion, governed by heat and mass transfer, will be modeled as steady state processes. A model for the shrinking droplet begins with the mass balance for a single droplet

$$\frac{dm_p}{dt} = -\dot{m}_c \tag{4.92}$$

The mass of the liquid droplet is m_p (where the p stands for particle) and the combustion rate is $\dot{m}_c$. The mass of the liquid droplet can be expressed in terms of its density and spherical geometry

$$m_p = \rho_l V = \frac{\pi}{6} \rho_l d_p^3 \tag{4.93}$$

Substituting Eq. (4.93) into the mass balance gives the expression

$$\frac{d(d_p)}{dt} = -\frac{2\dot{m}_c}{\pi \rho_l d_p^2} \tag{4.94}$$

This is a differential equation for the time rate of change of the droplet diameter. In order to solve this equation, we need a relation for the mass rate of combustion $\dot{m}_c$. The relation for the mass rate of combustion is determined by solving the equations of change for the mass fraction and temperature profiles as influenced by the droplet surface, the flame sheet, and the ambient environment. Even with these simplifying assumptions, this takes a bit of work.

The continuity, species continuity, and energy equations for a spherical geometry take the following form:

$$\frac{1}{r^2} \frac{d}{dr} \left(r^2 \rho v_r \right) = 0 \Rightarrow r^2 \rho v_r = \dot{m} = \text{constant} \tag{4.95}$$

$$\rho_g v_r \frac{dY_i}{dr} = \rho_g \mathcal{D} \left[\frac{1}{r^2} \frac{d}{dr} \left(r^2 \frac{dY_i}{dr} \right) \right] + \dot{\omega}_i'''; \quad i = \text{F, Ox} \tag{4.96}$$

$$v_r \frac{dT}{dr} = \frac{k_g}{\rho_g C_{pg}} \left[\frac{1}{r^2} \frac{d}{dr} \left(r^2 \frac{dT}{dr} \right) \right] + \frac{\dot{\omega}_F''' \Delta h_c}{\rho_g C_{pg}} \tag{4.97}$$

The solution of these equations depends on the boundary conditions at the liquid droplet surface, the flame sheet, and the ambient environment.

Recall at the beginning of this chapter I wrote the combustion reaction a one-step reaction in mass units (Eq. (4.14))

$$1\text{kg F} + s \text{ kg Ox} \rightarrow (s+1)\text{kg P} \tag{4.98}$$

The symbol F is fuel, s is the mass of oxidizer per unit mass of fuel, Ox is oxidizer, P is combustion products, and $(s+1)$ is the mass of products per unit mass of fuel. The reaction rates can be normalized using the stoichiometric coefficients in the following manner:

$$-\omega_F''' = -\frac{\omega_{Ox}'''}{s} = \frac{\omega_P'''}{(s+1)} \tag{4.99}$$

These equations govern the behavior of a burning liquid fuel droplet. In order to solve these equations, it is convenient to introduce a mathematical transformation known as the Shvab–Zeldovich transformation.

4.4.3 SOLUTION FOR TEMPERATURE AND CONCENTRATION PROFILES

The Shvab–Zeldovich transformation removes the source terms from the energy and species continuity equations. This simplifies the integration of the resulting differential equations. This presentation follows the derivation in Kanury's book (1975). The interested reader is encouraged to study the derivations presented in other combustion books because there are many different paths that can lead to the same final results. I especially recommend the presentations by Turns (2012, pp. 383–400), Kuo (2005, pp. 578–581), Williams (1985, pp. 52–62), and Law (2006, pp. 213–222).

We begin by multiplying the fuel species equation by Δh_c and add it to the energy equation:

$$\rho_g \alpha_g \frac{d}{dr} r^2 \frac{dC_{pg}T}{dr} - \dot{m}_s'' r_s^2 \frac{dC_{pg}T}{dr} + \rho_g D \frac{d}{dr} r^2 \frac{dY_F \Delta h_c}{dr} - \dot{m}_s'' r_s^2 \frac{dY_F \Delta h_c}{dr}$$
$$+ r^2 \left(\Delta h_c \omega_F''' - \Delta h_c \omega_F''' \right) = 0 \tag{4.100}$$

Assuming $Le = \alpha_g/D = 1$ and simplifying gives

$$\rho_g \alpha_g \frac{d}{dr} r^2 \frac{d(C_{pg}T + Y_F \Delta h_c)}{dr} - \dot{m}_s'' r_s^2 \frac{d(C_{pg}T + Y_F \Delta h_c)}{dr} = 0 \tag{4.101}$$

Dividing through by $\omega_F''' \Delta h_c + \Delta h_c(Y_{FS} - Y_{FR})$, we obtain

$$\rho_g \alpha_g \frac{d}{dr} r^2 \frac{db_{FT}}{dr} - \dot{m}_s'' r_s^2 \frac{db_{FT}}{dr} = 0 \tag{4.102}$$

$$b_{FT} = \frac{C_{pg}(T - T_\infty) + Y_F \Delta h_c}{\Delta h_{vap} + C_{pl}(T_S - T_R) + \Delta h_c(Y_{FS} - 1)} \tag{4.103}$$

Note that the quantity b_{FT} is dimensionless. The boundary conditions can now be transformed into nondimensional quantities. The boundary conditions at the droplet surface are based on the balance of mass and energy. The energy balance at the droplet surface

$$\dot{m}_S''\left[\Delta h_{vap} + C_{pl}(T_S - T_R)\right] = k_g\left[\frac{dT}{dr}\right]_{r=r_s} = \rho_g\alpha_g\left[\frac{d(C_{pg}T)}{dr}\right]_{r=r_s} \tag{4.104}$$

Multiplying the species mass balance at the droplet surface by Δh_c

$$\dot{m}_S''\Delta h_c(Y_{FS} - Y_{FR}) = \rho_g\mathcal{D}\left[\frac{d(\Delta h_c Y_F)}{dr}\right]_{r=r_s} \tag{4.105}$$

Adding the boundary conditions together and noting that $\alpha_g = \mathcal{D}$,

$$\dot{m}_S'' = \rho_g\alpha_g\left[\frac{db_{FT}}{dr}\right]_{r=r_s} \tag{4.106}$$

where b_{FT} has been previously defined. A similar process leads us to another dimensionless variable b_{OxT}

$$b_{OxT} = \frac{C_{pg}(T - T_\infty) + \Delta h_c(sY_{Ox} - Y_{Ox\infty})}{\Delta h_{vap} + C_{pl}(T_S - T_R)} \tag{4.107}$$

Finally, if we can assume that $\mathcal{D}_F = \mathcal{D}_{Ox}$, and assuming that the fuel droplet is pure ($Y_{FR} = 1$), then we get a third dimensionless variable b_{FOx}

$$b_{FOx} = \frac{Y_F - s(Y_{Ox} - Y_{Ox\infty})}{(Y_{FS} - 1)} \tag{4.108}$$

Using the Shvab–Zeldovich transformation, the three differential equations (energy and two species) and their associated boundary conditions have been transformed into one equation and one boundary condition. The dimensionless variables b_{FT}, b_{OxT}, and b_{FO} all satisfy the system

$$\rho_g\alpha_g\frac{d}{dr}r^2\frac{db}{dr} - \dot{m}_s''r_s^2\frac{db}{dr} = 0 \tag{4.109}$$

$$\dot{m}_S'' = \rho_g\alpha_g\left[\frac{db}{dr}\right]_{r=r_s} \tag{4.110}$$

Integrating these equations gives the radial profile and the mass flux at the surface

$$\ln\left[\frac{b_\infty - b_S + 1}{b - b_S + 1}\right] = \left(\frac{\dot{m}_s''r_s^2}{\rho_g\alpha_g}\right)\frac{1}{r} \tag{4.111}$$

$$\dot{m}_S'' = \frac{\rho_g\alpha_g}{r_s}\ln\left[\frac{b_\infty - b_R + 1}{b - b_S + 1}\right] \tag{4.112}$$

The expression for the $b(r)$ profile can be simplified by eliminating $\dot{m}_s''r_s^2/\rho_g\alpha_g$

$$(b - b_S + 1) = (B + 1)^{(1 - r_s/r)} \tag{4.113}$$

Here a new dimensionless variable, the Spalding transfer number, $B = b_\infty - b_S$, has been introduced (Spalding, 1953).

$$\dot{m}_S'' = \frac{\rho_g \alpha_g}{r_s} \ln(B + 1) \tag{4.114}$$

Eqs. (4.113) and (4.114) describe the temperature and concentration profiles, the flame location, and the mass burning rate that develop during the burning of a single liquid droplet. The Spalding transfer number is a dimensionless driving force for heat or mass transfer. It can take any of the following equivalent forms:

$$B = B_{OxT} = B_{FOx} = B_{FT} \tag{4.115}$$

$$B_{OxT} = b_{OxT\infty} - b_{OxTS} = \frac{sY_{Ox\infty} \Delta h_c + C_{pg}(T_\infty - T_S)}{\Delta h_{vap} + C_{pl}(T_S - T_R)} \tag{4.116}$$

$$B_{FOx} = b_{FOx} - b_{FOxS} = \frac{sY_{OxS} + Y_{FS}}{1 - Y_{FS}} \tag{4.117}$$

$$B_{FT} = b_{FT\infty} - b_{FTS} = \frac{Y_{FS} \Delta h_c + C_{pg}(T_\infty - T_S)}{\Delta h_{vap} + C_{pl}(T_S - T_R) + \Delta h_c(Y_{FS} - Y_{F\infty})} \tag{4.118}$$

The physical meaning of B can be grasped more readily if we assume that the temperature at the liquid surface is equal to the liquid's boiling point, $T_S = T_{bp}$, and recognize that the enthalpy of combustion and the enthalpy of vaporization are much larger than the sensible enthalpy values. The transfer number B_{OxT} can then be written in approximate form:

$$B_{OxT} \cong \frac{sY_{Ox\infty} \Delta h_c}{\Delta h_{vap}} \tag{4.119}$$

Kanury cautions that this approximation may be off by as much as 50% (reduced to 20% when taking the natural logarithm), so Eq. (4.88) should be used only for preliminary calculations where one is interested in order-of-magnitude calculations. Table 4.3 is a short list of B values taken from Kuo (2005, p. 581).

Table 4.3 Select Values of B for Liquid Fuels

Liquid Fuel	B
iso-Octane	6.41
Benzene	5.97
n-Heptane	5.82
Toluene	5.69
Automotive gasoline	~5.3
Kerosene	~3.4
Heavy fuel oil	1.7
Carbon	0.12

From Kuo, K., 2005. Principles of Combustion, second ed. John Wiley & Sons, p. 581.

Higher volatility fuels have higher B values. The combustion rate or mass burning rate of the fuel droplet can be calculated directly from Eq. (4.114) using any of the definitions of B. The choice of B will depend on which parameters are given and which are unknown.

Earlier we noted that a flame sheet encircles the fuel droplet. The location of the flame can be determined by the stoichiometric condition at the flame, $Y_{F,flame} = sY_{Ox,flame}$. Substituting this equality into the radial profile expression, Eq. (4.111), and using the results for B_{FOx} and b_{FOx} we obtain an expression for the flame position:

$$r_{flame} = \frac{\dot{m}_s'' r_s^2}{\rho_g \alpha_g} \ln(1 + sY_{Ox\infty}) \qquad (4.120)$$

This expression indicates that the flame radius increases as the combustion rate increases. An alternative expression can be obtained by using Eq. (4.120) Eq. (4.120) to eliminate $\dot{m}_s'' r_s / \rho_g \alpha_g$. The resulting expression is

$$r_{flame} = r_s \left[\frac{\ln(B + 1)}{\ln(1 + sY_{Ox\infty})} \right] \qquad (4.121)$$

Lower volatility fuels have lower B values which, in turn, result in smaller flame radii. In the limit of $B \to 0$, the flame radius shrinks to the droplet surface.

The temperature profile is contained in the solution for the radial profile of b. The temperature profile is extracted from the b profile by recognizing that there is no oxygen inside the flame zone ($r_s < r < r_{flame}$) and no fuel vapor outside the flame. Choosing $b_{OxT}(r)$, the solution is

$$\ln \left[\frac{b_{OxT\infty} - b_{OxTS} + 1}{b_{OxT} - b_{OxTS} + 1} \right] = \left(\frac{\dot{m}_s'' r_s^2}{\rho_g \alpha_g} \right) \frac{1}{r} \qquad (4.122)$$

Substituting the definition of b_{OxT}, rearranging and solving for temperature inside the flame zone gives

$$T(r) = T_S - \frac{h_{liquid}}{C_g} + \left[\frac{C_g(T_\infty - T_S) + sY_{Ox\infty} \Delta h_c + h_{liquid}}{C_g} \right] \exp \left(\frac{-\dot{m}_s'' r_s^2}{\rho_g \alpha_g r} \right) \qquad (4.123)$$

where $h_{liquid} = \Delta h_{vap} + C_l(T_S - T_R)$ is the total enthalpy of the liquid. Outside the flame zone ($r_{flame} < r < \infty$), one obtains

$$T(r) = T_S - \left(\frac{h_{liquid} + \Delta h_c}{C_g} \right) + \left[\frac{C_g(T_\infty - T_S) + Y_{FR} \Delta h_c + h_{liquid}}{C_g} \right] \exp \left(\frac{-\dot{m}_s'' r_s^2}{\rho_g \alpha_g r} \right) \qquad (4.124)$$

Finally, the temperature at the flame sheet can be calculated by substituting the flame radius equation into either temperature profile solution. The flame sheet temperature is

$$T_{flame} = T_S + \frac{C_{pg}(T_\infty - T_S) + sY_{Ox\infty} (\Delta h_c - h_{liquid})}{C_{pg}(1 + sY_{Ox\infty})} \qquad (4.125)$$

This is same as the adiabatic flame temperature at constant pressure. To see this, consider an energy balance on the burning droplet based on a unit mass of fuel. For each unit mass of fuel burnt, the heat released must raise the temperature of the air, raise the temperature of the fuel vapor, and vaporize and preheat a unit mass of liquid fuel. The energy balance is

$$sY_{Ox\infty} \Delta h_c = C_{pg}(T_{flame} - T_{\infty}) + sY_{Ox\infty} C_{pg}(T_{flame} - T_S) + sY_{Ox\infty} h_{liquid} \qquad (4.126)$$

Solving the energy balance for the flame temperature gives the same result as Eq. (4.126).

Similarly, one can solve for the mass fraction profiles of fuel and oxygen using the $b(r)$ solution. The fuel mass fraction is non-zero only inside the flame zone ($r_S < r < r_{flame}$). Selecting $b(r) = b_{FOx}$ and

$$\frac{b_{FOx} - b_{FOxS} + 1}{b_{FOx\infty} - b_{FOxS} + 1} = \frac{s(Y_{Ox} - Y_{Ox\infty}) - Y_F}{1 + sY_{Ox\infty}} + 1 = \exp\left(\frac{-\dot{m}_s'' r_s^2}{\rho_g \alpha_g r}\right) \qquad (4.127)$$

To get the fuel mass fraction profile, substitute $Y_{Ox} = 0$ and solve for Y_F

$$Y_F(r) = (1 + sY_{Ox\infty})\left[1 - \exp\left(-\frac{\dot{m}_s'' r_s^2}{\rho_g \alpha_g r}\right)\right] - sY_{Ox\infty} \qquad (4.128)$$

The oxygen mass fraction profile is obtained by setting $Y_F = 0$

$$Y_{Ox}(r) = Y_{Ox\infty} + \left(\frac{1 + sY_{Ox\infty}}{s}\right)\left[\exp\left(\frac{-\dot{m}_s'' r_s^2}{\rho_g \alpha_g r}\right) - 1\right] \qquad (4.129)$$

A plot of the temperature and mass fraction profiles is shown in Fig. 4.17.

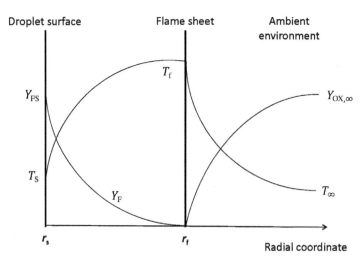

FIGURE 4.17

Temperature and mass fraction profiles for a burning liquid droplet.

This completes our discussion of the temperature and mass fraction profiles. We will now use these results to calculate burning times of droplets.

4.4.4 SOLUTION FOR BURNING TIME AND ILLUSTRATIVE RESULTS

The whole purpose of performing the detailed combustion analysis of a single burning fuel droplet is to calculate two global performance measures: the combustion rate and the burnout time. Recall that the mass balance for a burning droplet takes the form:

$$\frac{d(d_p)}{dt} = -\frac{2\dot{m}_c}{\pi \rho_1 d_p^2} = -\frac{K}{d_p^2} \tag{4.130}$$

This can be readily integrated with the initial condition that $d_p(t = 0) = d_{p0}$

$$d_p^2 = d_{p0}^2 - Kt \tag{4.131}$$

This is sometimes called the "diameter-squared" or "d-squared" law. The constant K is called the burning constant.

$$K = \frac{8\rho_g \alpha_g}{\rho_1} \ln(B+1) \tag{4.132}$$

The burnout time is the time for complete combustion, so $d_p = 0$. Thus, the burnout time is given by

$$t_b = \frac{d_{p0}^2}{K} = \frac{\rho_1 d_{p0}^2}{8\rho_g \alpha_g \ln(B + 1)} \tag{4.133}$$

For hydrocarbons burning in air, a reasonable order-of-magnitude estimate for the burning rate constant K is 10^{-6} m²/s (Kanury, 1975, pp. 178−179). The example below provides an illustration of how the burnout time varies with fuel properties and droplet diameter.

EXAMPLE 4.6

Calculate the burnout time for *iso*-octane, ethanol, and diesel fuel using the data in Kanury (1975, pp. 178−179). The burning constants are 1.44×10^{-6}, 9.3×10^{-7}, and 8.5×10^{-7} m²/s, respectively. Use droplet diameters of 1, 10, 30, 100, 300, and 1000 μm.

Solution

The solution is a straightforward application of the burnout time equation. The following table summarizes the results to three significant digits.

Burnout times in milliseconds for three fuels as a function of droplet diameter

Droplet Diameter (μm)	iso-Octane	Ethanol	Diesel Fuel
1	6.94×10^{-4}	1.08×10^{-3}	1.18×10^{-3}
10	6.94×10^{-2}	0.108	0.118
30	0.625	0.968	1.06
100	6.94	10.8	11.8
300	62.5	96.8	106
1000	694	1080	1180

The broad range in burnout times is due primarily to the diameter-squared effect. Fuel properties play a secondary role. To get a feel for the duration of the burnout time, consider the burnout times for 100-μm droplets. The human eye completes a full cycle of blinking in 100–400 ms. If you were observing the combustion of single 100-μm droplets, *the burnout time is so rapid that you could miss it if you blinked!*

4.4.5 SUMMARY

This section derived the relations for describing the behavior of a single liquid fuel droplet. This analysis is the basis for the d^2-law which says that the square of the droplet diameter shrinks linearly with time. It is important to note that similar shrinking droplet behavior has been observed experimentally with evaporating droplets and with burning droplets. This behavior confirms the inference that the combustion rate is diffusion limited. The burning liquid droplet is a useful analog for burning solid particles.

4.5 TRANSPORT PHENOMENA IN POROUS MEDIA

Piles of combustible dusts can, in certain circumstances, become susceptible to self-heating and thermal runaway. This can lead to the ignition of a flameless combustion wave, called a smolder wave that traverses the pile, thus emitting products of incomplete combustion. In a confined space, these products of incomplete combustion can accumulate to concentrations high enough to become toxic or even explosive. An additional hazard of smoldering is that if the smolder wave is exposed to a large increase in ventilation, it can transition to flaming combustion.

This section describes the physical structure of porous media and how this influences transport processes. A mathematical concept called local volume averaging provides the tools necessary to formulate the transport equations and constitutive relations for porous media. Local volume averaging leads to natural considerations of length scales and mixture properties. Typical magnitudes are presented based on the characteristics of the porous medium. We restrict ourselves to the energy equation because that is sufficient for characterizing the combustible dust hazards related to smoldering.

4.5.1 PHYSICAL DESCRIPTION OF A POROUS MEDIUM

Combustible dust deposits form a porous medium consisting of solid particles and void space. You can visualize the dust pile as a bed of spheres of different diameters. One cannot apply the governing equations transport phenomena within a porous medium directly because they are defined for a continuous medium. Moving from point to point within a porous medium from one end of the deposit to another, one would encounter solid or gas phase media in an erratic and unpredictable manner. To analyze the behavior of a porous medium, one must select a volume of space larger than a mathematical point so as to obtain smooth averages of the medium. This sampling volume is called a *representative elementary volume* (REV). It is convenient to think of REV as a sphere with a characteristic diameter d_{REV}.

In a sense, REV defines the size of a "point" within the porous medium. The properties of the porous medium are the average of the mixture properties at that point. By this technique, it is possible to transform a granular medium into a *pseudo-continuum*. Fig. 4.18 is an illustration of an REV selected from an arbitrary location within the dust deposit.

REV is like a filter in the sense that it smooths the measurement noise as one moves from point to point within the porous medium. The size of

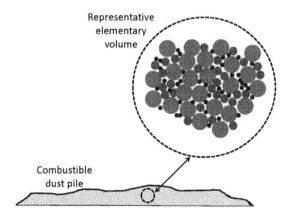

FIGURE 4.18

Diagram of a representative elementary volume for a combustible dust deposit.

the REV changes with the problem at hand. The REV size must be defined in relation to the microscopic characteristic length scale of the porous medium under consideration. The natural microscopic length scale for a combustible dust pile is its average particle diameter. The criterion for applicability of the REV concept has been described by Whitaker (1969, 1999, pp. 74−79) as

$$d_p \ll d_{REV} \ll L_0 \qquad (4.134)$$

This criterion states that the characteristic size of the REV must be much greater than a particle diameter and must be much smaller than the macroscopic length scale (eg, the thickness of the dust pile layer). The symbol $\ll$ implies that the ratio of the two sizes should be different by at least an order of magnitude. As a first approximation, the characteristic length for the REV of a dust accumulation is estimated as $100d_p$. The following example illustrates these relative magnitudes.

EXAMPLE 4.7

A manufacturer handles a combustible particulate material shaped as spheres with an average particle diameter of 10 mm. They store this material in a cylindrical silo with a diameter of 3 m. The handling of this material creates a combustible dust layer in the workspace. The thickness of the dust layer is typically around 5 mm. The dust particles have spherical shape with an average diameter of 50 μm.

You have been asked to evaluate the heat transfer characteristics in both the particulate material and the dust. As a first step, evaluate the size of the REV to determine if a pseudo-continuum model is appropriate.

Solution

The size of the REV must be much larger than the particle diameter but much smaller than the characteristic macroscopic length scale. For the particulate solid, REV is estimated as 100 times the average diameter of the particulate material: $d_{REV} = 100d_p = 100(10 \text{ mm}) = 1 \text{ m}$. The silo diameter is 3 m. The ratio of the silo diameter to the REV diameter is $L_0/d_{REV} = \frac{3 \text{ mm}}{1 \text{ mm}} = 3$. The ratio of 3 is too small, indicating that REV is too large to give a meaningful description of heat transfer behavior of the material in the silo. A pseudo-continuum model is not appropriate.

For the combustible dust layer, the size of the REV is estimated as $d_{REV} = 100 \, d_p = 100(50 \text{ μm}) = 5000 \text{ μm}$ or 0.5 mm. The ratio of the dust layer thickness to the size of the REV is $L_0/d_{REV} = \frac{5 \text{ mm}}{0.5 \text{ mm}} = 10$. This ratio is sufficiently large so that you can analyze the heat transfer characteristics of the dust layer using a pseudo-continuum model.

4.5.2 LOCAL VOLUME AVERAGING, LENGTH SCALES, AND MIXTURE PROPERTIES

Local volume averaging is the fundamental basis behind modern theories of multiphase systems. The basic concept is to define a volume average of any continuum field variable using the REV. The definition of a local volume average for any variable ψ is

$$\overline{\psi} = \frac{1}{V} \int_V \psi \, dV \tag{4.135}$$

The length scale of the averaging volume L_0 must be much larger than the particle diameter but much smaller than the characteristic dimension of the flow field. As a preliminary estimate, one may choose $L_0 \cong 100 \, d_p$.

The local volume average leads to the definition of average continuum quantities. Letting α_i represent the volume fraction of phase i, the mixture property ψ_m is defined by the relation:

$$\psi_m = \sum_{i=1}^{N} \alpha_i \psi_i \tag{4.136}$$

This relation holds for porous media that are homogeneous and isotropic. It can be used to calculate mixture thermodynamic properties like density and heat capacity. It can be used to calculate transport properties if a suitable description of the porous medium is available, but in many cases it is easier to measure the transport properties than it is to calculate them.

4.5.3 THE ENERGY TRANSPORT EQUATION FOR A POROUS MEDIUM

I am introducing only the energy equation for a porous medium because the self-heating, spontaneous ignition, and smolder wave propagation are to a first approximation adequately modeled by thermal theories. Furthermore, I neglect both forced and natural convection. Convection in porous media is an important topic in many fields of endeavor, but not so much in combustible dust hazard scenarios. We also neglect work done by pressure changes, viscous dissipation, and thermal radiation. For a more comprehensive discussion of heat transfer in porous media which includes consideration of these effects, I recommend the books by Kaviany (1995) and Nield and Bejan (2006).

To describe the heat transfer processes in a porous medium, we begin by writing separate thermal energy equations for the fluid phase and the solid phase. The variable α_g is the volume fraction of gas (usually air) and is often called the void fraction or the porosity. The solid fraction is denoted by $1 - \alpha_g$. The thermal energy equation for each phase takes the following form in rectangular coordinates:

$$\alpha_g \rho_g C_{pg} \frac{\partial T_g}{\partial t} = \alpha_g \frac{\partial}{\partial x}\left(k_g \frac{\partial T_g}{\partial x}\right) + \alpha_g \sum_{i=1}^{N} h_{f,i}^0 \omega_{g,i}'' \tag{4.137}$$

$$(1-\alpha_g)\rho_s C_s \frac{\partial T_s}{\partial t} = (1-\alpha_g)\frac{\partial}{\partial x}\left(k_s \frac{\partial T_s}{\partial x}\right) + (1-\alpha_g)\sum_{i=1}^{N} h_{f,i}^0 \omega_{s,i}''' \tag{4.138}$$

Each energy equation has an accumulation term, a heat conduction term, and a source term. For both phases, the term for heat conduction has been written with the expectation that the thermal conductivity is a function of position. This expectation arises from the recognition that heat conduction may be a function of the structure of the porous medium. An additional assumption is that the surface porosity is equal to the local porosity (interior to the porous body). This is an important assumption for the heat conduction term. We will return to this notion shortly. The source term denotes chemical reactions that occur within the REV. Thus, chemical reaction source term is written on a unit volume basis. More will be said about the chemical reaction source term later in this chapter.

Some simplification can be obtained if the gas and solid phases are in thermal equilibrium. If so, then the temperatures of the phases are equal, $T_g = T_s$. The two energy equations can be combined by adding them to obtain

$$(\rho C_p)_m \frac{\partial T}{\partial t} = \frac{\partial}{\partial x}\left(k_m \frac{\partial T}{\partial x}\right) + \sum_{i=1}^{N} h_{f,i}^0 \omega_{m,i}''' \tag{4.139}$$

The subscript m signifies the porous medium mixture. The mixture properties are defined as

$$(\rho C_p)_m = \alpha_g \left(\rho_g C_{pg}\right) + (1-\alpha_g)\left(\rho_s C_s\right) \tag{4.140}$$

The density and heat capacity are thermodynamic properties. Their calculation by the method of local volume averaging is a straightforward matter. A word of caution regarding the mixture density: it is not a constant. The mixture density is a function of the physical arrangement of the particles or packing density in the dust deposit. Deposits subjected to vibration will assume a more compact packing structure and will increase the mixture density. There are standardized methods for measuring the mixture density (often called the bulk density in the technical literature). As a rough rule of thumb, a dust deposit can be idealized as an assembly of randomly packed spheres of equal diameter. This configuration has a volume fraction of solid particles of approximately 0.6 with its complementary void fraction of 0.4. The theoretical density limit, assuming the particles are rigid, corresponds to a face centered cubic packing arrangement. This gives a maximum solid volume fraction of 0.741 and a porosity of 0.259 (Kaviany, 1995, pp. 20–24).

The mixture chemical reaction rate will in most cases become simplified to consider only the solid fraction because that is the fuel for the dust combustion reactions.

$$\omega_{m,i}''' = \alpha_g \omega_{g,i}''' + (1-\alpha_g)\omega_{m,i}''' \tag{4.141}$$

The calculation of the mixture thermal conductivity is a more complicated issue. Eq. (4.141) says that the mixture thermal conductivity is the weighted average of the two volume fractions.

$$Q = q_{\text{total}} \, A_{\text{total}} = q_g \, A_g + q_s \, A_s$$

$$Q = q_{\text{total}} \, A_{\text{total}} = -k_g A_g \left(\frac{\Delta T}{L}\right) - k_s A_s \left(\frac{\Delta T}{L}\right)$$

$$A_{\text{total}} = A_g + A_s, \qquad \alpha_g = \frac{A_g}{A_{\text{total}}}, \quad \alpha_s = \frac{A_s}{A_s}, \qquad \alpha_g + \alpha_s = 1$$

$$q_{\text{total}} = -\left[\alpha_g \, k_g + \left(1 - \alpha_g\right) k_s\right] \left(\frac{\Delta T}{L}\right) = -k_m \left(\frac{\Delta T}{L}\right)$$

FIGURE 4.19

Effective thermal conductivity of a porous medium with parallel structure.

$$k_m = \alpha_g k_g + (1 - \alpha_s)k_s \tag{4.142}$$

It was derived by assuming that the conductive heat flux through the fluid occurred simultaneously with the conductive heat flux through the solid. To use a heat transfer analogy, this is steady-state heat conduction with two heat fluxes occurring in parallel. The physical picture and the heat flux analysis are depicted in Fig. 4.19.

An alternative formulation can be derived if the two heat fluxes occur in series. This problem can be analyzed as a series of heat transfer resistances for steady-state heat conduction. In this case, the mixture thermal conductivity for heat fluxes occurring in series is given by the weighted harmonic mean k_H of the two thermal conductivities.

$$\frac{1}{k_H} = \frac{\alpha_g}{k_g} + \frac{\left(1 - \alpha_g\right)}{k_s} \tag{4.143}$$

The physical picture and the heat conduction analysis are shown in Fig. 4.20.

The parallel and series models for the effective thermal conductivity for a porous medium are presented here in the spirit of an intuitive argument. Neither of these results should be considered formal derivations.

In truth, neither scenario is likely to be an accurate model for predicting the effective thermal conductivity. The true effective thermal conductivity should be somewhere in between these two extreme models. Nield and Bejan argue that these two formulations place bounds on the magnitude of the mixture thermal

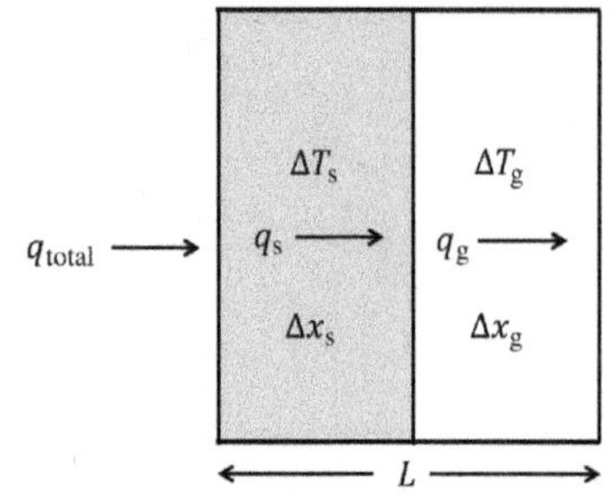

$$q_{total} = q_s = q_g$$

$$q_{total} = -k_m \left(\frac{\Delta T_{total}}{L} \right)$$

$$q_s = -k_s \left(\frac{\Delta T_s}{\Delta x_s} \right)$$

$$q_g = -k_g \left(\frac{\Delta T_g}{\Delta x_g} \right)$$

$$\Delta T_{total} = \Delta T_g + \Delta T_s = -\left(\frac{q_g \Delta x_g}{k_g} \right) - \left(\frac{q_s \Delta x_s}{k_s} \right) = -q \left(\frac{x_g}{k_g} + \frac{x_s}{k_s} \right)$$

$$q = -\frac{\Delta T_{total}}{\left(\frac{x_g}{k_g} + \frac{x_s}{k_s} \right)} = \frac{\Delta T_{total}/L}{\frac{1}{L} \left(\frac{x_g}{k_g} + \frac{x_s}{k_s} \right)} \qquad \alpha_g = \frac{x_g}{L}, \qquad 1 - \alpha_g = \frac{x_s}{L}$$

$$q = -k_H \left(\frac{\Delta T_{total}}{L} \right) \qquad \frac{1}{k_H} = \left(\frac{\alpha_g}{k_g} + \frac{1 - \alpha_g}{k_s} \right)$$

FIGURE 4.20

Effective thermal conductivity of a porous medium with series structure.

conductivity (2006, pp. 28–30). They argue that a better first estimate for practical purposes is the geometric mean of the thermal conductivities weighted by the volume fractions.

$$k_G = k_g^{\alpha_g} k_s^{1-\alpha_g} \tag{4.144}$$

Empirical tests of these model equations lend support to the upper and lower bound interpretations (Nield, 1991). Kaviany presents an extensive, formal analysis of this problem and lists several correlations for the mixture thermal conductivity (Kaviany, 1995, Chapter 3). Tavman et al. discuss a procedure for measuring the effective thermal conductivity for a powder (Tavman et al., 1998). The example below illustrates the calculation of the effective thermal conductivity for grain dust and will give you a feel for typical magnitudes.

EXAMPLE 4.8

Compare the performance of the three suggested models for the mixture thermal conductivity of a combustible dust sample against experimental measurements. Consider three cereal grains: wheat, sorghum, and corn. Compare the experimental measurements of effective thermal conductivity published by Chang et al. (1980) and compare them with calculations using the mixture, geometric, and harmonic mean models.

Solution

The table given below summarizes the calculations. The input data are obtained from a variety of sources, so some care must be taken to select data that are reasonably compatible. We begin with a table for the input data. The units on density are kg/m^3 and for thermal conductivity they are W/m-K.

Cereal Grain	Bulk Density[a]	Solid Density[b]	Air Density[c]	Porosity	Solid Fraction	k, air[c]	k, solid[d]
Wheat	280	1470	1.18	0.810	0.190	0.0261	0.177
Sorghum	430	1460	1.18	0.706	0.294	0.0261	0.177
Corn	510	1450	1.18	0.649	0.351	0.0261	0.177

[a]Chang, C.S., Lai, F.S., Miller, B.S., 1980. Thermal conductivity and specific heat of grain dust. Trans. ASAE 23, 1303–1306.
[b]Chang, C.S., 1988. Measuring density and porosity of grain kernels using a gas pycnometer. Cereal Chem. 65, 13–15.
[c]Kays, W.M., Crawford, M.E., 1980. Convective Heat and Mass Transfer, second ed. McGraw-Hill.
[d]Stroshine, R., 1998. Physical Properties of Agricultural Materials and Food Products. Purdue University.

Notice that the porosity values are much greater than the randomly packed spheres value of 0.4. This indicates that the powder samples were not subjected to a high degree of compaction. The porosity values were calculated from the mixture density relations (recall from Chapter 3 that $\rho_m = \alpha_s \rho_s + (1 - \alpha_s)\rho_g$)

$$\alpha_g = \frac{\rho_s - \rho_m}{\rho_s - \rho_g}, \alpha_s = 1 - \alpha_g$$

The mixture, geometric, and harmonic mean results are summarized in the following table:

Cereal Grain	k (meas)	k (m)	k (G)	k (H)
Wheat	0.0677	0.0705	0.0404	0.0315
Sorghum	0.0757	0.0949	0.0513	0.0355
Corn	0.0874	0.1082	0.0585	0.0382

None of the three models for the effective thermal conductivity predicted the measured values with great accuracy. This example suggests that the prediction of the effective thermal conductivity for a powder sample should be viewed with skepticism. If at all possible, measured values are to be preferred.

4.5.4 SUMMARY

There is an extensive body of literature on the subject of transport phenomena in porous media. This discussion was merely a brief survey that focused on those topics considered essential to evaluate the hazards of smoldering combustible dust. It was described how deposits of dust particles can form a porous body. Local volume averaging was introduced as a mathematical tool that permits the formulation of governing equations of change by describing a porous body as a pseudo-continuous medium. The energy equation was presented and the calculation of density, heat capacity, and the effective thermal conductivity was discussed. Due to the difficulty of describing the geometric structure of real porous bodies, in some cases it will be better if possible to work with measured properties instead of theoretical calculations.

4.6 DISPERSED MULTIPHASE FLOW

A dust explosion can be described as the passage of a flame through a cloud of solid particles suspended in air. The cloud of particles is a dispersed multiphase mixture, and the motion of the particles and the air suspending them is called a multiphase flow. Multiphase flow is the application of fluid dynamics principles to multiphase systems. The science of multiphase flow offers some important concepts for describing combustible dust phenomena. In particular, the flow of a dispersed cloud of particles is a useful model for dust deflagrations.

I begin with a description of the physical situation for the flow of dispersed particles in a gas stream and how the mixture flow behavior is affected by the solids concentration. Next is a description of single particle motion in a fluid and introduce two measures of particle motion behavior: the terminal settling velocity the response time. The concept of local volume averaging is again the primary tool for formulating and the necessary equations of change for the limiting condition of equilibrium flow. To describe dust deflagrations, we will need the mixture continuity, species continuity, and energy equations. In studying the behavior of porous media, in keeping with the literature we focused our attention on the gas volume fraction α_g. In the dispersed flow of particles, we will focus our attention on the volume fraction of solid particles α_s because the dust particles are the fuel of the multiphase mixture.

There is an extensive body of literature on the subject of multiphase transport phenomena. The discussion that follows is a biased survey that covers just those topics considered essential to evaluate the hazards of combustible dust deflagrations. A good overview of gas particle flows can be found in the book by Wallis (1969, Chapter 8). Two more modern books devoted entirely to the subject are Crowe et al. (2012) and Fan and Zhu (1998).

4.6.1 PHYSICAL DESCRIPTION OF DISPERSED MULTIPHASE FLOW

The flow of dispersed particles in air introduces additional sources of momentum transfer in the fluid or carrier phase. The relative magnitude of these effects

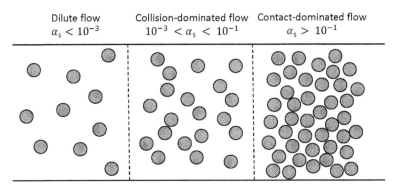

FIGURE 4.21

Flow regimes for dispersed multiphase flow (Crowe et al., 2012, p. 29).

depends on the particle concentration and the particle size. As a first-order classi-fication scheme, Crowe et al. divide gas particle motion into two limiting cases, dilute flow and dense flow (Crowe et al., 2012, pp. 26−29). In dilute flow, particle motion is dominated by the fluid forces of drag and lift and experience little momentum transfer from particle collisions. The flow of dust particles through the ductwork of a dust collection system is a dilute flow. In dense flows, momentum transfer by particle−particle collisions dominates particle motion. Dense flow can be further divided into collision-dominated flow and contact-dominated flow. Collision-dominated flow means that a particle−particle collision causes a change in particle trajectory. Contact-dominated flow means that parti-cles are in continuous contact and the contact forces are responsible for particle motion. A fluidized bed is an example of collision-dominated flow and the flow of bulk solids out of a storage silo is an example of contact-dominated flow. This range of flow behavior is illustrated in Fig. 4.21.

Crowe et al. use kinetic theory arguments to formulate order-of-magnitude criteria for classifying gas particle flows based on the volume fraction of solid α_s. They suggest the flow be classified as a dilute flow if $\alpha_s < 10^{-3}$, a collision-dominated flow if $10^{-3} < \alpha_s < 10^{-1}$, and contact-dominated if $\alpha_s > 10^{-1}$ (Crowe et al., 2012, pp. 26−29).

Additional insight into the dilute flow of particle mixtures can be obtained by considering two characteristics of the motion of a single particle: the terminal settling velocity and the response time. In each case, we will analyze the behavior of a spherical particle.

4.6.2 CHARACTERISTIC PARAMETERS FROM SINGLE PARTICLE BEHAVIOR

If a solid particle with an initial velocity of zero is released and allowed to fall freely, the particle will accelerate under the influence of gravity until the drag force is balanced by gravitational acceleration. At that point, the particle will reach a steady-state velocity called the terminal settling velocity. The terminal settling velocity is a characteristic velocity that one can use to evaluate the dynamic behavior of a particular combustible dust hazard scenario.

The derivation of an expression for the terminal settling velocity begins by writing a force balance on a single spherical particle $\sum F_i = m_p a_p$. In words,

$$\text{gravity force} - \text{buoyant force} - \text{drag force} = \text{mass} \times \text{acceleration} \tag{4.145}$$

At steady state, the acceleration term goes to zero. Substituting expressions for each of the forces gives

$$\rho_p \left(\frac{\pi}{6} d_p^3 \right) g - \rho_g \left(\frac{\pi}{6} d_p^3 \right) g - \frac{1}{2} C_D \rho_g v_{p,TSV}^2 \left(\frac{\pi d_p^2}{4} \right) \tag{4.146}$$

The subscripts p and g denote the particle and gas properties, respectively. The drag force coefficient, C_D, is defined in the usual manner and is based on the cross-sectional area of the particle. We will also need to introduce the definition of the Reynolds number based on the particle velocity Re_p.

$$C_D = \frac{F_{drag}/A_p}{1/2 \rho_g v_p^2}, \quad Re_p = \frac{\rho_g v_p d_p}{\mu} \tag{4.147}$$

Collecting terms and solving for the drag coefficient

$$C_D = \frac{4 g d_p}{3 v_{p,TSV}^2} \left(\frac{\rho_p - \rho_g}{\rho_g} \right) \tag{4.148}$$

As the particle falls, the air flows across the face and around the circumference of the particle. At very low speeds (very small particles), the flow around the sphere is laminar. This regime of flow is called the Stokes law region. It occurs at small Reynolds numbers, $Re_p < 1$. In the Stokes law region, the drag coefficient is a simple function of the Reynolds number, $C_D = 24/Re_p$. Substituting the expression for the drag coefficient and solving for the terminal settling velocity $v_{p,TSV}$:

$$v_{p,TSV} = \frac{\left(\rho_p - \rho_g \right) d_p^2 g}{18 \mu}, \quad v_{p,TSV} \cong \frac{\rho_p d_p^2 g}{18 \mu} \tag{4.149}$$

The second expression for the terminal settling velocity holds when $\rho_p \gg \rho_g$, a condition well satisfied for solid particles falling in air. For solid particles with a specific gravity of 1, the Stokes law region applies for particles $1 \, \mu m < d_p < 50 \, \mu m$. For particles smaller than 1 micron the assumption of continuum flow begins to falter and an adjustment must be made for velocity slip (the assumption in

Table 4.4 Drag Coefficient for Single Spherical Particles Over a Range of Reynolds Numbers

Region	Stokes	Intermediate	Newton's Law
Re_p	$Re_p < 1$	$1 < Re_p < 10^3$	$10^3 < Re_p < 2 \times 10^5$
C_D	$\dfrac{24}{Re_p}$	$\dfrac{24}{Re_p}\left(1 + 0.15Re_p^{0.687}\right)$	~ 0.44

continuum flow is that the gas velocity at the particle solid surface is zero). The adjustment is made with the Cunningham slip factor (Hinds, 1999, pp. 48–51). In most industrial settings, it is unlikely that particles in the submicron range will be a significant fraction of a combustible dust particle size distribution, so I will not discuss this further.

For larger particles, $d_p > 50 \, \mu m$, or those with a specific gravity greater than 1, the flow is outside the Stokes law region and a more complex value for the drag coefficient is required. At high Reynolds numbers ($Re_p > 10^3$), the drag coefficient assumes a limiting value of 0.44. This region is called the Newton's law region. Many different correlations have been proposed for the drag coefficient over a sphere. Table 4.4 summarizes one such recommendation (Hinds, 1999, pp. 42–46).

In the Stokes flow region, the terminal settling velocity can be calculated directly after computing the drag coefficient. In the Newton's law region, the drag coefficient is a constant and so, again, the terminal settling velocity can be computed directly.

$$v_{p,TSV} = 1.74 \left[\frac{g d_p \left(\rho_p - \rho_g \right)}{\rho_g} \right]^{1/2} \tag{4.150}$$

In the transition region, the drag coefficient is a function of the Reynolds number which contains the unknown velocity term. One must either solve by iteration or use the following correlation (Hinds, 1999, pp. 55–58):

$$v_{p,TSV} = \left(\frac{\mu}{\rho_g d_p} \right) \exp\left(-3.070 + 0.9935 J - 0.0178 J^2 \right) \tag{4.151}$$

$$J = \ln\left(C_D Re_p^2 \right) = \ln\left(\frac{4 \rho_p \rho_g g d_p^3}{3 \mu^2} \right) \tag{4.152}$$

One example of where the terminal settling velocity could be a useful parameter is in the evaluation of a dust dispersal scenario. The following example illustrates its use.

EXAMPLE 4.9

You wish to test the explosibility of four combustible dust samples in a 20-L sphere. Each dust sample has the same chemical composition but a different particle size distribution. The solid particle density of the material is $1500\,kg/m^3$. The mean diameters for the four samples are 1000, 300, 100, and $30\,\mu m$. The test requires that the powder is injected into the spherical chamber and then, after a time delay, a pyrotechnic match is actuated to ignite the cloud. The ignition time delay is set for 60 ms (ASTM E1226, 2012). A successful explosibility test requires a uniform suspension of the powder sample. You are curious if the suspension of the dust sample will be nonuniform due to gravitational settling. Calculate the terminal settling velocity and use this to calculate a time scale for settling, t_{TSV}, versus the ignition time delay. Assume the dispersal gas is air at 300 K and atmospheric pressure. The radius of a 20-L sphere is 0.168 m.

Solution

The time scale for gravitational settling can be defined as the radius of the 20-L sphere divided by the terminal settling velocity. The larger mean diameters are likely in the transition region of the drag coefficient. The smallest diameter particle is probably in the Stokes region. We calculate the velocities and then check the Reynolds number to verify that the correct drag coefficient relation was used. The density of air at the stated conditions is $1.1766\,kg/m^3$ and the viscosity is 1.853×10^{-5} Pa-s (Kays and Crawford, 1980, p. 388).

Parameter	$d_p = 30\ \mu m$	$d_p = 100\ \mu m$	$d_p = 300\ \mu m$	$d_p = 1000\ \mu m$
$C_D Re_p^2$	N/A	N/A	1.81×10^3	6.72×10^4
$v_{p,TSV}$ (m/s)	3.96×10^{-2}	0.440	1.57	5.07
Re_p	7.54×10^{-2}	0.838	29.9	322
Flow region	Stokes	Stokes	Transition	Transition
t_{TSV}(s)	4.24	0.201	0.107	0.031

The Stokes equation for settling velocity was used for the 30 and $100\,\mu m$ mean diameter samples. The Reynolds numbers computed from those velocities validate that selection. The velocities for the 300 and $1000\,\mu m$ diameter samples were calculated using the transition region equation. The Reynolds numbers verify that these particles are in the transition region.

The time scales for settling are much greater than the ignition delay time for the 30 and $100\,\mu m$ diameter samples. For these samples, the uniformity of the dust suspension should be unaffected by settling. The settling time scale of the $300\,\mu m$ diameter sample is comparable to the time scale of the delay, so the dust suspension may be somewhat nonuniform. For the largest mean particle diameter, the time scale for settling is approximately ½ the ignition delay time and, therefore, it can be expected that much of the sample will settle out of suspension prior to ignition. This means that the proposed test condition is most likely not appropriate for the 1000 μm sample.

Another useful characteristic of single particle motion is the momentum response time (Crowe, et al., 2012, pp. 24−26). The momentum response time is the time required for a gas traveling at a constant free stream velocity to accelerate a stationary particle to approximate the gas velocity (I will give a more specific definition shortly). The equation of motion for a spherical particle subject to fluid drag is

$$m_p \frac{dv_p}{dt} = \frac{1}{2} C_D \frac{\pi d_p^2}{4} \left(v_g - v_p\right) \left|v_g - v_p\right| \tag{4.153}$$

A relative Reynolds number is defined using the velocity difference or "slip" between the particle and the carrier gas

$$\text{Re}_{\Delta v} = \frac{\rho_p \left|v_g - v_p\right| d_p}{\mu} \tag{4.154}$$

Assuming Stokes flow, the relation $C_D = 24/\text{Re}_p$ can be substituted into the equation of motion and written as

$$\frac{dv_p}{dt} = \frac{1}{\tau_{\text{mom}}} \left(v_g - v_p\right), \ \tau_{\text{mom}} = \frac{\rho_d d_p^2}{18\mu} \tag{4.155}$$

The time constant τ_{mom} has units of time. The initial condition is that the particle velocity is initially zero. The solution for a steady-state gas flow is

$$v_p(t) = v_g \left[1 - \exp\left(-t/\tau_{\text{mom}}\right)\right] \tag{4.156}$$

One time constant corresponds to a particle velocity equal to 63% of the free stream gas velocity. The magnitude of the momentum time constant is proportional to the density of the particle: the more massive the particle, the slower it is to accelerate. But the particle diameter is a more significant parameter since the momentum time constant is proportional to the particle diameter squared; smaller particles are more rapidly accelerated. Table 4.5 demonstrates the effect of particle density and diameter on the momentum time constant. The densities chosen are for water, wheat grain dust, and aluminum.

A similar analysis can be used to explore the thermal response of a cold particle in a hot gas stream. Assume that the temperature of the particle is uniform. If the sole heat transfer mechanism between the particles and the gas is convection, then the energy equation for a single particle is

$$m_p C_s \frac{dT_p}{dt} = \text{Nu} \pi \, k_g \, d_p \left(T_g - T_p\right) \tag{4.157}$$

Table 4.5 Momentum Response Times in Milliseconds as a Function of Particle Diameter and Density

Particle Diameter (μm)	Particle Density (kg/m³)		
	1000	1400	2700
1	0.0030	0.0042	0.0081
10	0.30	0.42	0.81
100	30	42	81
1000	3000	4200	8100

where C_s is the heat capacity of the solid particle, Nu is the Nusselt number and the remaining symbols are as noted before. Assuming a spherical particle, the particle mass can be written in terms of its density and particle diameter. Collecting like terms yields

$$\frac{dT_p}{dt} = \frac{6\mathrm{Nu}\,k_g}{\rho_p C_s d_p^2}\left(T_g - T_p\right) \tag{4.158}$$

For convective heat transfer to spheres, the quantity Nu/2 goes to unity for small Reynolds numbers (Crowe et al., 2012, p. 25). Thus the energy equation simplifies to

$$\frac{dT_p}{dt} = \frac{1}{\tau_{\mathrm{therm}}}\left(T_g - T_p\right), \quad \tau_{\mathrm{therm}} = \frac{\rho_p C_s\, d_p^2}{12 k_g} \tag{4.159}$$

With the initial condition that the initial temperature of the particle is $T_{p,0}$, the solution to the differential equation is in the same form as for the momentum response time, and so the thermal response time has a similar meaning: it is the time it takes for the particle to reach 63% of the temperature change toward thermal equilibrium.

The relationship between the two response times can be revealed by taking their ratio:

$$\frac{\tau_{\mathrm{mom}}}{\tau_{\mathrm{therm}}} = \frac{2C_s}{3C_{p,g}\mathrm{Pr}}, \quad \mathrm{Pr} = \frac{C_{p,g}\,\mu}{k_g} \tag{4.160}$$

Over a fairly broad temperature range, the Prandtl number for air is approximately 0.7, so the thermal response time is of the same order of magnitude as the momentum response time.

4.6.3 LOCAL VOLUME AVERAGING AND MIXTURE PROPERTIES

We are interested in the flow of dispersed combustible dust particles in air. The analysis is easier if the dust suspension can be treated as a dilute flow. We return to our consideration of local volume averaging to develop criteria for dilute flow and to establish a length scale for the representative elementary volume. This task begins with the evaluation of the distance of separation between particles to satisfy dilute flow. You will notice that the previous discussion used the subscript "p" to denote particle properties and from this section onwards I will use the subscript "s" for solid. This is because in the previous section the discussion was about single particles. Now we will be discussing particle populations or, more succinctly, the solid phase.

When combustible dust deposits form a porous medium, the particles are in direct contact. In effect, the distance between adjacent particles is zero. This has a direct impact on the transport properties of the porous body. If the same combustible dust is dispersed as a cloud, the distance between particles is so great that the particles have little if any effect on adjacent particles. Consider a dust cloud consisting of equal-sized spheres. An order-of-magnitude estimate of the interparticle spacing of a dust cloud can be obtained from the following analysis. The volume fraction of solid is related to the number density of particles in the cloud by the following equation:

$$\alpha_s = N_p''' \left(\frac{\pi d_p^3}{6} \right), \quad N_p''' = \frac{N}{\mathcal{L}^3} \tag{4.161}$$

The number of particles per unit volume is designated as N_p'''. Consider a sampling volume with a characteristic length of $\mathcal{L}$ such that only one particle is in the volume ($N=1$). The characteristic length $\mathcal{L}$ is an order-of-magnitude estimate of the separation distance between particles in the dilute flow. The scaling argument for local volume averaging requires that $\mathcal{L}/d_p \gg 1$. To satisfy this inequality, I assign the ratio $\mathcal{L}/d_p = 10$. This restriction places a constraint on the maximum volume fraction

$$\alpha_s < \frac{\pi}{6} \left(\frac{d_p}{\mathcal{L}} \right)^3 < \frac{\pi}{6} \left(\frac{d_p}{10 d_p} \right)^3 < 5 \times 10^{-4} \approx 10^{-3} \tag{4.162}$$

Thus, our criterion for a dilute suspension is $\alpha_s < 10^{-3}$.

The characteristic length of the REV d_{REV} can be defined as $d_{REV} \gg \mathcal{L}$ and recall that the REV length scale must be much smaller than the length scale for flow ($d_{REV} \ll L_0$). These restrictions can be summarized as

$$d_p \ll \mathcal{L} \ll d_{REV} \ll L_0 \rightarrow \mathcal{L} \sim 10 d_p, \ d_{REV} \sim 100 d_p, \ L_0 \sim 1000 d_p \tag{4.163}$$

To use equations of change based on local volume averaging, the scaling constraints of Eq. (4.145) must be obeyed. For a powder sample with a mean particle diameter of 100 μm, the interparticle distance must be 1000 μm (1 mm), the characteristic length of the REV is 10 mm, and the length scale of the flow must be no smaller than 10 cm.

Returning to the definition of the local volume average of a variable ψ, the following density relations can be useful in the study of dilute dispersed particle flows. The mixture density is given by

$$\rho_m = \sum \alpha_i \rho_i = \alpha_s \rho_s + (1 - \alpha_s) \rho_g = \bar{\rho}_s + \bar{\rho}_g \tag{4.164}$$

In combustible dust flows, the quantity $\bar{\rho}_s = \alpha_s \rho_s$ is the dust concentration or, alternatively, it is the fuel concentration. It is most often reported in units of g/m^3. The volume averaged density can be related to mass fractions

$$Y_i = \frac{\int_{V_i} \rho_i dV}{\int_V \rho_m dV} = \frac{\rho_i V_i}{\rho_m V} = \frac{\alpha_i \rho_i}{\rho_m} = \frac{\bar{\rho}_i}{\rho_m} \rightarrow Y_i \rho_m = \alpha_i \rho_i \tag{4.165}$$

It is usually more convenient to calculate mixture properties from mass fractions. The above relation can be defined in terms of mass fractions as follows:

$$\sum_{i=1}^{2} \frac{\alpha_i}{\rho_m} = \sum_{i=1}^{2} \frac{Y_i}{\rho_i} \tag{4.166}$$

$$\sum_{i=1}^{2} \frac{\alpha_i}{\rho_m} = \frac{1}{\rho_m} \sum_{i=1}^{2} \alpha_i = \frac{1}{\rho_m} \tag{4.167}$$

Substituting Eq. (4.167) into (4.166) and solving for the mixture density gives

$$\rho_m = \left[\sum_{i=1}^{2} \frac{Y_i}{\rho_i} \right]^{-1} \tag{4.168}$$

This final relation allows the calculation of the mixture density from mass fractions.

4.6.4 GOVERNING EQUATIONS FOR EQUILIBRIUM FLOWS

Our interest in the governing equations for the dilute flow of dispersed combustible dust is limited to our desire to study dust deflagrations. Hence, we will only need the continuity equation, the energy equation, and the species continuity equations. At this stage of the presentation, it is assumed that the flow is in both velocity equilibrium (the local particle velocity is equal to the local gas velocity) and temperature equilibrium (the local particle temperature is equal to the local gas temperature).

These are the equations of change for steady planar flow for the equilibrium flow of a dispersed combustible dust mixture.

$$\frac{\partial}{\partial x} (\rho_m v_x) = 0 \text{ or } \dot{m}''_m = \rho_m v_x = \text{constant} \tag{4.169}$$

$$\rho_m C_{p,m} \left(v_x \frac{\partial T}{\partial x} \right) = k_m \frac{\partial^2 T}{\partial x^2} + Q \bar{\omega}''_F \tag{4.170}$$

$$\rho_m \left(v_x \frac{\partial Y_i}{\partial x} \right) = \rho_m \mathcal{D}_m \frac{\partial^2 Y_i}{\partial x^2} + \bar{\omega}'''_i \tag{4.171}$$

The subscript m denotes mixture properties. The advantage of assuming velocity and thermal equilibrium is that the governing equations of change take the same form as those for single phase flow. Of course, if the physical situation does not warrant these assumptions, then this form of the equations will not be appropriate for use.

The mixture thermal conductivity can be calculated from Maxwell's model for a dilute suspension of spheres (Kaviany, 1995, p. 129).

$$\frac{k_m}{k_g} = \frac{2\alpha_g + (3 - 2\alpha_g)\,\hat{k}}{3 - \alpha_g + \alpha_g \hat{k}}, \quad \hat{k} = \frac{k_s}{k_g} \tag{4.172}$$

The same model can be used for the calculation for the mixture diffusivity $\mathcal{D}_m$.

4.6.5 SUMMARY

The literature on dispersed multiphase flow is extensive. This section has covered just the basics of the fluid dynamics of dispersed particles as it relates to combustible dust deflagrations. As in our survey of porous media, the technique of local volume averaging has given us a procedure for calculating mixture properties. To simplify the form of the equations of change, it was assumed that the flow of particles was in both velocity and temperature equilibrium. Relaxing these assumptions requires more complex equations of change and will be discussed in Chapter 10.

4.7 THERMAL RADIATION

Thermal radiation is the transmission of heat by electromagnetic waves. Heat transfer by conduction or convection requires the presence of an intervening medium. Thermal radiation does not; radiant heat can be transferred from a hot surface to a cold surface even in a vacuum.

Thermal radiation can be an important mode of heat transfer in many combustion applications. Sooty diffusion flames from large hydrocarbon pool fires can emit 30% or more of their heat release as thermal radiation (Drysdale, 1997, p. 146). Depending on their relative size, pool fires can project hazardous levels of radiant heat sufficient to cause human injury through burn injuries. In gaseous flames, the magnitude of the thermal radiation heat flux is governed by the absorption characteristics of the combustion product gases or by the formation of soot. Thermal radiation is generally important in dust deflagrations because of the presence of high concentrations of fine particles.

In this section, I will give a brief introduction to thermal radiation in participating media from the viewpoint of Whitaker (1983) who derives the radiative transfer equation as a photon transport equation. I first introduce the radiative transport equation for a participating medium. I then present the equations for the emissive power of a black body and discuss the radiation processes of emission and absorption. I next introduce the concept of optical thickness and describe the calculation of the radiative properties of dust clouds based on the Mie theory of light scattering. The specific effects of thermal radiation on flame propagation will be discussed in later chapters. Comprehensive presentations on thermal radiation heat transfer

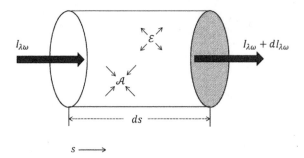

FIGURE 4.22

Control volume sketch for the radiative transfer equation.

in participating media can be found in the books by Özisik (1973), Siegel and Howell (1981), and Modest (2013).

4.7.1 THERMAL RADIATION IN PARTICIPATING MEDIA

All solid bodies emit electromagnetic radiation according to their temperature. Electromagnetic radiation exhibits both wave-like and particle-like behaviors (the wave particle duality of quantum mechanics). The choice of how to analyze its behavior is a matter of convenience. In the analysis of thermal radiation in a participating medium, it is convenient to think of thermal radiation as a stream of photons. As a stream of photons is emitted into a control volume, the photons are absorbed and scattered. The radiative transfer equation is an energy balance equation that accounts for the changes in photon energy as a function of absorption, emission, and interactions with solid surfaces. We will assume in this development that the participating medium is continuous, homogeneous, and isotropic; polarization can be neglected; and the medium is in local thermodynamic equilibrium. The physical situation is depicted in Fig. 4.22.

The radiative transfer equation at steady state is written as

$$\frac{dI_{\lambda\omega}}{ds} = \mathcal{E}_{\lambda\omega} - \mathcal{A}_{\lambda\omega} \tag{4.173}$$

The left-hand side is the change in radiation intensity $I_{\lambda\omega}$ along the path length s. $I_{\lambda\omega}$ is the radiant energy flux per unity wavelength λ per unit solid angle ω, $\mathcal{E}_{\lambda\omega}$ is the emission rate per unit volume, and $\mathcal{A}_{\lambda\omega}$ is the absorption per unit volume. This form of the radiative transfer equation has been written for monochromatic (single wavelength) radiation. The unit solid angle is defined in the usual way as shown in Fig. 4.23.

The most general expression for the radiative transfer equation is considerably more complex than the above equation. More information can be found in the radiation heat transfer books cited earlier.

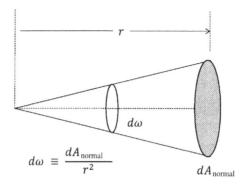

FIGURE 4.23

Definition of the solid angle ω.

An example application of the radiative transfer equation is to consider the attenuation of an incident beam of monochromatic thermal radiation as it travels through an absorbing and scattering medium. If we assume that the attenuation process can be modeled by a constant coefficient (called the extinction coefficient β_λ) with units of reciprocal length and is linear in the beam intensity, the radiative transfer equation for a one-dimensional geometry (coordinate x) becomes

$$\frac{dI_\lambda}{dx} = -A_\lambda = -\beta_\lambda I_\lambda \tag{4.174}$$

The boundary equation is $I_\lambda(x=0) = I_{\lambda 0}$. This equation can be integrated to yield

$$I_\lambda(x) = I_{\lambda 0}\exp(-\beta_\lambda x) \tag{4.175}$$

This is the familiar Lambert–Beer law (Whitaker, 1983, pp. 382–383). For a given value of path length x, larger values of the extinction coefficient lead to smaller values of the radiant intensity: the medium is said to have a greater attenuating effect on the radiant intensity.

The equation above gives rise to a useful dimensionless ratio called the *optical thickness*. The optical thickness τ_{opt} is the product of the absorption coefficient and the characteristic beam length for the system of interest, $\tau_{opt} = \beta_\lambda x_0$. Small values of the optical thickness imply small contributions of absorption or scattering to the radiant intensity: the participating medium approximates a *transparent* medium. This type of medium is called optically thin. Large values of the optical thickness imply large contributions of absorption or scattering to the local radiant intensity. This is called an optically thick medium and it approximates an *opaque* medium.

To calculate the intensity of the thermal radiation at the boundaries, we need to discuss black body radiation.

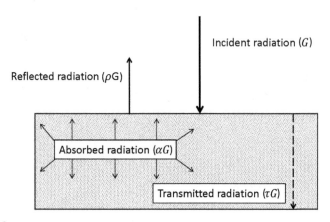

FIGURE 4.24

Surface energy balance for thermal radiation on a solid surface.

4.7.2 BLACK BODY RADIATION

All solid bodies emit thermal radiation. The degree to which a body interacts with its surroundings depends in part on the surface properties of the body. The incident radiation on a surface is either absorbed, reflected, or transmitted through the body. The surface properties relevant to thermal radiation are the absorptivity α, reflectivity ρ, and transmissivity τ, where each property is bounded between 0 and 1. Fig. 4.24 is a definition sketch for the fate of thermal radiation which strikes a solid surface.

A surface energy balance on the system yields

$$G = \alpha G + \rho G + \tau G \rightarrow \alpha + \rho + \tau = 1 \qquad (4.176)$$

If the transmissivity is zero, the body is said to be opaque. If both the reflectivity and transmissivity are zero, then the absorptivity is equal to 1 and the body is said to be black.

Black body radiation establishes an important reference case in the study of thermal radiation. The radiant intensity from a black body is independent of the direction of emission. Thus, the black body spectral intensity can be integrated over the hemispherical domain formed by the solid angle to give the simple relation for the spectral radiant heat flux

$$q_{b\lambda}(\lambda) = \pi I_{b\lambda}(\lambda) \qquad (4.177)$$

The spectral distribution for the black body emissive power is given by Planck's distribution

$$q_{b\lambda}(\lambda) = \frac{2\pi c^2 h}{\lambda^5 \left[\exp(ch/\lambda kT) - 1 \right]}. \qquad (4.178)$$

The constants in this equation are the speed of light c, Planck's constant h, and Boltzmann's constant k. The black body hemispherical emissive power is obtained by integrating over all wavelengths (Bird et al., 2002, pp. 493−495) and is given by the expression:

$$q_b = \sigma T^4 \tag{4.179}$$

The constant σ is the Stefan−Boltzmann constant. Its value is 5.67×10^{-8} W/m²-K⁴.

Now armed with the radiative transfer equation and the black body emissive power, we are ready to discuss the calculation of the radiative properties of dust clouds.

4.7.3 RADIATIVE PROPERTIES OF DUST CLOUDS

The relative significance of thermal radiation in a given dust deflagration depends on the dust concentration, the particle size distribution, and the size of the dust cloud. Since thermal radiation is a form of electromagnetic radiation, the computation of the effect of how particles interact with a source of thermal radiation is based on the study of light scattering by particles (van de Hulst, 1981). As in most physical phenomena, the calculation of the interaction of thermal radiation with dust particles is simplified as the length ratio of particle size versus wavelength approaches some asymptotic limit. Fig. 4.25 compares the spectrum of electromagnetic radiation with the potential range of particle diameters you may encounter in a combustible dust hazard scenario. The size range of nanoparticles

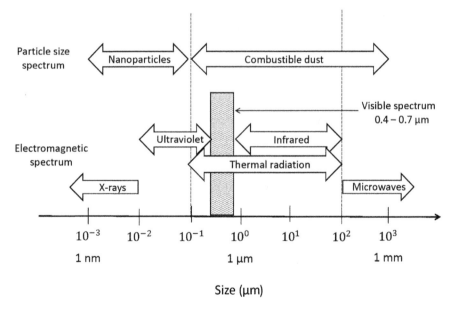

FIGURE 4.25

Spectrum of electromagnetic waves in comparison with the potential particle size range of combustible dusts.

is also shown in this figure as their importance in industrial products continues to grow.

The key message here is that the range of particle diameters of interest overlaps the range of wavelengths that span thermal radiation. This extensive overlap creates a challenge for the investigator interested in simple estimates for radiative properties of dust clouds. The formal analysis of the light scattering properties of combustible dust particles is based on the Mie theory of light scattering. The Mie theory involves solving Maxwell's equations to determine the absorption and scattering efficiency of particles based on their optical properties, particle diameter, and radiation wavelength. Mie organized his theory around two dimensionless parameters, the complex index of refraction m and the size parameter x:

$$m = n - ik, \; x = \pi d_p / \lambda \tag{4.180}$$

A comprehensive presentation of the Mie theory and various solutions for special cases can be found in van de Hulst (1981). Modest presents an up-to-date summary of the literature since van de Hulst's book (Modest, 2013, Chapter 12). Happily, we can avoid diving into the intricate details of Mie's theory as we will see that there is a limiting case that will suit most of our needs. One of the more important simplifications is that we can determine the absorption and scattering properties of a cloud of particles by adding the individual contributions of each particle.

The physical setting for light scattering by a single spherical particle is illustrated in Fig. 4.26. The incident radiation is intercepted by a particle. Some of the incident radiation is absorbed and the rest is scattered. A simple way to think of how the particle interacts with the incident radiation is that the particle intercepts a portion of the radiation equal to the projected area of the particle which is $\pi d_p^2 / 4$.

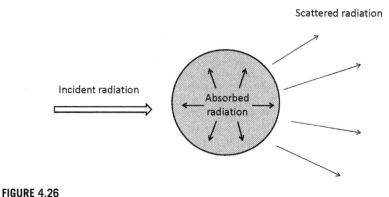

FIGURE 4.26

Absorption and scattering of incident thermal radiation by a spherical particle.

First define the dimensionless extinction efficiency factor as

$$Q_e = \frac{\text{radiant power absorbed and scattered by the particle}}{\text{radiant power geometrically incident on the particle}} \qquad (4.181)$$

The contributions of absorption and scattering are additive and are also expressed as dimensionless efficiency factors.

$$Q_e = Q_a + Q_s \qquad (4.182)$$

As van de Hulst notes, these are statements of the conservation of energy (van de Hulst, 1981, pp. 12–14). A monodisperse cloud of particles with a number concentration of particles per unit volume N_p has an extinction coefficient defined as

$$\beta_e = \left(\frac{\text{projected area}}{\text{particle}}\right)\left(\frac{\text{number of particles}}{\text{unitvolume}}\right)(\text{efficiency factor})$$

$$= \frac{\pi d_p^2 N_p Q_e}{4} \qquad (4.183)$$

The number concentration of particles is given by the expression

$$N_p = \frac{6\alpha_s}{\pi d_p^3} \qquad (4.184)$$

Mie's theory enables the investigator to calculate the efficiency factors for spherical particles of arbitrary size. The extraction of the efficiency factors involves somewhat tedious calculations or inspection of data plots. However, as noted by Hinds (1999, pp. 352–354), for particle diameters greater than 4 μm, the extinction efficiency asymptotically approaches a value of 2. Modest offers the criterion that $Q_e \to 2$ when $x = \pi d_p/\lambda \gg 1$ (Modest, 2013, p. 402). At the larger wavelengths of thermal radiation ($\lambda \cong 100\mu$m), Modest's criterion $Q_e \to 2$ corresponds to particle diameters on the order of $d_p \geq 32\mu$m. The asymptotic value $Q_e \to 2$ implies that the particle removes twice as much thermal radiation than what could be accomplished by simple geometric blocking. The factor of 2 arises from the combined effects of both absorption and scattering. Substituting the number concentration expression into the definition for the extinction coefficient, and setting $Q_e = 2$, the following simple relation is obtained:

$$\beta_e = \frac{3\alpha_s}{d_p} \qquad (4.185)$$

The radiative properties of a dust cloud are a function of particle concentration and particle diameter. Table 4.6 illustrates the relative magnitude of the trends for the extinction coefficient when $Q_e = 2$, $\alpha_s = 10^{-4}$, and the beam path length $x_0 = 3$m.

In this case, all three particle diameters result in an optically thick dust cloud.

The example below makes use of the extinction coefficient to examine the accuracy of an oft-quoted rule of thumb on how to identify a potentially hazardous concentration in a combustible dust cloud.

Table 4.6 Extinction Coefficient and Optical Thickness for Various Particle Diameters

d_p (μm)	β_e (m^{-1})	τ_{opt}
10	30	900
100	3	90
1000	0.3	9

EXAMPLE 4.10

Eckhoff quotes a rule of thumb about the optical thickness of combustible dust clouds (Eckhoff, 2003, p. 9). He states that an illuminated 25 W light bulb cannot be seen through a coal dust cloud with a length of 2 m and a dust concentration of 40 g/m^3 (roughly the MEC). Evaluate the plausibility of this statement by calculating the optical thickness of this dust cloud for three mean particle diameters: 10, 100, and 1000 μm. Assume the particle size distribution is monodisperse and ambient temperature and pressure.

Solution

This problem is a practical application of the Lambert–Beer solution. This model requires the calculation of the extinction coefficient and the optical thickness which, in turn, requires the calculation of the volume fraction of solid. The volume fraction of solid is calculated from the relation derived in the disperse multiphase flow section of this chapter.

$$Y_{coal}\rho_m = \alpha_{coal}\rho_{coal}$$

The coal dust mass fraction is calculated from

$$Y_{coal} = \frac{\bar{\rho}_{coal}}{\bar{\rho}_{coal} + \bar{\rho}_{air}} = \frac{0.04 \text{ kg/m}^3}{(0.04 + 1.1766)\text{kg/m}^3} = 0.0329$$

The density of air at 300 K and 1.01 bars is obtained from Kays and Crawford (1980, Appendix A).

$$\rho_m = \bar{\rho}_s + \bar{\rho}_g = 0.04 + 1.1766 = 1.22 \text{ kg/m}^3$$

The solid density of coal is estimated to be 1500 kg/m^3. Thus the volume fraction of coal dust at 40 g/m^3 is

$$\alpha_{coal} = \frac{Y_{coal}\rho_m}{\rho_{coal}} = \frac{0.0329(1.22)}{1500} = 2.68 \times 10^{-5}$$

The extinction coefficient is calculated from the relation $\beta_e = 3\alpha_s/d_p$ and the optical thickness from $\tau_{opt} = \beta_\lambda x_0$, where $x_0 = 2$m.

d_p (μm)	β_e (m^{-1})	τ_{opt}	$I(x=2)/I_0$
10	8.04	16.1	1.02×10^{-7}
100	0.804	1.61	2.00×10^{-1}
1000	0.0804	0.161	8.51×10^{-1}

For the smallest particle diameter, 10 μm, the claim is definitely verified. For the largest particles, 1 mm, the claim is not reliable. For the intermediate size of 100 μm, the light is attenuated by 80%, making the claim seem reasonable enough.

4.8 TURBULENCE

Turbulence is a ubiquitous phenomenon in fluid mechanics. Turbulence plays an important role in combustion due to its enhancement of heat transfer, mass transfer, and chemical reaction rates (Turns, 2012, Chapter 11). Although the importance of turbulence was recognized early on, it is only more recently that the effects of turbulence on dust explosion behavior and testing have been systematically explored (Amyotte et al., 1989). It will become apparent in this discussion that the effects of turbulence are not manifested at the length scale of individual particles. Instead, turbulence influences the dust combustion process at the scale of the flame thickness. Thus, there has been much effort directed at developing models for the turbulent burning velocity as a function of the laminar burning velocity. In this section, I will give a brief overview of some of the important characteristics of turbulence, describe the key length scales and their calculation for characterizing turbulent flows, and discuss the calculation of the turbulent burning velocity of premixed flames.

4.8.1 PHYSICAL DESCRIPTION OF TURBULENT FLOW

Turbulence is a flow phenomenon characterized by a three-dimensional, rotational, stochastically fluctuating, velocity field. In addition to velocity fluctuations, all of the continuum field variables may exhibit fluctuations about their mean value: density, species concentration, pressure, and temperature. The fluid dynamicist Osborne Reynolds recognized that these fluctuations could be represented mathematically by decomposing the instantaneous value of a variable into its mean value plus a fluctuating component (Turns, 2012, pp. 428–429). Thus the x-component of velocity can be represented as

$$v_x = \bar{v}_x + v_x'$$

(4.186)

The over bar represents the mean value of the velocity and the prime designates the fluctuating component. The mean value is obtained by calculating a time average over a representative time interval Δt sufficiently large to accommodate multiple fluctuation cycles but small enough to detect time-varying flows.

$$\bar{v}_x = \frac{1}{\Delta t} \int_0^{\Delta t} v_x(t)dt \tag{4.187}$$

The similarity to the concept of local volume averaging for multiphase systems should be apparent. It is frequently desired to find an average value of the fluctuating quantity. The root mean square value is the most common measure of central tendency used in turbulence measurement, where $v'_{rms} = \sqrt{(v')^2}$. A dimensionless fluctuating value is defined as the intensity $I_{rms} = v'_{rms}/\bar{v}$.

Turbulence is not a property of a fluid. It is a property of the flow. The Reynolds number is used to establish whether a flow is turbulent. The Reynolds number is a dimensionless number that compares the relative magnitude of inertial and viscous forces and is defined in the following way:

$$Re = \frac{\rho v \mathcal{L}}{\mu} \tag{4.188}$$

The specific definition of the characteristic length $\mathcal{L}$ depends on the flow geometry under consideration. The transition from laminar to turbulent flow tends to occur at large Reynolds numbers, but the definition of "large" is dependent on the flow geometry.

The nature of turbulence has been the subject of intensive research over the last 100-plus years. Until recently, the majority of this research has been based on the Reynolds decomposition of the flow variables into their mean and fluctuating components. While this is a very attractive approach conceptually, it leads to a mathematical difficulty called the turbulence closure problem. The turbulence closure problem is the result of the averaging process which generates new terms into the average equations of change. These new terms are unknown. These new terms must be modeled using additional physical insight. This task, called turbulence modeling, has generated an enormous body of literature (Pope, 2005). Some of the approaches to turbulence modeling will be addressed further in Chapter 10 I consider next a simpler approach based on the length scales of turbulence. The characteristic turbulent length scales are the basis for characterizing turbulent premixed flame behavior.

4.8.2 LENGTH SCALES IN TURBULENCE

A turbulent flow dissipates more mechanical energy than the comparable laminar flow counterpart. This is not just because the mean kinetic energy of the flow is greater, but also because the flow generates more lost work than a laminar flow. The lost work is essentially the kinetic energy extracted from the fluctuating flow field and dissipated as heat by viscous action. This transformation of kinetic energy into heat by viscous dissipation occurs through a process called the turbulent energy cascade. A turbulent flow consists of a broad distribution of swirling packets of fluid called eddies. The magnitude of the swirling motion is characterized by its vorticity. The turbulent cascade of energy is the mechanism by which the flow extracts mechanical energy from the large eddies and conveys it to the small eddies where the mechanical energy is dissipated into heat by viscous action (Tennekes and Lumley, 1980, Chapter 3). The full range of eddies that participate in this process is called the energy cascade.

It turns out that a given turbulent flow can be characterized to a certain degree by determining the characteristic sizes of these eddies. In a sense, we want to know the large, medium, and small sizes of the spectrum. The largest eddies are associated with the macroscale dimension of the flow field, the medium size is called the integral scale, and the smallest size is the Kolmogorov microscale. I introduce simple, intuitive definitions for these quantities below. A formal definition of these quantities would require an extensive discussion of turbulence theory. I direct the interested reader to the books by Tennekes and Lumley (1980) and Pope (2005) for more information on turbulence theory.

The largest eddy sizes are comparable with the flow macroscale $\mathcal{L}_{\text{flow}}$. The flow macroscale is usually a characteristic width of the flow field. In internal or confined flows, this dimension can be the width of the confinement. For example, in a 20-L spherical test chamber, the flow macroscale is the radius of the chamber.

The intermediate size scale of the energy spectrum is the integral scale $\mathcal{L}_{\text{int}}$. The integral scale, sometimes called the Taylor macroscale, is based on a statistical definition of the correlation distance between fluctuating velocities. Its value must generally be determined by measurement of the flow field of interest. Many investigators use an alternate measure for the intermediate size scale of the energy spectrum called the Taylor microscale $\mathcal{L}_\lambda$. The Taylor microscale is physically interpreted to be the largest size of eddies that begin to dissipate kinetic energy. Two turbulence Reynolds numbers can be defined from these length scales

$$\text{Re}_{\text{int}} = \frac{\rho v'_{\text{rms}} \mathcal{L}_{\text{int}}}{\mu}, \text{Re}_\lambda = \frac{\rho v'_{\text{rms}} \mathcal{L}_\lambda}{\mu} \tag{4.189}$$

The smallest length scale in a turbulent flow is the Kolmogorov microscale $\mathcal{L}_k$. This is the length scale where viscous dissipation occurs. It is defined by a relation derived from dimensional analysis

$$\mathcal{L}_k = \left[\frac{(\mu/\rho)^3}{\epsilon_0} \right]^{1/4} \tag{4.190}$$

Here the turbulent dissipation rate is related to the turbulent kinetic energy by the relation:

$$\epsilon_0 = \frac{3}{2} \frac{\left(v'_{rms}\right)^3}{\mathcal{L}_{int}}$$

(4.191)

The Kolmogorov length scale can be used to define a third turbulence number

$$Re_k = \frac{\rho v'_{rms} \mathcal{L}_k}{\mu}$$

(4.192)

The ratio of the length scales can be related by the following expressions (Turns, 2012, p. 434):

$$\frac{\mathcal{L}_k}{\mathcal{L}_{int}} = Re_{int}^{-3/4} \quad \text{and} \quad \frac{\mathcal{L}_\lambda}{\mathcal{L}_{int}} = Re_{int}^{-1/2}$$

(4.193)

One disadvantage of the turbulence parameters is that some measurement data are needed to perform the calculations. Given the very large body of literature on turbulence measurements, if you need some measurement data for a turbulence analysis, it is likely that you can find a publication that investigated a flow scenario similar to yours. In Chapter 8 I will summarize a number of studies that have been published on the turbulent flow inside combustible dust test chambers.

It is interesting to consider how the Kolmogorov length scale compares with typical combustible dust particle diameters. Consider a combustible dust with a mean particle diameter of 100 μm. We will use the study by Dahoe and his colleagues in which they measured the rms velocity fluctuations in a 20-L sphere during the preignition phase of dust dispersal using different dispersion devices (Dahoe et al., 2001a, 2001b). Similar studies by Pu et al. measured similar turbulence values with their characterization of dispersion in a 20-L sphere (Pu et al., 1988, 1990, 1991). Thus, the turbulent flow field during dust dispersal (preignition) in a 20-L sphere appears to be consistent in this regard.

As an example, one test condition Dahoe and his colleagues measured the fluctuating vertical velocity component at a dimensionless radial location equal to 0.36 from the center (along the horizontal plane) and at the elevation of the lowest point of the dispersion ring. Table 4.7 summarizes the measurements made at three distinct times (Dahoe et al., 2001a, p. 174). The integral scale turbulence number is calculated for each measurement, and the Kolmogorov length scale is computed for comparison with the mean particle diameter.

This time range represents a practical range of delay times *before* firing the igniter in the chamber. The lesson of Table 4.7 is that the Kolmogorov length scale, and hence the role of turbulence, varies dramatically with time. At short times, the particle diameter is much larger than the Kolmogorov length scale.

Table 4.7 Calculation of the Kolmogorov Length Scale for Different
Fluctuating Velocity Measurements During Dispersal in a 20-L sphere
(Dahoe et al., 2001a, p. 174)

Time (ms)	v'_{rms} (m/s)	Re_{int}	L_k (m)	L_k/d_p
10	25	4.79×10^4	8.04×10^{-6}	0.0804
50	4	1.89×10^3	3.18×10^{-5}	0.318
100	1	1.07×10^3	8.98×10^{-5}	0.898

This means that at this portion of the flow history the individual particles are
immersed in a smaller eddies. At long time scales, the Kolmogorov length scale
is comparable to the particle diameter. This type of length scale comparison
becomes more important during combustion.

4.8.3 BURNING RATES IN TURBULENT PREMIXED DUST FLAMES

I restrict this section to the consideration of gaseous flames. The application
of these ideas to dust deflagrations will be presented in Chapter 7 and
Chapter 8. From the beginning of the serious investigation of turbulent premixed flames, it has been remarked that one of the primary distinguishing
factors is the wrinkling of the flame surface (Kuo and Acharya, 2012, pp.
287−310). The wrinkling effect increases the flame surface area which in
turn increases the mass burning rate. The inference drawn is that the flame
still propagates at the laminar burning velocity. This simple conceptual model
for a turbulent premixed flame essentially says that the turbulent mass burning rate is given by

$$\dot{m}'''_{turbulent} = \rho_u S_u A_{turbulent} \text{ with } A_{turbulent} > A_{laminar} \tag{4.194}$$

indicating that the turbulent mass burning rate is greater than the laminar rate
only because of the difference in flame surface area.

Borghi developed a classification of different turbulent premixed combustion
regimes (Borghi, 1985). As the intensity of turbulence increases, or as the turbulent scale decreases, the wrinkling also increases. This range of behavior is called
the *wrinkled flame regime*. Eventually the wrinkling can become so severe that it
disrupts the continuity of the flame surface. At extreme levels of turbulence, the
flame becomes a nearly homogeneous reaction zone within its confining volume.
This is called the *distributed reaction regime*. The flame thickness increases and
once the flame surface continuity is disrupted the flame becomes a thicker region
of isolated flame packets called flamelets. This is called the *flamelets-in-eddies
regime* (Turns, 2012, pp. 459−460).

Given this description of the range of flame behavior, it is reasonable to
expect that the different regimes of behavior can be classified by comparing the

relative magnitudes of turbulence scale and fluctuating velocity $(\mathcal{L}_{int}, \mathcal{L}_k, v'_{rms})$ with the flame thickness and laminar burning velocity (δ, S_u). In fact, a large number of correlations have been proposed which take the following functional form:

$$S_{turbulent} = f\left(S_u, v'_{rms}, Re_{int}\right) \tag{4.195}$$

As an example, one of the simplest correlations is the earliest such correlation published by Damkohler in 1940 takes the form:

$$S_{turbulent} = S_u\left(1 + \frac{v'_{rms}}{S_u}\right) \tag{4.196}$$

Reviews of these many correlations with concomitant recommendations and limitations can be found in Andrews et al. (1975), Kuo and Acharya (2012, pp. 291−304), Dahoe et al. (2013), and Lipatnikov (2013, pp. 123−134 and 143−153).

One final parameter often used in turbulent premixed flame analysis is the Damköhler number Da. The Damköhler number is defined as the ratio of a characteristic flow time and a characteristic chemical reaction time. The Damköhler number can be defined in many different ways depending on the physical situation under consideration. For premixed turbulent combustion, the Damköhler number is defined in the following way:

$$Da = \frac{t_{flow}}{t_{chemical}} = \left(\frac{\mathcal{L}_{int}}{\delta}\right)\left(\frac{S_u}{v'_{rms}}\right) \tag{4.197}$$

In dimensionless form, some turbulent flame velocity correlations take the form:

$$\frac{S_{turbulent}}{S_u} = f(Re_{int}, Da) \tag{4.198}$$

This concludes our discussion, for now, on turbulent premixed combustion.

4.9 SUMMARY

This chapter has been just an overview of the transport phenomena topics needed for the study of combustible dust hazards. The equations of change for laminar reacting flows are the foundation for the investigation of combustible dust hazards. Three essential models for combustion analysis were introduced. The Chapman−Jouguet analysis for premixed flames (infinite kinetics) provided the opportunity to introduce the concepts of the deflagration and the detonation.

The second model, the premixed flame, considered the role of finite kinetics in the premixed flame and introduced the concepts of flame thickness and burning velocity. Both of these parameters are functions of the rate of chemical reaction and the thermal diffusivity. The premixed flame is a macroscale analog for a dust deflagration. The diffusion flame analysis for a liquid fuel droplet, the third combustion model, introduced the burnout time and mass burning rate for individual fuel droplets. The diffusion flame is the microscale analog for a single burning dust particle. Although developed for gaseous or liquid fuels, these three models form the basis for understanding dust fires and explosions.

To develop models for dust combustion, we needed to address multiphase systems, thermal radiation, and turbulence. Two aspects of multiphase systems were described, porous media and dispersed particle flows. The properties of porous media play an essential role in the development of models for the smoldering behavior of dust piles. The description of dispersed multiphase systems will provide the framework for analyzing dust deflagrations.

The roles of thermal radiation and turbulence in dispersed particle systems were also explored. The analysis of thermal radiation in a participating medium is important because clouds of fine particles absorb and scatter thermal radiation far more readily than ordinary gases like air. Turbulence is important in both the creation of combustible dust clouds as well as their combustion.

Armed with these tools, we are ready to begin our investigation of combustible dust hazards by considering smoldering phenomena.

REFERENCES

Amyotte, P., Chippett, S., Pegg, M.J., 1989. Effects of turbulence on dust explosions. Prog. Energy Combust. 14, 293–310.

Andrews, G.E., Bradley, D., Lwakabamba, S.B., 1975. Turbulence and turbulent flame propagation—a critical appraisal. Combust. Flame 24, 285–304.

Annamalai, K., Puri, I.K., 2007. Combustion Science and Engineering. CRC Press, Boca Raton, FL.

ASTM E1226, 2012. Standard Test Method for Explosibility of Dust Clouds. ASTM International, West Conshohocken, PA.

Bird, R.B., Stewart, W.E., Lightfoot, E.N., 2002. Transport Phenomena, second ed. John Wiley & Sons, New York.

Borghi, R., 1985. On the structure and morphology of turbulent premixed flames. In: Casci, C., et al., (Eds.), Recent Advances in the Aerospace Sciences. Plenum Press, New York, pp. 117–138.

Chang, C.S., 1988. Measuring density and porosity of grain kernels using a gas pycnometer. Cereal Chem. 65, 13–15.

Chang, C.S., Lai, F.S., Miller, B.S., 1980. Thermal conductivity and specific heat of grain dust. Trans. ASAE 23, 1303–1306.

Churchill, S.W., 1980. The Practical Use of Theory in Fluid Flow. Book 1: Inertial Flows. Etaner Press, Thornton, PA.

Crowe, C.T., Schwarzkopf, J.D., Sommerfield, M., Tsuji, Y., 2012. Multiphase Flows with Droplets and Particles, second ed. CRC Press, Boca Raton, FL.

Dahoe, A.E., Cant, R.S., Scarlett, B., 2001a. On the decay of turbulence in the 20-liter explosion sphere. Flow Turbul. Combust. 67, 159–184.

Dahoe, A.E., Cant, R.S., Scarlett, B., 2001b. On the transient flow in the 20-liter sphere. J. Loss Prev. Process Ind. 14, 475–487.

Dahoe, A.E., Skjold, T., Roekaerts, D.J.E.M., Pasman, H.J., Eckhoff, R.K., Hanjalic, K., et al., 2013. On the application of the Levenberg–Marquardt method in conjunction with an explicit Runge–Kutta and implicit Rosenbrock method to assess burning velocities from confined deflagrations. Flow Turbul. Combust. 91, 281–317.

Drysdale, D., 1997. An Introduction to Fire Dynamics, second ed. John Wiley & Sons, New York.

Eckhoff, R., 2003. Dust Explosions in the Process Industries, third ed. Gulf Professional Publishing, Elsevier, New York.

Fan, L.-S., Zhu, C., 1998. Principles of Gas–Solid Flows. Cambridge University Press, Cambridge.

Glassman, I., Yetter, R.A., Glumac, N.G., 2015. Combustion, fifth ed. Academic Press, New York.

Hinds, W.C., 1999. Aerosol Technology: Properties, Behavior, and Measurement of Airborne Particles, second ed. Wiley-Interscience, New York.

Ibarreta, A., Marr, K., Garner, S., O'Hern, S.C., Myers, T.J., 2015. On the use of laminar burning velocities in process safety. 11th Global Congress on Process Safety. American Institute of Chemical Engineers, Austin, TX, April 27–29.

Kanury, A.M., 1975. Introduction to Combustion Phenomena. Gordon and Breach Science Publishers, New York.

Kaviany, M., 1995. Principles of Heat Transfer in Porous Media, second ed. Springer-Verlag, Berlin.

Kays, W.M., Crawford, M.E., 1980. Convective Heat and Mass Transfer, second ed. McGraw-Hill, New York.

Krantz, W.B., 2007. Scaling Analysis in Modeling Transport and Reaction Processes: a Systematic Approach to Model Building and the Art of Approximation. John Wiley & Sons, New York.

Kuo, K., 2005. Principles of Combustion, second ed. John Wiley & Sons, New York.

Kuo, K., Acharya, R., 2012. Fundamentals of Turbulent and Multiphase Combustion. John Wiley & Sons, New York.

Law, C.K., 2006. Combustion Physics. Cambridge University Press, Cambridge.

Lee, J.H.S., 2008. The Detonation Phenomenon. Cambridge University Press, Cambridge.

Lewis, B., von Elbe, G., 1987. Combustion, Flames, and Explosions of Gases, third ed. Academic Press, New York.

Linan, A., Williams, F.A., 1993. Fundamental Aspects of Combustion. Oxford University Press, Oxford.

Lipatnikov, A., 2013. Fundamentals of Premixed Turbulent Combustion. CRC Press, Boca Raton, FL.

Markstein, G.H. (Ed.), 1964. Nonsteady Flame Propagation. Pergamon Press, Oxford.

Modest, M.F., 2013. Radiative Heat Transfer, third ed. Academic Press, New York.

Nettleton, M.A., 1987. Gaseous Detonations: their Nature, Effects and Control. Chapman and Hall, London.

Nield, D.A., 1991. Estimation of the stagnant thermal conductivity of saturated porous media. Int. J. Heat Mass Transfer 34, 1575–1576.

Nield, D.A., Bejan, A., 2006. Convection in Porous Media, third ed. Springer, Berlin.

Özisik, M.N., 1973. Radiative Transfer and Interactions with Conduction and Convection. Wiley-Interscience, New York.

Pope, S.B., 2005. Turbulent Flows. Cambridge University Press, Cambridge.

Pu, Y.K., Jarosinki, J., Tal, C.S., Kauffman, C.W., Sichel, M., 1988. The investigation of the feature of dispersion induced turbulence and its effects on dust explosions in closed vessels. Twenty-Second Symposium (International) on Combustion. The Combustion Institute, Elsevier.

Pu, Y.K., Jarosinski, J., Johnson, V.G., Kauffman, C.W., 1990. Turbulence effects on dust explosions in the 20-liter spherical vessel. Twenty-Third Symposium (International) on Combustion. The Combustion Institute, Elsevier.

Pu, Y.K., Li, Y.-C., Kauffman, C.W., Bernal, L.P., 1991. Determination of turbulence parameters in closed explosion vessels. Progress in Astronautics and Aeronautics: Dynamics of Deflagrations and Reactive Systems: Heterogeneous Combustion Volume 132, 107−123.

Siegel, R., Howell, J.R., 1981. Thermal Radiation Heat Transfer, second ed. McGraw-Hill, New York.

Sirignano, W.A., 2010. Fluid Dynamics and Transport of Droplets and Sprays, second ed. Cambridge University Press, Cambridge.

Slattery, J., 1981. Momentum, Energy, and Mass Transfer in Continua, second ed. Robert Krieger Press, Cambridge.

Spalding, D.B., 1953. The combustion of liquid fuels. Fourth Symposium (International) on Combustion. The Combustion Institute, Williams & Wilkins.

Spalding, D.B., 1979. Combustion and Mass Transfer. Pergamon, Oxford.

Strehlow, R.A., 1984. Combustion Fundamentals. McGraw-Hill, New York.

Stroshine, R., 1998. Physical Properties of Agricultural Materials and Food Products. Purdue University, West Lafayette, IN.

Tavman, S., Yolci, P., 1998. Thermal conductivity measurements of granular and powdered foods. In: Proceedings of the 3rd Karlsruhe Nutrition Symposium European Research towards Safer and Better Food, Karlsruhe, Germany, pp. 451−460.

Tennekes, H., Lumley, J.L., 1980. A First Course in Turbulence. The MIT Press, Cambridge, MA.

Thompson, P.A., 1972. Compressible-Fluid Dynamics. McGraw-Hill, New York.

Turns, S.R., 2012. An Introduction to Combustion, third ed. McGraw-Hill, New York.

van de Hulst, H.C., 1981. Light Scattering by Small Particles. Dover Publications, Mineola, NY.

Wallis, G.B., 1969. One-Dimensional Two-Phase Flow. McGraw-Hill, New York.

Whitaker, S., 1969. Advances in theory of fluid motion in porous media. Ind. Eng. Chem. 61, 14−28.

Whitaker, S., 1983. Fundamental Principles of Heat Transfer. Robert E. Krieger Publishing Company, Malabar, FL.

Whitaker, S., 1999. The Method of Local Volume Averaging. Kluwer Academic Publishers, Dordrecht, The Netherlands.

Williams, F.A., 1985. Combustion Theory, second ed. Benjamin/Cummings Publishing Company, Menlo Park, CA.

Zeldovich, Ya.B., Barenblatt, G.I., Librovich, V.B., Makhviladze, G.M., 1985. The Mathematical Theory of Combustion and Explosions, translated from Russian by McNeill, D. Consultants Bureau, Plenum Publishing Corporation, Washington, DC.

Smoldering phenomena

Under certain favorable conditions, combustible dust deposits can undergo smoldering. This is a very important hazard scenario to understand, because smoldering is oftentimes difficult to detect and may not reveal itself until it leads to a catastrophic event: exposure to a toxic atmosphere, a dust fire, or a dust explosion. The ignition energy required to initiate smoldering is much lower than the energy level required to ignite a flaming fire, and yet a smoldering fire can abruptly transition to a flaming fire (Rein, 2009). Thus, smoldering presents a hazard scenario that can progress without warning from a low hazard level to a high hazard level.

Smoldering is flameless combustion. Similar to a deflagration, smoldering propagates through a fuel bed as a combustion wave with a finite speed. Smoldering causes charring of the solid fuel and usually creates a glowing ember or coal; it is frequently accompanied by the generation of smoke. Smoldering can be initiated by an external ignition source or, for certain materials, by self-heating and spontaneous ignition. Fundamentally, smoldering requires that the fuel forms a porous char during combustion. Agricultural grain dusts, food ingredient dusts, wood dust, and coal dust have all demonstrated the ability to sustain smoldering combustion (Toong, 1983, p. 293). However, many other porous fuels also exhibit smoldering behavior including cellulose insulation, leaf litter in wildland settings, in situ coal beds, cotton fabric, and polyurethane foam. Despite the diversity of chemical compositions and physical forms of the fuels, the smoldering behavior of these disparate materials is remarkably similar to each other. They differ only in the details.

Smoldering is a very inefficient form of combustion. It generates much higher levels of products of incomplete production. For carbonaceous fuels, smoldering generates high levels of carbon monoxide, other partially oxidized species, and smoke (with particle diameters less than 1 μm in mean diameter) (Purser, 2002, pp. 2-83 to 2-171). These substances are flammable and can create explosive atmospheres in a confined space. Furthermore, carbon monoxide is a potent toxic compound. Thus, undisturbed smoldering piles of carbonaceous fuels can create both explosive and toxic atmosphere hazards.

If a smoldering dust pile is disturbed, it can lead to a dust deflagration in the form of a flash fire or an explosion (Palmer, 1973, pp. 273–297). Cross and Farrer (1982, p. 17) relate a case study where firefighters dislodged a smoldering layer of sawdust caked onto the wall of a dust collector. Upon dislodging the smoldering fuel, the material fell to the floor and erupted into a flash fire injuring three of the firefighters, one fatally. The CCPS/AIChE combustible dust book

(2005, pp. 117−124) presents a discussion of particulate fire scenarios commonly encountered in industry. Amyotte (2013, pp. 92−99) cautions against underestimating the hazard potential of dust layer ignition and fire. Eckhoff (2005, pp. 175−196) discussed several case studies involving gas and dust explosions caused by smoldering combustion in powder layers and deposits. Ohlemiller has reviewed the mathematical modeling of smoldering and has summarized key results from empirical studies (Ohlemiller, 1985, 2002). Rein (2009) reviewed the subject of smoldering and its occurrence in both natural and man-made environments.

One unfortunate feature of this chapter is the use of the Greek symbol δ to represent both characteristic lengths in discussions of smoldering phenomena and dimensionless numbers in the Semenov and Frank−Kamenetskii thermal runaway problems. This issue arises due to historical usage in the literature. The meaning of the symbol should be clear from its context.

5.1 PHYSICAL DESCRIPTION OF SMOLDERING

A combustible dust deposit can be visualized as a porous medium. The smolder wave is analogous to a deflagration wave, but there is an important difference. The structure of a deflagration wave (premixed flame) consisted of a preheat zone and a reaction zone. A smolder wave is different because it propagates not through a gaseous mixture but through a porous bed of fuel. The structure of a smolder wave is defined by the changes that occur in the solid matrix. At the simplest level of description, the thermal structure of a smolder wave consists of a pyrolysis zone and a char zone (Fig. 5.1) (Moussa et al., 1977). The temperature

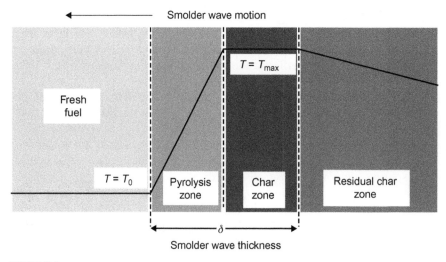

FIGURE 5.1

Thermal structure of a smolder wave.

rises in the pyrolysis zone, reaches a maximum, steady state value in the char zone, and decreases gradually in the residual zone.

The pyrolysis reactions are endothermic (in the absence of oxygen) and the char oxidation reactions are exothermic. Thus, there are a lot of different kinds of chemical reactions involved in smoldering including the pyrolysis of the solid fuel (which generates gas phase fuel vapor), homogeneous oxidation reactions with the fuel vapor, and heterogeneous char oxidation.

A smolder wave is a fuel-rich, ventilation-limited form of combustion. The combustion mechanism of smoldering is highly dependent on the rate and direction of the oxygen supply (Ohlemiller, 2002). The rate of the oxygen supply is largely a function of the permeability of the porous medium, the presence and direction of any airflow, and the orientation of the fuel bed with respect to gravity (horizontal or vertical). The direction of the oxygen supply—cocurrent with the smolder wave direction (forward smolder) or countercurrent to it (reverse smolder)—has a significant impact on the structure of the smolder wave (Ohlemiller, 1985, pp. 299–301). The direction of oxygen flow into the smolder wave is determined primarily by the geometry of the porous body, the degree of exposure of the porous body's boundary to the atmosphere, and the location of the ignition source. Faster smolder wave speeds are associated with greater oxygen flow rates.

In order for a dust deposit to smolder, it must satisfy the requirement that the fuel particles form a char upon heating (Drysdale, 1999, p. 275). Additional factors of the fuel particles that can influence the speed of propagation include the dust particle diameter and the depth or thickness of the deposit. In fact, Palmer and Tonkin found that these parameters are linked (Palmer and Tonkin, 1957). Palmer and Tonkin found that for a given dust there was a minimum dust layer thickness that would support the propagation of a smolder wave and that this minimum dust layer thickness increased with increasing particle diameter. This latter relationship is suggestive of the multiphase mixture rule of thumb that for the porous body to be treated as a pseudo-continuum, the characteristic thickness of the porous body δ_L has to be much greater than the particle diameter d_p, $\delta_L \gg d_p$.

5.2 IGNITION BY SELF-HEATING AND THERMAL RUNAWAY

Even at ambient temperature, a combustible dust is an exothermic, chemically reacting system. The Arrhenius temperature dependence for chemical reactions suggests that at ambient temperatures a combustible dust pile is slowly oxidizing. Such a body is said to be self-heating. If the rate of oxidation is slow enough, the heat of reaction is completely transferred to the surroundings by conduction, convection, and radiation. In part, the dissipation of heat depends on the geometry of the body, its thermal properties, and the ambient temperature. For a fixed ambient temperature, the rate of heat loss decreases with increasing pile size due to the thermal conductivity of the body. For a porous body, the size effect is amplified even more due to the presence of pore space filled with air, a natural thermal insulator. The finite thermal

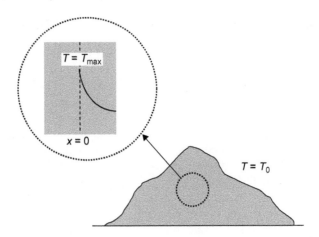

FIGURE 5.2

Depiction of a self-heating deposit of combustible dust.

conductivity of the body leads to the formation of a temperature field inside the body with the maximum temperature at the center of the body. The temperature monotonically decreases from the center of the body to the boundary of the body that is in contact with its surroundings. Fig. 5.2 illustrates the basic problem.

As the size of the body is increased, eventually the body is unable to dissipate the heat of reaction and the temperature increases with the maximum temperature occurring at the center of the body. The elevated temperature within the body causes the oxidation rate to increase which, in turn, increases the rate of heat generation. This causes a positive feedback loop with higher temperatures causing higher exothermic reaction rates which cause higher temperatures, etc. Eventually the center temperature becomes so large that it causes ignition of the fuel bed. This process of uncontrolled temperature rise is called a thermal runaway.

It turns out that for a given size and shape of a self-heating body there is a special temperature called the critical temperature where the rate of heat generation is just matched by the rate of heat loss. If the ambient temperature is below the critical temperature, the body is said to be in a subcritical state; if above the critical temperature, the body is in a supercritical state. Ignition occurs when the self-heating body achieves a supercritical state (Linan and Williams, 1993, pp. 86–91). Fig. 5.3 depicts the three thermal states of a self-heatingbody: subcritical, critical, and supercritical.

Note that at the critical state, there can be a significant maximum temperature at the center of the self-heating body (significant in the sense that the temperature can be high enough to cause discoloration or even charring at the center).

An alternative, equivalent thought experiment with self-heating bodies is to consider the ambient temperature fixed and to increase the size of the self-heating body. In this manner, you will proceed through the subcritical, critical, and supercritical states as the characteristic dimension of the body increases. In particular, the characteristic size of the self-heating body is the thinnest dimension for heat

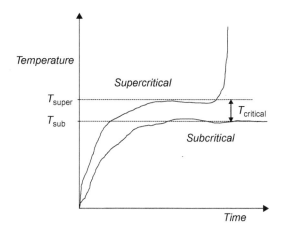

FIGURE 5.3

Temperature profiles for a self-heating body in subcritical and supercritical states.
The critical temperature occurs between the subcritical and supercritical states.

conduction. An important safety consideration to remember is that the critical temperature of a self-heating body decreases as its thickness increases: piles with larger thicknesses are more hazardous than thin piles.

In addition to the oxidation of a combustible dust solid, many factors can be responsible for facilitating the self-heating phenomenon: microbiological activity in organic materials, moisture absorption in cellulosic materials, and the presence of contaminants. Self-heating and thermal runaway is observed in a number of practical systems like linseed oil soaked rags, compost piles, and coal stockpiles. Bowes (1984) presents a comprehensive treatise on both theoretical and experimental aspects of self-heating and thermal runaway. A more up-to-date presentation of experimental methods is in the CCPS/AIChE combustible dust reference (CCPS/AIChE, 2005, Chapter 4).

There are two classical models for the analysis of spontaneous ignition phenomena: the Semenov model and the Frank−Kamenetskii model. These two models are discussed next followed by a brief survey of hot surface ignition.

5.2.1 SEMENOV MODEL

The Semenov model is based on the assumption of a uniform temperature within the self-heating body. For a solid material or a porous medium composed of combustible dust particles, this is akin to assuming that the thermal conductivity of the body is infinite. This model works better with fluids under agitation. We will explore this model as a simpler analog of the self-heating porous medium problem. In the next section we will consider the behavior of a self-heating body with finite thermal conductivity, the Frank-Kamenetskii problem. I will follow the presentation in Quintiere (2006, pp. 80−84; be mindful of errors in Quintiere's book which

I have corrected in this derivation). Additional information can be found in Glassman et al. (2015, pp. 368−373) and Varma et al. (1999, Chapter 3).

Consider a self-heating porous medium composed of a combustible dust. The volume and surface area of the body are fixed. The body is immersed in a constant ambient temperature T_∞. We are interested in exploring the temperature behavior of the system as influenced by heat generation and heat loss effects. I begin with the control volume form of the energy balance on the body.

$$\rho\, V\, C_{\text{mix}} \frac{dT}{dt} = \dot{q}'''_{\text{gen}} - \dot{q}'''_{\text{loss}} \tag{5.1}$$

$$\dot{q}'''_{\text{gen}} = \omega'''_{\text{F}}\, \rho\, V\, Q = A\, \rho\, V\, Q \exp\left(\frac{-E}{R_g\, T}\right) \tag{5.2}$$

$$\dot{q}'''_{\text{loss}} = h S_{\text{A}}(T - T_\infty) \tag{5.3}$$

Note that since the system of interest is a porous medium, the heat capacity C_{mix} is the mixture heat capacity defined as $C_{\text{mix}} = \rho_m^{-1}\left[\alpha_s\, \rho_s\, C_s + (1 - \alpha_s)\, \rho_{\text{air}} C_{\text{p,air}}\right]$. In the heat generation term, the form of the chemical reaction rate is a zeroth order reaction. Experience has shown that in many thermal runaway scenarios, the temperature behavior dominates the reaction rate term. The reactant conversion is so small that it has a negligible effect on the reaction rate in the early stages of the thermal runaway. The heat loss term is modeled with a convective heat transfer coefficient.

We invoke a mathematical insight developed by Semenov which enables us to better cope with the extremely nonlinear character of the Arrhenius term (Quintiere, 2006, pp. 80−84). Semenov's insight begins with the realization that the critical temperature will be approximately equal to the ambient temperature $T_c \cong T_\infty$. The reciprocal temperature that appears in the Arrhenius term can then be expanded with a first-order Taylor series about T_∞.

$$\frac{1}{T} \cong \frac{1}{T_\infty}\left[1 - \left(\frac{T - T_\infty}{T_\infty}\right) + \ldots\right] \tag{5.4}$$

$$\frac{E}{R_g\, T} \cong \frac{E}{R_g\, T_\infty}\left[1 - \left(\frac{T - T_\infty}{T_\infty}\right)\right] \tag{5.5}$$

Define the following dimensionless temperature variable θ and substitute into the Arrhenius expression:

$$\theta \equiv \frac{E}{R_g\, T_\infty}\left(\frac{T - T_\infty}{T_\infty}\right) \tag{5.6}$$

$$\exp\left(\frac{-E}{R_g\, T}\right) = \exp\left\{\left(\frac{-E}{R_g\, T_\infty}\right)\left[1 - \left(\frac{T - T_\infty}{T_\infty}\right)\right]\right\} = \exp\theta \exp\left(\frac{-E}{R_g\, T_\infty}\right) \tag{5.7}$$

Define a dimensionless time variable using a characteristic time based on the heat loss term

$$\hat{t} = \frac{t}{t_{\text{loss}}}, \quad t_{\text{loss}} = \frac{\rho\, V\, C_{\text{mix}}}{h\, S_{\text{A}}} \tag{5.8}$$

The energy balance equation in terms of the dimensionless variables becomes

$$\frac{d\theta}{d\hat{t}} = \delta \exp \theta - \theta \tag{5.9}$$

The Semenov parameter δ is defined as

$$\delta = \left[\frac{\rho V Q A \exp(-E/R_g T_\infty)}{h S_A T_\infty} \right] \left(\frac{E}{R_g T_\infty} \right) \tag{5.10}$$

The Semenov parameter is a dimensionless ratio of the heat generation rate to the cooling rate. In general terms, if the Semenov parameter is much smaller than 1, $\delta \ll 1$, the rate of cooling exceeds the rate of heat generation and the reaction is "quenched" (ie, the reaction rate becomes negligibly small). If the Semenov parameter becomes much greater than 1, $\delta \gg 1$, then the heat generation rate far exceeds the cooling rate and the result is a thermal runaway. The use of the Semenov parameter as a criterion for thermal stability will now be further refined.

Now that the energy balance is in dimensionless form, let us work our way toward its solution. In the case of thermal runaway analysis, it is not necessary to solve the energy balance to obtain the temperature history. Instead, we will evaluate the critical state of the self-heating body. This approach will give us a refined criterion for thermal stability. The critical state is the state at which the cooling rate just balances the heat generation rate. Finding the critical state is made all the more difficult because of the nonlinearity of the Arrhenius term. Fig. 5.4 shows how the nonlinear heat generation term competes with the heat loss term.

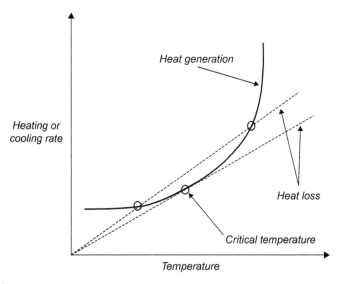

FIGURE 5.4

Heat generation and heat loss rate for the Semenov analysis.

Two conditions are required to determine the critical state:

$$\dot{q}''_{gen} = \dot{q}''_{loss} \quad \text{and} \quad \frac{d\dot{q}''_{gen}}{dT} = \frac{d\dot{q}''_{loss}}{dT} \tag{5.11}$$

In terms of the dimensionless variables, these conditions become

$$\delta_c \exp(\theta_c) = \theta_c \quad \text{and} \quad \delta_c \exp(\theta_c) = 1 \tag{5.12}$$

Solving for θ_c and δ_c yield the simple results

$$\theta_c = 1 \quad \text{and} \quad \delta_c = e^{-1} \cong 0.368 \tag{5.13}$$

Thus, a more refined criterion for thermal stability is that $\delta \leq 0.368$. If $\delta > 0.368$, a thermal runaway will result. If you know or can measure the properties of a self-heating body, for a specified volume and surface area you can use the critical Semenov parameter δ_c to calculate the critical ambient temperature.

The Semenov model can also be solved to give an estimate of the shortest possible time to a thermal runaway (Quintiere, 2006, pp. 127–129). This is obtained by assuming that the self-heating body is perfectly insulated so that it is an adiabatic system (zero heat loss). Returning to the dimensionless form of the energy balance, setting $\theta = 0$ is equivalent to setting $\dot{q}''_{loss} = 0$; thus

$$\frac{d\theta}{d\hat{t}} = \delta \exp \theta \rightarrow \int_0^{\hat{t}_{TR}} d\hat{t} = \frac{1}{\delta} \int_0^1 e^{-\theta} d\theta \rightarrow t_{TR} = \frac{(1 - e^{-1})}{\delta} \tag{5.14}$$

The limits of integration are based on the initial condition, $\theta(\hat{t}=0)=0$, and achievement of the critical condition, $\theta(\hat{t}=\hat{t}_{TR})=1$. The time to an adiabatic thermal runaway is

$$\hat{t}_{TR} = \frac{0.632}{\delta} \tag{5.15}$$

In dimensional terms, the adiabatic time to a thermal runaway is given by the equation:

$$t_{TR} = \left[\frac{\rho C_{mix} T_\infty}{QA \exp(-E/R_g T_\infty)}\right]\left(\frac{R_g T_\infty}{E}\right) \tag{5.16}$$

This is the minimum time for a self-heating body to achieve a thermal runaway. With heat losses, the time to thermal runaway will increase.

EXAMPLE 5.1

Calculate the Semenov parameter and the adiabatic time to thermal runaway for a self-heating body shaped as a cylinder with a radius of 1 m and a height of 5 m. The ambient temperature is 30°C. The material has the following properties: $\rho = 620$ kg/m³, $C_{mix} = 0.84$ kJ/kg-K, $A = 3 \times 10^6$ s⁻¹, $E = 70$ kJ/mol, $Q = 30$ MJ/kg, and $h = 10$ W/m²-K.

Solution

The solution to this example is a straightforward application of the Semenov model. The volume of the body is $\pi R^2 H = 5\pi \cong 15.7 \text{ m}^3$. The surface area, neglecting the top and bottom surfaces, is $\pi R H = 15\pi \cong 47.1 \text{ m}^2$.

First, to calculate the Semenov parameter we use the following equation:

$$\delta = \left[\frac{\rho V Q A \exp(-E/R_g T_\infty)}{h S_A T_\infty} \right] \left(\frac{E}{R_g T_\infty} \right)$$

Inserting the values given for the parameters

$$\delta = \left[\frac{620 \text{ kg/m}^3 \cdot 15.7 \text{ m}^3 \cdot 3.0 \times 10^6 \text{ J/kg} \cdot 3 \times 10^6 \text{ s}^{-1} \cdot \exp(-70,000/(8.314 \cdot 303))}{10 \text{ W/m}^2\text{K} \cdot 47.1 \text{ m}^2 \cdot 303 \text{ K}} \right]$$

$$\cdot \left(\frac{70,000}{8.314 \cdot 303} \right) = 14.6$$

The Semenov parameter $\delta \gg 1$, indicating that the body is supercritical and will experience a thermal runaway.

The next step is to calculate the adiabatic time to a thermal runaway.

$$t_{TR} = 0.632 \left[\frac{C_{mix} T_\infty}{Q A \exp(-E/R_g T_\infty)} \right] \left(\frac{R_g T_\infty}{E} \right)$$

Again inserting the values given for the various parameters:

$$t_{TR} = \left[\frac{0.632 \cdot (0.84 \text{ kJ/kgK}) \cdot 303 \text{ K}}{3 \times 10^6 \text{ J/kg} \cdot 15.7 \text{ m}^3 \cdot 3 \times 10^6 \text{ s} \cdot \exp(-70,000/(8.314 \cdot 303))} \right] \left(\frac{8.314 \cdot 303}{70,000} \right)$$

$$t_{TR} = 4.8 \times 10^{-2} \text{ s} = 48 \text{ ms}$$

The adiabatic time to a thermal runaway is extremely short. Heat losses will extend this time interval, but these results clearly indicate that this material should not be stored in such a large quantity.

5.2.2 FRANK−KAMENETSKII MODEL

The Frank−Kamenetskii model is based on the assumption that the self-heating body has a finite thermal conductivity. Thus, a temperature profile forms within the body. The exact nature of the temperature profile will depend on the size and shape of the body as well as the boundary conditions. We will consider only the simplest geometry, an infinite rectangular slab with a finite half-thickness of x_0, so that we can focus on the fundamental concepts. There is an extensive literature on the Frank–Kamenetskii model. Good summaries can be found in Beever (1995), Babrauskas (2003), and Annamalai and Puri (2007, pp. 621–631). The definitive reference on the subject is Bowes (1984).

The system of interest is self-heating porous medium shaped as a rectangular slab with a half-thickness of x_0 and of infinite extent in the other two coordinates (Fig. 5.5).

The energy equation for this problem requires terms for heat conduction and heat generation.

$$k_{eff} \frac{d^2 T}{dx^2} = -\rho Q A \exp\left(\frac{-E}{R_g T}\right) \tag{5.17}$$

Heat losses are accounted for through the boundary condition at the interface between the self-heating body and its surroundings. For simplicity, we assume that the temperature at the surface is equal to the ambient temperature T_∞. Thus the boundary conditions are

$$\frac{dT}{dx}(x = 0) = 0 \text{ (symmetry)} \quad \text{and} \quad T(x = x_0) = T_\infty \tag{5.18}$$

Introducing the temperature approximation for the Arrhenius term

$$\theta \equiv \frac{E}{R_g T_\infty} \left(\frac{T - T_\infty}{T_\infty}\right) \tag{5.19}$$

$$\exp\left(\frac{-E}{R_g T}\right) = \exp \theta \exp\left(\frac{-E}{R_g T_\infty}\right) \tag{5.20}$$

The dimensionless spatial variable $\hat{x} \equiv x/x_0$ is substituted into the energy equation with the dimensionless temperature variable giving the nondimensional energy equation

$$\frac{d^2 \theta}{d\hat{x}^2} = -\delta e^\theta \tag{5.21}$$

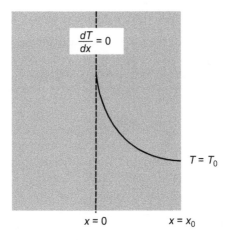

FIGURE 5.5

Rectangular slab geometry for a self-heating porous medium.

$$\delta = \left[\frac{\rho \, x_0^2 \, Q \, A \, \exp(-E/R_g \, T_\infty)}{k_{\text{eff}} \, T_\infty} \right] \left(\frac{E}{R_g \, T_\infty} \right) \tag{5.22}$$

The dimensionless parameter δ is the Frank−Kamenetskii parameter. It is the ratio of a heat generation term and the heat loss by conduction. The solution to this equation requires numerical methods. The complete details are available in Bowes (1984, pp. 24−34). The numerical solution yields two numerical results, the critical value $\delta_c = 0.878$ and the temperature at the center of the body $\theta_0 = 1.12$. Given the layer thickness and properties of a self-heating body, you can use the critical Frank−Kamenetskii parameter to calculate the critical ambient temperature. Just like the Semenov analysis, the critical ambient temperature represents the maximum temperature that the self-heating body can be exposed to without undergoing a thermal runaway.

Alternatively, with the ambient temperature fixed, you can calculate the critical half-thickness for a given geometry. As indicated by Beever, the range of critical half-thickness is bounded by the value for the isothermal boundary condition x_0 and by the adiabatic boundary condition where the critical half-thickness would be $x_0/2$ (Beever, 1995, p. 2-187).

The solution of the energy equation depends on the shape of the self-heating body. Table 5.1 summarizes the critical parameters for several geometries (from Beever, 1995, p. 2-181; Babrauskas, 2003, pp. 380−381). In each case, the boundary condition is isothermal with $T = T_\infty$.

Babrauskas (2003, p. 381) noted that the approximate formula for the critical Frank−Kamenetskii parameter of a shallow cone is useful because when poured onto a horizontal surface granular materials will assume the shape of a shallow cone.

Additional critical F−K parameters for different geometries have been summarized in Bowes (1984, pp. 71−123). For arbitrarily shaped, nonhollow bodies it has been shown that a sphere of equivalent volume has a lower critical ambient temperature. Therefore, if a solution for the critical parameters for a particular geometry of interest has not been published, substituting the critical parameters for a sphere of equivalent volume will yield a lower (more conservative) critical ambient temperature (Bowes, 1984, pp. 77−86).

Table 5.1 Critical Frank−Kamenetskii Parameters for Several Self-Heating Bodies (Beever, 1995, p. 2-181; Babrauskas, 2003, p. 381)

Geometry	Characteristic Dimension	δ_c	θ_0
Infinite rectangular slab	Thickness $2x_0$	0.878	1.12
Cube	Edge $2x_0$	2.52	1.89
Infinite cylinder	Radius r	2.00	1.39
Equicylinder	Radius r, height $2r$	2.76	1.78
Sphere	Radius r	3.32	1.61
Shallow cone	Base radius $= a$, height $= d$, $p = a/d$, $p \geq 0.5$	$1.2 + \dfrac{2}{p^2}$	N/A

EXAMPLE 5.2

Calculate the Frank–Kamenentskii parameters for the following accumulations of combustible dust. Assume that the ambient temperature is 30°C. The material has the following properties: $\rho = 620$ kg/m^3, $C_{mix} = 0.84$ kJ/kg-K, $A = 3 \times 10^6$ s^{-1}, $E = 70$ kJ/mol, $Q = 30$ MJ/kg, and $k_{eff} = 0.050$ W/m-K.

A. An accumulation of grain dust has been deposited on a large floor with a thickness of 3 mm. Will the accumulation self-heat to the point of ignition?
B. An accumulation of the same grain dust has assumed the shape of a conical pile with an angle of repose of 30°. The base diameter of the pile is 3 m. Will the accumulation self-heat to the point of ignition?

Solution

The task is to calculate the F–K parameter for each geometry and compare its value with the critical F–K value δ_c for that geometry.

For case A, we first calculate the F–K parameter:

$$\delta = \left[\frac{\rho \, x_0^2 \, Q \, A \, \exp(-E/R_g \, T_\infty)}{k_{eff} \, T_\infty} \right] \left(\frac{E}{R_g \, T_\infty} \right)$$

Inserting the values given for the parameters yields

$$\delta = \left[\frac{620 \text{ kg}/m^3 (0.003 \text{ m})^2 \cdot 3.0 \times 10^6 \text{ J/kg} \cdot 3 \times 10^6 \text{ s}^{-1} \cdot \exp(-70,000/(8.314 \cdot 303))}{0.050 \text{ W}/mK \cdot 303 \text{ K}} \right]$$

$$\cdot \left(\frac{70,000}{8.314 \cdot 303} \right) = 7.88 \times 10^{-2}$$

$$\delta = 0.0788 \ll 0.878 \text{(rectangular slab)} \rightarrow \text{subcritical}$$

For case B, we must first determine the critical F–K parameter for a cone with the specified geometry. First we need to calculate the height of the cone knowing the angle of repose: $d = a \tan \theta_r = 1.5$ m $\cdot \tan 30° = 0.866$ m. This give a value of $p = a/d = 1.5/0.866 = 1.73$.

For a shallow cone, $\delta_c = 1.2 + \frac{2}{p^2} = 1.87$. Substituting the values for the various parameters gives us

$$\delta = \left[\frac{620 \text{ kg}/m^3 (0.866 \text{ m})^2 \cdot 3.0 \times 10^6 \text{ J/kg} \cdot 3 \times 10^6 \text{ s}^{-1} \cdot \exp(-70,000/(8.314 \cdot 303))}{0.050 \text{ W}/mK \cdot 303 \text{ K}} \right]$$

$$\cdot \left(\frac{70,000}{8.314 \cdot 303} \right) = 6.57 \times 10^3$$

$$\delta = 6570 \gg 1.87 \text{(shallow cone)} \rightarrow \text{supercritical}$$

Thus, the thin layer is subcritical and would not be expected to undergo a thermal runaway. The conical pile, on the other hand, is supercritical and can be expected to undergo a thermal runaway.

5.2.3 **SURVEY OF ILLUSTRATIVE RESULTS ON IGNITION BEHAVIOR**

A wide range of studies have been published on self-heating and thermal run-away. These studies are essentially calorimetry studies, that is, the material under study is subjected to specified thermal conditions to observe the thermal response of the material. Bowes (1984, Chapters 5 and 6) gives a comprehensive discussion on calorimetric techniques for extracting Arrhenius kinetic parameters from self-heating experiments. Though it is a bit dated, Bowes has reviewed a broad range of studies on the self-heating of charcoal, coal, cellulose-based materials, agricultural grains, and oilseeds (Bowes, 1984, Chapters 8 and 9).

Agricultural grains, feed products, and dust are very susceptible to self-heating (Aldis and Lai (1979); Mills, 1989). In addition to the storage volume and ambient temperature, the self-heating behavior depends on many factors including moisture content of the material and the degree of contamination by mold spores. The greatest safety concern with these materials is the potential to undergo a thermal runaway. However, an additional concern is the potential economic loss caused by subcritical self-heating due to biological processes. Even if the material never goes supercritical, modest heating to temperatures on the order of 55°C can irreversibly damage grains. A common problem scenario with these materials is the creation of a cohesive mass that gets trapped in the storage bin. The stagnant material has a longer residence time in the silo than the free-flowing material and may go unnoticed for a full growing season. The stagnant material is an ideal candidate for causing self-heating problems. Aeration of storage bins is an accepted control to prevent or limit self-heating of the stored commodity.

The self-heating of coal in utility plant stockpiles is a well-known problem (Nelson and Chen, 2007). While supercritical coal piles present a serious hazard, even subcritical heating can lead to significant economic loss. Coal-fired power plants and coking facilities may maintain coal storage piles for as much as 90 days. These piles are normally stored outdoors due to their very large size (typically from 1000 to 7000 metric tonnes). Here, instead of aeration, compaction has been successfully used to limit self-heating by depriving the fuel of the oxygen it needs to promote self-heating and ultimately initiate combustion. One of the more important conclusions presented in this paper is that while there are many papers published with kinetic parameters for coals of various ranks and compositions, it is rare for investigators to report the uncertainty associated with these kinetic parameters. Thus, one may find themselves using kinetic data with little or no idea of its reliability.

Several investigators have published empirical studies on the hot surface ignition temperature of various combustible dusts. This experimental configuration is of practical importance in industrial settings where combustible dust may settle on electrical enclosures or equipment. Palmer and Tonkin (1957) published one of the earlier studies in which they examined the effect of layer thickness on the ignition temperature. They tested cork, wood, and coal dusts deposited in conical heaps with depths ranging from 25 to 125 mm on a temperature-controlled hot plate. Ignition was determined by the observation of smoldering at the top surface of the dust deposit. The observed ignition temperatures ranged from 210°C to 320°C. They also tested carbonized wood sawdust and noted that this material exhibited hot surface ignition temperatures approximately 30°C lower than untreated wood sawdust. Consistent with the Frank—Kamenetskii theory, thicker dust piles had lower ignition temperatures than thin piles; the data correlated well with ignition temperature (°C) versus the reciprocal thickness of the dust layer (cm). Palmer and Tonkin also noted in their results that for the same material, the ignition temperature increases with increasing average particle diameter.

Bowes and Townsend (1962) conducted hot surface ignition tests with beech sawdust and analyzed hot surface ignition data using the F—K self-heating model. They used a temperature-controlled hot plate and placed the dusts onto the heater as a flat layer. Dust layer thicknesses ranged from 3 to 25 mm. Ignition was determined by observing a temperature rise greater than the hot plate setpoint temperature. Tests were terminated when smoldering was observed at the top surface of the layers. Their results demonstrated that the thicker the dust layer, the lower the ignition temperature. With layer thicknesses of 3 mm or less, self-heating did not occur. Ignition temperatures ranged from 270°C to 370°C. Bowes and Townsend did not observe a clear trend in ignition temperature with changing particle diameter. They were able to correlate their data with the F—K model and extract Arrhenius kinetic parameters.

Miron and Lazzara (1988) performed hot surface ignition tests with fuel (coal, oil shale), agricultural materials (lycopodium spores, cornstarch, and a grain dust mixture consisting of corn and soybean dust), and brass powder (coated with stearic acid, $C_{17}H_{35}COOH$, to prevent oxidation). They noted the practical safety concern associated with hot surface ignition of combustible dusts, commenting on the U.S. federal regulation that the surface temperatures of permissible electrical equipment in underground coal mines must not exceed 150°C. Their experimental configuration was a flat layer of dust deposited onto a temperature-controlled hot plate. Ignition was determined to occur when the dust sample's central temperature attained 50°C or higher than the setpoint temperature of the hot plate. With the coal samples, ignition was accompanied by the emission of acrid smoke and gases. With the lycopodium spores, charring and the evolution of smoke was observed. The agricultural materials were more difficult to test. Both the cornstarch and the grain dust samples had a tendency to char and swell at the hot plate surface thus disrupting the temperature measurements (the thermocouples were displaced by the material motion). The cornstarch did not exhibit smoldering behavior, but the grain dust did in the form of glowing combustion. The brass powder self-heated to the point of ignition and transitioned to flaming combustion.

In their testing, Miron and Lazzara found that hot surface ignition temperatures decreased as particle size decreased, consistent with the results of Palmer and Tonkin. Ignition temperatures decreased and approached an asymptotic value with increasing depth of dust layer.

5.3 **THEORIES OF SMOLDER WAVE PROPAGATION**

Investigations into smolder wave propagation have revealed a rich diversity of smolder wave behavior (Ohlemiller, 1981, 2002). By smolder wave behavior, I mean the movement of the smolder front, the temperature and concentration profiles within the wave, and the interaction of the smolder wave with its surroundings. The range of smoldering behavior can be organized by four principal factors:

- Dimensionality of the smolder wave propagation: one-dimensional or multidimensional
- Direction of airflow compared to the direction of smolder wave propagation
- Orientation of the fuel bed with respect to gravity: horizontal or vertical
- Location and extent of bounding, impermeable surfaces.

This classification scheme is by no means unique.

The dimensionality of the smolder wave propagation is determined largely by two factors: the relative thickness of the fuel bed δ_L in comparison to the thickness of the smolder wave δ_S, and by the location of the ignition source. The smolder wave thickness depends somewhat on the particle size, but generally $\delta_S \approx 10$ mm. The fuel bed can be considered thin if $\delta_L < \delta_S$. Smolder wave propagation in thin fuel beds tends to be one dimensional. If $\delta_L \gg \delta_S$, the propagation tends to be multidimensional. If ignition occurs at the end of a thin fuel bed, the propagation tends to be one dimensional. If ignition occurs somewhere along the bed length (but not the edge) of a thick fuel bed, the smolder wave propagation is multidimensional. An additional consideration is that the smolder wave thickness must be small compared to any curvature of the smolder front (Ohlemiller, 1985, p. 300).

In the absence of airflow, natural convection and diffusion are the primary transport mechanisms for oxygen transport. With airflow, convection assumes a greater importance. If the airflow is in the opposite direction as the smolder wave motion, the configuration is called *reverse propagation* or *opposed propagation*. Air must flow through the unburnt fuel column toward the combustion zone. For an observer traveling with the smolder wave (a Lagrangian perspective), the flow of fresh fuel and oxygen into the wave is concurrent. The wave structure itself bears a certain resemblance to a gaseous deflagration wave. The pyrolysis and oxidation fronts are coupled and travel as one.

If the airflow is in the same direction as the motion of the smolder wave, it is called *forward propagation*. From the Lagrangian perspective, the flow of oxygen is countercurrent to the flow of fresh fuel. The wave structure for forward propagation is more complex than with reverse propagation. The pyrolysis and oxidation fronts form two distinct fronts (Rein, 2009). All other things being equal, forward propagation rates tend to be slower than reverse propagation (Ohlemiller, 2002). In real-world settings, smoldering tends to be multidimensional with characteristics of both forward and reverse propagations (Ohlemiller, 1985, p. 301).

The orientation of the fuel bed may have an impact on the significance of buoyancy and the hydraulic resistance to flow caused by the permeability of the porous medium. Shallow horizontal beds will present less resistance to vertical flow than tall vertical beds. Because smoldering produces product gases that are warmer than the ambient, the effect of buoyant heat and mass transfer is exerted upwards as the warm product gases tend to rise. In general, buoyant flows tend to hinder smoldering in downward propagation, but will tend to promote it in upwards propagation (Rein, 2009). In a vertical fuel bed, forward smoldering propagation is downwards and reverse smoldering propagation is upwards.

Finally, the location and extent of bounding, impermeable surfaces will generally have an important effect on the extent to which horizontal or vertical surfaces permit ingress and egress of fresh air and combustion products. Thus, the location and extent of impermeable surfaces will contribute to the development of one-dimensional or multidimensional smolder fronts.

The diversity of smolder wave behavior presents a formidable challenge for the investigator seeking insight into this phenomenon. Ohlemiller (1985) has presented a comprehensive discussion of physical and chemical phenomena contributing to smoldering including an insightful scaling analysis of the governing multiphase transport equations. In his analysis, he provides a hierarchy of modeling assumptions and discusses the relative merits of these different approaches. Overall, the smolder wave behavior is influenced by the porous structure of the fuel bed and its subsequent char formation. The behavior of individual particles exerts less influence on the physics than the properties of the porous medium.

In an excellent complementary article, Ohlemiller describes in detail experimental observations for 1D smolder propagation (forward and reverse), the properties of 2D smolder waves, and factors contributing to the transition from smoldering to flaming combustion (Ohlemiller, 2002). Rein (2009) has given a broad survey of smoldering phenomena including its importance in many wildland settings and reviews the more recent experimental and modeling literature.

We will now consider some very simple models of smoldering propagation.

5.3.1 1D SMOLDER WAVE PROPAGATION

In gaseous premixed flames, it is reasonable to approximate the mixture molar mass of the products to be equal to the mixture molar mass of the reactants. The accuracy of this assumption is based on the dilution of both mixtures by nitrogen. In a smoldering fuel bed, this assumption will not hold. The fuel solids may take up 60% by volume or more of the bed volume, and anywhere from 10% to 50% of the fuel mass may be consumed by the pyrolysis and oxidation reactions. Thus, you might be inclined to think that a thermal theory of smoldering will have little chance of success. It turns out, however, that thermal theories are good for at least an order of magnitude analysis of the problem. We shall consider two thermal models of smoldering, one based on a simple scaling argument and one based on a global energy balance.

We begin with Williams's calculation for the smolder wave speed based on a simple scaling argument as presented by Drysdale (1999, pp. 275–288). Williams's model is based on a global energy balance and uses the concept of an ignition temperature (Williams, 1977). The smoldering process is divided into three zones. Zone 1 is the pyrolysis zone characterized by a steep temperature rise from the ambient temperature, T_0, to the ignition temperature, T_{ign}. Zone 1 is where the emission of visible smoke occurs. Zone 2 is the charred zone. The temperature reaches is maximum value, T_{max}, glowing char occurs, and the emission of visible smoke ceases. Zone 3 is the residual porous char. The temperature begins to fall in this zone and glowing ceases. This description of smoldering zones is consistent with the observations of Moussa et al. (1977) who reported on smoldering experiments with α-cellulose and Leisch et al. (1984) who conducted smoldering experiments with grain dust and wood dust. Fig. 5.6 illustrates the physical system for the smoldering wave.

The basic premise is that the rate of propagation of a combustion wave is controlled by the rate of heat released by combustion. This energy balance is expressed in terms of fluxes. The energy balance is determined by equating the heat release rate with the enthalpy flux. Solving for the smolder wave velocity, v_{sm},

$$\dot{q}_c'' = \rho v_{sm}(\Delta h) \rightarrow v_{sm} = \frac{\dot{q}_c''}{\rho(\Delta h)} \tag{5.23}$$

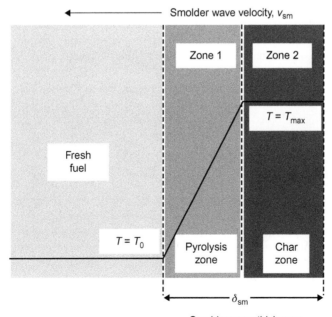

FIGURE 5.6

Definition sketch for Williams's scale analysis of smolder wave propagation.

In this equation, ρ is the fuel density and Δh is the enthalpy difference based on the difference between the ignition temperature and the initial temperature of the virgin fuel, and $\dot{q}_c''$ is the heat released by combustion at the smolder wave interface in Zone 2 expressed as a flux. If the ignition temperature is on the order of the maximum temperature of smoldering, $T_{ign} \approx T_{max}$, then the enthalpy difference can be expressed as a function of the heat capacity of the fuel mixture C_{mix} and the temperature difference:

$$\Delta h \cong C_{mix}(T_{max} - T_0) \tag{5.24}$$

If the heat transfer from Zone 2 to Zone 1 is by conduction alone, then the heat flux released by combustion can be approximated by

$$\dot{q}_c'' \cong \frac{k_{eff}(T_{max} - T_0)}{\delta_{sm}} \tag{5.25}$$

where k is the thermal conductivity of the fuel and δ_{sm} is the smolder wave thickness. Substituting these expressions into the equation for the smolder velocity, we obtain the simple result:

$$v_{sm} = \frac{k_{eff}(T_{max} - T_0)}{\delta_{sm}} \times \frac{1}{\rho_{mix} C_{mix}(T_{max} - T_0)} = \left(\frac{k_{eff}}{\rho_{mix} C_{mix}}\right)\frac{1}{\delta_{sm}} \tag{5.26}$$

$$v_{sm} = \frac{\alpha_{mix}}{\delta_{sm}} \tag{5.27}$$

The smolder velocity is equal to the thermal diffusivity of the fuel α divided by the smolder wave thickness. As an order of magnitude estimate, if $\alpha_{mix} \approx 10^{-7}$ m^2/s and $\delta_{sm} \approx 10^{-2}$ m, then a characteristic smolder velocity is 10^{-5} m/s. In comparison, a typical laminar burning velocity is 5×10^{-1} m/s and a typical detonation velocity is 10^3 m/s.

The determination of a steady-state smolder velocity using a global energy balance has been applied to both horizontal and vertical fuel bed configurations with and without forced air convection (Ohlemiller and Lucca, 1983; Dosanjh et al., 1987; Torero et al., 1993; Torero et al., 1994; Bar-Ilan et al., 2004). The energy balance requires adjustment depending on the presence and direction of forced convection. The orientation of the fuel bed (vertical or horizontal) has an effect only by its influence on buoyancy. As an example, let us consider a reverse smolder wave propagating in one direction with air flowing through the fuel bed from the opposite direction. A definition sketch is shown in Fig. 5.7.

A macroscopic energy balance over the smolder wave equates the heat released per unit mass of oxygen consumed in the smolder wave with the rise in temperature of the unburnt fuel and the incoming air (Rein, 2009; Dosanjh et al., 1987). The energy balance takes the following form:

$$\dot{m}_{total}'' C_{mix}(T_{max} - T_0) = \dot{m}_{ox}'' Q_{sm} - \dot{q}_{loss}'' \tag{5.28}$$

Here we have treated the solid and gas phases as a two-phase mixture with $C_{mix} = \rho_m^{-1}[\alpha_s \rho_s C_s + (1 - \alpha_s)\rho_{air} C_{p,air}]$, where α_g is the void volume fraction.

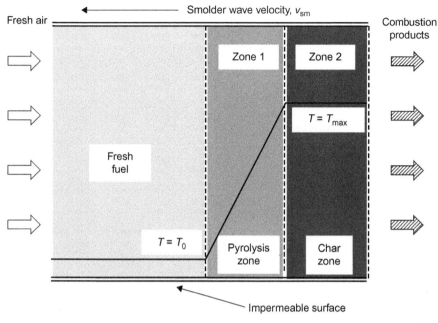

FIGURE 5.7

Reverse smolder wave propagation.

The chemical energy released by smoldering Q_{sm} is written on a per unit mass of oxygen consumed. The mass balance expressed as mass fluxes becomes

$$\dot{m}''_{total} = \dot{m}''_s + \dot{m}''_g \tag{5.29}$$

The smolder velocity is related to the mass flux of the smolder wave by the relation $\dot{m}''_{sm} = (1 - \alpha_g)\rho_s v_{sm}$. Neglecting heat losses and solving for the smolder velocity yields the following expression:

$$v_{sm} = \frac{\dot{m}''_{ox} Q_{sm}}{\rho_s C_s (1 - \alpha_g)(T_{max} - T_0)} \tag{5.30}$$

In real-world situations, one-dimensional smolder wave propagation is less common and multidimensional propagation is more common. We consider a simple two-dimensional wave problem next.

5.3.2 2D SMOLDER WAVE PROPAGATION

The rate of smolder wave propagation is determined by the rate of oxygen supply versus the rate of heat loss into the unburnt fuel. In two-dimensional smolder propagation, the smolder front develops a distinctive parabolic shape in the vertical profile. Gugan published an explanation of this phenomenon in cigarette

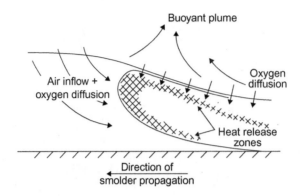

FIGURE 5.8

Ohlemiller's diagram showing the influx of oxygen into a two-dimensional smolder front.

Reproduced with permission from Ohlemiller, T.J., 1990. Smoldering combustion propagation through a permeable horizontal fuel layer. Combust. Flame 81, 341–353.

smoldering based on the relative rates of oxygen convection and diffusion. He demonstrated that the parabolic shape of the burning coal at the tip of the cigarette is due to these competing rates of oxygen supply (Gugan, 1966).

Ohlemiller (1981, 1990) applied this same reasoning to explain his observations of smolder wave propagation in a horizontal fuel bed. Ohlemiller observed that the vertical profile of the smolder wave assumed a parabolic shape as the wave traveled through a horizontal bed of cellulosic insulation. Following the example of Gugan, he was able to derive a simple model for the vertical profile of the smolder front. It is important to note that this analysis only gives the shape of the smolder wave profile; it does not give the smolder velocity.

Ohlemiller reasoned that the tip of the smolder wave (at the interface of the fuel bed surface and the air) could draw oxygen into it more easily than the deeper segment of the smolder wave. This variation in the oxygen supply led to the tip of the smolder wave traveling faster than the deeper portions of the wave. Eventually, the smolder wave achieved a steady velocity with the length of the front equal to approximately twice the layer height.

Ohlemiller realized that the two-dimensional smolder wave exhibited both forward and reverse smolder wave characteristics. Fig. 5.8 shows the influx of oxygen into a two-dimensional smolder front. The oxygen supply approaches the smolder front in two directions: in front of the wave (reverse smolder) and behind it (forward smolder). The exiting flow of combustion products from the smolder zone causes a buoyant plume to form above the smolder zone. This buoyant plume causes a convective flow of oxygen to enter the unburnt fuel in front of the smolder zone and draws the oxygen into the smolder wave where the oxygen is consumed. Simultaneously, oxygen diffuses through the charred zone and into the smolder wave.

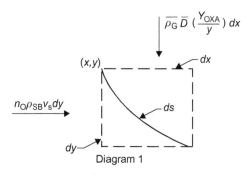

Diagram 1

FIGURE 5.9

Definition sketch for the diffusion analysis of the vertical profile of the smolder front.

The derivation of the smolder profile is based on the assumption that oxygen diffusion is the rate controlling step in smolder wave propagation. It is also assumed that the width of the smolder front is thin compared to the depth of the fuel layer. The oxygen fluxes are defined as follows (refer to Fig. 5.9).

Oxygen is supplied to the front of the smolder wave by convection and to the rear of the wave by diffusion. The convective flux flows across a differential area dy times the unit width of the fuel bed. The convective flux is given by the expression:

$$\text{Convective flux} = \rho_s\left(1 - \alpha_g\right)\left[\frac{0.23}{(F/A)_{st}}\right] v_{sm}\, dy \tag{5.31}$$

The bulk density of the solid fuel is represented by the product $\rho_s\left(1 - \alpha_g\right)$ and the factor $0.23/(F/A)_{st}$ is the mass of oxygen consumed per unit mass of fuel reacted. The differential length dy is located at depth y below the unburnt fuel bed surface.

The diffusive flux flows across a differential area dx times the unit width of the fuel bed. The diffusive flux is estimated using the expression:

$$\text{Diffusive flux} = \rho_g \mathcal{D}\left(\frac{\Delta Y_{ox}}{y}\right) dx \tag{5.32}$$

The density of the gas is ρ_g, the diffusivity of oxygen in the layer is $\mathcal{D}$, the oxygen mass fraction difference is ΔY_{ox}, and y is the depth of the bed at the smolder front location ds. The vertical profile of the smolder front is obtained by equating the convective and diffusive fluxes, and then integrating the resulting differential equation:

$$\rho_s\left(1 - \alpha_g\right)\left[\frac{0.23}{(F/A)_{st}}\right] v_{sm}\, dy = \rho_g \mathcal{D}\left(\frac{\Delta Y_{ox}}{y}\right) dx \tag{5.33}$$

$$x = \left[\frac{0.23\rho_s\left(1 - \alpha_g\right)v_{sm}}{2(F/A)_{st}\rho_g \mathcal{D}\Delta Y_{ox}}\right] y^2 \tag{5.34}$$

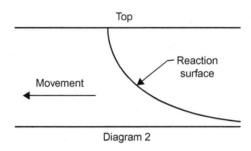

Diagram 2

FIGURE 5.10

Vertical profile of the smolder wave front.

The solution of the vertical profile is shown in Fig. 5.10. The foregoing analysis should be viewed as a very approximate model for the vertical profile of the two-dimensional smolder wave. An accurate calculation with this model would be difficult because there is little guidance on how to estimate the characteristic fluid properties and stoichiometry for the reaction.

The preceding analysis of smoldering was based on Ohlemiller's experiments with cellulosic insulation, a fibrous, fluffy material that bears only a passing resemblance to combustible dust. But this discrepancy in physical form does not rescind the validity of the observations. For example, an investigation of smoldering with horizontal layers of a cellulosic powder exhibited similar parabolic profiles of the smoldering wave front (Sato and Sega, 1989). Additionally, however, the tests by Sato and Sega revealed more complex interactions between the smolder wave, the unburnt powder, and the convective airflow above the fuel bed.

Finally, Palmer and Tonkin (1957) noted a quadratic relation in the breakthrough time as a function of bed depth for upwards propagation in vertical fuel beds. He interpreted these data as evidence that the rate of smoldering was controlled by the rate of oxygen diffusion.

5.4 CONFINED SMOLDERING

The generation rate and yield of combustion products depend on the ventilation condition. An overventilated condition (air present in amounts greater than the stoichiometric requirement) for the combustion of an organic fuel yields products of complete combustion, carbon dioxide and water vapor. An underventilated condition yields products of incomplete combustion (Purser, 2002). Smoldering combustion is an underventilated condition.

Smoldering generates of combustion products that are very different from those generated by flaming combustion. Compared to a flaming fire, smoldering generates a 10-fold increase in the proportion of CO to CO_2 generated, that is, the ratio of CO to CO_2 is on the order of one-tenth in a flaming fire and on the order of 1 in a smoldering fire (Rein, 2009). Carbon monoxide is both toxic and flammable. In an unconfined space with natural ventilation, the CO concentration is unlikely to pose a serious hazard to observers. If smoldering occurs in a confined space, and in the absence of forced ventilation with fresh air, then there is the potential for the CO concentration to rise to toxic or explosive levels.

The toxicity of CO is well established (NIOSH, 2015). The toxicity is directly related to the dose received where dose is defined as the total intake of the toxin. With CO, this is usually expressed as an exposure to a concentration level for a specified period of time. A concentration of 4000 parts per million by volume (ppmv) is fatal to humans in 30 minutes; a concentration of 5000 ppmv is fatal in 5 minutes (NIOSH, 2015). The specific physiological response to lower doses of CO will depend on a variety of circumstances including the medical condition of the person experiencing the exposure. Quintiere et al. (1982) were successful in building and testing a mathematical model for CO generation in a confined space. Their efforts were focused on the smoldering of upholstered furniture (specifically, the polyurethane foam cushion) in a residential setting. However, their methodology could be easily extended to industrial settings.

At higher concentrations, CO can form a flammable atmosphere. The flammable range of CO is 12.5–74%. Smoldering will produce a variety of products of incomplete combustion including smoke. These additional materials can contribute to either the toxicity or the flammability of the atmosphere.

The toxicity and flammability hazards of confined smoldering can affect not only the workers at an industrial facility but also the emergency responders such as firefighters. A number of industrial explosions at agricultural storage facilities (grain elevators) have been linked to smoldering fires (Ogle et al., 2014; Eckhoff, 2005, pp. 175–196). These hazardous conditions have also been encountered in smaller settings like privately owned farms (USFA, 1998). In many cases, the ignition source for an explosion has been the transition of the smoldering fire into a flaming fire.

5.5 TRANSITION FROM SMOLDERING TO A FLAMING FIRE OR AN EXPLOSION

It is difficult to predict if and when a smoldering fire will transition to a flaming fire. One of the greatest hazards of this transition is that a flaming fire is an excellent ignition source for a flammable atmosphere (Gummer and Lunn, 2003). Smolder waves are less capable as ignition sources for a flammable atmosphere unless the glowing char is exposed on the surface of the fuel bed. In contrast, flames can only occur on the surface of the fuel bed.

The transition from smoldering to flaming combustion is usually triggered by an increase in ventilation to the reaction zone. This can be accomplished by forced convection over the fuel surface. It is generally accepted that the transition to flaming can only occur in forward smolder propagation (Palmer, 1957; Ohlemiller, 2002). An alternative means of increasing ventilation is by dislodging and fluidizing the smoldering material. One way for this to occur is when smoldering material at a higher elevation is disturbed and dislodged followed by free fall. Another means of creating the sudden fluidization of smoldering material can occur during firefighting activities. The smoldering mass can be disrupted by the turbulent jet from a portable fire extinguisher or with a water hose stream. This latter scenario is especially hazardous because of the potential to disperse a cloud of unburnt combustible dust at the same time that a strong ignition source becomes available. This scenario can lead to a dust deflagration, turning a smaller hazard into a much larger hazard.

5.6 CONTROLLING SMOLDERING HAZARDS

The control of smoldering hazards with combustible dusts presents difficult challenges. From the perspective of the system safety hierarchy (see Chapter 1), the best strategy is to prevent smoldering from occurring. Depending on any number of operational or economic factors, this may not always be a practical strategy.

The strategies for hazard control can be organized according to the fire triangle. Considering the fuel component, a straightforward strategy is to limit the accumulation of self-heating solids. If the size of the accumulation can be kept smaller than the critical dimension, then the heat release by self-heating will be harmlessly dissipated to the environment. If a stockpile of self-heating material must be kept in a size greater than its critical dimension, then another approach with the fuel is to cool it below its critical temperature. The control of ambient temperature can be accomplished with refrigeration, but this introduces an expense beyond the simple storage of the material.

Another strategy is to control the oxidizer content of the storage atmosphere. This can be accomplished by using nitrogen or carbon dioxide gas to create an inert atmosphere. This strategy requires that the self-heating material is kept in a gas-tight enclosure or that the storage atmosphere be monitored and controlled to prevent the intrusion of oxygen into the enclosure. The oxygen content of the atmosphere does not need to be zero; it needs to be below the limiting oxygen concentration that will support combustion. Crowl (2003) provides a detailed analysis of how to determine the inert gas concentration required to prevent combustion.

The third leg of the fire triangle is the ignition source. Beyond simple self-heating, piles of combustible dust can be susceptible to ignition by external sources of heat. External ignition sources can be controlled through engineering controls like using electrical equipment of the proper design (so-called electrical classification for hazardous atmospheres) or by administrative controls like hot work safety and control of smoking materials.

If prevention fails, then the next strategy of controlling smoldering hazards is detection and response (Goforth and Ostrowski, 1987; USFA, 1998; OSHA, 2013). There are several potential indicators for smoldering combustion. These indicators include: carbon monoxide production, smoke production, odor, glowing embers, elevated temperatures, and production of charred material (Krause, 2009; Ogle et al., 2014). These indicators are not as reliable as those for a flaming fire. This is because the intensity of these indicators is dependent on both the rate smoldering, the configuration of the fuel bed, and environmental factors such as the size of the enclosure and the nature and magnitude of ventilation within the enclosure. For example, Palmer and Tonkin (1957) noted that the upwards propagation of a smolder wave in a deep fuel bed was not visibly evident until the wave approached the top surface of the bed. Ohlemiller (2002) indicated that smoke particles from smoldering may be 10–100 times the diameter of smoke particles generated by a flaming fire. Such large particles might be filtered out by the porous medium of the fuel bed and would, therefore, not be visible as a smoke emission.

Once smoldering has been detected, care must be taken in any corrective action to avoid making a hazardous situation worse. Corrective actions can include removing the unburnt fuel from the storage area at risk, but care must be taken to avoid disturbing the smolder zone. This removal activity may also require ventilation to limit the formation of an explosive atmosphere within the enclosure. Another strategy is to use inert gas to displace oxygen or to cover the smoldering mass with an impermeable cover. Depending on the size of the smoldering mass and its enclosure, extinguishing the fire by suffocating it may take several weeks as demonstrated in agricultural grain silos (Krause, 2009). Suppression of the smoldering reaction with water is a potential strategy, but the volume of water delivered must be of sufficient volume and flow rate that it floods the material, but must be delivered in a way so as to prevent the formation of combustible dust clouds. This can be a dangerous tactic to implement, and some guidance documents explicitly warn against it.

5.7 SUMMARY

This chapter examined some of the key features of smoldering combustion. This is an important topic for understanding the hazards of combustible dusts because smoldering can transition without warning from a low hazard level to a high hazard level. After providing a physical description of smoldering, we considered self-heating and spontaneous ignition as a mechanism for initiating a smolder wave. The Semenov and Frank–Kamenetskii models for a self-heating solid were presented. With the Semenov model, it is possible to calculate a time to thermal runaway. With the Frank–Kamenetskii model, results were given for the calculation of critical properties for various geometries.

Theories for smolder wave propagation were presented and illustrative results were discussed for one- and two-dimensional smolder waves. The hazards of confined smoldering behavior were investigated. Finally, the control of smoldering hazards was demonstrated with reference to the system safety hierarchy and the fire triangle.

In smoldering, the structure of the porous fuel bed has an effect on the initiation, propagation, and extinguishment of the combustion wave. Although this porous structure is a result of the interparticle contacts, the behavior of individual particles does not dominate this behavior. Dust deflagrations occur in dust suspensions where the interparticle spacing is on the order of several particle diameters. The behavior of dust deflagrations is more sensitive to the combustion behavior of the individual particle. In Chapter 6 we will turn our attention to the analysis of individual burning particles as our first step towards understanding dust explosion behavior.

REFERENCES

Aldis, D.F., Lai, F.S., 1979. Review of Literature Related to Engineering Aspects of Grain Dust Explosions (No. 1375). Department of Agriculture, Science and Education Administration, Washington, DC.

Amyotte, P., 2013. An Introduction to Dust Explosions. Butterworth Heinemann, Elsevier, New York.

Annamalai, K., Puri, I.K., 2007. Combustion Science and Engineering. CRC Press, Boca Raton, FL.

Babrauskas, V., 2003. The Ignition Handbook. Fire Science Publishers, Issaquah, WA.

Bar-Ilan, A., Rein, G., Walther, D.C., Fernandez-Pello, A.C., 2004. The effect of buoyancy on opposed smoldering. Combust. Sci. Technol. 176, 2027−2055.

Beever, P.F., 1995. Self-heating and spontaneous combustion, Section 2, Chapter 12. In: DiNenno, P.J., et al., (Eds.), SFPE Handbook of Fire Protection Engineering, second ed. National Fire Protection Association, Quincy, MA, pp. 2-180−2-189.

Bowes, P.C., 1984. Self-Heating: Evaluating and Controlling the Hazards. Elsevier, New York.

Bowes, P.C., Townsend, S.E., 1962. Ignition of combustible dusts on hot surfaces. Br. J. Appl. Phys. 13, 105−114.

CCPS/AIChE, 2005. Guidelines for Safe Handling of Powders and Bulk Solids. American Institute of Chemical Engineers. John Wiley & Sons, New York.

Cross, J., Farrer, D., 1982. Dust Explosions. Plenum Press, New York.

Crowl, D., 2003. Understanding Explosions. Center for Chemical Process Safety, American Institute of Chemical Engineers, John Wiley & Sons, New York.

Dosanjh, S.S., Pagni, P.J., Fernandez-Pello, A.C., 1987. Forced concurrent smoldering combustion. Combust. Flame 68, 131−142.

Drysdale, D., 1997. An Introduction to Fire Dynamics, second ed. John Wiley & Sons, New York.

Eckhoff, R., 2005. Explosion Hazards in the Process Industries. Gulf Publishing Company, New York.

Glassman, I., Yetter, R.A., Glumac, N.G., 2015. Combustion, fifth ed. Academic Press, New York.

Goforth, K.J., Ostrowski, C.R., 1987. Emergency Preplanning and Firefighting Manual: a Guide for Grain Elevator Operators and Fire Department Officials. National Grain and Feed Association.

Gugan, K., 1966. Natural smolder in cigarettes. Combust. Flame 10, 161−164.

Gummer, J., Lunn, G.A., 2003. Ignitions of explosive dust clouds by smouldering and flaming agglomerates. J. Loss Prevent. Process Ind. 16, 27−32.

Krause, U. (Ed.), 2009. Fires in Silos. Wiley-VCH, Berlin.

Leisch, S.O., Kauffman, C.W., Sichel, M., 1984. Smoldering combustion in horizontal dust layers. Twentieth Symposium (International) on Combustion. The Combustion Institute, Elsevier, New York.

Linan, A., Williams, F.A., 1993. Fundamental Aspects of Combustion. Oxford University Press, Oxford.

Mills, J.T., 1989. Spoilage and Heating of Stored Agricultural Products: Prevention, Detection, and Control. Agriculture Canada.

Miron, Y., Lazzara, C.P., 1988. Hot surface ignition temperatures of dust layers. Fire Mater. 12, 115−126.

Moussa, N.A., Toong, T.Y., Garris, C.A., 1977. Mechanism of smoldering of cellulosic materials. Sixteenth Symposium (International) on Combustion. The Combustion Institute, Elsevier, New York.

Nelson, M.I., Chen, X.D., 2007. Survey of experimental work on the self-heating and spontaneous combustion of coal, Reviews in Engineering Geology, Volume XVIII. Geological Society of America, pp. 31−83.

NIOSH, 2015. NIOSH Pocket Guide to Chemical Hazards. National Institute of Occupational Safety and Health, Atlanta, GA.

Ogle, R.A., Dillon, S.E., Fecke, M., 2014. Explosion from a smoldering silo fire. Process Saf. Progress 33, 94−103.

Ohlemiller, T.J., 1981. Smoldering Combustion Hazards of Thermal Insulation Materials. Report NBSIR 81-2350. National Bureau of Standards, U.S. Department of Commerce, Washington, DC.

Ohlemiller, T.J., 1985. Modeling of smoldering combustion propagation. Progress Energy Combust. Sci. 11, 277−310.

Ohlemiller, T.J., 1990. Smoldering combustion propagation through a permeable horizontal fuel layer. Combust. Flame 81, 341−353.

Ohlemiller, T.J., 2002. Smoldering combustion, Section 2, Chapter 9. In: DiNenno, P.J., et al., (Eds.), SFPE Handbook of Fire Protection Engineering, third ed. National Fire Protection Association, Quincy, MA, pp. 2-200−2-210.

Ohlemiller, T.J., Lucca, D.A., 1983. An experimental comparison of forward and reverse smolder propagation in permeable fuels beds. Combust. Flame 54, 131−147.

OSHA 3644, 2013. Firefighting Precautions at Facilities with Combustible Dust. Occupational Safety and Health Administration, Washington, DC.

Palmer, K.N., 1973. Dust Explosions and Fires. Chapman and Hall, London.

Palmer, K.N., Tonkin, P.S., 1957. The ignition of dust layers on a hot surface. Combust. Flame 1, 14−18.

Purser, D.A., 2002. Toxicity assessment of combustion products, Section 2, Chapter 6. In: DiNenno, P.J., et al., (Eds.), SFPE Handbook of Fire Protection Engineering, third ed. National Fire Protection Association, Quincy, MA, pp. 2-83−2-171.

Quintiere, J., Birky, M., McDonald, F., Smith, G., 1982. An analysis of smoldering fires in closed compartments and their hazard due to carbon monoxide. Fire Mater. 6, 99—110.

Quintiere, J.G., 2006. Fundamentals of Fire Phenomena. John Wiley & Sons, New York.

Rein, G., 2009. Smouldering combustion phenomena in science and technology. Int. Rev. Chem. Eng. 1, 3—18.

Sato, K., Sega, S., 1989. Smolder spread in a horizontal layer of cellulosic powder. Fire Safety Science 2, 87—96.

Toong, T.-Y., 1983. Combustion Dynamics: The Dynamics of Chemically Reacting Fluids. McGraw-Hill, New York.

Torero, J.L., Fernandez-Pello, A.C., Kitano, M., 1993. Opposed forced flow smoldering of polyurethane foam. Combust. Sci. Technol. 91, 95—117.

Torero, J.L., Fernandez-Pello, A.C., Kitano, M., 1994. Downward smolder of polyurethane foam. Fire Safety Science 4, 409—420.

USFA, 1998. Special Report: The Hazards Associated with Agricultural Silo Fires, USFA-TR-096. United States Fire Administration, Emmitsburg, MD.

Varma, A., Morbidelli, M., Wu, H., 1999. Parametric Sensitivity in Chemical Systems. Cambridge University Press, Cambridge.

Williams, F.A., 1977. Mechanisms of fire spread. Sixteenth Symposium (International) on Combustion. The Combustion Institute, Elsevier, New York.

Dust particle combustion models

6

At the microscopic level, a dust flame consists of a cloud of burning particles. The occurrence of a dust deflagration in an industrial setting involves phenomena at multiple length scales. Examining the physical and chemical processes involved in the combustion of a single particle can be considered the logical starting point for investigating the fundamental nature of a dust deflagration (Cloney et al., 2014). For example, Palmer (1973, pp. 181–190) described the potential insight into dust deflagrations that one could derive from single particle combustion studies. One specific example he gave was the problem of determining the relationship between dust flame speeds and particle burning times. Eckhoff (2003, pp. 251–265, 588–589) has provided a more recent survey of single particle combustion in the context of dust deflagrations. Amyotte has described the influence of chemical factors and particle size on dust explosibility (Amyotte, 2013, Chapters 5 and 6). This chapter will give us the opportunity to appreciate the complexity of solid fuel combustion. This complexity arises due to the physical and chemical processes involved in the combustion process. In addition, we will find that the microstructure of the particles can influence the combustion of many organic solids and metals, and that the microstructure can evolve over the course of the combustion event.

In this chapter, I first discuss the basic physical and chemical phenomena behind the combustion of single particles. An important concept—the overall reaction rate—is the foundation upon which much of single particle combustion analysis is based. The overall reaction rate incorporates into one term the rates of chemical reaction and mass transfer. The essential features of heat and mass transfer models for single particles will be introduced next. Both external and internal transport processes are considered. The heat and mass transfer models are used to evaluate the simplifying assumptions often invoked in single particle combustion models. These assumptions are explained in terms of the relevant time and length scales based on kinetic and transport processes. The particle diameter is a major influence on these characteristic scales.

Much of this information has been developed by observing how materials respond to heating. After describing a range of material behaviors, I will go on to describe the major classes of organic and inorganic solid fuels comprising combustible dusts. One of the more important observations from this discussion will be the fact that solid particles rarely retain their original physical form or chemical composition during combustion. I will close this section with a brief survey of some of the more popular experimental tools used for single particle combustion studies.

The next topic will introduce the three fundamental models for noncatalytic gas—solid reactions as applied to single particles: the shrinking particle model, the shrinking unreacted core model, and the progressive conversion model. These models are highly simplified descriptions of reaction and diffusion behavior that occur during combustion. Nevertheless, they are often successful in predicting a particle's mass combustion rate and the burnout time (time for complete combustion). The mass combustion rate and the burnout time are two of the more important attributes for characterizing single particle combustion. The performance equations can be used to analyze kinetic data to discern the relative importance of the transport and kinetic processes. The final topic of this section will cover extensions of these gas—solid kinetics models including higher order kinetics, nonisothermal effects, and microstructure development during reaction.

Section 6.3 presents a literature survey on single particle combustion for the key types of organic solid fuels: noncharring (completely volatile) organic solids, charring (partially volatile) organic solids, and nonvolatile organic solids. Then for each of these material types, we will explore the influence of particle size on combustion behavior, briefly survey the chemical kinetics investigations for these materials, and specifically consider the influence of chemical composition and particle microstructure on combustion behavior.

The particle size range of interest for combustible dust lies in the approximate range of 1—500 μm. Many of the studies that we will survey involve particle sizes that are much larger than this size range. By surveying a more comprehensive body of literature, it will become apparent that the relative importance of the physical and chemical processes involved in dust particle combustion is dependent on particle size.

Through this brief literature survey, we will discover the wide diversity of physical and chemical phenomena observed with these solid fuels and explore some of the limitations of single particle combustion modeling. We will find that organic solids that produce volatiles upon heating give rise to diffusion flame behavior. The d-squared law is typically a good first approximation for describing the particle size history of solid fuel particles during combustion. The method of analysis builds directly on the method used in Chapter 4 to analyze the combustion of liquid droplets. We will also briefly review models for organic solids that account for the influence of particle microstructure.

In Section 6.4, I will describe the combustion behavior of two classes of metal fuels, volatile metals and nonvolatile metals. The combustion of metals is complicated by the phase behavior of their metal oxides. We will explore models that account for this behavior with two combustible metals commonly found in industry: aluminum, a volatile metal and iron, a nonvolatile metal.

In this chapter, two important caveats must be borne in mind. First, the combustion models that I will present are all based on the assumption that the particle has a spherical shape and consists of a material which is continuous, homogeneous, and isotropic. The literature on these various solid fuels will portray a very different view: combustible dusts are usually composed of a variety of nonspherical particles with internal microstructures. The degree to which these departures from ideality are important will have to be evaluated on a case-by-case basis.

Second, the vast majority of the literature on solid fuels in particulate form involves investigations for achieving complete combustion in furnaces or other engineered systems. In contrast, in most accident settings, the combustion process is incomplete. This is a function of two characteristic scale factors: the size (length) scale of the particle and the timescale of the combustion process. The quantity of thermal energy liberated by combustion is directly related to the extent of fuel conversion. Thus, the fact that most accidental dust deflagrations exhibit incomplete combustion is an important consideration in evaluating the potential importance of particle size, shape, and microstructure.

Finally, to keep the size of this chapter reasonable, only steady state combustion will be considered. The important topic of particle ignition will not be covered.

6.1 SINGLE PARTICLE COMBUSTION PHENOMENA

In liquid droplet combustion, we saw that the droplet diameter decreased as it burned. The burnout time was a function of the square of the initial droplet diameter. This was called the d-squared law. It is tempting to assume that single particle combustion will follow the same pattern, ie, that the particle will shrink in size as it burns. It turns out the solid fuels exhibit a more complex range of behaviors. To place these behaviors into perspective, we need some background. Nevertheless, it will become evident that the investigation of single particle combustion builds directly on the diffusion flame model for liquid droplet combustion.

Chemical kinetics, the analysis of the network of physical and chemical steps which make up chemical reactions, is the first topic considered. The combustion of solid fuels generally requires both homogeneous and heterogeneous reactions. Homogeneous reactions occur within the fluid medium under study (we will neglect solid-state chemical reactions). The combustion reactions described in Chapter 4 were homogeneous reactions.

Heterogeneous reaction systems require consideration of mass transport processes. With heterogeneous reactions, a gaseous reactant—such as oxygen—must diffuse to the surface of the particle surface, adsorb onto the solid surface of the fuel particle, react with the solid to form a product, desorb the product from the surface, and diffuse back out into the free stream environment. Thus, the overall rate of reaction is dependent on both reaction and diffusion processes. To properly evaluate the rate of reaction in a heterogeneous system, the sequence of transport and reaction steps must be defined and quantified. This sequence of steps leads to the definition of an overall rate of reaction. This will be the subject of the next section.

The burning of individual particles of combustible dust depends significantly on the chemical composition and physical structure of the dust particle. The range of combustion behavior observed depends on how the solid material responds to external heating. One broad classification of behavior is organic versus inorganic. We will first consider organic solid fuels. There are two key questions to consider in this regard:

- How does the solid particle respond to the heat of the flame, does it vaporize (physical phenomenon) or pyrolyze (chemical phenomenon)?
- If the solid pyrolyzes, does it form a solid char or not?

The later discussion of single particle combustion behavior for organic solids will be structured into three fundamental types: noncharring solids, charring solids, and nonvolatile solids.

Since most inorganic combustible dusts are metals, I will concentrate most of the discussion of inorganic materials on metals. With metal combustion, there are three questions to consider:

- During combustion, does the solid metal melt and vaporize or does it remain a solid?
- Does the metal oxide product remain a vapor or does it condense?
- If the metal oxide product condenses, is it soluble with the molten metal?

The later discussion of single particle combustion behavior for metallic solids will be structured into two fundamental types: volatile metals and nonvolatile metals.

The final topic in this section will be an overview of experimental techniques used to investigate single particle combustion.

6.1.1 HETEROGENEOUS CHEMICAL REACTIONS AND THE OVERALL REACTION RATE CONCEPT

During combustion, combustible dust particles must interact with the ambient gas environment. Oxygen must diffuse toward the flame sheet and the combustion products must diffuse away from the flame sheet. Chemical reactions that occur exclusively within the bulk of a phase region are called homogeneous reactions. Chemical reactions that occur at an interface between two phases are called heterogeneous reactions.

Solid materials which form a solid char during combustion are an example of combustion mechanisms that include both homogeneous and heterogeneous chemical reactions. If the combustion mechanism involves any heterogeneous chemical reactions, this interaction takes the form of a sequence of steps involving diffusion, adsorption, and chemical reaction, each of which occurs at its own definite rate. The sequence of steps for the gas—solid interaction is depicted in Fig. 6.1.

The observed combustion rate, called the overall rate, is the combination of the rates of the individual steps. The rates of adsorption and desorption are

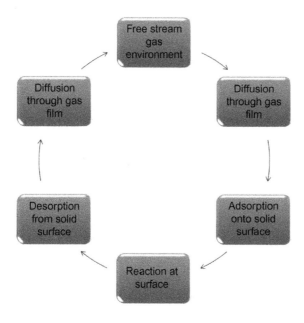

FIGURE 6.1

Sequence of diffusion, adsorption, and chemical reaction steps in a gas–solid combustion process.

generally much faster than the rates of diffusion or reaction and so they are usually neglected in the development of the overall rate expression.

Consider the following gas–solid chemical reaction:

$$A(g) + bB(s) \rightarrow C(g) \tag{6.1}$$

Here A, B, and C are chemical species, b is a stoichiometric coefficient, g indicates a gas phase species, and s denotes the solid reactant. Since the diffusion and reaction steps occur in series, then at steady state the overall rate for the combustion process can be expressed as a mass balance (in mole units) at the particle surface:

$$\frac{\dot{m}''_A}{\mathcal{M}_A} = \frac{\omega''_B}{\mathcal{M}_B} \tag{6.2}$$

This equation states that the mass transfer flux of oxygen across the gas film to the surface of the fuel particle (expressed in mole units) is equal to the surface rate of reaction (expressed in mole units). (Levenspiel, 1999, Chapters 17 and 25). The mass transfer rate can be modeled with a film mass transfer coefficient k_g. We will assume that the chemical reaction rate can be modeled as a linear function of the surface concentration of species A (we will discuss the relaxation of that assumption later in this chapter). We want to be able to calculate the rate of reaction using measureable quantities. We begin by writing the rate expressions in terms of their driving forces

$$\frac{\dot{m}_A''}{\mathcal{M}_A} = k_g(C_{A,\infty} - C_{A,s})$$
(6.3)

$$\frac{\omega_A''}{\mathcal{M}_A} = -b\,k_c\,C_{A,s}$$
(6.4)

The free stream molar concentration $C_{A,\infty}$ is a measurable quantity; the surface molar concentration $C_{A,s}$ is usually not measureable. At steady state, the rate of mass transfer through the boundary layer is equal to the chemical reaction rate so $\dot{m}_{A,g}'' = \omega_A''$ (note that the molar mass of A has canceled on both sides of the equation). Equating the rate expressions, we have

$$k_g(C_{A,\infty} - C_{A,s}) = bk_c C_{A,s}$$
(6.5)

This equation can be solved for the surface concentration in terms of the free stream concentration and the rate coefficients:

$$C_{A,s} = \left(\frac{k_g}{k_g + bk_c}\right)C_{A,\infty}$$
(6.6)

Substituting the expression for the surface concentration $C_{A,s}$ into Eq. (6.4) and simplifying eliminate the unknown surface concentration variable $C_{A,s}$ and yield the desired result:

$$\dot{m}_A'' = C_{A,\infty}\left[\frac{1}{(1/k_g) + (1/bk_c)}\right] = k_0 C_{A,\infty}$$
(6.7)

$$\frac{1}{k_0} = \frac{1}{k_g} + \frac{1}{bk_c}$$
(6.8)

The coefficient k_0 is the overall rate coefficient. Each rate coefficient can be thought of as conductance, so their reciprocals are resistances. Then the overall rate coefficient is a sum of additive resistances. This only holds true for linear rate expressions. The physical interpretation of the overall rate coefficient k_0 is readily apparent: if the mass transfer coefficient is much greater than the chemical rate coefficient, $k_g \gg bk_c$, then the rate of chemical reaction is the slower rate process and therefore controls the overall rate. The overall rate is said to be *kinetic controlled*. If the chemical reaction rate is much greater than the mass transfer rate, $bk_c \gg k_g$, then the mass transfer rate is the controlling process. The overall rate is said to be *mass transfer controlled*.

The overall rate concept will be the central focus of our discussion on noncatalytic gas–solid reactions.

EXAMPLE 6.1

Using Equation 6.8, compute the value of k_0 for the following four cases: $k_g = bk_c$; $k_g = 2 bk_c$; $k_g = 10 bk_c$; and $k_g = 100 bk_c$. Assume $bk_c = 1 \, m/s$. How large does k_g need to be to satisfy $k_g \gg bk_c$?

Solution

Solve Equation 6.8 for the overall rate coefficient k_0:

$$k_0 = \left[\frac{1}{k_g} + \frac{1}{bk_c} \right]^{-1}$$

Substituting the values for k_g gives the following values for the overall rate coefficient:

k_g	k_0
bk_c	0.50
$2 bk_c$	0.67
$10 bk_c$	0.91
$100 bk_c$	0.99

To satisfy the expression $k_g \gg bk_c$, the value of k_g required depends on the magnitude of acceptable error. If approximately 1% error is the maximum acceptable value, inspection of the table of values suggests that the value of the mass transfer coefficient must satisfy $k_g = 100 \, bk_c$.

6.1.2 TRANSPORT PHENOMENA IN SINGLE PARTICLE COMBUSTION

The combustion processes described in Chapter 4 and Chapter 5 demonstrate the interplay of thermodynamics, chemical kinetics, and transport processes. More specifically, as discussed in the previous section on overall reaction rates, the rate of combustion of a fuel depends on the competition between chemical reaction rates and transport rates. In this section, I will introduce some basic considerations of transport phenomena which play a role in single particle combustion. The relative magnitude of transport rates and reaction rates determines if a particular combustion process is diffusion controlled or kinetic controlled.

It turns out that the relative magnitudes of transport rates are not only largely determined by the physical properties of the ambient fluid but also directly influenced by the geometrical properties of the particle. Particle size, particle shape, and particle microstructure, especially porosity, can influence transport rates. In nonporous solids, heat and mass transfer occur at one single length scale: the particle diameter. Heat and mass transfer occur between the gaseous free steam environment and the particle surface.

In porous solids, heat and mass transfer occur at two length scales: the particle diameter (external transfer) and the pore diameter (internal transfer). Many of the combustible dusts encountered in the industrial realm are nonporous solids. However, there are many examples of organic solids that begin as nonporous materials, but, upon initiation of combustion, will gradually become porous solids. An important question for us to consider is whether the evolution of a porous structure in a particle is a significant factor in its combustion behavior.

The first point to consider is the natural time and length scales that arise in combustible dust flames. Then the heat and mass transfer interactions between particles and the free stream environment will be reviewed. The heat and mass transfer rates into the interior of a particle are considered next.

6.1.2.1 Typical time and length scales for dust deflagrations

The speed and thickness of a dust deflagration wave establishes an important timescale. In engineered systems, one of the primary goals of the combustion engineer is to achieve complete combustion of the fuel (or as close as realistically possible). An important observation in most dust deflagrations is that the combustion process is incomplete. Especially in accident settings, one usually finds unburnt fuel in the area where the deflagration occurred. This is a consequence of the timescale of the deflagration wave.

The timescale for a dust deflagration is the characteristic time for the flame to sweep through a cloud of fuel particles. The kinetic and transport processes responsible for the propagation of the flame must contribute to the combustion process within that timescale. Consider a dust deflagration with a burning velocity $S_u = 1$ m/s and a flame thickness $\delta = 0.01$ m. The timescale for the deflagration is $t_f = \delta/S_u = 0.01$ s $= 10$ ms. With a flame thickness of $\delta = 0.1$ m, the timescale becomes $t_f = 0.1$ s $= 100$ ms. So the characteristic timescale for a dust deflagration will normally fall within a range of $10-100$ ms. The relative contribution of external and internal transport processes to single particle combustion can be gauged by comparison of with these timescales.

6.1.2.2 External heat and mass transfer

The external transport processes of interest are convective heat and mass transfer from the ambient gas stream to the external surface of the particle. Convective transport is a boundary layer process. A well-accepted correlation for convective heat transfer to a spherical particle is the Ranz–Marshall correlation (Crowe et al., 2012, pp. 104–105; Bird et al., 2002, p. 438; Szekely et al., 1976, 46–47).

$$Nu = 2 + 0.60\,Re^{1/2}\,Pr^{1/3} \tag{6.9}$$

$$Nu = \frac{hd_p}{k}; \quad Re = \frac{\rho v d_p}{\mu}; \quad Pr = \frac{\mu C_p}{k} \tag{6.10}$$

The Nusselt number Nu, the Reynolds number Re, and the Prandtl number Pr are all defined in terms of the free stream fluid properties. The characteristic length in the Nusselt number and Reynolds number is the particle diameter. A similar correlation is used for convective mass transfer (Szekely et al., 1976, pp. 10−13; Levenspiel, 1999, pp. 577−581).

$$Sh = 2 + 0.6\ Re^{1/2}\ Sc^{1/3} \tag{6.11}$$

$$Sh = \frac{k_g d_p}{\mathcal{D}}; \quad Sc = \frac{\mu}{\rho \mathcal{D}} \tag{6.12}$$

The Sherwood number Sh and the Schmidt number Sc are defined in terms of the free stream fluid properties and the characteristic length is the particle diameter.

As discussed in Chapter 4 thermal radiation is another mode of heat transfer that can influence particle temperature. We will neglect thermal radiation effects at the microscale level. In Chapter 7, thermal radiation effects will be an important consideration at the macroscale level.

6.1.2.3 Internal heat and mass transfer

Internal transport processes refer to the heat and mass transfer that occurs within the particle. The rates of transport may be affected by the phase distribution or porous structure of the particle. For solid particles of a single phase, the particle body can be treated as continuous, homogeneous, and isotropic. The mechanism of heat transfer to the particle interior is thermal conduction. The characteristic timescale for thermal conduction for a given particle size is given by the equation $t_{conduction} = r_p^2/\alpha$, where the thermal diffusivity is given by $\alpha = k/\rho C_p$. Because of the spherical symmetry of the particle, the characteristic length for the particle is the radius, not the diameter. To understand the order of magnitude of the conduction time, we will consider two materials, anthracite coal to represent an organic solid and pure aluminum to represent a metallic solid (neglecting the influence of any oxide layer). The room temperature thermal diffusivities for these materials are calculated as follows (property data from Incropera and DeWitt, 1990, Appendix A):

$$\alpha(\text{anthracite}) = \frac{k}{\rho C_p} = \frac{0.26\ W/m \cdot K}{(1350\ kg/m^3)(1260\ J/kg \cdot K)} = 1.53 \times 10^{-7}\ m^2/s \tag{6.13}$$

$$\alpha(\text{aluminum}) = \frac{k}{\rho C_p} = \frac{237\ W/m \cdot K}{(2702\ kg/m^3)(903\ J/kg \cdot K)} = 9.71 \times 10^{-5}\ m^2/s \tag{6.14}$$

The thermal diffusivity of aluminum is 600 times greater than the value for anthracite. Table 6.1 presents some characteristic values of the conduction time scale for a range of particle sizes. For ease of comparison with other timescales, these values are reported in milliseconds.

This calculation demonstrates that the thermal conduction timescale for metal particles is much shorter than the timescale for a dust deflagration. In the size range of interest for combustible dust hazards, metal particles are likely to be in thermal equilibrium with the free stream gas. The same will be true for only the smaller particle size range for organic particles (perhaps $d_p < 10\ \mu m$).

Table 6.1 Thermal Conduction Time-scale for Anthracite and Aluminum as a Function of Particle Size

Particle Radius (μm)	t (Anthracite) (ms)	t (Aluminum) (ms)
1000	6540	10.3
100	65.4	0.103
10	0.654	0.00103
1	0.00654	0.0000103

Since this calculation is based on only two materials and the thermal properties were evaluated at room temperature, these generalizations should be used cautiously.

We are now in position to determine the relative importance of internal and external heat transfers to a particle. The Biot number is defined as $\mathrm{Bi} = h d_\mathrm{p}/k_\mathrm{m}$. The Biot number appears to have the same functional form as the Nusselt number, but it has a different interpretation. The Nusselt number is an indicator of the relative importance of convection and conduction through the fluid to the particle surface. The Biot number is an indicator of the relative importance of convection from the fluid to the particle surface versus the heat conduction within the particle. At high Biot numbers, $\mathrm{Bi} \gg 1$, heat conduction within the particle is the controlling heat transfer resistance and there is a significant temperature gradient within the particle. At low Biot numbers, $\mathrm{Bi} \ll 1$, the particle is isothermal and convection from the fluid is the controlling resistance.

In principle, a similar calculation can be done for diffusion. The characteristic timescale for diffusion is $t_\mathrm{diffusion} = r_\mathrm{p}^2/\mathcal{D}$. Solid state diffusion of oxygen into a solid matrix is an unlikely mechanism of combustion in normal circumstances. Even the largest diffusion coefficients in solids are very small compared to the thermal diffusivities described above. The difficulty is that the order of magnitude of the diffusion coefficient in the solid state is very sensitive to the choice of diffusing molecule, the microstructure of the solid matrix, and the temperature of the solid. As an example for determining the order of magnitude, one of the largest values of diffusion coefficients in metals is on the order of 10^{-12} m^2/s (Bird et al., 2002, p. 518). Even for a particle with a radius of 1 μm, the diffusion time is 1000 ms, far too large for solid state diffusion to be a factor in all but the most unusual of cases.

In Chapter 4 there was a discussion of models for the calculation of the thermal conductivity of a porous medium. In considering the mass transfer processes internal to a particle, mass diffusion in a porous medium is an important consideration. The simplest model that has been demonstrated to be useful in many applications is to consider a modified form of Fick's law of diffusion. The essential idea here is that the value of the diffusion coefficient in a gaseous medium becomes diminished by the structure of the porous medium. In other words, instead of diffusion occurring in free space, it now occurs in the free space defined by the enclosing surfaces of the porous medium. Estimating the value of transport coefficients in a porous medium is a very challenging problem because one must define the structure of the porous medium and then determine how it impacts the diffusion process. I will present only one simple model for the effective diffusivity or diffusion coefficient. Additional information can be found in the literature (Szekely et al., 1976, pp. 23–33; Froment and Bischoff, 1979, pp. 159–175; Whitaker, 1999, Chapter 1).

Maxwell derived a model for the effective diffusivity which though originally intended for the physical system of a dilute suspension of spheres has been determined to be useful for porous media as well (Whitaker, 1999, pp. 43–46). This is similar to the result for the effective thermal conductivity that was presented in Chapter 4. The model equation is

$$\frac{\mathcal{D}_m}{\mathcal{D}} = \frac{2\alpha_g}{\alpha_g(3 - \alpha_g)} \tag{6.15}$$

In this equation, $\mathcal{D}_m$ is the mixture diffusivity (the effective diffusivity in the porous medium), $\mathcal{D}$ is the binary diffusivity in air, and α_g is the volume fraction of gas in the porous medium (the porosity). As an example, a typical porosity for a porous medium composed of a random packing of sphere is $\alpha_g = 0.4$; the effective diffusivity is calculated from Maxwell's model as $(\mathcal{D}_m/\mathcal{D}) = 0.77$. Typical values for binary diffusivities of gases are on the order of 10^{-4} m²/s. Using the expression for the diffusion timescale $t_{diffusion} = r_p^2/\mathcal{D}$, porous particles with a radius on the order of 1 mm could experience significant internal diffusion effects. However, this order-of-magnitude calculation is based on the simple effective diffusivity model cited above. The ultimate test of the potential significance of internal diffusion is in the conduct of single particle combustion experiments.

6.1.3 THE RESPONSE OF SOLID FUELS TO EXTERNAL HEATING

When a solid particle burns, it will exhibit a flame near or at the particle surface. A portion of the heat liberated by the combustion process is transported to the particle surface. Most combustion reactions occur in the gas phase. Thus the heat of the flame must volatilize the solid material so that the fuel vapor can diffuse toward the flame. Although there are a few rare examples of an organic solid that melts and vaporizes during combustion, most organic solids do not. Instead the heat of the flame causes the solid to decompose (pyrolyze) and generate volatile matter (Di Blasi, 1993; Beyler and Hirschler, 2002). The original parent material may not fully decompose and so it may form a char (carbon-rich solid) and/or ash (inorganic residue).

Pyrolysis generates a distribution of decomposition products. The number and variety of decomposition products is usually a function of the heating rate. Hence, the set of decomposition products generated at a slow heating rate is usually different from the set of products generated by a fast heating rate. Thousands of papers have been published on the study of pyrolysis of organic materials. Before using a particular study to assist in the investigation of combustible dust behavior, it is important to check that the conditions of the study are sufficiently similar to the expected heating rate encountered in a particular combustible dust hazard scenario.

Metals can undergo phase changes such as solid–solid transitions, melting, or boiling. Since metals are elements (or mixtures of elements), they do not undergo chemical decomposition in the way that organic solids pyrolyze. The products of metal combustion in air, metal oxides (and sometimes metal nitrides), exhibit a range of behaviors including diffusion away from the burning particle and into the free stream, condensation onto the metal surface of the burning particle, or dissolution into the metal forming a mixture (Dreizin, 2000).

6.1.4 INTRODUCTION TO EXPERIMENTAL METHODS FOR SINGLE PARTICLE COMBUSTION

There is a vast literature on the experimental investigation of single particle combustion. Most of this research was motivated by studies to advance the state of knowledge in solid fuels for use in industrial combustors and rocket engines. In this section, we will first briefly cover some of the experimental methods used in single particle combustion studies.

There are a number of calorimetry methods that monitor the response of materials to heating (Brown, 2001). These instruments usually test a small bulk sample of the material and not necessarily a single particle. Two of the more commonly used instruments are differential scanning calorimetry (DSC) and thermogravimetry (TG). Calorimetric methods are very useful, but one must bear in mind their limitations.

In DSC, a sample of material is heated at a prescribed rate of temperature rise and the material temperature is monitored and compared with a reference. If an endothermic or exothermic reaction occurs, the instrument supplies more or less heat to the sample, respectively, and records the heat input. The DSC record indicates the temperature at which these thermal events occur. DSC is useful for identifying phase transitions and decomposition reactions and their enthalpies. An important shortcoming of DSC is the small size of the sample that is used (frequently on the order of milligrams) and the limitation of the maximum rate of temperature rise that can be accomplished with the instrument.

TG is an instrument that heats a sample in a controlled atmosphere (air or inert gas) and records the weight change in the sample as a function of temperature. TG is a popular tool for investigating the pyrolysis of materials and other chemical reactions. It suffers from some of the same limitations of DSC. The kinetic analysis of TG data is presented in Brown's book (2001, Chapter 10).

Many experimental studies have been performed with single particles heated in furnaces. These studies employ one of two methods of presenting the particle to the thermal environment: a static particle configuration (eg, a particle suspended on a wire) and a falling particle technique (allowing the particle to fall through a heated tube or drop-tube furnace). In the 1950s, these studies recorded the particle reaction with still photography or videography. Particle size data could be extracted from the captured images. Later, particles were mounted on microbalances to monitor weight change with time as well as the photographic methods. These single particle experiments have created a large literature with experimental data that allow direct comparison with gas–solid kinetic models derived in the next section.

Finally, a number of single particle combustion observations have been obtained by studying flame propagation through dust clouds in both unconfined and confined environments. These studies are addressed primarily in Chapter 7 and Chapter 8.

6.2 NONCATALYTIC GAS–SOLID REACTION MODELS

Let us begin by considering the simplest of combustion models for combustible dust particles. In this simple model, the combustible dust particle is identified as the fuel and the oxygen in the surrounding air is the oxidizer. To burn, oxygen must diffuse to the particle (droplet) surface, react with the fuel, and the product must diffuse into the free stream environment. This basic kinetic scheme is called a noncatalytic gas–solid reaction model. The term gas–solid is a bit misleading. The particle can be a liquid droplet as long as it maintains a well-defined size and shape. The term noncatalytic signifies that we will only consider homogeneous or heterogeneous chemical reactions in which the solid actively participates in the reaction. In a catalytic reaction, the catalyst speeds up the reaction rate but is not consumed as a reactant. This distinction is significant in that so much modeling effort has gone into the study of heterogeneous systems with catalytic reactions (Aris, 1975; Levenspiel, 1999, Chapters 17–22; Carberry, 2001, Chapters 5, 8, and 9). In our application of combustible dust, catalytic reactions rarely dominate the combustion process.

Noncatalytic gas–solid reaction models incorporate the reaction and mass transfer steps into an overall reaction rate that describes the depletion of fuel as the combustion reaction progresses. The combustion process may result in changes in particle size, shape, and microstructure. We will consider the simplest geometry, spherical particles, as a means of investigating the significance of particle size on the combustion process.

Intuitively, one might expect that the particle diameter will shrink as the combustion reaction progresses because the reaction is consuming the fuel particle. This was indeed the case when we investigated the diffusion flame model for liquid fuel droplets. With solid fuels, however, the changes in particle geometry can be more complicated. The particle diameter may shrink, stay the same, or grow as the combustion reaction progresses. The process of pyrolysis in charring organic solids can lead to the formation of char structures. Solid fuels with gaseous combustion products only, such as pure carbon, will shrink in particle diameter as the reaction proceeds. Volatile metals may exhibit the same behavior. But charring organic solids may maintain a constant diameter, or even swell in size, as the pyrolysis reactions generate volatile gaseous products for combustion. Likewise, some nonvolatile metals may experience a constant particle diameter history, or even growth, if the combustion products are condensed phases which deposit on the parent metal particle surface.

Particle shape exerts little influence on steady combustion behavior if the solid fuel melts or pyrolyzes without charring. Such particles tend to assume a spherical shape during combustion. Charring organics and some metals may preserve a nonspherical shape during combustion. There are few studies on the effect of particle shape on combustion behavior. When possible, the effects of particle shape will be described in terms of available empirical studies.

The topic of particle microstructure will arise later when we consider specific types of fuels. A good summary of gas—solid reaction models that can incorporate microstructural changes can be found in the monograph by Szekely and his coauthors (Szekely et al., 1976). Ramachandran and Doraiswamy have also reviewed the subject (Ramachandran and Doraiswamy, 1982). Both charring organic solids and metallic solids can experience microstructural changes during combustion.

We are specifically interested in the burning of combustible dust particles. To avoid introducing too much complexity into our modeling efforts, we will examine the reaction kinetics of spherical particles only. Spherical particles represent the simplest geometric model for combustible dusts. We will consider three kinetic models: the shrinking particle model, the shrinking unreacted core model, and the progressive conversion model. We will then briefly review some of the extensions of these models that have been proposed in the literature.

Before we begin our study of gas—solid reaction models, we will take a brief detour to consider the relationship between fractional conversion and particle radius.

The fractional conversion x_B is the fraction of the limiting reactant that has been converted to products. It can be related directly to the shrinking core radius (Levenspiel, 1999, p. 572).

$$1 - x_B = \frac{\text{volume of unreacted core}}{\text{total particle volume}} = \left(\frac{r_c}{r_0}\right)^3 = \hat{r}_c^3 \qquad (6.16)$$

An additional relation can be derived for the change in particle size as a function of fractional conversion. In general, a particle may grow, shrink, or maintain a constant particle size during reaction (Carberry, 2001, p. 317). The moles remaining of solid B can be related to the initial particle radius r_0 by the equation:

$$n_B = \frac{4\pi \rho_B}{3 \, \mathcal{M}_B} (r_0^3 - r_c^3) \qquad (6.17)$$

The moles of product C generated by the chemical reaction can be related to the radius of the particle r_p at any instant by the equation:

$$n_C = \frac{4\pi\rho_C}{3\,\mathcal{M}_C}(r_p^3 - r_c^3) \tag{6.18}$$

An additional relation that is needed is from the reaction stoichiometry: $n_B = (b/c)n_C$. Equating the expressions, $n_B = n_C$ and solving for the particle radius r_p yields the following expression:

$$\frac{r_p}{r_0} = \left[\frac{c\,\mathcal{M}_C\rho_B}{b\,\mathcal{M}_B\rho_C} + \left(1 - \frac{c\,\mathcal{M}_C\rho_B}{b\,\mathcal{M}_B\rho_C}\right)\left(\frac{r_c}{r_0}\right)^3\right]^{1/3} \tag{6.19}$$

The significance of this expression becomes more obvious when written in dimensionless form

$$\hat{r} = \left[\beta + (1-\beta)\hat{r}^3\right]^{1/3} \tag{6.20}$$

$$\hat{r} = \frac{r_c}{r_0}, \quad \beta = \frac{c\,\mathcal{M}_C\rho_B}{b\,\mathcal{M}_B\rho_C} \tag{6.21}$$

In terms of fractional conversion, the particle radius varies with reaction progress as

$$\hat{r} = [1 + (1-\beta)x_B]^{1/3} \tag{6.22}$$

If $\beta > 1$, the particle expands upon reaction; if $\beta < 1$, the particle shrinks; and if $\beta = 1$, the particle size remains constant. The key finding to remember here is that the particle radius may change with time as the reaction consumes the solid reactant. The factors which govern this transformation are the solid densities of the reactant and product, their molar masses, and their stoichiometric coefficients.

EXAMPLE 6.2

Using Equation 6.22, calculate the value of the dimensionless particle core radius at 50% conversion for three cases: $\beta = 0.8$, 1.0, and 1.2.

Solution

This calculation illustrates the physical interpretation of the coefficient β as a particle size parameter. Substituting the value for the conversion $x_b = 0.5$, Equation 6.22 takes the form:

$$\hat{r} = [1 - 0.5\,(1 - \beta)]^{1/3}$$

For a shrinking particle with $\beta = 0.8$, $\hat{r} = 0.966$. For a particle of constant size with $\beta = 1.0$, $\hat{r} = 1$. For an expanding particle with $\beta = 1.2$, $\hat{r} = 1.03$.

6.2.1 SHRINKING PARTICLE MODEL

The shrinking particle model has already been introduced as the d-squared law for liquid droplet combustion. An important distinction is that the analysis for liquid droplet combustion was derived for a diffusion flame in the gas phase. The primary assumption for the diffusion flame analysis was that the combustion reaction in the gas phase was assumed to be infinitely fast. Thus, the combustion rate was limited by the rates of fuel vapor and oxygen mass transfer and by the rate of heat transfer.

Our analysis of shrinking particle kinetics is simplified by three assumptions: (1) the particle is isothermal throughout the reaction; (2) mass transfer across the gas film boundary layer surrounding the particle can be described by a linear mass transfer coefficient; and (3) the chemical reaction kinetics is described by an irreversible, first-order reaction that is dependent only on the gas phase concentration of species A. The overall rate depends on the diffusion of species A across the gas film boundary layer and followed by the chemical reaction at the surface of the particle. It is assumed that the diffusion of gaseous product (species D) does not influence the overall rate of reaction. The kinetic process results in the shrinking and gradual disappearance of the solid particle. Fig. 6.2 depicts over time the progress of the reaction front and the disappearance of the particle.

The chemical reaction to be considered is

$$A(g) + bB(s) \rightarrow c\,C(s) + dD(g) \tag{6.23}$$

Here $A, B, C,$ and D are chemical species; b, c, d are the stoichiometric coefficients, g indicates a gas phase species, and s denotes the solid reactant or product.

The essential feature of a single burning particle is that it shrinks in size as it burns. The time duration for complete particle combustion is called the burnout time. This feature is captured by an integral mass balance on the particle combined with some simple geometry. The objective of this analysis is to derive an expression for the burnout time. We will consider two limiting cases for the overall reaction rate: diffusion control through the gas film and kinetic control.

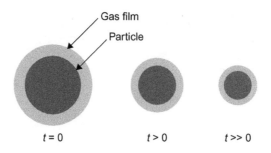

Gas film

Particle

$t = 0$ $t > 0$ $t \gg 0$

FIGURE 6.2

The shrinking particle model.

6.2.1.1 Gas film diffusion control

If the rate of chemical reaction is much greater than the rate of diffusion across the gas film, then the overall reaction rate is controlled by diffusion. The gas phase concentration of species A declines from its free stream value of $C_{A,\infty}$ to a value of zero at the solid surface of the particle, $C_{A,c} = 0$. The gas film is assumed to have a constant thickness of δ surrounding the particle. Fig. 6.3 shows the concentration profile of species A when the gas film resistance is controlling.

For diffusion through the gas film, we will follow Levenspiel's example and use the Frossling correlation for calculating the mass transfer coefficient of species A in the ambient free stream environment (Levenspiel, 1999, p. 577−581).

$$\text{Sh} = \frac{k_g d_p X_{A,\infty}}{\mathcal{D}} = 2 + \text{Sc}^{1/3}\text{Re}^{1/2} = 2 + 0.6\left(\frac{\mu}{\rho\mathcal{D}}\right)^{1/3}\left(\frac{\rho v d_p}{\mu}\right)^{1/2} \tag{6.24}$$

In this equation, the three dimensionless groups—Sh, Sc, Re—are the Sherwood, Schmidt, and Reynolds numbers, respectively. The other symbols have been defined elsewhere. Levenspiel noted the following asymptotic limits for this correlation:

$$k_g \approx \frac{1}{d_p} \quad \text{for small } d_p \text{ and } v \tag{6.25}$$

$$k_g \approx \left(\frac{v}{d_p}\right)^{1/2} \quad \text{for large } d_p \text{ and } v \tag{6.26}$$

We will use these two limits to evaluate the overall rate when gas film diffusion is controlling.

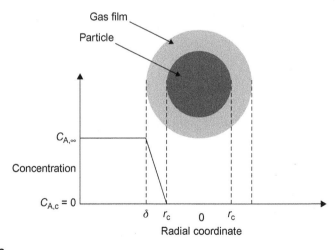

FIGURE 6.3

Concentration profile for species A when the gas film resistance is controlling.

To derive the burnout time, we begin with a mass balance on the particle written in units of moles. Since the chemical reaction occurs at the particle surface, we will normalize the overall reaction rate with the surface area of the particle S_p.

$$-\frac{1}{S_p}\frac{dn_B}{dt} = -\frac{1}{4\pi r_c^2}\frac{dn_B}{dt} = bk_g\left(C_{A,\infty} - C_{A,c}\right) = bk_g C_{A,\infty} \tag{6.27}$$

From geometry, we know the moles of B can be obtained by

$$n_B = C_B V = C_B\left(\frac{4\pi}{3}r_c^3\right) \tag{6.28}$$

The molar density C_B is calculated by dividing the mass density ρ_B by the molar mass $\mathcal{M}_B$. The radius of the particle, r_c, is designated by the subscript "c" as in "core," which is the terminology that will be used later in the description of the shrinking core model. The differential dn_B is calculated as

$$dn_B = d(C_B V) = 4\pi r_c^2 C_B dr_c \tag{6.29}$$

Substituting the expression for dn_B into the mass balance gives us the expression

$$-C_B\frac{dr_c}{dt} = bk_g C_{A,\infty} \tag{6.30}$$

The small particle limit corresponds to small Reynolds numbers (Stokes flow regime, $Re \rightarrow 0$). The mass transfer coefficient can be estimated as $k_g = \mathcal{D}/r_c$. The mass balance can now be integrated with the initial condition $r_c(t=0) = r_0$.

$$t = \frac{C_B r_0^2}{2b\mathcal{D}C_{A,\infty}}\left[1 - \left(\frac{r_c}{r_0}\right)^2\right] \tag{6.31}$$

The burnout time τ occurs when the particle disappears and is found by setting $r_c = 0$.

$$\tau = \frac{C_B r_0^2}{2b\mathcal{D}C_{A,\infty}} \tag{6.32}$$

The particle radius history can be cast in dimensionless form

$$\hat{t} = 1 - \hat{r}_c^2, \quad \text{with} \quad \hat{t} = t/\tau, \quad \hat{r}_c = r_c/r_0 \tag{6.33}$$

In terms of fractional conversion, the dimensionless particle radius history becomes

$$\hat{t} = 1 - (1-x_B)^{2/3} \tag{6.34}$$

Equation (6.33) is equivalent to the d-squared law derived for the diffusion flame surrounding a liquid fuel droplet. To see the equivalence, simply invert the equation to get it into the following form:

$$r_c^2 = r_0^2 - Kt \tag{6.35}$$

$$K = \left(\frac{8b\,\mathcal{M}_B}{\rho_B}\right)\mathcal{D}\,C_{A,\infty} \quad \text{shrinking solid particle, diffusion control} \tag{6.36}$$

$$K = \frac{8\rho_g\alpha_g}{\rho_l}\ln(B+1) \quad \text{liquid droplet diffusion flame} \tag{6.37}$$

As a result of invoking the simple device of a mass transfer coefficient at low Reynolds numbers, significantly less effort was involved in deriving the results for the diffusion-controlled shrinking particle reaction as compared with the diffusion flame analysis for liquid droplets.

At large particle sizes, the Reynolds number is large; the limiting form of the mass transfer correlation gives $k_g = (constant)/(r_c)^{1/2}$. Substituting this relation for k_g into the particle mass balance and integrating gives the result:

$$t = (\text{constant})\left[1 - \left(\frac{r_c}{r_0}\right)^{3/2}\right] \tag{6.38}$$

which gives an equation for the burnout time:

$$\tau = \frac{(\text{constant})r_0^{3/2}}{C_{A,\infty}} \tag{6.39}$$

In terms of fractional conversion, we obtain

$$\frac{t}{\tau} = 1 - (1 - x_B)^{1/2} \tag{6.40}$$

Because we have used the simplifying device of the large particle limit, the burnout time and performance equations cannot be used to calculate exact values. However, they can be used to compare different particle combustion scenarios.

6.2.1.2 Kinetic control

If the mass transfer rate is much faster than the chemical reaction, then the overall rate is kinetic controlled. In this limiting behavior, the gas phase concentration of species A is constant throughout the gas film. The concentration of species A then goes to zero at the solid surface, $C_{A,c}$. Fig. 6.4 shows the concentration profile of species A when the overall rate is kinetic controlled.

To calculate the burnout time, we write the mass balance on the particle in molar units.

$$-\frac{1}{S_p}\frac{dn_B}{dt} = -\frac{1}{4\pi r_c^2}\frac{dn_B}{dt} = bk_c C_{A,\infty} \tag{6.41}$$

We use the stoichiometric relation to transform the left-hand side into a time rate of change of the particle radius

$$dn_B = d(C_B V) = 4\pi r_c^2 C_B dr_c \tag{6.42}$$

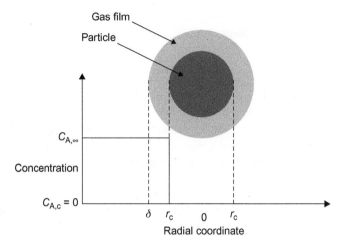

FIGURE 6.4

Concentration profile of species A when the chemical reaction is the rate-determining step.

This relation is substituted into the mass balance and the resulting differential equation is integrated with the initial condition $r_c(t = 0) = r_0$.

$$t = \frac{C_B r_0}{b k_c C_{A,\infty}} \left[1 - \left(\frac{r_c}{r_0} \right) \right] \tag{6.43}$$

The burnout time τ occurs when the particle radius is zero or $r_c = 0$.

$$\tau = \frac{C_B r_0}{b k_c C_{A,\infty}} \tag{6.44}$$

Casting the particle radius history into dimensionless terms yields

$$\hat{t} = 1 - \hat{r} \quad \text{with} \quad \hat{t} = t/\tau, \quad \hat{r}_c = r_c/r_0 \tag{6.45}$$

This equation can be written in terms of fractional conversion

$$\hat{t} = 1 - (1 - x_B)^{1/3} \tag{6.46}$$

Formally, these results only hold for a first-order irreversible chemical reaction. The same formalism can be used for a more general kinetics model by using a Taylor series expansion of the more general kinetics equation and retaining only the linear term.

6.2.1.3 Combination of resistances

Thus far we have derived performance equations for the shrinking particle model with reference to the limiting cases of diffusion- or kinetic-controlled reaction rates. Since both rate processes are linear, the resistances can be combined such that an intermediate case can be considered. The resistances can be combined in the following way. First, it is recognized that these resistances occur in series in the manner described in Section 6.1.1. This means that the rate of diffusion must equal the rate of reaction. For the shrinking particle model, the combined resistances take the following form:

$$-\frac{1}{4\pi r_c^2}\frac{dn_B}{dt} = \left[\frac{1}{1/k_g + 1/bk_c}\right]C_{A,\infty} = k_{total}C_{A,\infty} \tag{6.47}$$

An alternative way to think of the combination of resistances is to consider the time duration implied by the action of the resistance. The time needed to achieve a given level of conversion is equal to the sum of each individual time to achieve that level of conversion as if each resistance was acting alone on the overall rate (Levenspiel, 1999, pp. 579—581).

$$t_{total} = t_{film\ diffusion} + t_{kinetic} \tag{6.48}$$

Similarly, for the burnout time, the combined resistances give

$$\tau_{total} = \tau_{film\ diffusion} + \tau_{kinetic} \tag{6.49}$$

If limiting condition prevails, then one of the two terms on the right-hand side of Eq. (6.48) or (6.49) is negligibly small. If both resistances are significant, then each of the two terms on the right-hand side must be calculated.

EXAMPLE 6.3

Compare the dimensionless times for a shrinking particle to achieve 10%, 50%, 90%, and 99% conversion for a diffusion-controlled reaction (Stokes limit) and a kinetic-controlled reaction.

Solution

This is a straightforward application of Equation 6.46: $\hat{t} = 1 - (1 - x_B)^{1/3}$. Substituting the values of the conversion gives the dimensionless time of reaction for a shrinking particle.

x_B	$\hat{t}$
0.10	0.035
0.50	0.206
0.90	0.536
0.99	0.785

The dimensionless time for reaction rises slowly due to its cube root dependence on conversion.

6.2.1.4 Summary of the shrinking particle model

The shrinking particle model is the simplest noncatalytic gas–solid reaction model. This model is a simple analog for gasification reactions. The essential behavior of shrinking particle kinetics can be characterized by two limiting cases: diffusion-controlled combustion and kinetic-controlled combustion. In the limit of small particle sizes, the diffusion-controlled shrinking particle is equivalent to the d-squared law derived for liquid droplet diffusion flames. However, it is important to note that no flame has been explicitly introduced through this model. The reaction process is simply described by a first-order irreversible chemical reaction. In the kinetic-controlled limit, the shrinking of the particle is described by a function that is linear in particle size. Finally, it was shown how to calculate the particle burnout time if both rate processes are comparable in magnitude.

6.2.2 SHRINKING UNREACTED CORE MODEL

In the shrinking unreacted core model, it is assumed that the reaction rate is much faster than the rates of diffusion. Therefore, the reaction occupies a thin zone called the reaction front. As the reaction front passes through the solid particle, it leaves behind a porous solid product layer. In pyrolysis and combustion processes, this product layer would be an ash layer. The reaction front begins at the outer surface of the particle and proceeds radially inwards. It is assumed that the unreacted material (the fuel) is impermeable to gas diffusion.

Yagi and Kunii are believed to be the first investigators to use the shrinking unreacted core model to analyze particle combustion (Yagi and Kunii, 1955). Their approach used the particle mass balance approach presented in this section. Shen and Smith derived a more general shrinking core model with analytical solutions for the concentration profile and allowed for the more general case with finite diffusion and reaction rates. They also expanded this work by accounting for nonisothermal effects (Shen and Smith, 1965). Ishida and Wen developed a more general model that could be reduced in limiting cases to either the shrinking unreacted core approximation or the progressive conversion approximation (Ishida and Wen, 1968; Wen, 1968; Ishida and Wen, 1971).

The reaction front requires oxygen for the reaction. Thus oxygen must diffuse through the gas film to the particle surface. Then the oxygen must diffuse radially inwards through the porous product layer to the reaction front where it is consumed. Gaseous products then diffuse outwards through the product layer and then through the gas film surrounding the particle and into the free stream environment.

As the reaction front proceeds toward the particle center, the fuel volume continuously decreases hence, the name of the model, the shrinking unreacted core model. We will assume that the particle diameter is constant throughout the reaction. Fig. 6.5 depicts over time the progress of the reaction front and the generation of the porous product layer.

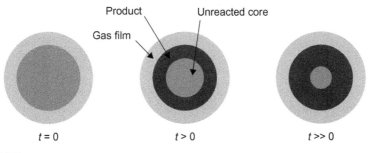

Product   Unreacted core

Gas film

$t = 0$ $t > 0$ $t \gg 0$

FIGURE 6.5

The shrinking unreacted core model with constant particle diameter.

The reaction to be considered is written as

$$A(g) + bB(s) \rightarrow cC(s) + dD(g) \qquad (6.50)$$

Here A, B, C, and D are chemical species; b, c, d are the stoichiometric coefficients; g indicates a gas phase species; and s denotes the solid reactant or product. The conceptual model displayed in Fig. 6.8 is based on the assumption that the particle diameter remains constant. Following Levenspiel (1999, pp. 570–576), the derivation that follows will invoke that assumption for simplicity. This is in fact true only if the bulk density of the product layer is equal to the bulk density of the unreacted core. This assumption can be relaxed (Wen, 1968; Szekely et al., 1976, pp. 73–88; Carberry, 2001, pp. 310–329).

Consider a spherical fuel particle immersed in air of sufficient volume that the depletion of oxygen by the combustion reaction is negligible. The ambient environment is constant in temperature and pressure. The particle is assumed to be isothermal throughout the course of the reaction. This is a strong assumption and is invoked only because it greatly simplifies the analysis. Whether the resulting model is useful will depend on the characteristics of a given application.

The objective is to derive an expression for the burnout time, ie, the time duration required for complete combustion of the fuel particle. We must account for three rate processes: the diffusion of oxygen through the boundary layer (gas film) surrounding the particle, the diffusion through the porous product layer, and the chemical reaction rate. We will examine the case when each of these three processes controls the overall reaction rate.

6.2.2.1 Gas film diffusion control

If the gas film diffusion rate is the slowest step, then the gaseous reactant A is consumed at the outer surface of the product layer. The concentration of A is at a maximum in the free stream environment and is zero at the particle surface. The concentration of A in the free stream environment is $C_{A,\infty}$, the concentration of A at the particle surface (the exterior surface of the product layer) is $C_{A,p}$, and the concentration of A at the surface of the shrinking core is $C_{A,c}$. In molar concentration units, the concentration of species A at the external surface of the particle's product layer is $C_A(r = r_p) = C_{A,p} = C_{A,c} = 0$. Thus the concentration gradient through the gas film is at its maximum value: $\Delta C_A = C_{A,\infty} - C_{A,p} = C_{A,\infty}$. Fig. 6.6 shows the concentration profile for species A when the gas film resistance is the rate-determining step. The thickness of the gas film is δ (measured from the exterior surface of the particle), the radius of the product layer is r_p, and the radius of the shrinking core is r_c. Since the radius of the particle is assumed to be constant throughout the reaction, I will designate the particle radius as $r_p = r_0$ in the derivation that follows.

We wish to observe the progress of the reaction and determine the time required for complete reaction. We begin by writing a mass balance on the unreacted particle in molar units. From the reaction stoichiometry, we know that $dn_A = bdn_B$. Since we are observing a heterogeneous chemical reaction, it is convenient to write the mass balance normalized by the exterior surface area of the particle.

$$-\frac{1}{S_p}\frac{dn_B}{dt} = -\frac{1}{4\pi r_0^2}\frac{dn_B}{dt} = -\frac{b}{4\pi r_0^2}\frac{dn_A}{dt} = bk_g\left(C_{A,\infty} - C_{A,p}\right) = bk_g C_{A,\infty} \qquad (6.51)$$

The quantity $bk_g C_{A,\infty}$ is a constant. Using the relation for the volume of a sphere, we can write the change in the number of moles of B as a change in the particle radius:

$$n_B = C_B V_c, \quad -dn_B = -C_B dV_B = -C_B d\left(\frac{4\pi}{3}r_c^3\right) = -4\pi C_B r_c^2 dr_c \qquad (6.52)$$

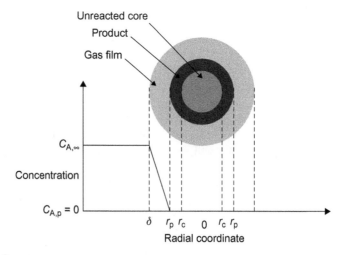

FIGURE 6.6

Concentration profile for species A for the shrinking unreacted core model when the gas film resistance is controlling.

$$-\frac{1}{S_p}\frac{dn_B}{dt} = -\frac{C_B r_c^2}{r_0^2}\frac{dn_B}{dt} = bk_g C_{A,\infty} \tag{6.53}$$

The molar concentration of B is related to its mass density by the relation $C_B = \rho_B/\mathcal{M}_B$, where $\mathcal{M}_B$ is the molar mass of B. This linear differential equation can be integrated directly with the initial condition that $r_c(t=0) = r_0$.

$$t = \frac{C_B r_0}{3bk_g C_{A,\infty}}\left[1 - \left(\frac{r_c}{r_0}\right)^3\right] \tag{6.54}$$

The burnout time τ, which is the time required for complete combustion, is found by setting $r_c = 0$.

$$\tau = \frac{C_B r_0}{3bk_g C_{A,\infty}} \tag{6.55}$$

When the overall reaction is controlled by the rate of gas film diffusion, the burnout time is linearly proportional to the initial particle radius. Finally, the history for the unreacted shrinking core can be written in dimensionless terms

$$\hat{t} = 1 - \hat{r}_c^3, \quad \text{with} \quad \hat{t} = t/\tau, \quad \hat{r}_c = r_c/r_0 \tag{6.56}$$

In terms of fractional conversion, the dimensionless particle history assumes the simple result

$$\hat{t} = x_B \tag{6.57}$$

6.2.2.2 Product layer diffusion control

If the diffusion of species A through the product layer is the slowest step, then the concentration profile of species A is a constant in the gas film and changes across the porous product layer. Thus, $C_{A,p} = C_{A,\infty}$ at the interface of the gas film and the particle exterior surface, and $C_{A,c} = 0$ at the interface of the product layer and the reactant. Fig. 6.7 shows the concentration profile for species A when the rate-determining step is the diffusion through the product layer.

To observe the reaction progress over time, we write a mass balance in molar units on the particle.

$$-\frac{dn_A}{dt} = 4\pi r^2 \dot{m}_A'' = 4\pi r_p^2 \dot{m}_{A,p}'' = 4\pi r_c^2 \dot{m}_{A,c}'' = \text{constant} \tag{6.58}$$

We invoke Fick's law for binary diffusion as the constitutive equation for the mass flux $\dot{m}_A''$.

$$\dot{m}_A'' = \mathcal{D}_e \frac{dC_A}{dr} \tag{6.59}$$

$$-\frac{dn_A}{dt} = 4\pi r^2 \mathcal{D}_e \frac{dC_A}{dr} = \text{constant} \tag{6.60}$$

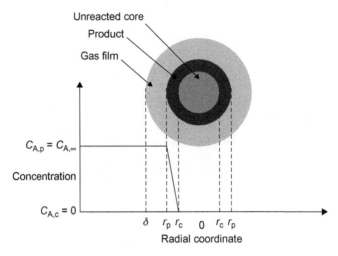

FIGURE 6.7

Concentration profile for species A for the shrinking unreacted core model when the product layer diffusion resistance is controlling.

The effective diffusivity $\mathcal{D}_e$ is the binary diffusivity of species A (oxygen in our application) as modified by the structure of the porous medium that constitutes the product layer. This differential equation is first integrated in the radial direction across the product layer thickness

$$- \frac{dn_A}{dt} \int_{r_0}^{r_c} \frac{dr}{r^2} = 4\pi \mathcal{D}_e \int_{C_{A,p}=C_{A,\infty}}^{C_{A,c}=0} dC_A \tag{6.61}$$

$$- \frac{dn_A}{dt} \left(\frac{1}{r_c} - \frac{1}{r_0} \right) = 4\pi \mathcal{D}_e C_{A,\infty} \tag{6.62}$$

The equation gives the relationship between the time rate of change in the particle at each point within the product layer. The stoichiometry of the reaction allows us to relate the change in moles of A with the change in the thickness of the product layer.

$$- dn_B = - b\, dn_A = - 4\pi C_B r_c^2 dr_c \tag{6.63}$$

Substituting Eq. (6.63) into (6.62), we get the differential equation:

$$- C_B r_c^2 \frac{dr_c}{dt} \left(\frac{1}{r_c} - \frac{1}{r_0} \right) = b\mathcal{D}_e C_{A,\infty} \tag{6.64}$$

This equation is now integrated over time

$$- C_B \int_{r_0}^{r_c} \left(\frac{1}{r_c} - \frac{1}{r_0} \right) r_c^2 dr_c = b\mathcal{D}_e C_{A,\infty} \int_0^t dt \tag{6.65}$$

Integrating, collecting like terms, and simplifying, the solution is

$$t = \frac{C_B r_0^2}{6 b \mathcal{D}_e C_{A,\infty}} \left[1 - 3 \left(\frac{r_c}{r_0} \right)^2 + 2 \left(\frac{r_c}{r_0} \right)^3 \right] \tag{6.66}$$

The desired result for the time for complete conversion of B is found by setting $r_c = 0$.

$$\tau = \frac{C_B r_0^2}{6 b \mathcal{D}_e C_{A,\infty}} \tag{6.67}$$

When diffusion through the product layer controls the overall rate, the burnout time is a function of the square of the initial particle radius. The expression for the history of the unreacted core radius can now be written in dimensionless form:

$$\hat{t} = 1 - 3\hat{r}^2 + 2\hat{r}^3 \quad \text{with} \quad \hat{t} = t/\tau, \quad \hat{r}_c = r_c/r_0 \tag{6.68}$$

In terms of fractional conversion, the dimensionless equation becomes

$$\hat{t} = 1 - 3(1 - x_B)^{2/3} + 2(1 - x_B) \tag{6.69}$$

6.2.2.3 Kinetic control

If the chemical reaction rate is the slowest step, then the concentration profile of species A is constant across the gas film and the product layer and then changes abruptly at the reactant–product interface. Fig. 6.8 depicts the concentration profile for kinetic control.

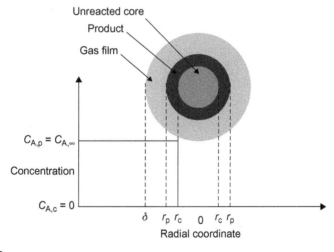

FIGURE 6.8

Concentration profile for species A for the shrinking unreacted core model when the chemical reaction rate is the controlling step.

To predict the dynamic behavior of the solid–gas reaction under kinetic control, we begin with a mass balance in molar units on the particle. The overall rate in this case is proportional to the surface area of the unreacted core. Thus, the mass balance on the particle takes the following form:

$$-\frac{1}{4\pi r_c^2}\frac{dn_A}{dt} = k_c C_{A,\infty} \tag{6.70}$$

In this analysis, the chemical reaction is assumed to be an irreversible reaction and is first order in the gas concentration of species A. We need to transform the left-hand side of this equation into an expression for the time rate of change in the radius of the unreacted core.

$$-\frac{b}{4\pi r_c^2}\frac{dn_A}{dt} = -\frac{1}{4\pi r_c^2}\frac{dn_B}{dt} = -C_B\frac{dr_c}{dt} = bk_c C_{A,\infty} \tag{6.71}$$

Integrating this equation with the initial condition that $r_c(t=0) = r_0$ and solving for t yields an expression for the history of the unreacted core radius

$$t = \frac{C_B r_0}{bk_c C_{A,\infty}}\left[1 - \left(\frac{r_c}{r_0}\right)\right] \tag{6.72}$$

The burnout time is found by setting $r_c = 0$

$$\tau = \frac{C_B r_0}{bk_c C_{A,\infty}} \tag{6.73}$$

And the unreacted core history for kinetic control can now be written in dimensionless form

$$\hat{t} = 1 - \hat{r} \quad \text{with} \quad \hat{t} = t/\tau, \quad \hat{r}_c = r_c/r_0 \tag{6.74}$$

In terms of fractional conversion, the solution becomes the linear relation

$$\hat{t} = 1 - (1 - x_B)^{1/3} \tag{6.75}$$

The relations for kinetic control of the shrinking unreacted core model are identical to the relations for kinetic control of the shrinking particle model.

6.2.2.4 Combination of resistances

There are three performance equations for the shrinking unreacted core model based on three potential sources of controlling resistance: gas film diffusion, product layer diffusion, and reaction kinetics. As before, because the rate equations for each resistance is linear, they can be combined into a single rate equation.

$$\frac{dn_B}{dt} = \frac{1}{\left[\dfrac{1}{k_g} + \left(\dfrac{r_0}{r_c}\right)\dfrac{(r_0 - r_c)}{\mathcal{D}} + \left(\dfrac{r_0}{r_c}\right)^2\dfrac{1}{k_c}\right]} bC_{A,\infty} \tag{6.76}$$

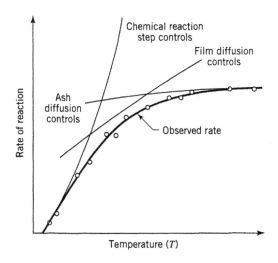

FIGURE 6.9

Relative importance of the rates of gas film diffusion, product layer (ash) diffusion, and chemical reaction on the overall reaction rate as a function of temperature.

Figure reproduced with permission from Levenspiel, O., 1999. Chemical Reaction Engineering, third ed.
John Wiley & Sons, New York. p. 584.

As with the shrinking particle model, an alternative way to think of the combination of resistances is to consider the time duration implied by the action of the resistance. The time needed to achieve a given level of conversion is equal to the sum of each individual time to achieve that level of conversion as if each resistance was acting alone on the overall rate (Levenspiel, 1999, pp. 579—581).

$$t_{\text{total}} = t_{\text{film diffusion}} + t_{\text{ash diffusion}} + t_{\text{kinetic}} \qquad (6.77)$$

In terms of the burnout time the relation is

$$\tau_{\text{total}} = \tau_{\text{film diffusion}} + \tau_{\text{ash diffusion}} + \tau_{\text{kinetic}} \qquad (6.78)$$

The relative importance of these three resistances as a function of temperature is illustrated in Fig. 6.9.

At lower temperatures the chemical reaction rate is the limiting resistance. As the temperature increases, gas film diffusion becomes the limiting rate. At high temperatures, it is the product layer (ash) that is the controlling resistance.

6.2.2.5 Summary of the shrinking unreacted core model

The shrinking unreacted core model is the next step in developing a kinetics model for noncatalytic gas—solid reactions. In its simplest form, the particle diameter remains constant while a nonporous solid is transformed into a porous product. The product layer forms at the outer surface of the particle and grows radially inwards. An analysis of the rate processes reveals two diffusion-controlled cases—through the gas film or through the porous solid product—and one kinetic-controlled case. When the overall rate is controlled by the diffusion through the product layer, the burnout time is related to the square of the particle diameter. When the overall rate is controlled by either the gas film diffusion or by the rate of chemical reaction, the burnout time is proportional to a linear function of the particle diameter. Thus, simple burning time measurements will not by themselves be sufficient to determine the rate-controlling step if the material conforms to the shrinking unreacted core model.

EXAMPLE 6.4

Compare the dimensionless times for a particle obeying the unreacted shrinking core model to achieve 10%, 50%, 90%, and 99% conversion for a diffusion-controlled reaction (Stokes limit), ash layer diffusion-controlled reaction, and a kinetic-controlled reaction.

Solution

This problem is a straightforward application of Equations 6.57, 6.69, and 6.75:

$$\text{Gas diffusion control}: \hat{t} = x_B$$
$$\text{Ash layer diffusion control}: \hat{t} = 1 - 3(1 - x_B)^{2/3} + 2(1 - x_B)$$
$$\text{Kinetic control}: \hat{t} = 1 - (1 - x_B)^{1/3}$$

Substituting the values of the conversion gives the dimensionless time of reaction for a shrinking particle.

x_B	$\hat{t}$ Gas Diffusion	$\hat{t}$ Ash Layer Diffusion	$\hat{t}$ Kinetic
0.10	0.10	0.0035	0.035
0.50	0.50	0.110	0.206
0.90	0.90	0.554	0.536
0.99	0.99	0.881	0.785

The dimensionless time for reaction varies widely with conversion depending on which process is rate-controlling.

6.2.3 PROGRESSIVE CONVERSION MODEL

In the progressive conversion model, it is assumed that the chemical reaction rate varies continuously along the particle radius. The progressive conversion model is the classic reaction–diffusion model for porous solid particles. In the realm of porous particles, the progressive conversion model is the basis for most heterogeneous catalytic kinetics studies in single catalyst pellets (Aris, 1975). Ausman and Watson derived the progressive conversion model for application to the

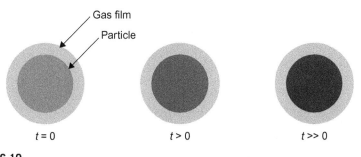

FIGURE 6.10

The progressive conversion model for a particle of constant diameter.

combustion of carbon deposited on the surface of porous catalysts (Ausman and Watson, 1962). Ishida and Wen developed a more general model that could be reduced in limiting cases to either the shrinking unreacted core approximation or the progressive conversion approximation (Ishida and Wen, 1968; Wen, 1968; Ishida and Wen, 1971). Because the applicability of the progressive conversion model to a combustible dust deflagration is somewhat uncertain, and because the mathematical analysis is quite a bit more involved, we will skip over the derivation and simply discuss a couple of key points about the model. The interested reader is directed to the references for more information.

The conceptual model for progressive conversion is illustrated in Fig. 6.10. The particle is assumed to be a solid fuel particle composed of a single chemical species. A gas film is assumed to surround the particle. The gas film is the boundary layer for oxygen diffusion from the free stream to the particle surface. The oxygen then diffuses into the particle interior and reacts with the solid. In general terms, there is an oxygen concentration profile within the particle with the maximum concentration at the particle surface. The oxygen concentration profile is monotonically decreasing toward the center of the particle.

The relevance of the progressive conversion model for combustible dust deflagrations is difficult to ascertain. For solid state particles, the diffusion timescale is too slow for a dust deflagration. For porous particles, the diffusion timescale may be sufficiently small to make this reaction model a potential candidate. This generalization will be examined more carefully later in this chapter when we examine the major findings from empirical studies on single particle combustion. What is perhaps more important is that this generalized approach to the reaction–diffusion problem for a gas–solid reaction reduces to the limiting forms of the unreacted shrinking core model and to the shrinking particle model.

6.2.4 EXTENSIONS OF THE NONCATALYTIC GAS–SOLID KINETICS MODELS

The three gas–solid kinetics models presented in this section are greatly simplified portrayals of real combustion behavior. However, these models are useful in the sense that they can offer a reasonable representation of the burnout time for a combustible dust particle. The shrinking particle model and the shrinking unreacted core model have enjoyed much use in combustion analysis. The progressive conversion model has been less successful. Three important limitations to these model as developed in this section are that the models assume isothermal behavior, that the reaction kinetics can be represented as a first-order irreversible reaction, and that the transport and reaction processes are unaffected by changes in the particle microstructure. There have been many contributions in the literature to relax these assumptions. We will only touch briefly on these three topics.

The consideration of more formal chemical kinetic phenomena will lead to nonlinear models. Wen analyzed the behavior of the shrinking unreacted core model with an nth order rate law (Wen, 1968). Although the numerical results will vary depending on the form of kinetic model and its rate constants, the basic behavior of the shrinking particle and shrinking unreacted core models is essentially unchanged. Szekely and coauthors warn in their text that calculations of burnout time (time to complete conversion in their text) may be in serious error if the linear model is used for nonlinear kinetics (Szekely et al., 1976, pp. 89–91).

Nonisothermal reaction behavior can lead to unstable behavior (Wen and Wang, 1970). It has been observed that the shrinking unreacted core model is sometimes unstable at the reaction approaches complete conversion. Unstable behavior can manifest itself as a rough or uneven reaction front or as a sudden change in the rate-controlling step. If the reaction is endothermic, there will be one stable steady state. If the reaction is exothermic, multiple steady states are possible (Wen and Wang, 1970; Szekely et al., 1976, pp. 93–101).

There are two basic concepts that have been implemented to model microstructural changes in gas–solid reactions: the grain model and the pore model (Ramachandran and Doraiswamy, 1982). In the grain model, the individual particle is assumed to be composed of a compacted mass of very small grains. The compacted mass of grains form macropores that allow gas to diffuse into the particle and contact the individual grains. The solid–gas reaction is assumed to occur at the individual grain level. There has been some success in using these models to investigate microstructural changes in particles caused by sintering (Szekely et al., 1976, pp. 125–167).

In the pore model, a pore distribution network is assumed for the particle. The gas–solid reaction occurs at the pore walls (Petersen, 1957; Szekely et al., 1976, pp. 109–125). These models have enjoyed some success in describing changes in pore size distribution caused by gas–solid reactions.

6.2.5 SUMMARY OF NONCATALYTIC GAS–SOLID REACTION MODELS

Our analysis of single particle combustion will be based on the two idealized gas–solid kinetics models, the shrinking particle and shrinking unreacted core models. To the extent possible, the performance of the model will be tested against the experimental results produced by single particle combustion tests. We will examine the limiting behavior of diffusion- and kinetic-controlled burning. Diffusion-controlled burning is characterized by chemical reaction rates that are much faster than the diffusion rates of combustion reactants and products. In a general sense, larger particles at higher temperatures experience diffusion-controlled burning. In kinetic-controlled burning, the rate of chemical reaction determines the rate of combustion. This limit is most often associated with smaller particles and colder temperatures. The equations for calculating burnout time, the particle size history, and the conversion history are summarized for the shrinking particle and shrinking unreacted core models in Table 6.2.

Table 6.2 Summary of Key Performance Equations for the Shrinking Particle and Shrinking Unreacted Core Models

Limiting Case	Performance Equation	Shrinking Particle	Shrinking Unreacted Core
Gas film diffusion, small particle, Re = 0	Burnout time	$\tau = \dfrac{C_B r_0^2}{2b\mathcal{D}C_{A,\infty}}$	$\tau = \dfrac{C_B r_0}{3bk_g C_{A,\infty}}$
	Radius history	$\dfrac{t}{\tau} = 1 - \left(\dfrac{r_c}{r_0}\right)^2$	$\dfrac{t}{\tau} = 1 - \left(\dfrac{r_c}{r_0}\right)^3$
	Conversion history	$\dfrac{t}{\tau} = 1 - (1-x_B)^{2/3}$	$\dfrac{t}{\tau} = x_B$
Gas film diffusion, large particle, v_{gas} = constant	Burnout time	$\tau = \dfrac{(\text{const})r_0^{3/2}}{C_{A,\infty}}$	N/A
	Radius history	$\dfrac{t}{\tau} = 1 - \left(\dfrac{r_c}{r_0}\right)^{3/2}$	N/A
	Conversion history	$\dfrac{t}{\tau} = 1 - (1-x_B)^{1/2}$	N/A
Product layer diffusion	Burnout time	N/A	$\tau = \dfrac{C_B r_0^2}{6b\mathcal{D}_e C_{A,\infty}}$
	Radius history	N/A	$\dfrac{t}{\tau} = 1 - 3\left(\dfrac{r_c}{r_0}\right)^2 + 2\left(\dfrac{r_c}{r_0}\right)^3$
	Conversion history	N/A	$\dfrac{t}{\tau} = 1 - 3(1-x_B)^{2/3} + 2(1-x_B)$
Kinetic control	Burnout time	$\tau = \dfrac{C_B r_0}{bk_c C_{A,\infty}}$	$\tau = \dfrac{C_B r_0}{bk_c C_{A,\infty}}$
	Radius history	$\dfrac{t}{\tau} = 1 - \dfrac{r_c}{r_0}$	$\dfrac{t}{\tau} = 1 - \dfrac{r_c}{r_0}$
	Conversion history	$\dfrac{t}{\tau} = 1 - (1-x_B)^{1/3}$	$\dfrac{t}{\tau} = 1 - (1-x_B)^{1/3}$

We will consider primarily the effect of particle size on combustion behavior. The typical particle size range of concern for combustible dust particles is 1 μm for the lower bound and 500 μm (0.5 mm) for the upper bound. That is only a guideline, and there will always be exceptions to that statement. Generally, particles larger than 500 μm are difficult to ignite. They may participate in a dust deflagration, but they will not dominate the behavior. The lower limit is admittedly somewhat arbitrary. Small particles are easier to ignite and will burn faster. Small particles are dangerous, but the smaller the particle, the greater the cohesive strength of the powder deposit. So really small particles (submicron sizes) tend to form agglomerates and may be difficult to suspend into a cloud (Eckhoff, 2009). The lower limit of particle size is really a matter of which particle sizing technology is being used.

Particle size in particular is important because it establishes a timescale for reaction. This can be compared with the typical timescales of a dust deflagration, 10−100 ms. The burnout time for a single particle should not exceed the dust deflagration timescale (in an order-of-magnitude sense). Particle sizes that exceed the deflagration timescale are unlikely to achieve complete combustion. This is an important inference because engineering studies of single particle combustion are typically designed to achieve complete combustion (at least as an ideal). Hence, the published engineering literature on single particle combustion must be viewed with the perspective that any given study may be better viewed as an ideal reference case and not necessarily what one will encounter in an investigation of a combustible dust accident.

When possible, we will also explore the effect of particle shape, but there are fewer experimental studies available for that purpose. The potential effects of particle microstructure will complete the discussion.

6.3 SINGLE PARTICLE COMBUSTION MODELS FOR ORGANIC SOLIDS

In this section, we will survey single particle combustion studies on organic solids. In general, the combustion of solid particles is more complicated than the combustion of liquid droplets. One combustion model is not sufficient because different types of solid fuels can exhibit fundamentally different combustion behaviors. Since transport processes are scale dependent, there is often an interplay between diffusional and kinetic controls with particle diameter acting as a key parameter. Therefore, I have organized this discussion of combustion phenomena according to how the solid responds to external heating.

The combustion behavior of organic solids can be classified according to their propensity to generate volatile matter and form a solid char. It is convenient to divide organic materials into three categories:

- Noncharring, completely volatile solids
- Charring, partially volatile solids
- Nonvolatile solids.

This classification of organic solids is based loosely on the proximate analysis of the material which includes moisture, volatile matter, fixed carbon, and ash (see Chapter 2) (Speight, 2005). The volatile component of the combustible dust is the primary fuel in a dust deflagration.

Experimental data for the single droplet combustion of organic liquids will be considered first to establish a benchmark for the performance of a diffusion-controlled shrinking particle model. This benchmark will be our reference case for the comparison of single particle combustion models for each fuel type. The specific objective of this section is to determine how well a simple gas–solid kinetics model is able to predict the progress of combustion and the burnout time. We will find that for combustible dust particles in the particle size range of interest—1–500 μm—the shrinking particle kinetics model provides an adequate representation of most forms of organic or metallic solids.

6.3.1 ORGANIC LIQUIDS

The combustion of liquid fuel droplets was discussed in Chapter 4. The diffusion flame analysis which led to the development of the d-squared law can be thought of as a diffusion-controlled shrinking particle model. In many of the references to be cited in this chapter, this model is often referred to as the *oil drop model*. Godsave appears to have published one of the earlier accounts of the d-squared law (Godsave, 1949). In his 1953 article, Godsave presented a discussion of his experiments with 16 hydrocarbon liquids (Godsave, 1953). Fig. 6.11 is a direct photograph (no supplemental lighting used) of four of the liquid fuels tested alongside a coal gas flame for comparison. The variation in luminosity with fuel type is evident from the photograph.

Godsave's experimental apparatus involved suspending a single pendant droplet from a silica fiber, igniting the droplet, and recording the burning behavior with photography. The images were captured by photographing the droplets in silhouette by a cinematograph camera. By using strong background lighting, he was able to wash out the image of the flame making it easier to record and measure the droplet diameters (Fig. 6.12).

The initial droplet diameters were on the order of 1.5 mm. In the 1953 paper, he gave a plot of single droplet combustion data from tests using three liquid fuels: benzene, tertiary butyl benzene, and tertiary amyl benzene. He plotted the square of the droplet diameter versus time and observed an excellent fit of the data to the model equation (Fig. 6.13).

He obtained the burning rate constant K (in his paper he called it the evaporation constant) by fitting the data to the d-squared law:

$$d_p^2 = d_{p,0}^2 - Kt \qquad (6.79)$$

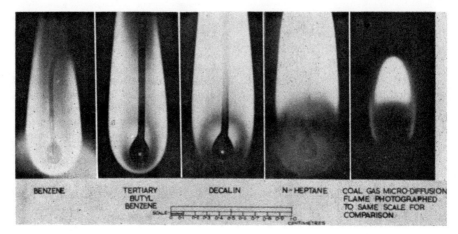

FIGURE 6.11

Photographs of single droplet diffusion flames for four different hydrocarbon liquids.

Reproduced with permission from Godsave, G.A.E., 1953. Studies of the combustion of drops in a fuel spray—the burning of single drops of fuel. In: Fourth Symposium (International) on Combustion. The Combustion Institute, Elsevier, New York. pp. 818–830.

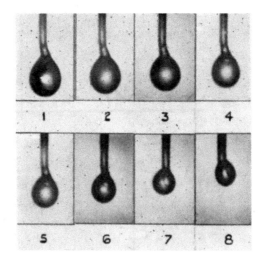

FIGURE 6.12

Silhouette photograph of burning petroleum ether droplet, original diameter 1.5 mm.

Reproduced with permission from Godsave, G.A.E., 1953. Studies of the combustion of drops in a fuel spray—the burning of single drops of fuel. In: Fourth Symposium (International) on Combustion. The Combustion Institute, Elsevier, New York. pp. 818–830.

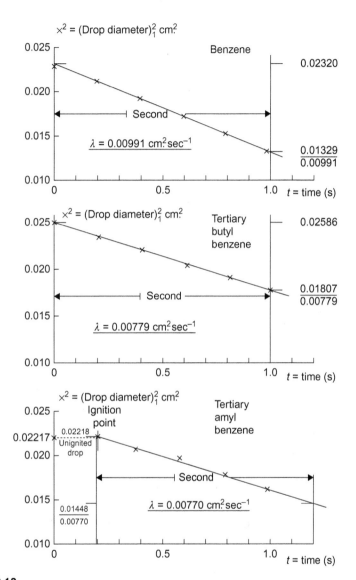

Godsave estimated the experimental uncertainty in determining the value of the burning rate constant as approximately $\pm 5\%$. Table 6.3 summarizes the burning rate constants that he obtained for 13 liquid hydrocarbons.

He commented in his paper that for typical organic liquids similar to those he tested, one would predict that the burnout time for a 100-μm droplet would be on the order of 10 ms.

Goldsmith and Penner (1954) presented a derivation for the diffusion flame analysis of single droplet combustion (very similar to the one featured in Chapter 4) and demonstrated that the d-squared law followed directly from the analysis. They demonstrated the ability of the diffusion flame analysis to predict the burning rate constant using combustion data published by others. They also indicated that the flame radius predicted by the diffusion flame analysis was larger than the measured values by a factor of 3. They attributed the discrepancy to the enhancement of heat transfer due to the natural convection caused by the combustion process.

Goldsmith (1956) performed single droplet experiments at various oxygen concentrations with *n*-heptane, ethanol, benzene, and toluene. Droplets were suspended on the tip of a quartz fiber. He again affirmed the goodness of fit of the combustion data to the d-squared law. He estimated his experimental uncertainty

Table 6.3 Burning Rate Constants Experimental Values from Single Droplet Combustion Tests (Godsave, 1953)

Chemical	Chemical Formula	Molar Mass	Burning Constant ($\mu m^2/ms^a$)
Petroleum ether	N/A	N/A	987
Kerosene	N/A	N/A	964
Diesel fuel	N/A	N/A	795
Ethanol	C_2H_6O	46.068	815
Benzene	C_6H_6	78.108	972
Ethyl benzene	C_8H_{10}	106.160	865
o-Xylene	C_8H_{10}	106.160	794
p-Xylene	C_8H_{10}	106.160	769
Isopropyl benzene	C_9H_{12}	120.186	783
tert-Butyl benzene	$C_{10}H_{14}$	134.212	774
tert-Amyl benzene	$C_{11}H_{16}$	148.238	782
n-Heptane	C_7H_{16}	100.198	967
Iso-Octane	C_8H_{18}	114.224	950

[a]*Nonstandard, but convenient units.*

as ±1%. Goldsmith observed that combustion experiments with certain fuels in enriched oxygen atmospheres caused the formation of a brittle, carbonaceous residue. The residue was formed when burning aromatic fuels like benzene and toluene.

These early experiments validate the basic features of the diffusion flame analysis and the d-squared law for liquid fuel droplets. The concept of these single droplet combustion tests seems easy: suspend a fuel droplet on a fiber, ignite and record the decreasing droplet diameter. But getting good, reproducible data is rarely easy. We shall see that combustible dust particles can introduce their own experimental difficulties.

6.3.2 NONCHARRING ORGANIC SOLIDS: SYNTHETIC POLYMERS

Some organic solids will melt and vaporize upon heating. Many organic polymers will pyrolyze and volatilize during combustion. This behavior is identical to the combustion of liquid droplets. Because these materials form little or no char, they are essentially completely volatile.

Polymer microstructure (crystallinity) does not influence the combustion process. Instead, it is the molecular structure of the polymer that influences combustion and the potential formation of char. The primary determinant in charring behavior of polymers is cross-linking. Greater cross-link densities favor charring behavior. Fillers and additives may also affect combustion behavior and contribute to the formation of char.

Noncharring organic solids are completely volatile (Beyler and Hirschler, 2002). Upon the application of external heat, noncharring organic solids generate volatile matter until the original mass has been fully consumed. Noncharring organic solids can be modeled using the diffusion flame model developed for liquid droplets. Synthetic polymers that melt and flow upon heating are called thermoplastics. If heated to a sufficiently high temperature, many thermoplastics will pyrolyze, ignite, and burn without charring. Thermosets are polymers that have cross-links that connect the polymer chains together. Generally speaking, thermosets are charring organic solids. In this section, we will review the combustion behavior of thermoplastics. Only a small selection of papers will be reviewed here with the main goal of establishing the best fit for the single particle combustion model. Most of these earlier studies were performed with spherical particles with diameters exceeding 1 mm which may be less relevant to combustible dust hazards. However, there are later studies that used particles less than 500 μm in diameter, a size range that is more relevant to our interests.

6.3.2.1 Combustion rate law

One of the earlier papers that investigated the combustion behavior of thermoplastics was by Essenhigh and Dreier (1967). They studied the combustion behavior of 11 polymers using spheres measuring 1−2 mm in diameter. Particles were suspended on a silica fiber, ignited, and allowed to burn to completion. The shrinking diameter of the spheres during combustion was recorded by photography. Each of the spheres burned with a diffusion flame that surrounded the sphere. The flame radius was offset from the polymer sphere surface. The nylon sphere maintained its spherical shape. The other polymers deformed slightly during melting and combustion, but still maintained essentially curvilinear profiles.

Most of the polymers burned like liquid droplets and conformed to the d-squared law. The phenolic polymer and polycarbonate burned in two stages. In the first stage, they burned like a liquid droplet, but later, in the second stage, they formed a char. The char continued to burn but at a different rate than in the first stage. During the heating and ignition of the particles, some exhibited swelling up to 1.5 times the initial diameter. The pyrolysis reaction appeared to be limited to the surface of most of the polymers. The polystyrene–butadiene copolymer exhibited volumetric pyrolysis: bubbles formed within the sphere and ruptured at the surface.

Burning rate constants were determined for all 11 polymers. The values obtained indicated that the polymers were burning faster than the typical hydrocarbon liquid droplets. This was attributed to the observation that the flame was much closer to the polymer surface leading to enhanced rates of heat transfer. Table 6.4 summarizes the burning rate constants obtained from their experiments.

Comparing the polymer burning rate constants with the hydrocarbon liquids, the polymer constants are typically two to three times larger.

Waibel and Essenhigh reported on an extension of this work by performing single particle combustion tests at different concentrations of oxygen ranging from 10% to 70% (Waibel and Essenhigh, 1973). They tested nylon, polystyrene, and polymethyl methacrylate (PMMA) particles with spherical diameters ranging

Table 6.4 Summary Polymer Burning Rate Constants (Essenhigh and Dreier, 1967)

Polymer	C/H Atomic Ratio	Burning Rate Constant ($\mu m^2/ms$)
Polyethylene	0.500	2510
Polypropylene	0.500	2480
Nylon 6 (Zytel)	0.549	3400
Polymethyl methacrylate (Lucite)	0.625	2750
Polystyrene–butadiene	0.666	2270
Polyvinyl acetate	0.666	3940
Cellulose acetate	0.714	16,400
Polystyrene	1.00	1740
Polycarbonate	1.14	8620
Phenolic resin (novolak)	1.14	1240
Polyamide resin (Versamid)	N/A	3370

from 1 to 2 mm. They recorded the combustion data with photography and by weighing the change in particle mass during combustion. They found that the polymers obeyed the d-squared law over the full range of oxygen concentrations tested. In their study, they also derived a model for particle combustion based on the pyrolysis reaction process. Raghunandan and Mukunda (1977) performed single particle combustion experiments with polystyrene (spherical particle diameters of 2, 3, 4, and 5 mm) and derived a similar model to Waibel and Essenhigh. They recorded the progress of the combustion reaction by measuring the change in particle mass over time and measured the central temperatures of the particles during combustion (Raghunandan and Mukunda, 1977).

There has been some interest in the behavior of burning polymer particles in a microgravity environment (Yang et al., 2008). Although the behavior of particle combustion in microgravity is beyond the scope of this book, this study is mentioned because they also performed experiments in terrestrial gravity. Their test program used mostly PMMA spherical particles at several diameters (2, 2.5, 3, 3.2, 4.8, and 6.4 mm), but they also performed some experiments with polypropylene and polystyrene. They noted that in terrestrial gravity the progress in combustion followed the d-squared law. Interestingly, in microgravity the bubbling of the polymer during combustion gave rise to sputtering and ejection of material sufficient to cause impulsive motion of the particles while suspended on a wire. Thus, in microgravity, the burning particles did not obey the d-squared law.

Two studies of particular interest investigated the burning behavior of spherical particles in a size range more relevant to combustible dust hazards (Panagiotou and Levendis, 1994; Panagiotou et al., 1994). In one study, Panagiotou and his colleagues investigated the single particle burning behavior of four thermoplastics: polystyrene, polyvinyl chloride (PVC), polyethylene, and polypropylene (Panagiotou and Levendis, 1994). The spherical particles ranged in size from 53 to 300 μm. Fig. 6.14 shows the size and appearance of typical unburnt particles used in the investigation.

Instead of suspending particles on a fiber, they used a drop-tube furnace and recorded the combustion process with video photography and optical pyrometry. The nonchlorinated polymers all followed the d-squared law, although there was evidence that the combustion process was at least partially controlled by the finite rate of pyrolysis. The PVC underwent a charring reaction early on in the combustion process and deviated considerably from the d-squared law. Fig. 6.15 shows the diffusion flame surrounding a burning polystyrene particle with an initial particle diameter of 60 μm. The flame diameter attained a maximum value of 1.5 mm.

They operated the furnace at a constant temperature and with varying levels of oxygen. Their experimentally determined burning rate constants for polystyrene, polyethylene, and polypropylene are summarized in Table 6.5.

The burning rate constants for PVC are not shown because the data did not fit the d-squared law. The burning rate constants derived from this experimental

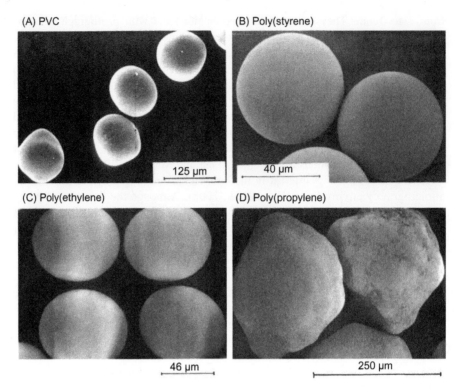

FIGURE 6.14

Typical unburnt polymer particles used in a single particle combustion study by Panagiotou and Levendis.

Reproduced with permission from Panagiotou, T., Levendis, Y., 1994. A study on the combustion characteristics of PVC, poly(styrene), poly(ethylene), and poly(propylene) particles under high heating rates. Combust. Flame 99, 53–74.

FIGURE 6.15

High-speed video frames of a burning polystyrene particle. The time between two consecutive frames is 1 ms (left to right).

Reprinted with permission from Panagiotou, T., Levendis, Y., 1994. A study on the combustion characteristics of PVC, poly(styrene), poly(ethylene), and poly(propylene) particles under high heating rates. Combust. Flame 99, 53–74.

Table 6.5 Summary of Burning Rate Constants for Three Polymers Under Various Combustion Conditions (Panagiotou and Levendis, 1994)

Polymers	Combustion Conditions	Initial Diameter (μm)	Burning Rate Constant (μm^2/ms)
Polystyrene	1373 K, air	60	550
Polystyrene	1373 K, 50% O_2	60	710
Polystyrene	1373 K, O_2	60	650
Polystyrene	1473 K, air	250	800
Polyethylene	1373 K, air	71	560
Polyethylene	1473 K, air	212	500
Polypropylene	1473 K, air	250	520

program are similar to those determined for organic liquids, but different from the values report for polymers by Essenhigh and Dreier (1967). It is unknown whether the discrepancy is due to the difference in experimental methods or due to the difference in particle sizes. Nevertheless, the body of results indicates that noncharring organic solids conform to the d-squared law.

6.3.2.2 Combustion mechanism

Beyler and Hirschler (2002) identify four mechanisms of polymer thermal decomposition. Two of these mechanisms are directed at the main chain of the polymer: chain scission and cross-linking. Chain scission is the process of breaking a chemical bond on the main chain, thus creating two smaller chains. As the chain scission process continues, the product is typically a distribution of oligomers ranging from 1 to 10 monomer units. As the thermal decomposition proceeds, it is this pool of liquid oligomers that volatilizes and burns in the flame.

Cross-linking is the union of two main chains by a chemical bond. Cross-linking usually leads to char formation, but that is not the only mechanism that creates char. Another mechanism for char formation is side chain elimination.

The other two mechanisms for the thermal decomposition of polymers operate on the side groups or side chains along the main chain. They are side chain elimination and side chain cyclization. Side chain elimination involves the stripping away of functional groups attached to the side of the main polymer chain. The side group that was removed from its main chain position reacts with other side groups forming a low molecular weight liquid. In cyclization, the broken main chain reacts with itself and forms a cyclic structure on the main chain. This process can also be the mechanism for forming a carbonaceous char.

A notable feature of noncharring polymers is the generation of bubbling and volatile release during combustion. These observations were noted in a number of studies using macroscopic PMMA samples that were typically 10 mm thick or more with bubble diameters as large and 1 mm (Kashiwagi and Ohlemiller, 1982). The bubbles transported volatiles from the interior of the molten polymer to the surface.

Oxygen facilitated the chain scission process in the polymer melt, but the depth of oxygen penetration was limited to the diameter of the bursting bubbles at the polymer surface (Kashiwagi, 1994; Brown and Kashiwagi, 1996). Although they were unable to record the size of the bubbles, investigators have reported observing the formation of bubbles in particles on the order of $1-3$ mm in diameter (Essenhigh and Dreier, 1967; Raghunandan and Mukunda, 1977; Yang et al., 2000). For particle sizes of interest in combustible dust hazards, the potential role of bubble formation has not yet been determined.

6.3.2.3 Effects of cross-linking

While there are many synthetic polymers that do not form a char while burning, there are many that do. Two industrially common polymers are known char-formers: polyvinyl chloride and polyacrylonitrile. This appears to be a function of the specific chemical structure of the main chain (Beyler and Hirschler, 2002). Panagiotou and Levendis (1994) observed the formation of char during single particle combustion experiments with PVC particles. They determined that PVC does not obey the d-squared law over the entire lifetime of the particle.

   Another factor that leads to char formation is cross-linking (Panagiotou et al., 1994). In their 1994 paper, Panagiotou et al. describe tests that they performed with polystyrene spheres that had varying degrees of cross-linking. They conducted two campaigns of single particle combustion experiments: particles with diameters in the range from 47 to 63 μm with varying degrees of cross-linking (ranging from 0% to 25%) and particles with an 8% degree of cross-linking and diameters ranging from 47 to 350 μm. They also tested styrene monomer droplets in the size range of $400-500$ μm. They found that the noncross-linked polystyrene particles burned as a liquid droplet, but the cross-linked particles followed a char oxidation period after the burnout of volatile material. With the set of particles of constant degree of cross-linking, particles larger than 150 μm followed the d-squared law with volatiles escaping from a thin layer at the surface of the particle. Smaller particles seemed to lose volatiles from the depth of the particle interior. It was estimated that most of the particle mass (perhaps as much as 98%) burned during the volatile combustion phase. The flame temperatures were similar to those obtained during the combustion of styrene droplets. As the degree of cross-linking increased, the amount of soot generated in the flame increased and the time required for total burnout increased. As expected, the time for total burnout increased with increasing particle size.

EXAMPLE 6.5

Compare the burnout times for a droplet of iso-octane and a polyethylene particle for three diameters: 25, 100, and 500 μm. Assume both obey the d-squared law. Use burning constants $K = 950(\mu m)^2/ms$ for iso-octane and $K = 500(\mu m)^2/ms$ for polyethylene.

Solution

The burnout time for the d-squared law is given by the equation $\tau = d_{p,0}^2/K$. The table below summarizes the calculations.

$d_{p,0}$, µm	Iso-octane τ, ms	Polythylene τ, ms
25	0.66	1.25
100	10.5	20
500	263	500

As explained in Section 6.1.2, the time scale for a typical dust deflagration is on the order of 10–100 ms. The burnout calculations indicate that 25 and 100 µm particles would react to complete conversion in a deflagration. The 500 µm particles would react only partially.

6.3.3 CHARRING ORGANIC SOLIDS: COAL

Coal is a metamorphic rock. It is the organic residue of fossilized prehistoric vegetation. Depending on its exposure to heat and pressure over time, coal progresses through a series of transformations generally starting with lignite and progressing to bituminous and ultimately to anthracite coal. This classification is called the rank of the coal. Lignite is considered a low rank coal and anthracite is a high rank coal. As the rank of the coal increases the carbon content increases, the hydrogen content decreases, and the mechanical strength of the coal increases. ASTM has developed one of the more commonly accepted classification schemes for the rank of coal, ASTM D388 (ASTM, 1998). The rank of coal is based on the quantity of fixed carbon, volatile matter, and its heat of combustion. The elemental composition of several common organic solid fuels and types of coal is given in Table 6.6.

Coal is an important combustible dust because it is used as a solid fuel in many utilities to generate electricity. It is also an important feedstock for

Table 6.6 Approximate Elemental Composition of Common Organic Solid Fuels in Mass Percent

Materials	Carbon (%)	Hydrogen (%)	Oxygen (%)
Wood (cellulose)	44	6	50
Peat	59	6	35
Lignite	71	5	24
Subbituminous coal	74	5	21
Bituminous coal	84	5	11
Anthracite	94	3	3
Graphite	100	–	–

Adapted from Edgar, T.F., 1983. Coal Processing and Pollution Control. Gulf Publishing, New York. p. 23.

manufacturing coke, an essential material for the steel industry. While roughly 60% of the coal in the United States comes from aboveground mining, the remaining 40% comes from underground mines (Crelling et al., 2010). An underground mine is the perfect setting for a dust explosion: combustible dust is everywhere and the tunnels form long, one-dimensional channels that provide a degree of confinement rarely seen in built structures. One safeguard used in underground mines is to deploy limestone rock dust to cover all nonworking surfaces within the mine (Amyotte, 2006). Recent surveys of the coal particle size distribution in underground mines indicate that the finest size fraction, minus 200 mesh or $d_p < 75\ \mu m$, make up 30–40% of the coal dust in the mines (Sapko et al., 2007). These very fine dusts are easily lofted and settle on top of the rock dust thereby diminishing its effectiveness (NIOSH, 2006).

Coal will pyrolyze upon exposure to external heating forming three products: gaseous volatiles, solid char, and ash. Because some of the parent material becomes char, coal is only partially volatile. Furthermore, the precise yield of volatiles versus char is dependent on the heating rate. If coal is reacted (oxidized) to completion, both the volatile matter and the char will contribute to the total heat release of combustion. The volatile matter forms a "cloud" around the particle and burns first. The oxidation of the char material follows. The combustion of volatiles occurs much faster than char oxidation (about 10 times faster). For particle sizes smaller than 200 μm, the timescale for devolatilization and combustion is on the order of 100 ms while the timescale for char oxidation and burnout is on the order of 1000 ms (Smith, 1982). Depending on the heating rate, these two oxidation processes may occur simultaneously (Saastamoinen et al., 1993; Veras et al., 1999).

Coal has a small amount of inorganic content. The combustion process generates ash. If the ash particles are transported into the free stream gas, they have little effect on the rate of combustion. If the ash particles coalesce onto the surface of the burning particle, the ash can form a porous layer which will impede combustion (analogous to the shrinking unreacted core model).

In coal dust deflagrations, coal rarely burns to completion. One of the earliest observations of coal residue following a coal mine explosion was reported by the chemist Michael Faraday and the geologist Charles Lyell in their published report on the investigation of the explosion at the Haswell Collieries (Faraday and Lyell, 1845). They reported finding large deposits of coke (devolatilized coal) in the region of the mine associated with the explosion. They collected samples of the coke and compared it with unburnt coal from the same mine. They found that much of the volatile content had been driven out of the coke. George Rice of the U.S. Geological Survey made similar observations (Rice, 1910). He reported the proximate analysis of coke samples with unburnt coal and demonstrated comparable changes in volatile matter (decreased by 25–50%) and fixed carbon (25% decrease).

Coal is perhaps the only combustible dust that has been studied so extensively in postexplosion environments. More recent studies by the U.S. Bureau of Mines (later incorporated into the National Institute of Occupational Safety and Health or NIOSH) performed coal dust explosion testing in the laboratory and performed controlled dust explosion experiments in a full-scale mine (Hertzberg et al., 1982; Ng et al., 1983; Cashdollar et al., 2007; Man et al., 2009). Using modern laboratory techniques and instrumentation like electron microscopy, these investigators were able to establish conclusively the important role of volatiles combustion and, to a lesser degree, char oxidation in accidental coal dust deflagrations. In the context of incident investigations, they have also demonstrated that coke formation is potential indicator of flame travel during an accidental deflagration. Real-time sampling and analysis of both gas and solid particulate matter indicate that the dust deflagration process in a coal mine occurs in the fuel-rich regime and indicated that some degree of char oxidation likely occurred (Conti et al., 1988, 1991).

Coal particles can take on a variety of sizes and shapes. Some typical examples are shown in Fig. 6.15 (Cashdollar, 2000). The unburnt particles have a rough and angular shape. After a dust deflagration, the particles have become larger and more rounded due to the softening, swelling, and devolatilization that occurred during combustion (Fig. 6.16).

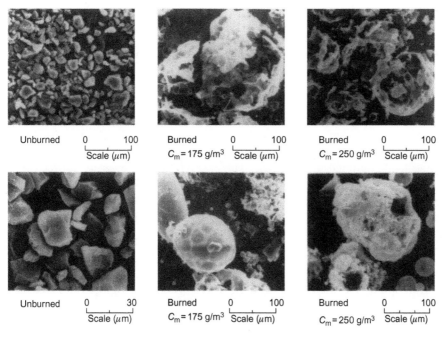

Unburned   0 —— 100
             Scale (μm)

Burned   0 —— 100
$C_m = 175$ g/m^3 Scale (μm)

Burned   0 —— 100
$C_m = 250$ g/m^3 Scale (μm)

Unburned   0 —— 30
             Scale (μm)

Burned   0 —— 100
$C_m = 175$ g/m^3 Scale (μm)

Burned   0 —— 100
$C_m = 250$ g/m^3 Scale (μm)

FIGURE 6.16

Variation in the particle sizes and shapes of a Pittsburgh coal dust sample before and after combustion.

Reprinted with permission from Cashdollar, K.L., 2000. Overview of dust explosibility characteristics, J. Loss Prev. Process Ind. 13, 189–199.

The combustion of volatile material relies almost exclusively on homogeneous chemical reactions. The combustion of the volatile matter can be modeled using the diffusion flame model developed for liquid droplets. Char oxidation, on the other hand, relies on a network of heterogeneous and homogeneous chemical reactions. The devolatilization process typically creates a porous char material (Laurendeau, 1978; Smith, 1982; Simons, 1983). The porous structure evolves during the combustion process. The pore size distribution and the specific surface area have a dominant effect on the oxidation rate of the char material. The combustion of the char material requires a new mathematical model, the nonvolatile organic solid combustion model which will be discussed later in the section on nonvolatile organic solids.

6.3.3.1 Combustion rate laws for volatiles and for char

Some of the earliest studies on single particle combustion for coal were published by Nusselt in 1916 and 1924 (Essenhigh and Fells, 1960). Nusselt's model was based on a diffusion flame analysis and predicted a d-squared law for the shrinking coal particle. In their paper, Essenhigh and Fells discussed the resistances in series model for solid particle combustion. For a set of 10 bituminous coals with a particle size range from 775 to 4000 μm, they obtained typical burning rate constants for volatiles combustion that ranged from 740 to 2200 μm^2/ms. For coal char residues with particle sizes from 300 to 4000 μm, they found that char oxidation approximately followed the d-squared law with burning rate constants on the order of $47-100 \mu m^2$/ms (Table 6.7). To a first approximation, their data indicate that volatiles combustion was an order of magnitude faster than char oxidation.

The effect of oxygen on burnout times was studied by Beeston and Essenhigh (1963) using a single particle combustion apparatus. Their experiments were conducted with a bituminous coal in the particle size range from approximately 400 to 1900 μm. They observed that during volatiles combustion a flickering, luminous flame surrounded the particle. This was followed by a steady glowing of the particle during char oxidation. Burnout times decreased with increasing oxygen concentrations. They posited the d-squared law for diffusion-controlled combustion and a d-linear law for kinetic-controlled combustion.

In another study, Essenhigh and colleagues analyzed the relationship between diffusion-controlled and kinetic-controlled combustions and concluded that the burning of coal char particles with a size smaller than 100 μm was kinetic

Table 6.7 Summary of Burning Rate Constants for Volatiles and Char Derived from 10 Bituminous Coals (Essenhigh and Fells, 1960)

Coal Components	Initial Diameter Range (μm)	Burning Rate Constant (μm^2/ms)
Volatiles	775–4000	740–2200
Char	300–4000	47–100

controlled (Essenhigh et al., 1965). They demonstrated how to estimate the values of adsorption and kinetic parameters from combustion data.

Kanury discussed the combustion of solid particles in terms of the diffusion flame model (Kanury, 1975, pp. 195−216). He offered the interpretation that as the volatile matter content decreases to zero the Spalding mass transfer number ($B = sY_{Ox\infty}$) decreases and the flame radius approaches the particle radius. Kanury suggested that kerosene droplets could be considered an analog for coal volatiles (100% volatile) and carbon particles an analog for coal char (0% volatile). He compared the burnout times for kerosene droplets and carbon particles of the same diameter and concluded that a kerosene droplet burned approximately 13 times faster than a carbon particle. If a coal particle had a high volatiles content, it should burn at a rate approaching that of a kerosene droplet. If a coal particle had a low volatile content, it should burn with a rate approaching a carbon particle. These observations are consistent with the generalization that volatiles combustion is approximately an order of magnitude greater than char oxidation.

Laurendeau reviewed the coal char combustion literature and concluded that chemical kinetics became the rate-controlling step only for particle sizes less than 1 μm (Laurendeau, 1978). Laurendeau also concluded that external diffusion control occurred for particles larger than 100 μm. In the size range from 1 to 100 μm, he asserted that both diffusion and kinetics controlled the combustion rate of coal char. A number of chemical kinetics studies were summarized in his paper. Smith also reviewed a number of chemical kinetics studies for coal char oxidation (Smith, 1982). He cautioned that several studies indicated that coal particles smaller than 100 μm could exhibit asymmetric jetting behavior during devolatilization. The d-squared law is a consequence of the spherical diffusion flame model, so while the spherical model works well in general, there may be exceptions when considering the behavior of jetting particles.

It is important to realize that most of the combustion studies on coal particles were performed with the goal of understanding how best to achieve higher levels of combustion efficiency (i.e., complete combustion). Complete combustion in a combustible dust accident is rare. Thus, if you are interested in calculating the burnout time of a coal particle in a dust deflagration scenario, I recommend that you assume that only the volatile matter contributes to the combustion process.

6.3.3.2 Devolatilization and combustion of large coal particles

While the shrinking particle model describes well the combustion behavior of coal particles less than 1 mm (1000 μm), studies with larger particles have demonstrated the formation of a porous product layer (usually ash) that is consistent with the shrinking unreacted core model. Although these larger particle sizes are unlikely to contribute to a dust deflagration, these studies are important in that they demonstrate that the kinetic model for particle combustion is size-dependent. Some studies have focused on the ash layer as the product, others have focused on the char layer (devolatilized coal) as the product. Abdel-Hafez conducted bituminous coal combustion tests with 3 mm coal particles in a fluidized bed of sand particles (Abdel-Hafez, 1988). He found that the shrinking unreacted core model fit the experimental results well.

Chern and Hayhurst performed single stationary particle tests with a set of six different bituminous coal (Chern and Hayhurst, 2004). Their objective was to determine if the shrinking core model described devolatilization in larger coal particles. They formed spherical particles with a diameter of 14 mm and suspended each particle individually in a basket and immersed into a bed of hot sand. They found that devolatilization indeed proceeded as a shrinking core process with the devolatilized coal as the product layer and the native coal as the unreacted core.

A study by Everson et al. demonstrated the shrinking core model for coal combustion using a bituminous coal (Everson et al., 2015). In their study, they used coal particles with 20, 30, and 40 mm diameters. They found that the combustion process followed the shrinking core model with an ash layer as the product layer and the unreacted coal at the core. Fig. 6.17 illustrates the accumulation of the ash layer (product layer) on the coal particle.

Thus, the shrinking core model can mimic the progression of the devolatilization front as it progresses into the interior of the coal particle, and it can also simulate the growth of the ash layer. In summary, the shrinking unreacted core model has been shown to be a useful model for large coal particles ($d_p > 3$ mm). However, this is a size range that is not likely to be important in the investigation of combustible dust hazards. Therefore, the shrinking particle model will likely be the predominant model in evaluating the combustible dust hazards of coal.

6.3.3.3 Transformation of particle microstructure

During devolatilization, the microstructure of coal particles can change dramatically. These changes can be a factor in the rate of devolatilization and char

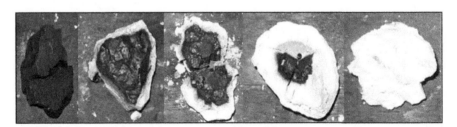

FIGURE 6.17

Sections of a 30-mm bituminous particle at various stages of conversion from 0% to 100% (from left to right) in a combustion furnace at 1000°C.

Reproduced with permission from Everson, R.C., Neomagus, H.W.J.P., van der Merwe, G.W., Koekemoer, A., Bunt, J.R., 2015. The properties of large coal particles and reaction kinetics of corresponding chars. Fuel 140, 17–26.

oxidation. One of the important features of devolatilization is the increase in porosity of the coal particles. The increase in porosity is due to the heat release caused by the combustion of volatiles. The heating of the coal particle leads to softening of the char and bubble nucleation and growth. Porosity formation in devolatilized coal ("coke") is a potential indicator of devolatilization and combustion in the course of an accidental deflagration. The magnitude of change in the devolatilized char particle is directly related to the volatiles content of the native coal.

With regards to observations of the burnt coal dust residue, the NIOSH (formerly U.S. Bureau of Mines) investigators have demonstrated a strong correlation between observations obtained from single particle combustion studies, laboratory explosibility studies, and experimental mine explosion tests (Ng et al., 1983; Cashdollar, 1996; Cashdollar et al., 2007; Man et al., 2009). The particle microstructure was observed via scanning electron microscopy. A consistent observation that has been corroborated over many types of bituminous coal samples with various combustion environments is the transformation of sharp, angular particle shapes (unburnt coal) to smooth, rounded particles (Ng et al., 1983). Many of the burnt coal particles exhibit blow holes or form cenospheres. The NIOSH investigators also developed a simple field test based on density separation in denatured ethanol: coal particles sink while coked particles float (Cashdollar et al., 2007; Man et al., 2009).

While coal tends to have some degree of porosity, devolatilization increases the total pore volume as well as the pore size distribution (Simons, 1983; Tsai and Scaroni, 1987). Much of the porosity structure is believed to be the result of the woody cellular structure from the progenitor materials from which the coal has been created (Wildman and Derbyshire, 1991). During devolatilization coal particles swell, and swelling causes porosity changes. Swelling is caused by coal softening and bubble nucleation during devolatilization process (Dakic et al., 1989; Yu et al., 2002, 2003, 2007).

Examination of the devolatilized char structures has provided important evidence for the devolatilization process and has helped explain the relationship between volatiles content and dust explosibility parameters (Hertzberg et al., 1982, 1988). Certain observations are a recurring theme in the study of devolatilized char structures: the bulk density decreases with devolatilization, the true density increases, the pore volume increases, particles often swell and then fragment during devolatilization, and particles often exhibit blow holes (cenospheres) (Ng et al., 1983; Rastogi et al., 1996). There is compelling evidence that small burning coal particles of high volatility can emit jets of volatile material (Smith, 1982). It seems reasonable to associate the formation of cenospheres with this jetting behavior.

EXAMPLE 6.6

Consider a spherical bituminous coal particle with a diameter of 300 μm. Calculate the burnout times for volatiles and char. Assume the combustion of both the volatiles and char obey the d-squared law with $K(volatile) = 1500(\mu m)^2/ms$ and $K(char) = 50(\mu m)^2/ms$. Neglect particle swelling.

Solution

It is assumed that volatiles combustion occurs first and goes to completion before char oxidation begins. The burnout time is calculated as $\tau = d_{p,0}^2/K$. The burnout time for volatiles combustion is $\tau(volatiles) = 60\ ms$. The burnout time for char oxidation the is $\tau(char) = 1800\ ms$. If the typical time scale for a coal dust deflagration is between 10 and 100 ms, then it is unlikely that char oxidation will contribute appreciably to the propagation of the deflagration wave.

6.3.4 CHARRING ORGANIC SOLIDS: BIOMASS

Biomass is defined generally as plant material or plant-derived material that is intended for utilization in some form of energy conversion. Biomass includes wood, herbaceous materials, and agricultural crop wastes (Jones et al., 2014). In the context of this book, I will also use the term biomass to include residues derived from biomass such as grain dust. Biomass usually does not refer to human food or animal feed.

Like coal, biomass can be characterized in terms of proximate analysis and ultimate analysis. Neither analysis is a reliable basis for the classification of biomass, but both types of analysis are valuable as a means of understanding the basic combustion properties of different types of biomass. For example, herbaceous biomass (leaves, grassy stems) typically has a higher ash content than woody biomass. Table 6.8 presents data that are representative of the differences between woody and herbaceous biomass.

Table 6.8 Comparison of the Proximate and Ultimate Analyses Values for a Woody Biomass and a Herbaceous Biomass (Jones et al., 2014)

Fuel	Proximate Analysis			Ultimate Analysis			
	Volatile Matter	Fixed Carbon	Ash	C	H	O	N
Wood pine chips	80.0	19.4	0.6	52.1	6.1	41.44	0.3
Wheat straw	71.89	17.8	10.32	46.35	6.29	47.92	4.04

Proximate analysis is weight percent on a dry basis.
Ultimate analysis is weight percent based on a dry ash-free basis.

Biomass is usually also defined as a lignocellulosic material. Lignocellulosic materials are composed of three naturally occurring polymers: cellulose, hemicellulose, and lignin. Cellulose and hemicellulose are both polymers composed of carbohydrates and lignin is an aromatic polymer. Different forms of biomass will have varying levels of these three polymers. Wood for example is approximately 50% cellulose, 25% hemicellulose, and 25% lignin (Beyler and Hirschler, 2002). However, there is considerable variability in the composition of different wood species and even between trees of the same species (Jones et al., 2014; Ross, 2010, Chapter 1). These polymers form fibers that ultimately create the microstructure and the structure of the plant body (Mettler et al., 2012). A good summary of the physical and chemical properties of wood is summarized by Ragland and Aerts (1991).

Biomass includes a variety of plant-based materials including plant structures and vegetative residues like leaves, twigs, stems, branches, bark, and sawdust. Four examples of the varied structures of biomass are shown in Fig. 6.18: sugar bagasse, pine sawdust, torrefied (a type of thermal treatment resulting in partial pyrolysis), pine sawdust and olive residue (the residue that remains after extracting the oil) (Riaza et al., 2014a,b). Row (a) shows the materials in their

FIGURE 6.18

Four examples of biomass material.

Reproduced with permission from Riaza, J., Khatami, R., Levendis, Y.A., Alvarez, L., Gil, M.V., Pevida, C., et al., 2014a. Single particle ignition and combustion of anthracite, semi-anthracite, and bituminous coals in air and simulated oxy-fuel conditions. Combust. Flame 161, 1096–1108; Riaza, J., Khatami, R., Levendis, Y.A., Alvarez, L., Gil, M.V., Pevida, C., et al., 2014b. Combustion of single biomass particles in air and in oxy-fuel conditions. Biomass Bioenergy 64, 162–174.

as-received condition. The second row, row (b), is a sample of each material after grinding and sieving into a size range of 75−150 μm. Row (c) are the SEM images of the ground and sieved fractions.

Typical biomass materials exhibit high levels of physical and chemical variabilities (Kenney et al., 2013). A portion of this variability arises due to the natural biological variation that arises in nature. The other main contributor to physical variability is the harvesting, handling, and mechanical processing that occurs during its use. This variability introduces uncertainty into the characterization of a biomass material. It is important when reviewing biomass data to bear in mind that mean values of a physical or chemical parameter by themselves may not be adequate for describing the distribution of a property.

I include grain dust, an agricultural grain residue, within the class of biomass materials. Grain dust is a significant source of dust deflagration accidents (Eckhoff, 2009; Eckhoff, 2003, pp. 20−25; CSB, 2006). Corn, sorghum, wheat, soybeans, rice, cornstarch, and barley represent 80% of the grains most likely to be involved in a dust explosion in the United States (Maness and Sargent, 2002). Grain dust is not simply small particles of its parent grain. Grain dust typically includes starch granule fragments, lignocellulosic debris from the stems and stalks of its parent grain, soil grains, and fungal spores (Martin and Sauer, 1976; Martin, 1981; Parnell et al., 1986; Goynes et al., 1986; Dashek et al., 1986). The presence of fungal spores is a source of a respiratory hazard for workers who handle grain and are exposed to its fines. Depending on the location of the sample, grain dust tends to have large percentages of particles under 100 μm in diameter and may contain as much as 10% by weight of respirable particles (the respirable range is defined as 0.01 μm $< d_p <$ 10 μm; Hinds, 1999, Chapter 11).

The microstructure of biomass arises out of its biological function (Ross, 2010). Biomass exhibits a diverse range of microstructural features that can be described generally as *fibrous* (Jones et al., 2014). Much of the biomass microstructure is retained after pyrolysis and char formation (Wildman and Derbyshire, 1991; Downie et al., 2009; Avila et al., 2011; Chen et al., 2013). The residual structure created during the pyrolysis of plant material may affect the release of volatiles. Fig. 6.19 illustrates how charred wood particles have retained their fibrous structure after being pyrolyzed in a drop-tube furnace at a temperature of 950°C (Chen et al., 2013).

There are characteristic length scales associated with the microstructure that can influence the evolution of the char microstructure. The exact nature of the microstructure depends on the biomass source: wood, bark, herbaceous materials, oil seeds, starches, or other agricultural residues. The chemical composition plays a complimentary role in determining the persistence of microstructural features during devolatilization and char oxidation.

While there are similarities between coal and biomass combustion, they do exhibit some very important differences (Riaza et al., 2014a,b). These differences include:

- Biomass particles are less dense than coal, so for the same size particles biomass has less mass and less heating value

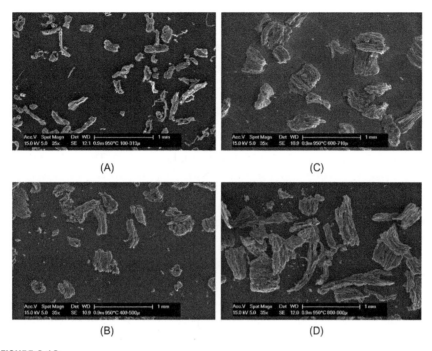

FIGURE 6.19

SEM micrographs of beech wood char particles after fast pyrolysis at a temperature of 950°C: (A) 350 μm, (B) 500 μm, (C) 700 μm, and (D) 800 μm.

Reproduced with permission from Chen, L., Dupont, C. Salvador, S., Grateau, M., Boissonet, G., Schweich, D., 2013. Experimental study on fast pyrolysis of free-falling millimetric biomass particles between 800°C and 1000°C. Fuel 106, 61–66.

- Biomass tends to have much higher volatile contents (as much as 80%) and less fixed carbon than coal (volatile content rarely >45%)
- Biomass flames tend to be less sooty than coal flames
- Biomass chars tend to be more reactive than coal chars.

6.3.4.1 Combustion rate laws for volatiles and for char

Although the key results of the shrinking particle and shrinking unreacted core models were presented earlier in this chapter, a paper by Kanury deserves citation here because he derived several models for biomass particle combustion (Kanury, 1994). He presented ignition models for thermally thin and thermally thick fuels and derivations for diffusion-controlled and kinetic-controlled combustions for both flaming combustion and char oxidation. His results are consistent with the gas—solid reaction models presented earlier with the primary difference that his models include theoretical expressions for the burn rate constants.

In surveying the biomass combustion literature, it is interesting to note that the investigation of single particle combustion in the context of combustible dust ($d_p < 500\ \mu m$) is a more recent development. Austin and his colleagues studied the combustion behavior of single corncob particles with diameters ranging from 300 to 1800 μm (Austin et al., 1991; Austin et al., 1996). The corncob residue was selected as an analog for grain dust because he could select nearly spherical particles and nonspherical particles for testing. Austin thus considered both the influence of particle size and shape on the burnout time for volatiles and char combustion. Austin conducted a series of single particle combustion tests by inserting a static particle into a constant temperature furnace. The changing particle diameter was recorded by video camera and the images were analyzed to calculate the instantaneous particle diameter. For the particle size range tested, he found a good fit of the burnout times with the d-squared law by using a modified definition for the mean particle size. The burn rate constant for volatiles, K, was calculated from the burnout time:

$$K = d_{p,0}^2/\tau \tag{6.80}$$

Fig. 6.20 shows a plot of burnout time data as a function of particle size. Note that Austin calculated the coefficient K_v from the slope of the line; $K_v = 1/K$, the reciprocal of the burn rate constant. In convenient units, the burn rate constant for volatiles combustion is $K = 1470\ \mu m^2/ms$.

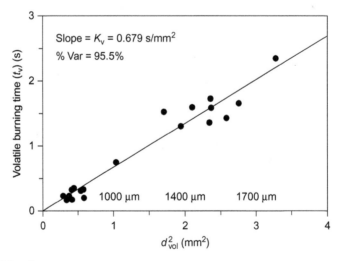

FIGURE 6.20

Burnout time for volatiles combustion of corncob particles as a function of particle diameter.

Reprinted with permission from Austin, P.J., Kauffman, C.W., Sichel, M., 1991. Combustion of single nonspherical cellulosic particles. Prog. Astronaut. Aeronaut. 152, 136–154.

Austin typically observed two distinct stages in the combustion process: first a nearly spherical diffusion flame would surround the particle (volatiles combustion) and eventually transition to char combustion. The spherical shape of the volatiles combustion flame was noted to be independent of the particle shape. The diameter of the volatiles diffusion flame was approximately three times the particle diameter. During volatiles combustion the particle diameter shrank to approximately 90% of its initial value. Note that this shrinkage behavior during volatiles combustion is different from that observed with coal.

Austin observed a fairly distinct transition from volatiles combustion to char oxidation. He observed that char ignition was uniform for the spherical char particles, but for nonspherical char particles the corners of the particle would ignite first and then the ignition wave would sweep across the char surface. The char combustion data demonstrated a good fit to the d-squared law (Fig. 6.21).

In Fig. 6.21, the abscissa of the graph is an effective particle size parameter and has units of diameter squared. The parameter is calculated by multiplying two different measures of the particle diameter. The first diameter, designated as $d_{V/A} = 6(\text{particle volume/particle surface area})$ and $d_{a.ave} = (L + W + H)/3$, where L, W, H refer to the principal directions of length, width, and height of a nonspherical particle. Again, Austin calculated the reciprocal of the burn rate constant from the slope of the line. The burn rate constant for char combustion in convenient units is $K = 424 \ \mu m^2/ms$.

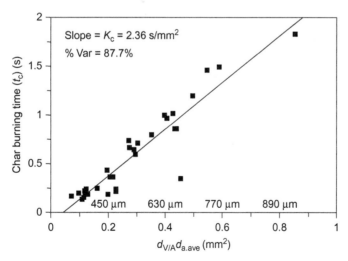

FIGURE 6.21

Burnout time for char combustion of corncob particles as a function of particle diameter.

Reprinted with permission from Austin, P.J., Kauffman, C.W., Sichel, M., 1991. Combustion of single nonspherical cellulosic particles. Prog. Astronaut. Aeronaut. 152, 136–154.

More recent studies of biomass particle combustion have tended to correlate combustion data with particle mass rather than particle diameter. This is likely due to the highly irregular shapes often encountered with biomass particles. One study of particular interest was by Riaza and his colleagues in which they obtained high quality images of spherical diffusion flame formation surrounding irregularly shaped particles (Riaza et al., 2014a,b). Fig. 6.22 shows spherical flames during volatiles combustion for bagasse at different oxygen concentrations.

Burning times for volatile combustion and char oxidation were correlated to particle mass by Mason et al. (2015). In their study, they tested three types of wood—pine, eucalyptus, and willow—at a range of particle diameters from 0.5 to 4 mm. They found that volatile burnout times correlated to particle mass

(A) Sugar cane bagasse

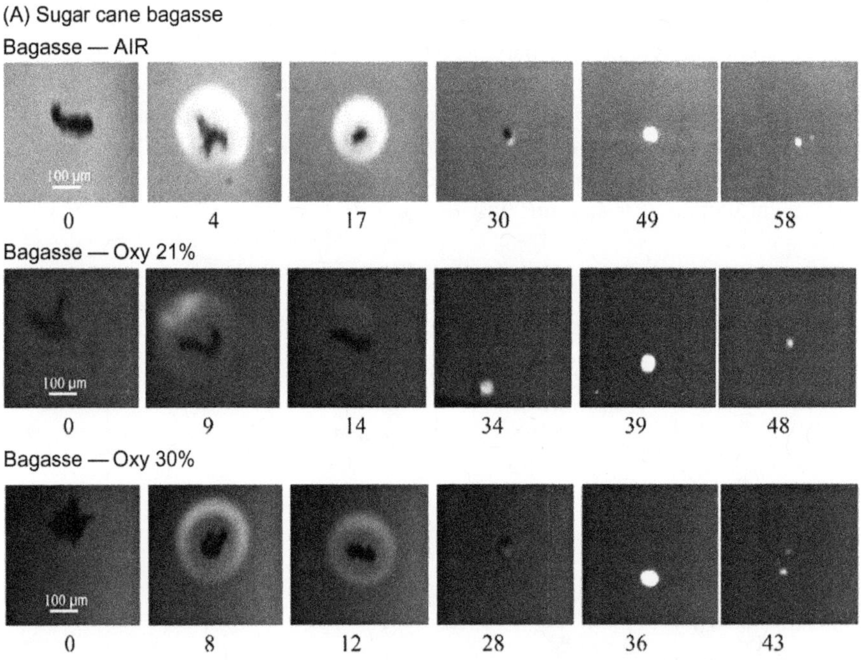

FIGURE 6.22

Spherical diffusion flame surrounding burning particle at three different oxygen concentrations. Volatile combustion in first three frames followed by char oxidation. Numbers underneath each photograph represents the burning time in milliseconds.

Reproduced with permission from Riaza, J., Khatami, R., Levendis, Y.A., Alvarez, L., Gil, M.V., Pevida, C., et al., 2014a. Single particle ignition and combustion of anthracite, semi-anthracite, and bituminous coals in air and simulated oxy-fuel conditions. Combust. Flame 161, 1096–1108; Riaza, J., Khatami, R., Levendis, Y.A., Alvarez, L., Gil, M.V., Pevida, C., et al., 2014b. Combustion of single biomass particles in air and in oxy-fuel conditions. Biomass Bioenergy 64, 162–174.

raised to the 0.6 power. This translates roughly to correlating to particle diameter $m_p^{0.6} \approx (\rho \, d_p^3)^{0.6} \approx d_p^{1.8}$. This correlation is suggestive of the d-squared law.

In summary, the d-squared law has been tested with biomass materials, and there is some empirical support for its validity. However, there are fewer such studies compared to coal. One possible reason is the greater diversity in particle shapes favoring a more elongated, fibrous profile that arise in pulverized biomass compared to the more near-spherical shapes of pulverized coal.

As a tentative guideline, biomass particles with a diameter smaller than 1 mm (1000 μm) appear to obey the shrinking particle model. For volatiles combustion, there is not at this time a well-defined range of particle diameters that specify diffusion-controlled versus kinetic-controlled combustion. The state of affairs for char oxidation is better understood, but from the point of view of prediction, one must know the fuel source for the char and have some knowledge of the char microstructure.

6.3.4.2 Devolatilization and combustion of large biomass particles

For biomass particles larger than 1 mm ($d_p > 1000$ μm), the shrinking core model tends to give a better representation of the combustion behavior. As with coal particle combustion, some of these studies have focused their investigation on the char layer as the product, and some have focused it on the ash layer. Since combustible dust hazards are typically associated with particles smaller than this size range, the shrinking core model is less relevant than the shrinking particle model. As with coal particle studies, the value of these larger particle size biomass studies is that they confirm that the appropriate selection of the combustion model depends on the particle size.

Maa and Bailie published one of the earliest modeling efforts of wood combustion using the shrinking core model (Maa and Bailie, 1973). They concluded in their paper that the combustion was kinetic controlled for particles smaller than 1 mm and heat transfer controlled for particles larger than 30 mm. Ouedraogo et al. performed a modeling study on combustion data derived from wood particles ranging from 10 to 75 mm (Ouedraogo et al., 1998). They found that the consideration of blowing effects in the boundary layer were an important element of their mass transfer model. Galgano and Di Blasi incorporated finite rate kinetics into their model formulation and compared their model with that developed by Maa and Bailie (Galgano and Di Blasi, 2003).

There is a large body of work that has attempted more fundamental approaches to the modeling of the transport and kinetic processes involved in single particle combustion. This is an extremely challenging objective and could easily distract our attention away from combustible dust hazards. Thus, only a few of these studies are identified here to provide a sense of the activity in this field.

Simmons and Ragland conducted stationary single particle combustion experiments with wood cubes measuring 10 and 20 mm (Simmons and Ragland, 1985). They found that both boundary later resistances (heat transfer and oxygen mass transfer) and internal heat conduction were controlling resistances. In an additional set of experiments with cubes with sizes of 5, 10, 15, 20, and 25 mm, they found that the volatiles burnout time and char burnout timescaled with particle size raised to the 1.5 power (Simmons and Ragland, 1986). Di Blasi and his coworkers have published several papers exploring the role of finite kinetics pyrolysis (Di Blasi, 1996, 2002, 2008; Branca and Di Blasi, 2003). A significant challenge ahead is to find ways to simplify the complex pyrolysis reaction networks and kinetic expressions so that they can more easily be incorporated into reactor models. Momeni and his colleagues explored the effect of particle size and shape on burnout times (Momeni et al., 2013a,b). They found that spherical particles had faster burnout times compared to cylinders of the same diameter. Burnham et al. survey the global chemical kinetics of cellulose pyrolysis and give recommendations for kinetic parameters (Burnham et al., 2015).

6.3.4.3 Transformation of particle microstructure

Observing the burning of large pieces of wood ($d_p > 10$ cm), one commonly sees the formation of fissures and cracks in the charred material (Fig. 6.23).

These characteristic features are caused by a combination of the thermal degradation, devolatilization, and microstructural changes which result in shrinkage. The shrinkage causes the formation of cracks and fissure development in the charred material (Roberts, 1971; Zicherman and Williamson, 1981; Shen et al., 2009).

FIGURE 6.23

Typical char formation on a flat slab of Douglas fir wood. Top and side views of char fissures and cracks; box denotes the termination of a fissure. Scale is in inches.

Reprinted with permission from Zicherman, J.B., Williamson, R.B., 1981. Microstructure of wood char. Part 1: Whole wood. Wood Sci. Technol. 15, 237–249.

The development of the char zone on a wood slab is shown in Fig. 6.24 (Roberts, 1971).

The shrinkage of the charred material causes stresses in the pyrolysis zone separating the char from the native wood. Theses stresses cause cracks to form in the char thereby creating channels which are a preferential pathway for the release of volatiles (Zicherman and Williamson, 1981; Drysdale, 1999, pp. 182–190; Shen et al., 2009). The characteristic cracks and fissures in charred wood specimens suggest a range of characteristic length scales. In particular, it is useful to know if there is a specimen size limit below which these cracks and fissures may not form. Asked another way, are these features relevant to the behavior of biomass-derived combustible dust?

Kuo and Hwang performed single particle combustion experiments with stationary rosewood particles in a constant temperature furnace (Kuo and Hwang, 2003). They observed the appearance of cracks and fissures in charred wood spheres of both 20 and 50 mm diameter. They noted that the cracks and fissures appeared early in the flaming combustion (volatiles combustion) of the 50 mm spheres, but appeared near the end of volatile combustion for the 20 mm spheres (Fig. 6.25).

The development of this type of char structure with 50 mm diameter spheres has also been noted in other investigations using oak, rosewood, and pine (Kuo and Hsi, 2005; Hsi and Kuo, 2007). In the study by Kuo and Hsi, they observed that the anisotropy of the wood grain had a measurable influence on ignition behavior.

Mukunda and coworkers performed a number of single particle combustion experiments with teak wood particles in a constant temperature furnace (Mukunda et al., 1984). They tested spheres with diameters of 10, 15, 20, and 25 mm. They reported that none of the spheres fractured and there were no

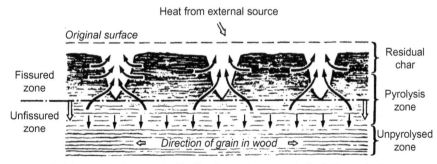

Solid arrows indicate probable directions of movement of volatile products

FIGURE 6.24

Conceptual model of wood char development.

Reprinted with permission from Roberts, A.F., 1971. Problems associated with the theoretical analysis of the burning of wood. In: Thirteenth Symposium (International) on Combustion. The Combustion Institute, Elsevier, New York. pp. 893–903.

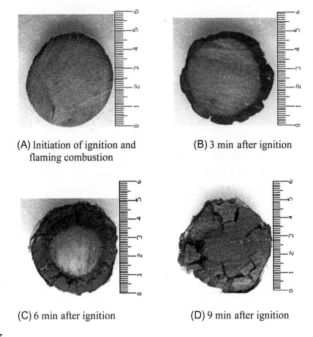

(A) Initiation of ignition and flaming combustion

(B) 3 min after ignition

(C) 6 min after ignition

(D) 9 min after ignition

FIGURE 6.25

Char pattern development in 50 mm diameter rosewood spheres with the formation of characteristic cracks and fissures.

Reprinted with permission from Kuo, J.T., Hwang, L.H., 2003. Mass and thermal analysis of burning wood spheres. Combust. Sci. Technol. 175, 665–693.

observable cracks in the char surface. The fact that Mukunda and coworkers did not observe cracking in the char surface for their wood spheres, but Kuo and Hsi did, serves as a reminder about the complex challenges of characterizing the combustion behavior of biomass. The difference in wood type, biological variation, the effect of grain anisotropy, and differences in their experimental apparatus may all be contributing factors for these disparate observations.

Several investigators have reported that small particles heated rapidly tend to lose their original fibrous structure while large particles heated slowly tend to retain it. Cetin et al. conducted both stationary particle and falling particle experiments with radiata pine, eucalyptus, and sugar cane bagasse to simulate softwood, hardwood, and agricultural residue (Cetin et al., 2004). The particle diameters ranged from 50 to 2000 µm, with the specific size range tested depending on which of three reactors were used: a wire mesh reactor for high heating rates (9500 °C/s), a tubular reactor for low heating rates (~10 °C/s), and a drop-tube reactor for extremely high heating rates (~10^5 °C/s). In sum, they found that slowly heated particles largely retained the fibrous microstructure of the parent material. Particles heated extremely fast exhibited swollen, spherical shapes with

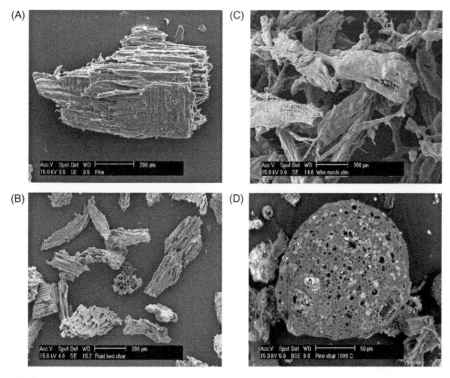

FIGURE 6.26

Parent pine sawdust particle (A) and charred particles caused by low (B), high (C), and extremely high heating rates (D).

Reprinted with permission from Cetin, E., Moghtaderi, B., Gupta, R., Wall, T.F., 2004. Influence of pyrolysis conditions on the structure and gasification reactivity of biomass chars. Fuel 83, 2139–2150.

little or no fibrous structure. Fig. 6.26 shows SEM images of pine sawdust particles exposed to varying heating rates.

The interpretation offered by the investigators was that slow heating rates allowed the volatiles to escape the particle through the natural porosity of the parent particle microstructure. At extremely high heating rates, the particle melts and the release of volatiles causes the particle to swell into a sphere. At intermediate heating rates (the tubular reactor), the particles exhibited a smaller degree of melting (sharp edges became rounded) and the amount of swelling was small. Similar progressions in char particle morphology were observed for the eucalyptus and bagasse materials. Other investigators have observed similar trends (Biagini et al., 2008; Dall'Ora et al., 2008; Biagini et al., 2009; Fisher et al., 2012). Alternatively, for a similar thermal exposure, larger particles will better retain their fibrous microstructure and smaller particles will tend to lose it through melting and swelling (Chen et al., 2013).

As a whole, these studies indicate that microstructural feature will impose little if any effect on biomass-derived combustible dust.

EXAMPLE 6.7

Consider a spherical particle of pine dust with a diameter of 300 μm. Calculate the burnout times for volatiles and char. Assume the combustion of both the volatiles and char obey the d-squared law with $K(volatile) = 1470 \ (\mu m)^2/ms$ and $K(char) = 424 \ (\mu m)^2/ms$.

Solution

It is assumed that volatiles combustion occurs first and goes to completion before char oxidation begins. The burnout time is calculated as $\tau = d_{p,0}^2/K$. The burnout time for volatiles combustion is $\tau(volatiles) = 61 \ ms$. The burnout time for char oxidation the is $\tau(char) = 212 \ ms$. If the typical time scale for a biomass dust deflagration is between 10 and 100 ms, then volatiles oxidation will dominate the combustion process and char oxidation may contribute to a lesser degree to the propagation of the deflagration wave.

6.3.5 NONVOLATILE ORGANIC SOLIDS

Solid carbon has numerous sources and uses. Solid carbon is available naturally and synthetically in a bewildering variety of forms. The International Union for Pure and Applied Chemistry (IUPAC) has defined the terminology for 118 different forms of solid carbon (Fitzer et al., 1995). Graphite, coal char, biochar, activated carbon, and carbon black can be classified generally as solid carbon. The classification "solid carbon" is approximate as it can be exceedingly difficult to drive out all volatile matter from organic solids. A better classification is nonvolatile organic solids. From the point of view of combustible dust hazards, nonvolatile organic solids are a useful benchmark for devolatilized carbonaceous solid fuels such as cross-linked polymers, coal, and biomass.

Nonvolatile organic solids are essentially carbon. But there are many different forms of carbon depending on the source material composition, the source material microstructure, and the degree of pyrolysis achieved by thermal treatment. The resulting material, called char, may be mostly carbon in composition but it may still have some hydrogen, oxygen, and nitrogen in its solid matrix. Depending on the source material, there may also be inorganic salts that comprise the ash component.

In addition to a complex chemical composition, the particle microstructure may include porosity with a wide range of pore diameters. The char may also exhibit atomic scale features like crystallites, edge defects, and vacancies. These solid state features have been shown to have a dramatic effect on the reactivity of the solid.

Although the specific surface area of some nonvolatile organic solids may be quite large, it can be expected that it will have a lesser impact on the combustion rate of a cloud of particles in a dust deflagration due to the short time frame of an accidental deflagration. In engineered systems like furnaces, the residence time of the particles can be controlled and it is possible in such circumstances to drive the combustion reaction towards completion.

6.3.5.1 Combustion kinetics for solid carbon

In the discussion on charring organic solids, it was shown that for both coal and biomass char oxidation follows the d-squared law. For particle sizes larger than 1 mm, the combustion process appears to be diffusion controlled (Essenhigh et al., 1965; Essenhigh and Fells, 1960). Due to the simpler chemical composition of the fuel, it is possible to discuss char oxidation in terms of specific chemical reactions.

Once the volatile matter has been driven out of the carbonaceous solid, further chemical change can only occur by char oxidation. Therefore, the kinetic model for the oxidation process will require at least one heterogeneous chemical reaction to describe the char oxidation process. Early studies of carbon combustion established the importance of heterogeneous chemical reactions in the oxidation process. Smith and Gudmundsen studied the combustion behavior of 5 mm diameter spheres shaped from electrode grade carbon (Smith and Gudmundsen, 1931). They noted the shrinking particle character of the global reaction process, observed that both the surface temperatures and the specific surface reaction rate increased as the particle size decreased. Experiments were conducted with both dry air and moist air. The combustion rate was found to be greater with dry air, but moist air gave a higher surface temperature.

Tu et al. conducted experiments to determine the relationship between the combustion rate and air velocity (Tu et al., 1934). They used brush carbon spheres with a diameter of 25 mm. From their experiments, they concluded that air velocity was a key variable in determining if carbon combustion was diffusion controlled or kinetic controlled. Essenhigh et al. later re-analyzed the data published by Tu et al. and were able to refine some of the observations of the combustion process and noted that combustion of char particles smaller than 100 μm was kinetic controlled (Essenhigh et al., 1965).

Davis and Hottel experimented with 25 mm carbon spheres and demonstrated that under certain conditions the combustion process involved a bluish carbon monoxide flame sheet surrounding the carbon particle (Davis and Hottel, 1934). Parker and Hottel studied the structure of the gas film (chemical composition and temperature) using a microsampling technique (Parker and Hottel, 1936). They found that at high temperatures there was a higher concentration of carbon dioxide and very little carbon monoxide produced at the carbon surface. This implies that carbon dioxide is reduced by the carbon to become carbon monoxide. The carbon monoxide then reacts with the oxygen to form carbon dioxide at the flame sheet. Thus, some of the carbon dioxide diffuses out into the free stream gas and some of it diffuses towards the carbon particle surface (Kanury, 1975, pp. 202−205).

Three heterogeneous chemical reactions are known to play an important role in the oxidation of carbon. These reactions are stoichiometric relations; they are not elementary reactions.

$$\text{Heterogeneous, exothermic reaction I}, \omega_I'' : \quad 2C + O_2 \rightarrow 2CO \qquad (6.81)$$

$$\text{Heterogeneous, endothermic reaction II}, \omega_{II}'' : \quad C + CO_2 \rightarrow 2CO \qquad (6.82)$$

$$\text{Heterogeneous, exothermic reaction III}, \omega_{III}''' : \quad C + H_2O \rightarrow CO + H_2 \qquad (6.83)$$

These chemical reactions work in concert to convert the carbon fuel into carbon dioxide. The diffusion reaction model bears similarity to the diffusion flame analysis for liquid droplets. Perhaps the biggest difference is that in the liquid droplet model the fuel vapor must be evaporated from the droplet surface by the heat of the flame sheet. In the carbon particle model, a heterogeneous chemical reaction is responsible for the gasification of the solid (Kanury, 1975, Chapter 6; Turns, 2012, Chapter 14; Laurendeau 1978; Saxena, 1990; Annamalai and Ryan, 1993).

We will consider two different models for diffusion-controlled char oxidation, the one-film model and the two-film model, and one model for kinetic-controlled combustion. The one-film model assumes that oxidation occurs at the char surface with no flame; instead, depending on the particle temperature, the particle will probably glow due to the heat of reaction released at the particle surface. The maximum temperature in the one-film model is at the particle surface. The glowing phenomenon is called incandescence and is commonly observed in single particle combustion experiments.

The two-film model is a refinement of the one-film model; the two-film model includes both heterogeneous and homogeneous chemical reactions to describe the char oxidation process. In the two-film model, a flame sheet surrounds the solid particle some distance from the particle surface (Kanury, 1975, Chapter 6; Turns, 2012, Chapter 14). The maximum temperature in the two-film model is at the flame sheet. Fig. 6.27 depicts the general concentration profiles for the one-film and two-film models.

The one-film and two-film models are ideal cases which are relevant at different particle size ranges. The combustion of carbon particles smaller than $50-100\,\mu m$ is described by the one-film model. The combustion of carbon particles larger than $1000\,\mu m$ (1 mm) is described by the two-film model. Another refinement is the continuous film model which bridges the one-film and two-film models (Turns, 2012, p. 448). The continuous film model envisions a diffuse reaction zone within the particle's boundary layer. The combustion of carbon particles in the size range from 100 to $1000\,\mu m$ is described by the continuous film model (Laurendeau, 1978, p. 257; Annamalai and Ryan, 1993, p. 420). Since these particle size ranges are order-of-magnitude estimates, you will find different investigators giving different ranges (but still all within the same order of magnitude). The kinetic-controlled model is relevant when mass transfer limitations can be overcome. The guidance one usually finds for kinetic-controlled particle

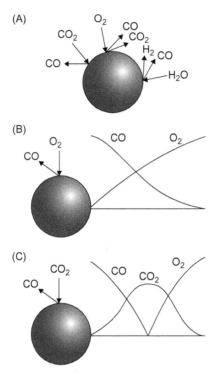

FIGURE 6.27

Carbon combustion models: (A) the heterogeneous chemical reactions involved in oxidation, (B) concentration profiles for the one-film model, and (C) concentration profiles for the two-film model.

Reprinted with permission from Annamalai, K., Ryan, W., 1993. Interactive processes in gasification and combustion—II. Isolated carbon, coal and porous char particles. Prog. Energy Combust. Sci. 19, 383–446.

combustion is that it is limited to lower particle temperatures and smaller particle sizes (Annamalai and Ryan, 1993, pp. 415–423).

This presentation for the one-film and two-film models generally follows the development by Kanury (Kanury, 1975, Chapter 6). An alternative approach can be found in the books by Turns (Turns, 2012, Chapter 14) and Annamalai and Puri (Annamalai and Puri, 2007, Chapter 9).

6.3.5.2 One-film combustion model

We consider a spherical particle of radius suspended in air. In the one-film model, the particle is consumed by a heterogeneous reaction; no homogeneous reactions occur in the gas phase. The chemical reaction is represented by

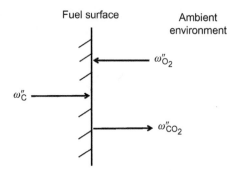

FIGURE 6.28

Surface control volume at the particle surface showing reaction rates for the one-film carbon particle combustion model.

$$C(s) + O_2 \rightarrow CO_2 \tag{6.84}$$

The consumption of the fuel particle requires oxygen to diffuse from the free stream toward the particle surface. The reaction rates for this system are illustrated in Fig. 6.28.

The one-film model assumes there is only one chemical reaction and it is a heterogeneous reaction.

$$\varphi C(s) + O_2 \rightarrow (\varphi + 1)CO_2 \tag{6.85}$$

The coefficient φ is the mass ratio of carbon to oxygen for the stoichiometric condition. Do not confuse this stoichiometric representation (consistent with Kanury's notation) with the one used in Chapter 4. In Chapter 4, we defined the coefficient s as the mass ratio of oxidizer to fuel ($s = 1/\varphi$). The conservation of mass requires that the reaction rates are related: $w_c'' + w_{O_2}'' = w_{CO_2}''$.

We begin the analysis of the one-film model with the continuity equation written in spherical coordinates.

$$\frac{1}{r^2}\frac{d}{dr}(r^2 \rho_g v_r) = 0 \Rightarrow r^2 \rho_g v_r = r^2 \dot{m}'' = constant \tag{6.86}$$

To model this system, we need to solve the species equation for oxygen written in spherical coordinates.

$$\dot{m}'' \frac{dY_{O_2}}{dr} = \rho_g \mathcal{D}_{O_2}\left[\frac{1}{r^2}\frac{d}{dr}\left(r^2 \frac{dY_{O_2}}{dr}\right)\right] \tag{6.87}$$

The solution of the oxygen species equation requires two boundary conditions. The first boundary condition is the free stream concentration of oxygen in air and the second arises by restricting ourselves to diffusion-controlled combustion.

$$Y_{O_2}(r \rightarrow \infty) = Y_{O_2,\infty} \tag{6.88}$$

$$w''_{O_2,r_p} = \left[\rho_g D_{O_2} \frac{dY_{O_2}}{dr} \right]_{r=r_p} - w''_{C,r_p} Y_{O_2,r_p} \tag{6.89}$$

The stoichiometry of the reaction gives the following mass balance relation:

$$\varphi w''_{O_2,r_p} = w''_{C,r_p} \tag{6.90}$$

Substituting Eq. (6.90) into Eq. (6.89) gives a revised flux boundary condition:

$$\left[\rho_g D_{O_2} \frac{dY_{O_2}}{dr} - \dot{m}''_{C,r_p} Y_{O_2,r_p} \right]_{r=r_p} = \frac{w''_{C,r_p}}{\varphi} \tag{6.91}$$

Integrating the oxygen species equation and invoking the flux boundary condition

$$r^2 \rho_g D_{O_2} \frac{dY_{O_2}}{dr} - w''_{C,r_p} r_p^2 Y_{O_2} = \frac{w''_{C,r_p} r_p^2}{\varphi} \tag{6.92}$$

This can be rearranged into a more useful form:

$$r^2 \rho_g D_{O_2} \frac{dY_{O_2}}{dr} - \left[w''_{C,r_p} r_p^2 \left(Y_{O_2} + \frac{1}{\varphi} \right) \right] = 0 \tag{6.93}$$

Integrating again and applying the second boundary condition gives an expression for the oxygen concentration profile

$$\ln \left[\frac{Y_{O_2} + 1/\varphi}{Y_{O_2,\infty} + 1/\varphi} \right] = - \left(\frac{w''_{C,r_p} r_p^2}{\rho_g D_{O_2}} \right) \frac{1}{r} \tag{6.94}$$

For the one-film model it is assumed that $Y_{O_2}(r = r_p) = 0$. Thus, Eq. (6.94) becomes

$$\frac{w''_{C,r_p} r_p}{\rho_g D_{O_2}} = \ln \left[\frac{Y_{O_2,\infty} + 1/\varphi}{1/\varphi} \right] = \ln \left(\varphi Y_{O_2,\infty} + 1 \right) = \ln(B_{of} + 1) \tag{6.95}$$

where the Spalding mass transfer number for the one-film model is defined as $B_{of} = \varphi Y_{O_2,\infty}$. This expression can be recast to calculate the mass burning rate

$$\dot{m}_{C,r_p} = w''_{C,r_p} \left(4\pi r_p^2 \right) = 4\pi r_p \rho_g D_{O_2} \ln(B_{of} + 1) \tag{6.96}$$

Referring to the definition of B in Chapter 4, for liquid droplet diffusion flames, this definition for the transfer number implies that the flame is at the surface of the particle. For the carbon oxidation reaction, the value of the stoichiometric coefficient is $\varphi = 12$ kg C/16 kg $O_2 = 0.75$ and for air $Y_{O_2,\infty} = 0.238$, so the transfer number for carbon is $B_{of} = 0.179$.

The burnout time and the particle diameter history are calculated from the expression derived in Chapter 4,

$$t_b(\text{one} - \text{film}) = \frac{\rho_p d_0^2}{8\rho_g D_{O_2} \ln(B_{of} + 1)} = \frac{d_0^2}{K_{of}} \tag{6.97}$$

$$d_p^2(t) = d_0^2 - K_{of}t; \quad K_{of} = \left(\frac{8\rho_g D_{O_2}}{\rho_p}\right) \ln(B_{of} + 1); \quad B_{of} = \varphi Y_{O_2, \infty} \tag{6.98}$$

These expressions provide the necessary tools for calculating the behavior of carbon particles for the one-film model assuming diffusion-controlled combustion.

If the temperature of the burning particle cannot be specified (eg, by measurement), then it can be calculated in a manner similar to the temperature calculation done for liquid droplet diffusion flames. The books by Kanury and Turns illustrate the method (Kanury, 1975, pp. 210−211; Turns, 2012, pp. 457−459).

6.3.5.3 Two-film combustion model

The two-film model assumes that the rate-determining step in combustion is the heterogeneous reaction that converts solid carbon and carbon dioxide into carbon monoxide (Kanury, 1975, pp. 211−213; Turns, 2012, pp. 459−465). The carbon monoxide reacts with oxygen in a flame sheet surrounding the carbon particle to create carbon dioxide. The control volume identifying the species mass flows is depicted in Fig. 6.29.

The two-film combustion model is based on two chemical reactions, one heterogeneous and one homogeneous (the subscript p stands for particle and the subscript g stands for gas):

$$\varphi_p C(s) + CO_2 \rightarrow \left(1 + \varphi_p\right) CO \tag{6.99}$$

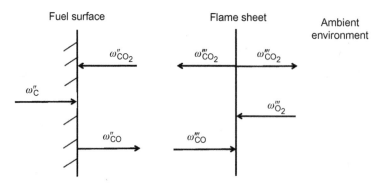

FIGURE 6.29

Surface control volumes at the particle surface and the flame sheet showing the reaction rates for the two-film model of carbon combustion.

$$\varphi_{\rm g} {\rm CO} + {\rm O}_2 \rightarrow \left(1 + \varphi_{\rm g}\right){\rm CO}_2 \tag{6.100}$$

The analysis begins with the species equations for oxygen and carbon dioxide.

$$\dot{m}'' \frac{dY_{\rm O_2}}{dr} = \rho_{\rm g} \mathcal{D}_{\rm O_2} \left[\frac{1}{r^2} \frac{d}{dr}\left(r^2 \frac{dY_{\rm O_2}}{dr}\right)\right] - \omega_{\rm O_2}''' \tag{6.101}$$

$$\dot{m}'' \frac{dY_{\rm CO_2}}{dr} = \rho_{\rm g} \mathcal{D}_{\rm CO_2} \left[\frac{1}{r^2} \frac{d}{dr}\left(r^2 \frac{dY_{\rm CO_2}}{dr}\right)\right] - \omega_{\rm CO_2}''' \tag{6.102}$$

The reaction rates for oxygen and carbon dioxide are related by stoichiometry $\omega_{\rm O_2}''' = -\omega_{\rm CO_2}'''/\left(1+\varphi_{\rm g}\right)$. The two species equations can be combined into one by multiplying the CO_2 species equation by the factor $\left(1+\varphi_{\rm g}\right)^{-1}$, and adding them together (assuming equal diffusivities). Introducing a new variable, $\tilde{Y} = Y_{\rm O_2} + Y_{\rm CO_2}/\left(1+\varphi_{\rm g}\right)$, the transformed species equation is, with some rearrangement,

$$\frac{d}{dr}\left(r^2 \rho_{\rm g} \mathcal{D} \frac{d\tilde{Y}}{dr}\right) - \left(\omega_{\rm C,r_p}''' r_0^2\right)\frac{d\tilde{Y}}{dr} = 0 \tag{6.103}$$

Integrating this equation once gives the intermediate result

$$\left(r^2 \rho_{\rm g} \mathcal{D} \frac{d\tilde{Y}}{dr}\right) - \left(\omega_{\rm C,r_p}''' r_0^2\right)\tilde{Y} = \text{constant} \tag{6.104}$$

The value of the constant is determined from the application of the boundary condition $Y_{\rm O_2}(r = r_0) = 0$:

$$\text{constant} = \frac{1}{\left(1 + \varphi_{\rm g}\right)} \left[r_0^2 \rho_{\rm g} \mathcal{D} \frac{dY_{\rm CO_2}}{dr} - \omega_{\rm C,r_p}''' r_0^2 Y_{\rm CO_2}\right]_{r=r_0} \tag{6.105}$$

Recognizing the $\omega_{\rm CO_2}''' = \omega_{\rm C,r_p}'''/\varphi_{\rm p}$, the value of the constant simplifies to the following expression:

$$\text{constant} = \frac{\omega_{\rm C,r_p}''' r_0^2}{\varphi_{\rm p}\left(1 + \varphi_{\rm g}\right)} \tag{6.106}$$

Substituting Eq. (6.106) into Eq. (6.105) gives

$$\left(r^2 \rho_{\rm g} \mathcal{D} \frac{d\tilde{Y}}{dr}\right) - \left(\omega_{\rm C,r_p}''' r_0^2\right)\tilde{Y} = \frac{\omega_{\rm C,r_p}''' r_0^2}{\varphi_{\rm p}\left(1 + \varphi_{\rm g}\right)} \tag{6.107}$$

Integrating this equation requires a boundary condition for $\tilde{Y}(r \to \infty)$ which is actually two boundary conditions: $Y_{O_2}(r \to \infty) = Y_{O_2,\infty}$ and $Y_{CO_2}(r \to \infty) = 0$. Upon integration and application of the boundary conditions, we obtain

$$\ln\left[\frac{\tilde{Y} + \dfrac{1}{\varphi_p\left(1 + \varphi_g\right)}}{Y_{O_2,\infty} + \dfrac{1}{\varphi_p(1 + \varphi_g)}}\right] = \left(\frac{\omega''_{C,r_p} r_0^2}{\rho_g D}\right)\frac{1}{r} \tag{6.108}$$

Application of the boundary condition at the particle surface, $Y_{O_2}(r = r_0) = 0; Y_{CO_2}(r = r_0) = 0; \therefore \tilde{Y}(r = r_0) = 0$. Solving for the mass burning rate of carbon yields

$$\frac{\omega''_{C,r_p} r_0}{\rho_g D} = \ln\left[Y_{O_2,\infty} \varphi_p\left(1 + \varphi_g\right) + 1\right] = \ln(B_{tf} + 1) \tag{6.109}$$

where the Spalding transfer number for the two-film model is defined as $B_{tf} = Y_{O_2,\infty} \varphi_p\left(1 + \varphi_g\right)$. Substituting the values for the stoichiometric parameters, one obtains $\varphi_g = 2(28 \text{ kg CO})/32 \text{ kg O}_2 = 1.75; \varphi_p = 12 \text{ kg C}/44 \text{ kg CO}_2 = 0.273; Y_{O_2,\infty} = 0.238(\text{air}); \therefore B_{tf} = 0.179$. This is the same value of the one-film transfer number.

$$t_b(\text{two} - \text{film}) = \frac{\rho_p d_0^2}{8\rho_g D \ln(B_{tf} + 1)} = \frac{d_0^2}{K_{tf}} \tag{6.110}$$

$$d_p^2(t) = d_0^2 - K_{tf} t; \quad K_{tf} = \left(\frac{8\rho_g D}{\rho_p}\right)\ln(B_{tf} + 1); \quad B_{tf} = Y_{O_2,\infty} \varphi_p\left(1 + \varphi_g\right) \tag{6.111}$$

These are the equations needed to predict diffusion-controlled carbon combustion for the two-film model. Typical calculated values for the burn rate constant for carbon particles are $= 44 \ \mu m^2/ms$ (Kanury, 1975, p. 208) and $K = 63 \ \mu m^2/ms$ (Annamalai and Ryan, 1993, p. 397).

6.3.5.4 Kinetic-controlled combustion model

Several good references provide detailed discussions on kinetic-controlled combustion of carbon particles (Walker et al., 1959; Laurendeau, 1978; Smoot and Pratt, 1979, Chapter 9; Smith, 1982; Smoot and Smith, 1985, Chapter 4; Annamalai and Ryan, 1993). There are many subtleties to consider in this field of investigation that spans nearly 50 years, and many of these subtleties are related to the particle microstructure which will be the topic of the next section. I will simply summarize the recommendations of Annamalai and Ryan.

For kinetic-controlled combustion, the particle diameter shrinks linearly with time. The burn rate constant is related to the Arrhenius parameters describing the heterogeneous carbon oxidation reaction. Thus,

$$d_p(t) = d_0 - K_{kin} t; \quad K_{kin} = \left(\frac{2\rho_g Y_{O_2,\infty}}{\rho_p}\right) A \exp\left(\frac{-E}{R_g T}\right) \tag{6.112}$$

The recommended values for the Arrhenius parameters are $A = 450$ m/s; $E = 66,400$ kJ/kmol. These are the equations needed to predict kinetic-controlled carbon combustion.

6.3.5.5 Effect of particle microstructure

The volatiles in a char-forming organic fuel burn faster than the char by an order of magnitude. For combustible dust, the majority of the hazard comes from volatiles combustion. This does not mean that the contribution of char oxidation can be neglected. In fact, dust explosion testing has shown that nonvolatile organic fuels can support a deflagration wave if exposed to a sufficiently strong ignition source (Ng et al., 1983; van der Wel et al., 1991; Denkevits and Dorofeev, 2006; Myers and Ibarreta, 2013a).

Devolatilization of charring solids tends to create porous solid structures. The porosity is formed as the volatile matter escapes the carbon-rich solid. The porosity distribution defines the internal surface area of the particle, and it is not unusual for the internal surface area to account for the majority of the specific surface area of the particle as measured by gas adsorption methods. In an uncontrolled industrial environment, porous particles can collect and store airborne volatile contaminants. Therefore, even if all of the original volatile matter is driven out of the porous particles, the foreign contaminants may become devolatilized in a deflagration wave.

In addition to porosity, carbon solids often possess microstructural features which can enhance combustion rates. The microstructure may be derived from the original structure of the parent material (biomass, coal), or it may be created by thermal or chemical processing. These microstructural features include crystallite size and orientation, surface concentration of basal plane edges, surface concentration of edge defects, and the presence of catalytic impurities (Walker et al., 1991; Contescu et al., 2012; Badenhorst and Focke, 2013). A great deal of research effort has gone into developing an understanding of how microstructure affects reactivity, but there is much work yet to be done (Simons, 1983; Walker et al., 1991; Pierson, 1993, Chapter 3; Hurt, 1998; Antal and Grønli, 2003; Marsh and Rodriguez-Reinoso, 2006, Chapter 5; Biagini et al., 2008; Avila et al., 2011). This is a very active area of research with polymer char oxidation (Levendis et al., 1989) and coal and biomass char oxidation (Smith, 1982; Wildman and Derbyshire, 1991; Hurt, 1998; Lu et al., 2000, 2002; Avila et al., 2011; Levendis et al., 2011). It is difficult to summarize the breadth and depth of investigations on the effect of particle microstructure on char oxidation rates. Suffice it to say that it should not be assumed that a nonvolatile organic solid will be too difficult to ignite and participate in a dust deflagration. For any given material, it will most likely be easier to test it in the laboratory rather than try to predict its combustion behavior a priori.

EXAMPLE 6.8

It has been noted in this chapter that the typical time scale for a dust deflagration is between 10 and 100 ms. Consider a spherical particle of carbon. Assume that it obeys the d-squared law with $K(carbon) = 50(\mu m)^2/ms$. What particle diameter of carbon is implied by a burnout time of 100 ms? To 10 ms?

Solution

The particle diameter for a given burnout time is calculated as $d_{p,0} = \sqrt{K\tau}$. A burnout time of 100 ms corresponds to a carbon particle diameter of $d_{p,0} = 71\ \mu m$. A burnout time of 10 ms corresponds to a particle diameter of $d_{p,0} = 22\ \mu m$. With no volatiles to contribute to the combustion process, a dust deflagration of carbon particles must consist of smaller particles. Using the Tyler sieve series designation (Chapter 2), only the particles passing a No. 200 mesh ($d_{p,0} < 75\ \mu m$) would have a burnout time comparable to a deflagration time scale.

6.3.6 SUMMARY OF ORGANIC SOLID COMBUSTION

We have considered the combustion behavior of several types of organic fuel particles. Although our interest is in combustible dust hazards, the first type considered was organic liquid fuels or the so-called "oil drop" model; this is the classic diffusion flame model for liquid droplets. The oil drop model has the same functional form as the shrinking particle model. The analogy between the oil drop model and the combustion of volatiles from a solid particle was recognized early in the history of combustion research and its utility is still evident today. The next topic was noncharring, completely volatile solids. The example of this type of fuel was certain synthetic polymers like polymethyl methacrylate, polystyrene, and polyethylene. The next fuel considered was charring, partially volatile solids. Two examples were considered: coal and biomass. Rate laws for the combustion of volatiles and the resulting char were explored. Finally, we discussed the combustion of nonvolatile organic solids. Chars derived from coal and biomass are two examples of nonvolatile organic solids.

Generally speaking, the combustion rate of volatiles is 10 times faster than char oxidation. In the context of combustible dust hazards, volatiles combustion will be the primary source of heat release. Char oxidation will also contribute, but usually to a lesser degree. But even nonvolatile organic solids may pose a combustible dust hazard depending on its ability to collect airborne volatile contaminants, the fineness of its particle size, the extent of its porosity, or due to other microstructural features.

6.4 SINGLE PARTICLE COMBUSTION MODELS FOR METALLIC SOLIDS

Metals found in nature are usually partly or fully oxidized. Manufactured metal particulates are usually in their reduced state (at least initially). They are either manufactured that way—like atomized metallic powders—or are created as a byproduct of a manufacturing process—like shavings, filings, and particulate debris. The degree of their reactivity will depend on their composition, the presence or absence of an oxide coating on the surface of the particle, the presence of additives, or the presence of contaminants.

Metals do not decompose upon heating like organic materials do. Metals do undergo phase transitions upon heating. Phase transitions have a profound influence on the combustion process for metals. Metals exhibit a diversity of physical and chemical steps in the combustion process. For example, unlike organic solids, metals frequently interact with their combustion products and influence the combustion reaction mechanism. Metal combustion mechanisms can be classified by the volatility of the metal with respect to the flame temperature, the volatility of the metal oxides, and the solubility of any of the metal oxide species in the parent metal (Yetter and Dryer, 2001; Dreizin, 2000).

The following guidelines summarize the basic types of metal combustion. We begin by considering the properties of the metal and three reference temperatures: the melting point of the metal (T_{bm}), the melting point of the metal oxide (T_{box}), and the adiabatic combustion temperature at constant pressure (T_{ad}) (Grosse and Conway, 1958; Glassman, 1960; Markstein, 1967; Yetter and Dryer, 2001). Table 6.9 summarizes three metal combustion scenarios based on these three temperatures.

Low temperature oxidation occurs when the reaction temperature is less than the boiling points of either the metal or its metal oxide. Surface combustion occurs when the flame temperature is on the order of the melting point of the metal oxide, but less than the melting point of the metal. If the flame temperature is greater than the melting point of the metal, but less than the melting point of the metal oxide, the combustion reaction will occur in the gas phase.

Now consider the metal oxide product. If the flame temperature is greater than the metal oxide melting point, then there will be a liquid aerosol that will either be transported away from the metal surface or it will be transported to and condense on the metal surface. If the flame temperature is less than the metal oxide melting point, then the metal oxide will form a solid aerosol.

Table 6.9 Metal Combustion Regimes

Metal Combustion Regime	Temperature Criteria
Low temperature oxidation	$T_{ad} < T_{bm} < T_{box}$
Surface combustion	$T_{ad} \sim T_{box} < T_{bm}$
Vapor phase combustion	$T_{bm} < T_{ad} \leq T_{box}$

Glassman identified four key observations about metal combustion (Glassman, 1960):

- Flame temperatures are limited by the boiling point of the metal oxides
- If the boiling point of the metal oxide exceeds the boiling point of the metal, then combustion occurs in the vapor phase. If the boiling point of the metal oxide is less than the boiling point of the metal, then combustion occurs at the metal surface
- The high flame temperatures of metal combustion lead to significant thermal radiation fluxes
- Factors important in determining steady combustion behavior may not be as important in ignition behavior.

He suggested a postulate to explain why the flame temperatures from metal combustion were limited by the boiling point of the metal. His postulate was that the enthalpy of combustion was less than the enthalpy of vaporization or decomposition for the metal oxides.

The combustion mechanism for metallic solids oftentimes involves a combination of both homogeneous and heterogeneous reactions. Numerous research studies have investigated the fundamental kinetics of metal combustion. Much of this work has been motivated by the use of metals as energetic additives to rocket propellants. Chemical mechanisms have been elucidated using sophisticated spectroscopic techniques to identify the intermediate chemical species formed during combustion. In the combustion of organic solids, the pyrolysis step produces a multitude of chemical species that can be difficult to identify in a reproducible way. In metal combustion, on the other hand, the number of intermediate species involved in the oxidation process is much smaller and more reproducible making the investigation of metal combustion mechanism a bit simpler than it is for organics (note that I said *simpler*, not *simple*).

There are many metals and alloys that can present a combustible dust hazard (Cashdollar and Zlochower, 2007; Eckhoff, 2003, pp. 138−140). Most of these metal dusts have very specialized uses (eg, energetic additives or rocket propellants) and are not in common use in industry. Therefore, I have selected two metals for investigation: aluminum, a volatile metal, and iron, a nonvolatile metal. Both of these metals are readily found in industrial use as both particulate solid products and as scrap or residue produced by manufacturing processes. By choosing these two metals, I am able to cover the range of gas−solid kinetics behavior with aluminum exhibiting shrinking particle behavior and iron exhibiting shrinking core behavior.

6.4.1 VOLATILE METALLIC SOLID: ALUMINUM

Aluminum dust has been a recognized combustible dust hazard for nearly 100 years (Price and Brown, 1921; Mason and Taylor, 1937; Cassel et al., 1949;

Ogle et al., 1988; Myers, 2008; Castellanos et al., 2014). There are many studies on the single particle combustion behavior of aluminum and numerous publications analyzing the chemical kinetics and transport processes behind it. With this surfeit of literature, I have had to be selective. For more information, Beckstead has written two comprehensive reviews on the aluminum particle combustion literature (Beckstead, 2002, 2005).

Aluminum particles are fabricated in a wide range of particle size distributions and, generally speaking, two basic shapes: spherical (atomized) and flake (ball-milled). Aluminum particles have a thin oxide layer that coats the particle. The stoichiometry of the combustion reaction is

$$2Al(s) + \frac{3}{2}O_2 \rightarrow Al_2O_3(s) \qquad (6.113)$$

Aluminum particles burn with a bright, white light. The combustion behavior of aluminum particles can be modeled as a diffusion flame/shrinking droplet model (Kanury, 1975; Ballal, 1983). The boiling point of aluminum is 2740 K ($\sim$2470°C) and the boiling point for the oxide Al_2O_3 is 3253 K (2980°C) (Ballal, 1980). The flame temperature approaches 3800 K ($\sim$3530°C) in oxygen and 3500 K ($\sim$3230°C) in air, so the aluminum particle is a molten droplet and the aluminum oxide combustion product is volatilized forming a smoke halo surrounding the burning droplet (Grosse and Conway, 1958; Glassman et al., 2015, pp. 494–495).

6.4.1.1 Combustion rate law

The earliest studies on aluminum particle combustion postulated the diffusion flame/shrinking particle model which gives rise to the d-squared law (Friedman and Maček, 1962; Kuehl, 1965; Prentice, 1970). However, nonideal combustion behavior such as incomplete combustion and particle fragmentation was frequently observed (Friedman and Maček, 1963; Maček, 1963; Kuehl, 1965; Wilson and Williams, 1971). A recurring observation in these studies was that aluminum particle combustion followed the d-squared law until particle fragmentation or the onset of other nonideal behavior. Wilson and Williams indicated that the burn rate constant for aluminum was in the range of $K = 140-400 \, \mu m^2/ms$. However, they also noted that the data were better correlated with an exponent of 1.5 for the combustion burn rate law. Wilson and Williams also observed that the flame diameter was 3–5 times the particle diameter. Yetter and Dryer reported on experiments that yielded burn rate constant $K = 400 \, \mu m^2/ms$ (Yetter and Dryer, 2001, p. 465).

Dreizin used a novel atomization technique to produce uniform, monosized particles with diameters of 85, 120, 165, and 190 μm (Dreizin, 1996). He ignited the particles and measured the temperature during combustion. He quenched the particles at different times and measured the internal particle composition. He fit his measured particle diameter data to the d-squared law (Fig. 6.30).

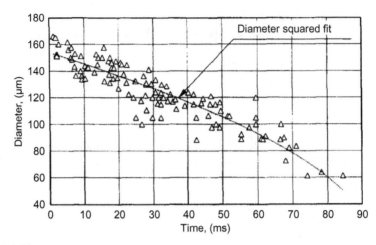

FIGURE 6.30

Aluminum particle diameter-squared versus burnout time.

Reprinted with permission from Dreizin, E.L., 1996. Experimental study of stages of aluminum particle combustion in air. Combust. Flame 105, 541–546.

Dreizin also indicated the important role that phase equilibrium played in modifying the combustion behavior of aluminum particles and its influence on the development of condensed aluminum oxide on the particle surface.

6.4.1.2 Combustion behavior of aluminum particles

Some of the earliest investigations on aluminum particle combustion noticed non-ideal behavior such as incomplete combustion, asymmetric burning caused by the presence of a condensed oxide cap, transparent oxide shells, and particle fragmentation (Friedman and Maček, 1962, 1963; Prentice, 1970; Kuehl, 1965; Maček, 1963).

Law developed a diffusion flame model for metal particles taking into account motion and condensation of metal oxides (Law, 1973). A number of other modeling efforts have since been published that attempt to describe the nonideal combustion behavior. Bucher et al. used enhanced spectroscopic techniques to measure the concentrations of intermediates in the flame zone of aluminum particles burning in air (Bucher et al., 1996). They also captured images of the burning particles showing a range of combustion behavior including formation of the aluminum oxide smoke halo surrounding the droplet with subsequent spinning, jetting of aluminum, and fragmentation (Fig. 6.31).

They subsequently extended their study of aluminum combustion to several other oxidizing environments (Bucher et al., 2000).

Dreizin identified three distinct phases of particle combustion behavior: (1) steady radiant emission stage, (2) a period of higher but oscillatory intensity, and (3) a period of diminishing radiant emission. He has offered arguments for

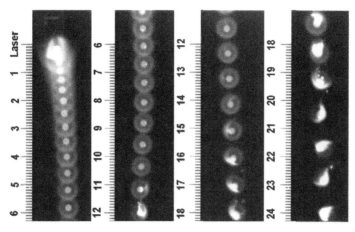

FIGURE 6.31

Luminosity images for four burning aluminum particles (210 μm diameter) in free fall. Scale on the left indicates distance from point of ignition by laser. Images show the full range of combustion behavior.

Reprinted with permission from Bucher, P., Yetter, R.A., Dryer, F.L., Parr, T.P., Hanson-Parr, D.M., Vicenzi, E.P., 1996. Flame structure measurement of single, isolated aluminum particles burning in air. In: Twenty-Sixth Symposium (International) on Combustion. The Combustion Institute, Elsevier, New York. pp. 1899–1908.

the role that phase equilibrium plays in modifying combustion behavior during the life cycle of the burning particle (Dreizin, 1996, 1999a, 1999b, 2000).

Beckstead has presented a comprehensive review of the aluminum particle literature and has critiqued the d-squared law (Beckstead, 2002, 2005). He developed an empirical fit to the particle diameter history data comprising almost 400 data points from 10 different sources and concluded that an exponent between 1.5 and 1.8 gave a better fit to the data than the d-squared law. Beckstead reported on a comprehensive reaction−diffusion model developed with his colleague Liang and, based on simulations with the model and comparisons with empirical data, reported that the model supported the conclusion that the exponent should be 1.88 (Beckstead, 2002).

Des Jardin et al. have developed a more comprehensive model for aluminum particle combustion that does not depend on the assumption of diffusion-controlled combustion and does not restrict the reaction zone to a flame sheet (Des Jardin et al., 2005). Yang and his colleagues have published a particle combustion model with finite kinetics which they claim gives good agreement with experimental measurements (Yang and Yoon, 2010).

6.4.2 NONVOLATILE METALLIC SOLID: IRON

The combustible dust hazard presented by iron is perhaps less well known compared to aluminum. Although there is an early reference to the potential combustibility of iron powder, the authors of the reference admitted that they had been unsuccessful in confirming the report (Price and Brown, 1921, p. 8). Iron (or steel) powder is certainly far less reactive in air than many other metals (Cashdollar, 1994). A more recent accident has demonstrated the hazard of finely divided iron powder. This accident involved fine iron dust that had been treated in a furnace with a hydrogen atmosphere to remove surface scale from the particles (CSB, 2011). While very fine aluminum dust represents a severe dust explosion hazard, finely divided iron dust has a much lower hazard potential. This is because iron is a nonvolatile metal and so its combustion relies on a heterogeneous reaction. The adiabatic flame temperature for iron combustion in air is 2285 K ($\sim$2000°C) and the boiling point of iron is 3023 K (2750°C). There is no gas phase diffusion flame in iron combustion. Instead, the particle glows as the heterogeneous reaction between iron and oxygen proceeds and the product layer accumulates on the droplet surface. The stoichiometry of the combustion reaction is

$$Fe(l) + \frac{1}{2}O_2 \rightarrow FeO(l) \tag{6.114}$$

The stoichiometry relation is written at the reaction temperature. As the droplet is quenched the other iron oxides, FeO reacts to form Fe_2O_3 and Fe_3O_4 (Hirano et al., 1983).

6.4.2.1 Combustion rate law

The literature on iron dust flames is quite limited. Much of the literature on iron combustion is based on experiments in pure oxygen atmospheres. There are a small number of investigations that have focused on the combustion behavior of iron particles in air. One such study has specifically investigated the gas—solid kinetics of iron particle combustion in air and found that the shrinking unreacted core model fits the data reasonably well with the combustion product, iron oxide, growing as the product layer around a shrinking unreacted iron particle (Sun et al., 2000). The shrinking unreacted core model for iron particle combustion was based on the following interpretation of the gas—solid reaction (Sun et al., 2000):

- Reaction at the air—iron oxide interface
  - Diffusion through the gas film surrounding the particle
  - Adsorption of oxygen molecules onto the oxide surface
  - Dissociation of the oxygen molecule into oxygen atoms
- Diffusion of the oxygen atoms through the iron oxide layer
- Reaction at the metal oxide boundary.

The reaction at the air—iron oxide interface was assumed to be the rate-controlling step. The shrinking core model is illustrated in Figure 6.32.

The kinetic expression for the oxygen consumption reaction was derived from previous studies of iron combustion in oxygen (Hirano et al., 1983; Hirano and Sato, 1993).

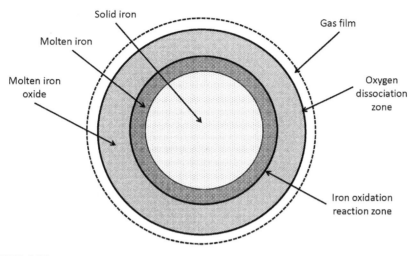

FIGURE 6.32

Shrinking unreacted core model for iron particle combustion.

The combustion reaction was postulated to be kinetic controlled which resulted in a d-linear rate law. Using the nomenclature of Sun et al.,

$$t_b = \frac{b\rho R_0}{3\dot{m}_{00}} \tag{6.115}$$

where b is mass ratio of oxygen to iron, ρ is the density of solid iron, R_0 is the initial particle radius, and $\dot{m}_{00}$ is the initial oxygen mass consumption rate.

6.4.2.2 Combustion behavior of iron particles

Sun et al. performed unconfined flame tests to discern the combustion mechanism. Two iron powders were tested had a particle diameter range of 1−3 μm and 2−4.5 μm. The burnout times ranged from 5 to 30 ms. The burnt particles had a rougher surface and the particle diameters of the burnt particles were larger than the unburnt diameters by approximately 20%. The growth of the particle diameter is consistent with the fact that the density of the iron oxide is lower than the density of iron. The heterogeneous nature of the combustion reaction is evident in Fig. 6.33 from another experimental study of flame propagation in iron dust clouds by the same investigators (Sun et al., 1998).

The gas−solid kinetic model for the combustion of iron has been the subject of several investigations (Hirano et al., 1983; Hirano and Sato, 1993). Much of this work has been motivated by concern over the safety of using iron or steel in oxygen service (Lanyi, 2000; Ward and Steinberg, 2009; Cox et al., 2014).

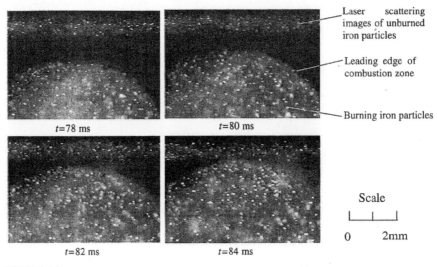

FIGURE 6.33

Flame propagation in a cloud of iron dust. Iron dust concentration of 1.05 kg/m^3; time is the burning time since ignition.

Reprinted with permission from Sun, J.-H., Dobashi, R., Hirano, T., 1998. Structure of flames propagating through metal particle clouds and behavior of particles. In: Twenty-Seventh Symposium (International) on Combustion. The Combustion Institute, Elsevier, New York. pp. 2405–2411.

6.4.3 SUMMARY OF METAL COMBUSTION

Metallic solids are less prevalent as a combustible dust hazard in comparison with organic solids, but they can present equivalent or even greater hazards due to the very high reaction temperatures. Metals can exhibit shrinking particle or shrinking unreacted core kinetics. One might be tempted to think that the chemical mechanism for metal combustion might be simpler in comparison with charring organic solids. But the phenomenon of metal combustion depends on the formation and interaction of various metal oxides which can lead to complex combustion behavior.

6.5 THE RELEVANCE OF SINGLE PARTICLE COMBUSTION TO COMBUSTIBLE DUST HAZARDS

Regardless of the combustible dust hazard—explosion, flash fire, or smoldering—the fundamental event occurring in the microscale is the combustion of a single particle. In the next chapter we will see that in some cases the structure of an unconfined dust flame is composed of individual burning particles. In other cases the dust flame is composed of clusters of particles burning together in a combined manner. Single particle combustion behavior provides a reference point against which one can develop insight into the nature of the specific type of combustion wave of interest.

In this chapter we have seen that the d-squared model for diffusion-controlled single particle combustion works remarkably well for a broad diversity of organic and metallic solid fuels. The factors which fundamentally determine how fast a particle burns are the following:

- Initial diameter
- Chemical composition
- Volatiles content (generated by vaporization or pyrolysis)
- Microstructure

The determination of the particle's initial diameter is simply a matter of measurement (Chapter 2). Particle diameter is a primary determinant governing the relative magnitudes of transport and kinetic rate processes. Empirical studies have shown that the overall combustion rate of larger charring organic solid particles often obey shrinking unreacted core kinetics with an ash layer, a char layer, and an unburnt core. But for particles in the size range of greatest interest in combustible dust hazards—particle diameters less than 500 μm—the combustion rate of organic solids follows approximately shrinking particle kinetics (the d-squared law). Some metallic solids like iron exhibit shrinking unreacted core kinetics regardless of the particle diameter.

Kanury pointed out that the diffusion flame analysis developed for liquid droplets (the d-squared law) could be adapted to solid fuels with a suitable modification of the transfer number B_{OxT} [Kanury, 1975, Chapter 6]. Ballal argued persuasively that the Spalding transfer number B is a useful correlating factor or diffusion-controlled combustion of gas, liquid, and solid fuels [Ballal, 1983].

$$B_{OxT} = \frac{sY_{Ox\infty}\,\Delta h_c + C_{pg}(T_\infty - T_S)}{\Delta h_{vap} + C_{pl}(T_s - T_R)} \tag{6.116}$$

Thus the calculation of the transfer number contains the chemical factors. It requires the determination of the relevant thermodynamic properties and a stoichiometric coefficient. For most fuels the magnitude of the enthalpies of combustion and vaporization are much greater than the sensible enthalpy, and one can simplify the expression for the transfer number to

$$B_{OxT} \cong \frac{sY_{Ox\infty}\,\Delta h_c}{\Delta h_{vap}} \tag{6.117}$$

For pyrolyzing solids, the enthalpy of vaporization is replaced by the enthalpy of gasification (pyrolysis).

A final approximation for diffusion-controlled heterogeneous combustion is formulated from the observation that in heterogeneous combustion the flame standoff is zero, i.e., the flame is at the particle surface or $r_{flame} = r_s$.

Table 6.10 Characteristic Burn Rate Constants and B Values for Select Solid Fuels

Solid Fuel	Burning Rate Constant $(\mu m)^2/ms$	B Value	Source
Polyethylene	2500	1.38	Petrella, 1979
Coal, bituminous	1500 (volatiles) 75 (char)	0.174	Ballal, 1983
Wood	1500 (volatiles) 400 (char)	1.5	Petrella, 1979
Carbon	50	0.174	Ballal, 1983
Aluminum	250	0.42	Ballal, 1983
Iron	1.20	0.0659	Bidabadi and Mafi, 2013

$$r_{flame} = \frac{r_s \, ln(B+1)}{ln(1 + sY_{Ox\infty})} \to B_{OxT} \cong sY_{Ox\infty} \tag{6.118}$$

Table 6.10 is a summary of some representative solid fuels with typical burn rate constants K and their transfer numbers B_{OxT}.

One observation to draw from this table is that greater B_{OxT} values imply greater K values. Note the very small value for the transfer number for iron which is consistent with our understanding that iron burns heterogeneously.

Finally we have seen that solid particles often possess a solid state microstructure which can influence the rate of combustion. Furthermore, the particle microstructure can change during combustion. Particle microstructure tends to exhibit a hierarchy of structural features. These microstructural features tend to be of greater importance for engineered combustion systems where the objective is to achieve complete combustion with minimal formation of pollutants. For most combustible dusts, the effects of particle microstructure are of secondary importance because the extent of combustion is rarely complete.

EXAMPLE 6.9

Calculate the burnout time for a 50 µm particle and a 500 µm particle for each of the following fuels: polyethylene, bituminous coal, wood, carbon, aluminum, and iron. Use the data in Table 6.10. For coal and wood assume that only volatiles combustion is relevant.

Solution

The burnout time is calculated as $\tau = d_{p,0}^2/K$.

Fuel	$\tau(d_{p,0} = 50 \ \mu m)$, ms	$\tau(d_{p,0} = 500 \ \mu m)$, ms
Polyethylene	1.0	100
Bituminous coal	1.7	167
Wood	1.7	167
Carbon	50	5000
Aluminum	10	1000
Iron	2080	208,000

The long burnout times for the non-volatile solids, carbon and iron, are especially striking.

6.6 SUMMARY

This chapter briefly surveyed the immense scientific literature on single particle combustion. We began with a discussion of the physical and chemical phenomena important to solid particle combustion, and how the combustion of organic solids differed from metallic solids.

We then considered the topic of noncatalytic gas-solid reactions. The shrinking particle and shrinking unreacted core models were introduced as the unifying model for single particle combustion, and two extremes of combustion behavior, diffusion-controlled and kinetic-controlled, were described. In the particle size range of interest for combustible dust, particle diameters less than 500 μm, most solid fuels follow shrinking particle kinetics.

The next section surveyed the literature on single particle combustion studies. Some of the early studies on liquid droplet combustion were reviewed and the ability of the d-squared law to fit the measurement data was demonstrated. One key finding from this survey was that the greater the volatility of the solid fuel, the faster the rate of combustion. The d-squared law was shown to act as a suitable combustion model for completely volatile, partially volatile, and non-volatile organic solids. Particle microstructure can cause deviations from the d-squared law, but the deviations tend to be a second-order effect in most cases. Kinetic-controlled combustion can be reasonably described by a linear function of particle diameter. A rule of thumb based solely on particle diameter is not always a reliable method for determining if a particular combustible dust will burn with diffusion-controlled or kinetic-controlled combustion. With that caution in mind, the combustion of particle diameters smaller than 10 μm tend to be kinetically-controlled.

The combustion behavior of partially volatile (charring) organic solids tends to depart from shrinking particle kinetics with increasing particle size. Depending on the specific material, as the particle diameter increases to somewhere in the range of 1 to 5 mm, the particle begins to exhibit shrinking unreacted core behavior. With increasing particle size, particle microstructure also becomes a more important influence on combustion behavior.

Metallic solids exhibit a broader range of behavior than organic solids. With metallic solids it may be difficult to predict their combustion behavior without empirical data to guide the investigation. We considered two examples of metallic combustible dusts, aluminum, a volatile metal, and iron, a nonvolatile metal. The d-squared law gives acceptable burnout time predictions for diffusion-controlled combustion of volatile metals like aluminum. But aluminum can also exhibit anomalous behavior, like fragmentation or jetting, which will cause deviations from d-squared behavior. Iron exhibits shrinking unreacted core kinetics and follows d-linear behavior.

A final point to take away from this chapter is that the Spalding transfer number B is a powerful tool for correlating the combustion behavior of a diverse set of fuels in the vapor, liquid, and solid state. In examining the combustion data described in this chapter, it is again emphasized that the more volatile the solid fuel, the faster it will burn.

Particle size and volatility are the primary determinants of single particle combustion behavior. Particle microstructure, which is important in engineered combustion systems, tends to be less important in accidental deflagration behavior. Armed with this understanding of single particle combustion, we are now ready to explore unconfined dust flame propagation.

REFERENCES

Abdel-Hafez, A.H., 1988. Simplified overall rate expression for shrinking-core bituminous char combustion. Chem. Eng. Sci. 43, 839–845.

Amyotte, P., 2006. Solid inertants and their use in dust explosion prevention and mitigation. J. Loss Pre. Process Ind. 19, 161–173.

Amyotte, P., 2013. An Introduction to Dust Explosions. Butterworth Heinemann, Elsevier, New York.

Annamalai, K., Puri, I.K., 2007. Combustion Science and Engineering. CRC Press, Boca Raton, FL.

Annamalai, K., Ryan, W., 1993. Interactive processes in gasification and combustion—II. Isolated carbon, coal and porous char particles. Prog. Energy Combust. Sci. 19, 383–446.

Antal Jr., M.J., Grønli, M., 2003. The art, science, and technology of charcoal production. Ind. Eng. Chem. Res. 42, 1619–1640.

Aris, R., 1975. The Mathematical Theory of Diffusion and Reaction in Permeable Catalysts. Volume I. The Theory of the Steady State. Clarendon Press, Oxford.

ASTM D388, 1998. Standard Classification of Coals by Rank. ASTM International, West Conshohocken, PA.

Ausman, J.M., Watson, C.C., 1962. Mass transfer in a catalyst pellet during regeneration. Chem. Eng. Sci. 17, 323–329.

Austin, P.J., Kauffman, C.W., Sichel, M., 1991. Combustion of single nonspherical cellulosic particles. Prog. Astronaut. Aeronaut. Vol. 152, 136–154.

Austin, P.J., Kauffman, C.W., Sichel, M., 1996. Ignition and combustion of cellulosic dust particles. Combust. Sci. Technol. 112, 187–198.

Avila, C., Pang, D.H., Wu, T., Lester, E., 2011. Morphology and reactivity characteristics of char biomass particles. Bioresour. Technol. 102, 5237–5243.

Badenhorst, H., Focke, W., 2013. Comparative analysis of graphite oxidation behavior based on microstructure. J. Nucl. Mater. 442, 75–82.

Ballal, D.R., 1980. Ignition and flame quenching of quiescent dust clouds of solid fuels. Proc. Royal Soc. Lond. A369, 479–500.

Ballal, D.R., 1983. Flame propagation through dust clouds of carbon, coal, aluminium and magnesium in an environment of zero gravity. Proc. Royal Soc. Lond. A385, 21–51.

Beckstead, B.W., 2002. A summary of aluminum combustion. Internal Aerodynamics in Solid Rocket Propulsion. RTO/VKI Special Course, Rhode-Saint-Genese, Belgium.

Beckstead, B.W., 2005. Correlating aluminium burning times. Combust. Explos. Shock Waves 41, 533–546.

Beeston, G., Essenhigh, R.H., 1963. Kinetics of coal combustion: the influence of oxygen concentration on the burning-out times of single particles. J. Phys. Chem. 67, 1349–1355.

Beyler, C.L., Hirschler, M.M., 2002. Thermal decomposition of polymers, Section 1, Chapter 7. In: DiNenno, P.J., et al., (Eds.), SFPE Handbook of Fire Protection Engineering, third ed. National Fire Protection Association, Quincy, MA, pp. 1-110–1-131.

Biagini, E., Narducci, P., Tognotti, L., 2008. Size and structural characterization of lignin-cellulosic fuels after the rapid devolatilization. Fuel 87, 177–186.

Biagini, E., Simone, M., Tognotti, L., 2009. Characterization of high heating rate chars of biomass fuels. Proc. Combust. Inst. 32, 2043–2050.

Bird, R.B., Stewart, W.E., Lightfoot, E.N., 2002. Transport Phenomena, second ed. John Wiley & Sons, New York.

Branca, C., Di Blasi, C., 2003. Global kinetics of wood char devolatilization and combustion. Energy Fuels 17, 1609–1615.

Brown, J.E., Kashiwagi, T., 1996. Gas phase oxygen effect on chain scission and monomer content in bulk poly(methyl methacrylate) degraded by external thermal radiation. Polym. Degrad. Stabil. 52, 1–10.

Brown, M.E. (Ed.), 2001. Introduction to Thermal Analysis: Techniques and Applications. second ed. Kluwer Academic Publishers, Dordrecht.

Bucher, P., Yetter, R.A., Dryer, F.L., Parr, T.P., Hanson-Parr, D.M., Vicenzi, E.P., 1996. Flame structure measurement of single, isolated aluminum particles burning in air. Twenty-Sixth Symposium (International) on Combustion. The Combustion Institute Elsevier, New York.

Bucher, P., Ernst, L., Dryer, F.L., Yetter, R.A., Parr, T.P., Hanson-Parr, D.M., 2000. Detailed studies on the flame structure of aluminum particle combustion. Prog. Astronaut. Aeronaut. 185, 689–722.

Burnham, A.K., Zhou, X., Broadbelt, L.J., 2015. Critical review of the global chemical kinetics of cellulose thermal decomposition. Energy Fuels 29, 2906–2918.

Carberry, J.J., 2001. Chemical and Catalytic Reaction Engineering. Dover Publications, Mineola, NY.

Cashdollar, K.L., 1994. Flammability of metals and other elemental dust clouds. Process Saf. Prog. 13, 139–145.

Cashdollar, K.L., 1996. Coal dust explosibility. J. Loss Prev. Process Ind. 9, 65–76.

Cashdollar, K.L., 2000. Overview of dust explosibility characteristics. J. Loss Prev. Process Ind. 13, 189–199.

Cashdollar, K.L., Zlochower, I.A., 2007. Explosion temperatures and pressures of metals and other elemental dust clouds. J. Loss Prev. Process Ind. 20, 337–348.

Cashdollar, K.L., Weiss, E.S., Montgomery, T.G., Going, J.E., 2007. Post-explosion observations of experimental mine and laboratory coal dust explosions. J. Loss Prev. Process Ind. 20, 607–615.

Cassel, H.M., das Gupta, A.K., Guruswamy, S., 1949. Factors affecting flame propagation through dust clouds. Third Symposium on Combustion, Flames and Explosions. William Wilkins and Co., Elsevier.

Castellanos, D., Carreto-Vazquez, V.H., Mashuga, C.V., Trottier, R., Mejia, A.F., Mannan, M.S., 2014. The effect of particle size polydispersity on explosibility characteristics of aluminum dust. Powder Technol. 254, 331–337.

Cetin, E., Moghtaderi, B., Gupta, R., Wall, T.F., 2004. Influence of pyrolysis conditions on the structure and gasification reactivity of biomass chars. Fuel 83, 2139–2150.

Chen, L., Dupont, C., Salvador, S., Grateau, M., Boissonet, G., Schweich, D., 2013. Experimental study on fast pyrolysis of free-falling millimetric biomass particles between 800°C and 1000°C. Fuel 106, 61–66.

Chern, J.-S., Hayhurst, A.N., 2004. Does a large coal particle in a hot fluidised bed lose its volatile content according to the shrinking core model? Combust. Flame 139, 208–221.

Cloney, C.T., Amyotte, P.R., Khan, F.I., Ripley, R.C., 2014. Development of an organizational framework for studying dust explosion phenomena. J. Loss Prev. Process Ind. 30, 228–235.

Contescu, C.I., Guldan, T., Wang, P., Burchell, T.D., 2012. The effect of microstructure on air oxidation resistance of nuclear graphite. Carbon 50, 3354–3366.

Conti, R.S., Zlochower, I.A., and Sapko, M.J. Rapid (Grab) Sampling During Full-Scale Explosions—Microscopic and Analytical Evaluation. Report of Investigations 9192, U.S. Department of the Interior, Bureau of Mines, 1988.

Conti, R.S., Zlochower, I.A., Sapko, M.J., 1991. Rapid sampling of products during coal mine explosions. Combust. Sci. Technol. 75, 195–209.

Cox, B.L., Dee, S.J., Hart, R.J., Morrison, D.R., 2014. Development of a steel component combustion model for fires involving pure oxygen systems. Process Saf. Prog. 33, 299–304.

Crelling, J.C., Hagemann, H.W., Sauter, D.H., Ramani, R.V., Vogt, W., Leininger, D., et al., 2010. Coal. Ullman's Encyclopedia of Industrial Chemistry. Wiley-VCH Verlag Gmbh & Co, New York.

Crowe, C.T., Schwarzkopf, J.D., Sommerfield, M., Tsuji, Y., 2012. Multiphase Flows with Droplets and Particles, second ed. CRC Press, Boca Raton, FL.

CSB, 2006. Investigation Report: Combustible Dust Hazard Study. U.S. Chemical Safety and Hazard Investigation Board, Washington DC.

CSB, 2011. Case Study: Metal Dust Flash Fires and Hydrogen Explosion—Hoeganaes Corporation, Gallatin, TN. U.S. Chemical Safety and Hazard Investigation Board, Washington DC.

Dakic, D., van der Honing, G., Valk, M., 1989. Fragmentation and swelling of various coals during devolatilization in a fluidized bed. Fuel 68, 911–916.

Dall'Ora, M., Jensen, P.A., Jensen, A.D., 2008. Suspension combustion of wood: influence of pyrolysis conditions on char yield, morphology, and reactivity. Energy Fuels 22, 2955–2962.

Dashek, W.V., Olenchock, S.A., Mayfield, J.E., Wirtz, G.H., Wolz, D.E., Young, C.A., 1986. Carbohydrate and protein contents of grain dusts in relation to dust morphology. Environ. Health Perspect. 66, 135–143.

Davis, H., Hottel, H.C., 1934. Combustion rate of carbon: combustion at a surface overlaid with stagnant gas. Ind. Eng. Chem. 26, 889–892.

Denkevits, A., Dorofeev, S., 2006. Explosibility of fine graphite and tungsten dusts and their mixtures. J. Loss Prev. Process Ind. 19, 174–180.

Des Jardin, P.E., Felske, J.D., Carrara, M.D., 2005. Mechanistic model for aluminum particle ignition and combustion in air. J. Propul. Power 21, 478–485.

Di Blasi, C., 1993. Modeling and simulation of combustion processes of charring and non-charring solid fuels. Prog. Energy Combust. Sci. 19, 71–104.

Di Blasi, C., 1996. Kinetic and heat transfer control in the slow and flash pyrolysis of solids. Ind. Eng. Chem. Res. 35, 37–46.

Di Blasi, C., 2002. Modeling intra- and extra-particle processes of wood fast pyrolysis. AIChE J. 48, 2386–2397.

Di Blasi, C., 2008. Modeling chemical and physical processes of wood and biomass pyrolysis. Prog. Energy Combust. Sci. 34, 47–90.

Downie, A., Crosky, A., and Munroe, P. Physical properties of biochar. In: Lehmann, J., Joseph, S. (Eds.), Biochar for Environmental Management: Science and Technology, London, GB: Earthscan, pp. 13–32, 2009.

Dreizin, E.L., 1996. Experimental study of stages of aluminum particle combustion in air. Combust. Flame 105, 541–546.

Dreizin, E.L., 1999a. Internal heterogeneous processes in aluminum combustion. Fifth International Microgravity Combustion Workshop, NASA/CP—1999-208917. National Aeronautics and Space Administration.

Dreizin, E.L., 1999b. On the mechanism of asymmetric aluminum particle combustion. Combust. Flame 117, 841–850.

Dreizin, E.L., 2000. Phase changes in metal combustion. Prog. Energy Combust. Sci. 26, 57–78.

Drysdale, D., 1999. An Introduction to Fire Dynamics, second ed. John Wiley & Sons, New York.

Eckhoff, R., 2003. Dust Explosions in the Process Industries, third ed. Gulf Professional Publishing, Elsevier, New York.

Eckhoff, R., 2009. Understanding dust explosions—the role of powder science and technology. J. Loss Prev. Process Ind. 22, 105–116.

Edgar, T.F., 1983. Coal Processing and Pollution Control. Gulf Publishing, New York.

Essenhigh, R.H., and Dreier, W.L. Combustion behavior of thermoplastic polymer spheres burning in quiescent atmospheres of air. AIAA Paper No. 67-103, AIAA 5th Aerospace Sciences Meeting, New York, 1967.

Essenhigh, R.H., Fells, I., 1960. Combustion of liquid and solid aerosols. Discuss. Faraday Soc. 30, 208–221.

Essenhigh, R.H., Froberg, R., Howard, J.B., 1965. Combustion behavior of small particles. Ind. Eng. Chem. 57, 32–43.

Everson, R.C., Neomagus, H.W.J.P., van der Merwe, G.W., Koekemoer, A., Bunt, J.R., 2015. The properties of large coal particles and reaction kinetics of corresponding chars. Fuel 140, 17–26.

Faraday, M., Lyell, C., 1845. On the subject of the explosion in the Haswell Colleries and on the means of preventing similar accidents. Philos. Mag. 26, 16–35.

Fisher, E.M., Dupont, C., Darvell, L.I., Commandre, J.-M., Sadddawi, A., Jones, J.M., et al., 2012. Combustion and gasification characteristics of chars from raw and torrefied biomass. Bioresour. Technol. 119, 157–165.

Fitzer, E., Köchling, K.-H., Boehm, H.P., Marsh, H., 1995. Recommended terminology for the description of carbon as a solid. Pure Appl. Chem. 67, 473–506.

Friedman, R., Maček, A., 1962. Ignition and combustion of aluminum particles in hot ambient gases. Combust. Flame 6, 9–19.

Friedman, R., Maček, A., 1963. Combustion studies of single aluminum particles. Ninth Symposium (International) on Combustion. The Combustion Institute, New York, Elsevier.

Froment, G.F., Bischoff, K.B., 1979. Chemical Reactor Analysis and Design. John Wiley & Sons, New York.

Galgano, A., Di Blasi, C., 2003. Modeling wood degradation by the unreacted-core-shrinking approximation. Ind. Eng. Chem. Res. 42, 2101–2111.

Glassman, I., 1960. Combustion of metals: physical considerations. Solid Propellant Rocket Research, Progress in Astronautics and Aeronautics 253–258.

Glassman, I., Yetter, R.A., Glumac, N.G., 2015. Combustion, fifth ed. Academic Press, New York.

Godsave, G.A.E., 1949. Combustion of droplets in a fuel spray. Nature. 164, 708–709.

Godsave, G.A.E., 1953. Studies of the combustion of drops in a fuel spray—the burning of single drops of fuel. Fourth Symposium (International) on Combustion. The Combustion Institute Elsevier, New York.

Goldsmith, M., 1956. Experiments on the burning of single drops of fuel. Jet Propul. 26, 172–178.

Goldsmith, M., Penner, S.S., 1954. On the burning of single drops of fuel in an oxidizing atmosphere. Jet Propul. 24, 245–251.

Goynes, W.R., Ingber, B.F., Palmgren, M.S., 1986. Microscopical comparison of cotton, corn, and soybean dusts. Environ. Health. Perspect. 66, 125–133.

Grosse, A.V., Conway, J.B., 1958. Combustion of metals in oxygen. Ind. Eng. Chem. 50, 663–672.

Hertzberg, M., Cashdollar, D.L., Ng, D.L., Conti, R.S., 1982. Domains of flammability and thermal ignitability for pulverized coals and other dusts: particle size dependences and microscopic residue analyses. Nineteenth Symposium (International) on Combustion. The Combustion Institute Elsevier, New York.

Hertzberg, M., Zlochower, I.A., and Edwards, J.C. Coal Particle Pyrolysis Mechanisms and Temperatures. Bureau of Mines Report of Investigations, RI 9169, U.S. Department of the Interior, 1988.

Hinds, W.C., 1999. Aerosol Technology: Properties, Behavior, and Measurement of Airborne Particles, second ed. Wiley-Interscience, New York.

Hirano, T., Sato, J., 1993. Fire spread along structural metal pieces in oxygen. J. Loss Prev. Process Ind. 6, 151–157.

Hirano, T., Sato, K., Sato, Y., Sato, J., 1983. Prediction of metal fire spread in high pressure oxygen. Combust. Sci. Technol. 32, 137–159.

Hsi, C.-L., Kuo, J.T., 2007. Mass and thermal characteristics of single wooden spheres burning in high temperature air streams, November 23–24. Twenty-Fourth National Symposium Chung Yuan Christian University. Chinese Mechanical Engineering Society, Taoyuan, Chungli, Taiwan

Hurt, R.H., 1998. Structure, properties, and reactivity of solid fuels. Twenty-Seventh Symposium (International) on Combustion. The Combustion Institute Elsevier, New York.

Incropera, F.P., DeWitt, D.P., 1990. Fundamentals of Heat and Mass Transfer, third ed. John Wiley & Sons, New York.

Ishida, M., Wen, C.Y., 1968. Comparison of kinetic and diffusional models for solid–gas reactions. AIChE J. 14, 311–317.

Ishida, M., Wen, C.Y., 1971. Comparison of zone-reaction model and unreacted-core shrinking model in solid-gas reactions—I. Isothermal analysis. Chem. Eng. Sci. 26, 1031–1041.

Jones, J.M., Lea-Langton, A.R., Ma, L., Pourkashanian, M., Williams, A., 2014. Pollutants Generated by the Combustion of Solid Biomass Fuels. Springer, Berlin.

Kanury, A.M., 1975. Introduction to Combustion Phenomena. Gordon and Breach Science Publishers, New York.

Kashiwagi, T., Ohlemiller, T.J., 1982. A study of oxygen effects on nonflaming transient gasification of PMMA and PE during thermal irradiation. Nineteenth Symposium (International) on Combustion. The Combustion Institute. Elsevier, New York.

Kashiwagi, T., 1994. Polymer combustion ad flammability—role of the condensed phase. Twenty-Fifth Symposium (International) on Combustion. The Combustion Institute, Elsevier, New York.

Kenney, K.L., Smith, W.A., Gresham, G.L., Westover, T.L., 2013. Understanding biomass feedstock variability. Biofuels 4, 111–127.

Kuehl, D.K., 1965. Ignition and combustion of aluminum and beryllium. AIAA J. 3, 2239–2247.

Kuo, J.T., Hsi, C.-L., 2005. Pyrolysis and ignition of single wooden spheres heated in high-temperature streams of air. Combust. Flame 142, 401–412.

Kuo, J.T., Hwang, L.H., 2003. Mass and thermal analysis of burning wood spheres. Combust. Sci. Technol. 175, 665–693.

Lanyi, M.D., 2000. Discussion on steel burning in oxygen (from a steelmaking metallurgist's perspective),". In: Steinberg, T.A., et al., (Eds.), Flammability and Sensitivity of Materials in Oxygen-Enriched Atmospheres: Ninth Volume, ASTM STP 1395. American Society for Testing and Materials, West Conshohocken, PA.

Laurendeau, N.M., 1978. Heterogeneous kinetics of coal char gasification and combustion. Prog. Energy Combust. Sci. 4, 221–270.

Law, C.K., 1973. A simplified theoretical model for the vapor-phase combustion of metal particles. Combust. Sci. Technol. 7, 197–212.

Levendis, Y.A., Flagan, R.C., Gavalas, G.R., 1989. Oxidation kinetis of monodisperse spherical carbonaceous particles of variable properties. Combust. Flame 76, 221–241.

Levendis, Y.A., Joshi, K., Khatami, R., Sarofim, A.F., 2011. Combustion behavior in air of single particles from there different coal ranks and from sugarcane bagasse. Combust. Flame 158, 452–465.

Levenspiel, O., 1999. Chemical Reaction Engineering, third ed. John Wiley & Sons, New York.

Lu, L., Sahajawalla, V., Harris, D., 2000. Characteristics of chars prepared from various pulverized coals at different temperatures using drop-tube furnace. Energy & Fuels 14, 869–876.

Lu, L., Kong, C., Sahajwalla, V., Harris, D., 2002. Char structural ordering during pyrolysis and combustion and its influence on char reactivity. Fuel 81, 1215–1225.

Maa, P.S., Bailie, R.C., 1973. Influence of particle sizes and environmental conditions on high temperature pyrolysis of cellulosic material—I (Theoretical). Combust. Sci. Technol. 7, 257–269.

Maček, A., 1963. Fundamentals of combustion of single aluminum and beryllium particles. Eleventh Symposium (International) on Combustion. The Combustion Institute. Elsevier, New York.

Man, C.K., Cashdollar, K.L., Zlochower, I.A., Green, G.M., 2009. Observations of post-explosion dust samples from an experimental mine. 6th US National Combustion Meeting. The Combustion Institute, University of Michigan, Ann Arbor, MI.

Maness, J.E., Sargent, L.M., 2002. Storage and handling of grain mill products, Section 9, Chapter 7. In: Cote, A.E., et al., (Eds.), Fire Protection Handbook, twentieth ed. National Fire Protection Association, Quincy, MA, pp. 9-85–9-103.

Markstein, G.H., 1967. Heterogeneous reaction processes in metal combustion. Eleventh Symposium (International) on Combustion. The Combustion Institute, Elsevier, New York.

Marsh, H., Rodriguez-Reinoso, R., 2006. Activated Carbon. Elsevier, New York.

Martin, C.R., 1981. Characterization of grain dust properties. Trans. ASAE 24, 738−742.

Martin, C.R., Sauer, D.B., 1976. Physical and biological characteristics of grain dust. Trans. ASAE 19, 720−723.

Mason, P.E., Darvell, L.I., Jones, J.M., Pourdashanian, M., Williams, A., 2015. Single particle flame-combustion studies on solid biomass fuels. Fuel 151, 21−30.

Mason, R.B., Taylor, C.S., 1937. Explosion of aluminum powder dust clouds. Ind. Eng. Chem. 29, 626−631.

Mettler, M.S., Vlachos, D.G., Dauuenharer, P.J., 2012. Top ten fundamental challenges of biomass pyrolysis for biofuels. Energy Environ. Sci. 5, 7797−7809.

Momeni, M., Yin, C., Kaer, S.K., Hansen, T.B., Jensen, P.A., Glarborg, P., 2013a. Experimental study on effects of particle shape and operating conditions on combustion characteristics of single biomass particles. Energy Fuels 27, 507−514.

Momeni, M., Yin, C., Kaer, Hvid, S.L., 2013b. Comprehensive study of ignition and combustion of single wooden particles. Energy Fuels 27, 1061−1072.

Mukunda, H.S., Paul, P.J., Srinivasa, U., Rajan, N.K.S., 1984. Combustion of wooden spheres—experiments and model analysis. Twentieth Symposium (International) on Combustion. The Combustion Institute, Elsevier, New York.

Myers, T.J., 2008. Reducing aluminum dust explosion hazards: case study of dust inerting in an aluminum buffing operation. J. Hazard. Mater. 159, 72−80.

Myers, T.J. and Ibarreta, A., 2013. Assessing the hazard of marginally explosible dust, American Institute of Chemical Engineers, 2013 Spring Meeting, 9th Global Congress on Process Safety, San Antonio, TX, April 28—May 1.

Ng, D.L., Cashdollar, K.L., Hertzberg, M., Lazzara, P., 1983. Electron Microscopy Studies of Explosion and Fire Residues, Bureau of Mines Information Circular 8936. U.S. Department of the Interior.

NIOSH, 2006. Float Coal Dust Explosion Hazards. U.S. Department of Health and Human Services, Centers for Disease Control and Prevention, National Institute for Occupational Safety and Health, DHHS (NIOSH) Publication No. 2006−125.

Ogle, R.A., Beddow, J.K., Chen, L.-D., Butler, P.B., 1988. An investigation of aluminum dust explosions. Combust. Sci. Technol. 61, 75−99.

Ouedraogo, A., Mulligan, J.C., Cleland, J.G., 1998. A quasi-steady shrinking core analysis of wood combustion. Combust. Flame 114, 1−12.

Palmer, K.N., 1973. Dust Explosions and Fires. Chapman and Hall, London.

Panagiotou, T., Levendis, Y., 1994. A study on the combustion characteristics of PVC, poly(styrene), poly(ethylene), and poly(propylene) particles under high heating rates. Combust. Flame 99, 53−74.

Panagiotou, T., Levendis, Y., Delicatsios, M.A., 1994. Combustion behavior of (poly) styrene particles of various degrees of crosslinking and styrene monomer droplets. Combust. Sci. Technol. 103, 63−84.

Parker, A.S., Hottel, H.C., 1936. Combustion rate of carbon: study of gas−film structure by microsampling. Ind. Eng. Chem. 28, 1334−1341.

Parnell Jr., C.B., Jones, D.D., Rutherford, R.D., Goforth, K.J., 1986. Physical properties of five grain dust types. Environ. Health. Perspect. 66, 183−188.

Petersen, E.E., 1957. Reaction of porous solids. AIChE J. 3, 443–448.

Pierson, H.O., 1993. Handbook of Carbon, Graphite, Diamond, and Fullerenes: Properties, Processing and Applications. Noyes Publications, Park Ridge, NJ.

Prentice, J.L., 1970. Combustion of pulse-heated single particles of aluminum and beryllium. Combust. Sci. Technol. 1, 385–398.

Price, D.J. and Brown, H.H. (assisted by H.R. Brown and H.F. Roethe). Dust Explosions—Theory and Nature of, Phenomena, Causes and Methods of Prevention. National Fire Protection Association, Quincy, Boston, MA, 1921.

Raghunandan, B.N., Mukunda, H.S., 1977. Combustion of polystyrene spheres in air. Fuel 56, 271–276.

Ragland, K.W., Aerts, D.J., 1991. Properties of wood for combustion analysis. Bioresour. Technol. 37, 161–168.

Ramachandran, P.A., Doraiswamy, L.K., 1982. Modeling of noncatalytic gas–solid reactions. AIChE J. 28, 881–900.

Rastogi, S., Klinzing, G.E., Proscia, W.M., 1996. Morphological characterization of coal under rapid heating devolatilization. Powder Technol. 88, 143–154.

Riaza, J., Khatami, R., Levendis, Y.A., Alvarez, L., Gil, M.V., Pevida, C., et al., 2014a. Single particle ignition and combustion of anthracite, semi-anthracite, and bituminous coals in air and simulated oxy-fuel conditions. Combust. Flame 161, 1096–1108.

Riaza, J., Khatami, R., Levendis, Y.A., Alvarez, L., Gil, M.V., Pevida, C., et al., 2014b. Combustion of single biomass particles in air and in oxy-fuel conditions. Biomass Bioenergy 64, 162–174.

Rice, G.S., 1910. The Explosibility of Coal Dust. U.S. Geological Survey, Bulletin 425, Washington, DC.

Roberts, A.F., 1971. Problems associated with the theoretical analysis of the burning of wood. Thirteenth Symposium (International) on Combustion. The Combustion Institute, Elsevier, New York.

Ross, R.J., editor, 2010. Wood Handbook—Wood as an Engineering Material. General Technical Report FPL-GTR-190, Forest Products Laboratory, U.S. Department of Agriculture.

Saastamoinen, J.J., Aho, M.J., Linna, V.L., 1993. Simultaneous pyrolysis and char combustion. Fuel 72, 599–609.

Sapko, M.J., Cashdollar, K.L., Green, G.M., 2007. Coal dust particle size survey of US mines. J. Loss Prev. Process Ind. 20, 616–620.

Saxena, S.C., 1990. Devolatilization and combustion characteristics of coal particles. Prog. Energy Combust. Sci. 16, 55–94.

Shen, J., Smith, J.M., 1965. Diffusional effects in gas–solid reactions. I&EC Fundamentals 4, 293–301.

Shen, D.K., Gu, S., Luo, K.H., Bridgwater, A.V., 2009. Analysis of wood structural changes under thermal radiation. Energy Fuels 23, 1081–1088.

Simmons, W.W., Ragland, K.W., 1985. Single particle combustion analysis of wood, Chapter 41. In: Overend, R.P., et al., (Eds.), Fundamentals of Thermochemical Biomass Conversion. Elsevier, New York, pp. 777–792.

Simons, G.A., 1983. The role of pore structure in coal pyrolysis and gasification. Prog. Energy Combust. Sci. 9, 269–290.

Smith, D.F., Gudmundsen, A., 1931. Mechanism of combustion of individual particles of solid fuels. Ind. Eng. Chem. 23, 277–285.

Simmons, W.W., Ragland, K.W., 1986. Burning rate of millimeter sized wood particles in a furnace. Combust. Sci. Technol. 46, 1–15.

Smith, I.W., 1982. The combustion rates of coal chars: a review. Nineteenth (International) Symposium on Combustion. The Combustion Institute, Elsevier, New York.

Smoot, L.D., Pratt, D.T. (Eds.), 1979. Pulverized Coal Combustion and Gasification: Theory and Applications for Continuous Flow Processes. Springer, Berlin.

Smoot, L.D., Smith, P.J., 1985. Coal Combustion and Gasification. Plenum Press, New York.

Speight, J.G., 2005. Handbook of Coal Analysis. Wiley-Interscience, New York.

Sun, J.-H., Dobashi, R., Hirano, T., 1998. Structure of flames propagating through metal particle clouds and behavior of particles. Twenty-Seventh Symposium (International) on Combustion. The Combustion Institute, Elsevier, New York.

Sun, J.-H., Dobashi, R., Hirano, T., 2000. Combustion behavior of iron particles suspended in air. Combust. Sci. Technol. 150, 99–114.

Szekely, J., Evans, J.W., Sohn, H.Y., 1976. Gas–Solid Reactions. Academic Press, New York.

Tsai, C.-Y., Scaroni, A.W., 1987. The structural changes of bituminous coal particles during the initial stages of pulverized-coal combustion. Fuel 66, 200–206.

Tu, C.M., Davis, H., Hottel, H.C., 1934. Combustion rate of carbon: combustion of spheres in flowing gas streams. Ind. Eng. Chem. 26, 749–757.

Turns, S.R., 2012. An Introduction to Combustion, third ed. McGraw-Hill, New York.

van der Wel, P.G.J., Lemkowitz, S.M., Scarlett, B., van Wingerden, C.J.M., 1991. A study of particle factors affecting dust explosions. Part. Part. Sys. Character. 8, 90–94.

Veras, C.A.G., Saastamoinen, J., Carvalho Jr., J.A., Aho, M., 1999. Overlapping of the devolatilization and char combustion stages in the burning of coal particles. Combust. Flame 116, 567–579.

Waibel, R.T., Essenhigh, R.H., 1973. Combustion of thermoplastic polymer particles in various oxygen atmospheres: comparison of theory and experiment. Fourteenth Symposium (International) on Combustion. The Combustion Institute, Elsevier, New York.

Walker Jr., P.L., Rusinko Jr., F., Austin, L.G., 1959. Gas reactions of carbon. Adv. Catal. Vol. XI, 133–221.

Walker Jr., P.L., Taylor, R.L., Ranish, J.M., 1991. An update on the carbon–oxygen reaction. Carbon 29, 411–421.

Ward, N., Steinberg, T., 2009. The rate-limiting mechanism for the heterogeneous burning of cylindrical iron rods. J. ASTM Int. 6 (6).

Wen, C.Y., 1968. Noncatalytic heterogeneous solid fluid reaction models. Ind. Eng. Chem. 60, 34–54.

Whitaker, S., 1999. The Method of Local Volume Averaging. Kluwer Academic Publishers, Dordrecht.

Wildman, J., Derbyshire, F., 1991. Origins and functions of macroporosity in activated carbons from coal and wood precursors. Fuel 70, 655–661.

Wilson, R.P., Williams, F.A., 1971. Experimental study of the combustion of single aluminum particles in O_2-Ar. Thirteenth Symposium (International) on Combustion. The Combustion Institute, Elsevier, New York.

Yagi, S., Kunii, D., 1955. Studies on combustion of carbon particles in flames and fluidized beds. Fifth Symposium (International) on Combustion. The Combustion Institute, Elsevier, New York.

Yang, H., Yoon, W., 2010. Modeling of aluminum particle combustion with emphasis on the oxide effects and variable transport properties. J. Mech. Sci. Technol. 24, 909–921.

Yang, J.C., Hamins, A., Donnelly, M.K., 2000. Reduced gravity combustion of thermoplastic spheres. Combust. Flame 120, 61–74.

Yang, Y.B., Sharifi, V.N., Swithenbank, J., Ma, L., Darvell, L.I., Jones, J.M., et al., 2008. Combustion of a single particle of biomass. Energy Fuels 22, 306–316.

Yetter, R.A., Dryer, F.L., 2001. Metal particle combustion and classification, Chapter 6. In: Ross, H.D. (Ed.), Microgravity Combustion: Fire in Free Fall. Academic Press, New York, pp. 419–478.

Yu, J.-L., Strezov, V., Lucas, J., Liu, G.-S., Wall, T., 2002. A mechanistic study on char structure evolution during coal devolatilization—experiments and model predictions. Proc. Combust. Inst. 29, 467–473.

Yu, J.-L., Strezov, V., Lucas, J., Wall, T., 2003. Swelling behavior of individual coal particles in the single particle reactor. Fuel 82, 1977–1987.

Yu, J.-L., Lucas, J.A., Wall, T.F., 2007. Formation of the structure of chars during devolatilization of pulverized coal and its thermoproperties: a review. Prog. Energy Combust. Sci. 33, 135–170.

Zicherman, J.B., Williamson, R.B., 1981. Microstructure of wood char: Part 1: whole wood. Wood Sci. Technol. 15, 237–249.

Unconfined dust flame propagation

Having examined the combustion behavior of single dust particles, we advance to unconfined burning dust clouds. Our theoretical frame of reference will be the one-dimensional laminar flame. The goal in this chapter will be to characterize the combustion behavior of these flames in terms of the burning velocity, flame thickness, lower flammability limit, minimum ignition energy, and quenching diameter. Ideally, we would like to relate the combustion attributes of the dust flame to the equivalence ratio (or dust concentration) and the particle size of the combustible dust. Additionally, we will find that many of the insights derived from the study of single particle combustion will inform our investigation of unconfined dust flames.

The combustion attributes associated with unconfined flame propagation have direct relevance to the evaluation of combustible dust hazards. For example, a flash fire is an unconfined dust flame. The flame temperature and thermal radiation directly correlate with the damage potential of a flash fire. The burning velocity for a dust flame is one of the input parameters needed for some of the explosion development models to be introduced in Chapter 8. Thus, our motivation for studying unconfined flame propagation has both theoretical and practical significance.

I begin the first section with a discussion of the one-dimensional laminar dust flame. After a brief survey of experimental methods, we will learn how dust flames differ from premixed gas flame and mist flames. In some experimental studies, dust flames seem to be composed of individual burning particles while other studies indicate a more diffuse flame structure with groups of particles burning in unison. Some potential heuristics indicating when group combustion is favored over single particle combustion will be introduced. A simple model for the burning velocity and flame thickness of heterogeneous flames due to Williams will be derived. Then some of the complicating factors in heterogeneous flame behavior will be examined including velocity slip, turbulence, and thermal radiation.

The next four sections will present models for laminar dust flame behavior. The thermal theory of flame propagation is discussed first since the thermal theory lends itself more readily to simple analytical solutions for the burning velocity. Next we will introduce the method of Ballal and Lefebvre which is based on a consideration of characteristic time scales. The Ballal—Lefebvre model incorporates the diffusion flame model for single particle combustion. As a step toward greater sophistication, the third section on flame propagation briefly discusses a mathematical technique called activation energy asymptotics. Ignition and quenching of dust flames will be the topic of Section 7.6.

Next we will survey some of the empirical studies on heterogeneous flame behavior including aerosol mists, organic solids (with an emphasis on coal and biomass), and metallic solids (aluminum and iron). This survey will present characteristic results for a range of real materials and will serve as a reminder that the mathematical models introduced earlier are merely approximations for describing the complex and fascinating behavior of dust flames. Following the presentation on heterogeneous flame studies, we will review the literature on accidental unconfined flame propagation or flash fires. To understand the potential impact of a flash fire on people, we will need to understand the injury potential of flash fires. Finally, we will briefly review methods for controlling flash fire hazards.

7.1 THE ONE-DIMENSIONAL LAMINAR DUST FLAME

In Chapter 4, I introduced the concept of a premixed gas flame and described some of its characteristics. A premixed dust flame exhibits many of the same features as the gas flame: it has a burning velocity and a flame thickness, both of which are measureable in principle. The deflagration time scale can be calculated from the burning velocity and the flame thickness. In a premixed dust flame, the fuel does not form a molecular mixture with air, it forms a two-phase mixture with discrete particles. The presence of discrete particles introduces a new length scale into the problem of flame propagation, and with it an overall reaction rate of the particle consisting of both transport and kinetic processes.

We will begin our study of unconfined dust flames with a brief survey of the experimental methods and measurements used by investigators to study dust flames. The term *unconfined* is used here to signify that the flame propagates at constant pressure. One of the greatest challenges in performing dust flame propagation experiments is the difficulty of creating a uniform dust dispersion. We restrict our attention to two types of flames: standing flames and propagating flames. Standing dust flames have been established in various types of burners. Propagating flames have been created in confined channels, inside combustion chambers, and in unbounded dust cloud configurations. Every apparatus devised thus far has both strengths and weaknesses as an experimental platform for studying dust flames.

A final consideration for dust flames has to do with the phenomenon of group combustion. Depending on the volatility of the fuel, the fuel concentration, and the particle size distribution, combustible dust particles may burn individually with each particle surrounded by its own flame sheet or as a cluster of particles surrounded by a single flame. We will discuss some of the literature published on this subject and introduce suggested criteria that enable one to predict the phenomenon of group combustion.

7.1.1 EXPERIMENTAL METHODS

A number of different techniques have been developed for the measurement of the laminar burning velocity in premixed gas systems (Andrews and Bradley, 1972a,b; Rallis and Garforth, 1980; Lewis and von Elbe, 1987, pp. 226−301; Law, 2006, pp. 263−275). These techniques have been adapted to the study of dust flames with the primary challenge being the creation of a uniform dust−oxidizer mixture. Creating a *nonuniform* cloud of combustible dust is relatively easy. Simply visualize tossing a handful of wheat flour up into the air (in the absence of ignition sources) and you can imagine the formation of the dust cloud as the particles first disperse and then begin to settle under the influence of gravity. The problem with this method is that there is no control over the dust concentration at either a macroscale or microscale.

Creating a *uniform* dust cloud for combustion studies is a serious experimental challenge. Four basic configurations have been used for dust cloud combustion studies: burners for stationary flames, channels for propagating flames, open chambers, and freely propagating flames. Dust flame propagation measurements are often disturbed by buoyancy effects. To isolate this effect on flame formation and propagation, a number of studies have been performed in microgravity environments. The performance of experiments in a microgravity environment offers additional technological challenges for the experimentalist.

In this chapter, we restrict ourselves to *unconfined* (constant pressure) flames, so we will not consider closed combustion chambers or bombs; we will cover those when we investigate *confined* flame propagation in Chapter 8. Stationary unconfined dust flames are established using a burner. A burner is a device for anchoring a flame to a fixed position. Several investigators have developed burners for stationary flames (Cassel et al., 1948; Friedman and Maček 1963; Horton et al., 1977; Jarosinski et al., 1986; Shoshin and Dreizin, 2002). This list is simply small set of examples; many more will be introduced when we discuss various heterogeneous flame studies. The basic principle for the dust burner is essentially the same in each case. Fig. 7.1 denotes a simplified schematic for a combustible dust burner based on the Bunsen burner concept where the laminar burning velocity is related to the flow of the unburnt dust−air mixture through the flame cone by the continuity relation $S_u = v_u \sin \theta$ (Turns, 2012, pp. 261−262).

While generally all based on the same concept, the technological details varied reflecting yet another clever device for fluidizing the powder or for making the flame more amenable to combustion measurement and diagnostics.

Channels, conduits, and open chambers have been a popular means for investigating propagating flames (Ballal and Lefebvre, 1981; Joshi and Berlad, 1986; Proust and Veyssière, 1988; Chen et al., 1996; Gao et al., 2015a,b). The channel is typically a cylindrical tube oriented vertically. A variety of mechanisms have been developed to feed the powder into the tube and ignite it. Some designs have incorporated specific features to ensure a uniform dispersion of the dust.

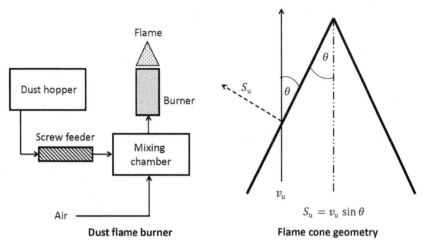

FIGURE 7.1

Schematic of typical dust flame burner.

It is especially difficult to produce a uniform dust cloud for freely propagating flames. Some investigators have tried using balloons to create a container of negligible mass and strength so that the confinement will disappear upon ignition of the cloud (Skjold et al., 2013; Julien et al., 2015a). Other investigators have developed impulsive dispersal methods. Holbrow and colleagues have generated fireballs by venting dust deflagrations from a vessel (Holbrow et al., 2000) Stern and his colleagues have used a cylindrical chamber with a lid that is detached from the cylinder upon injection and ignition of the dust cloud (Stern et al., 2015a,b).

Finally, a number of investigators have devised equipment suitable for microgravity experiments (Berlad, 1981; Ballal, 1983a; Law and Faeth, 1994; Ross, 2001; Goroshin et al., 2011). Experiments in microgravity are not a matter of intellectual curiosity; they serve an extremely important purpose by permitting an independent evaluation of the role of buoyancy in combustion behavior (Ronney, 1998). Gravity exerts its influence on dust flames in two contrary ways: the upward expansion of heated, less dense gas and the downward settling of dust particles.

One essential feature of the scientific method is the independent manipulation of experimental variables in a way that allows one to observe the effect that the variable has on the behavior of the system. Microgravity experiments permit one to directly observe the effect of buoyancy on dust flame combustion. Microgravity environments can be established by using drop towers, aircraft following parabolic trajectories, or on orbital spacecraft. The design of these experimental apparatuses requires a good deal of engineering to assemble the combustion device and instrumentation into a rugged, reliable automated package.

As a final note, it has been noted by a number of investigators that measured laminar burning velocities are a function of the size of the burner used in the experiments (Smoot and Horton, 1977). One important factor in this phenomenon is the Markstein length of the combustible dust (Dahoe et al., 2002). The Markstein length is a parameter that permits one to evaluate the effect of the curvature of the flame front. Flat flames are the reference geometric model. As the flame becomes increasingly curved, the burning velocity will either increase or decrease depending on the magnitude and direction of the flame curvature. For example, in a stationary Bunsen burner flame, the curvature is considered to be concave with respect to the unburnt mixture and the burning velocity is faster than if it were a flat flame. The influence of the flame curvature on the burning velocity is related to the Markstein length $\mathcal{L}_M$ by the relation:

$$S_u = S_u^0 \left(1 + \frac{\mathcal{L}_M}{\kappa} \right) \tag{7.1}$$

where S_u^0 is the laminar burning velocity of a flat flame and κ is the curvature of the flame. Dahoe and his colleagues measured the laminar burning velocity of cornstarch flames over a range of dust concentrations. Their analysis of the Markstein length of dust flames appears to be the only investigation of its kind thus far (Dahoe et al., 2002). Additional information on flame stretch, curvature, and the Markstein length can be found in the references (Tseng et al., 1993; Kuo, 2005, pp. 471−496; Law, 2006, Chapter 10).

7.1.2 FLAMES: PREMIXED GASES, AEROSOL MISTS, AND DUST CLOUDS

For any given fuel−air system, the characteristics of premixed gas flames—burning velocity and flame thickness—depend primarily on the equivalence ratio. Dispersing fuel droplets or particles into the air is in one sense a premixed fuel−air mixture. But with aerosol mists and dust clouds, the "granularity" of the fuel introduces a length scale, the droplet/particle diameter that influences the overall reaction rate for combustion. The discrete nature of the fuel droplets/particles introduces transport and kinetic processes that are slower than their molecular counterpart. Furthermore, a cloud of solid particles can introduce new transport phenomena—such as thermal radiation and turbulent particle−fluid interactions—that were not significant in premixed gas combustion and are less significant in aerosol mists. Flame propagation characteristics in aerosol mists and dust clouds depend not only on the equivalence ratio but also on the particle size distribution (Palmer, 1973, Chapter 6).

There is a conceptual advantage to examine the flame propagation behavior of premixed gases, aerosol mists, and dust clouds. As we progress through these different fuel systems, transport and kinetic processes become slower and more complicated. The volatile component of the fuel is decreasing leading to different patterns of fuel consumption. In premixed gas flames, fuel lean mixtures usually burn the fuel to completion; only in fuel rich conditions is any fuel left over. In aerosol mists and dust clouds, incomplete combustion is the rule rather than the exception. Again, particle size plays a key role in determining the degree to which unburnt fuel is left behind the deflagration wave (Eckhoff, 2003, pp. 265−294).

In Chapter 4, we discussed the combustion of individual liquid droplets and investigated the theoretical properties of the single droplet diffusion flame. In Chapter 6, we explored some of the early literature which established the general applicability of the diffusion flame model for single droplets and single particles. It was shown that there is a direct correspondence of liquid droplet combustion with diffusion-controlled combustion of solid particles. In a similar way, we shall see that studying flame propagation in aerosol mists will give us insight into the combustion behavior of dust clouds. Equally important, we will find that knowledge of single particle combustion behavior will lend us insight into the combustion of dust clouds.

7.1.3 SINGLE PARTICLE VERSUS GROUP COMBUSTION

In Chapter 6 we explored the combustion behavior of a wide range of organic and metallic combustible dusts from the point of view of the single particle. Our intent in providing that background was to foster insight into dust cloud combustion processes. The value of single particle combustion studies was recognized early on by combustion scientists, but they also realized that there was the possibility, if not probability, that an array of burning particles might exert an influence on the overall combustion process. For example, a single fuel particle burning in a large but finite volume of air has no competitors for oxygen. If the same fuel particle is placed in a dense cloud of similar fuel particles with the same finite volume of air, the subject particle must now compete with its neighboring particles for oxygen (Bryant, 1971a). On the other hand, clouds of combustible dust particles have a lower autoignition temperature than single particles, so the group phenomenon can both hinder and promote combustion behavior (Cassel and Liebman, 1959). The interaction of the particles in the gasification or combustion process is generally called *group combustion*.

Cassel and Liebman used a particle spacing criterion to infer the significance of group interactions, and a number of investigators have followed suit (Cassel and Liebman, 1959; Eckhoff, 2003, pp. 8−9; Crowe et al., 2012, pp. 21−23). The essential argument is a geometric one and is based on the number density of particles in the cloud. We will assume a monosized distribution of spherical particles. Recall the definition for the volume fraction of particles α_p in a dust cloud:

$$\alpha_p = \frac{\text{volume of particles}}{\text{volume of cloud}} = \frac{\pi}{6} n_p \, d_p^3 \qquad (7.2)$$

In this equation, n_p is the number density of particles and d_p is the particle diameter. Assume that the particle spacing in the cloud is uniform. Since we are only interested in an order of magnitude estimate of particle spacing, the exact geometrical shape of the dust cloud is not important. Therefore, we assume that the cloud has a cubical shape and its volume can be calculated as $V_{cloud} = L^3$, where L is the length of an edge of the cube. The interparticle spacing ℓ_p is calculated as the length per unit particle or $n_p^{-1/3}$

$$\ell_p = n_p^{-1/3} = \left[\frac{\pi}{6} \left(\frac{d_p^3}{\alpha_p} \right) \right]^{1/3} \tag{7.3}$$

$$\frac{\ell_p}{d_p} = \left(\frac{\pi \, \rho_p}{6 \, \rho_m} \right)^{1/3} \cong \left(\frac{\pi}{6\alpha_p} \right)^{1/3} \tag{7.4}$$

The symbol ρ_p represents the dust particle (solid) density and ρ_m is the mass concentration of the dust mixture. I have also used the approximation that the volume fraction of particles is essentially $\alpha_p \cong \rho_m / \rho_p$.

The interparticle spacing is a convenient reference length to gauge the potential interaction between adjacent particles due to transport processes or reaction kinetics. Perhaps the most obvious comparison to make is to compare the flame diameter with the interparticle spacing. As a thought experiment, consider two identical candle flames. As the candle flames are brought close to each other, they will eventually merge into one flame. That is the type of interaction that is described as group combustion. Some typical values for the interparticle spacing are presented in Table 7.1 for coal, aluminum, and iron dust.

The significance of the interparticle spacing calculations is made apparent by comparing with the expected flame standoff distance for each material. During diffusion-controlled combustion of bituminous coal (volatiles), the flame standoff is estimated at $d_f/d_p = 14$, for aluminum $d_f/d_p = 2 - 5$, and for iron $d_f/d_p = 1$ (recall from Chapter 6 that iron combustion is via a kinetic-controlled heterogeneous reaction). On the basis of the interparticle spacing, one would conclude that coal can undergo group combustion near stoichiometric and richer dust concentrations, but aluminum and iron do not.

More formal analyses of group combustion are available for droplet arrays, streams, and aerosol mists (Annamalai and Ryan, 1992; Sirignano, 2010; Kuo and Acharya, 2012b; Sirignano, 2014). Similarly, there are more formal analyses of particle arrays, streams, and dust clouds (Annamalai et al., 1994). Energy and mass transport into a cloud of particles takes longer than it does with a single

Table 7.1 Typical Values of Interparticle Spacing Ratio at Different Dust Concentrations for Three Combustible Dusts

Dust Concentration (g/m³)	ℓ_p/d_p, coal dust, $\rho_p = 1400$ kg/m³	ℓ_p/d_p, aluminum dust, $\rho_p = 2700$ kg/m³	ℓ_p/d_p, iron dust, $\rho_p = 7800$ kg/m³
30	36	45	64
300	17	21	30
3000	7.8	9.7	14

particle. In much of the group combustion literature, the group combustion models are typically normalized by some function of a single particle combustion model.

For the purposes of this book, the interparticle spacing concept will be sufficient in most cases. We will consider one aspect of group combustion with a simple model of oxygen depletion during combustion.

7.1.4 DEFLAGRATION WAVE EFFICIENCY

In this section, I introduce the concept of deflagration wave efficiency. In many theoretical studies of dust cloud combustion, the investigator assumes that the deflagration time scale t_{wave} (flame thickness divided by the laminar burning velocity) is equal to the burnout time of a single particle τ.

$$t_{wave} \equiv \frac{\delta}{S_u} = \tau \qquad (7.5)$$

I call this an *efficient* deflagration wave. This equality will hold only in the limit of a very dilute dust cloud with very small particles or, equivalently, it will only hold in the limiting condition that can be described as very fuel lean or at a high level of excess oxygen. Recall from Chapter 6, that the typical length and time scales for a laminar dust deflagration are a flame thickness $\delta \cong 0.01-0.1$ m and $t_{wave} \cong 100-10$ ms, respectively, if the burning velocity is $S_u = 1$ m/s.

In reality, we should expect that the deflagration time scale will be shorter than the burnout time or $t_{wave} < \tau$. This inference is based on the empirical observation that significant levels of unburnt fuel are typically found in deflagration tests (see Chapter 6). The finite rate of transport and kinetic processes at the particle length scale explains the slower particle burnout time. In the single particle analysis, the fuel particle was surrounded by an infinite expanse of air at a constant ambient temperature. The group combustion model teaches us that when the system becomes a cloud of particles, the air is no longer infinite; instead, the individual particles begin to exert an influence on the combustion process for their nearest neighbors. One method for gauging the impact of a changing oxygen environment is to consider the combustion of a well-mixed dust cloud in a finite volume.

EXAMPLE 7.1

Consider a monodisperse cloud of combustible dust. Upon ignition, a steady one-dimensional deflagration wave sweeps through the cloud with a laminar burning velocity $S_u = 1$ m/s and a flame thickness $\delta = 0.01$ m. Assume the dust deflagration wave is efficient and that the particle combustion is diffusion controlled. Calculate the particle diameter required for the deflagration to be efficient for the following combustible dusts: polyethylene

$(K = 2500 \ \mu m^2/ms)$, bituminous coal $(K_{volatiles} = 1500 \ \mu m^2/ms)$, aluminum $(K = 250 \ \mu m^2/ms)$, and iron $(K = 1.20 \ \mu m^2/ms)$.

Solution

For an efficient deflagration wave, $t_{wave} = \tau$. The time scale for the deflagration wave is $t_{wave} = \delta/S_u = (0.01 \ m)/(1 \ m/s) = 10$ ms. Thus, the burnout time for the particles τ is 10 ms.

For diffusion-controlled combustion, the burnout time is given by the expression $\tau = d_{p,0}^2/K$. The particle diameter corresponding to an efficient dust deflagration $d_{p,0}^*$ is simply $d_{p,0}^* = \sqrt{K\tau}$. The table below summarizes the calculations for the four combustible dusts.

Dust	$K \ (\mu m^2/ms)$	$d_{p,0}^* \ (\mu m)$
Polyethylene	2500	158
Bituminous coal	1500	122
Aluminum	250	50
Iron	1.20	3.5

The reader should keep in mind that these results apply only to a deflagration wave with the specified flame thickness and laminar burning velocity.

7.1.5 THE WELL-MIXED REACTOR MODEL FOR DUST CLOUD COMBUSTION

I will present one model to illustrate the effect of diminishing oxygen concentration on the burnout time of combustible dust particles. This formulation is based on the work of Bryant (for the kinetic-controlled combustion; for diffusion-controlled combustion, he credits a publication in German by Nusselt in 1924) (Bryant, 1971a). The objective is to examine the impact of oxygen consumption on single particle combustion in both diffusion-controlled and kinetic-controlled conditions in a well-mixed finite volume. It is assumed that the population of particles is monodisperse. Because we assume that the particles and the oxygen are well mixed (no spatial gradients in particle or oxygen concentration), we can use the single particle model to describe the combustion process. It is further assumed that the combustion rate is not affected by the temperature of the cloud; this assumption allows us to ignore the energy balance calculation.

The physical system is a cloud of monodisperse particles dispersed in air. The volume of the cloud V, and thus the number of moles of oxygen, are finite. It is assumed that all particles are ignited simultaneously and, at any given instant in time, have burned to the same level of conversion. A mass balance on oxygen gives the following algebraic relation:

$$C_{ox}(t) = C_{ox,0} - \frac{4\pi}{3}\left(r_0^3 - r_p^3\right)\frac{\rho_p N}{V(F/O)_{st}}$$

(7.6)

where C_{ox} is the molar concentration of oxygen, r_0 is the initial particle radius, r_p is the instantaneous particle radius (in other words, $r_p = r_p(t)$), ρ_p is the particle density, N is the number of particles in the cloud, and $(F/O)_{st}$ is the mass ratio of fuel required for complete combustion of a unit mass of oxygen.

Next Bryant defined an oxygen–fuel equivalence ratio $\mathbb{E}$

$$\mathbb{E} = \frac{C_{ox,0}V(F/O)_{st}}{\left(\frac{4\pi}{3}\right)r_0^3\rho_p N}$$

(7.7)

Combining these two equations to eliminate the volume V, the mass balance on oxygen can be written as

$$C_{ox}(t) = \frac{C_{ox,0}}{\mathbb{E}}\left[\left(\frac{r_p}{r_0}\right)^3 + \mathbb{E} - 1\right]$$

(7.8)

We will now use this equation to evaluate the impact of a diminishing oxygen concentration on both diffusion-controlled and kinetic-controlled combustions. For the case of diffusion-controlled combustion of a single particle, we assume that the mass transfer coefficient for the oxygen flux can be approximated as the film coefficient, $k_g = D/r_p$. The fuel mass balance written for a single particle undergoing diffusion-controlled combustion is

$$\frac{dm_p}{dt} = -\frac{D S_p C_{ox}(t)}{r_p}$$

(7.9)

The equation for the instantaneous oxygen concentration is substituted into the particle mass balance and integrating with the initial condition $r_p(t=0) = r_0$:

$$\int_0^\tau dt = -\left(\frac{\rho_p}{DC_{ox,0}}\right)\int_{r_0}^0 \left[\frac{\mathbb{E}}{\left(\frac{r_p}{r_0}\right)^3 + \mathbb{E} - 1}\right] r_p dr_p$$

(7.10)

The solution of this integral gives the burnout time for diffusion-controlled combustion in a diminishing oxygen atmosphere:

$$\tau_{cloud} = \left(\frac{\rho_p r_0^2}{DC_{ox,0}}\right)\left(\frac{\mathbb{E}}{3(\mathbb{E}-1)^{1/3}}\right)\ln\left[\frac{1 - (\mathbb{E}-1)^{1/3} + (\mathbb{E}-1)^{2/3}}{\left(1 + (\mathbb{E}-1)^{1/3}\right)^2}\right]$$

$$+ 2\sqrt{3}\left(\tan^{-1}\left\{\frac{2 - (\mathbb{E}-1)^{1/3}}{\sqrt{3}(\mathbb{E}-1)^{1/3}}\right\} + \tan^{-1}\left(\frac{\sqrt{3}}{3}\right)\right)\right]$$

(7.11)

This result tells us that the burnout time of the cloud undergoing diffusion-controlled combustion is equal to the product of the burnout time for a single particle times a modifying factor:

$$\tau_{\text{cloud}} = \left(\frac{\rho_p \, r_0^2}{DC_{\text{ox},0}} \right) f_{\text{diff}}(\mathbb{E}) = \tau_p f_{\text{diff}}(\mathbb{E}) \tag{7.12}$$

Following the same procedure, the burnout time for a dust cloud undergoing kinetic-controlled combustion in an oxygen diminishing atmosphere is given by the expression:

$$
\begin{aligned}
\tau_{\text{cloud}} = & \left(\frac{\rho_p \, r_0}{k_c C_{\text{ox},0}} \right) \left(\frac{\mathbb{E}}{6(\mathbb{E}-1)^{2/3}} \right) \left\{ \ln \left[\frac{(\mathbb{E}-1)^{2/3} + 2(\mathbb{E}-1)^{1/3} + 1}{(\mathbb{E}-1)^{2/3} - (\mathbb{E}-1)^{1/3} + 1} \right] \right. \\
& \left. + 2\sqrt{3} \, \tan^{-1} \left[\frac{2 - (\mathbb{E}-1)^{1/3}}{\sqrt{3}(\mathbb{E}-1)^{1/3}} \right] + 2\sqrt{3} \, \tan^{-1} \left(\frac{\sqrt{3}}{3} \right) \right\}
\end{aligned} \tag{7.13}
$$

Again, this result says that the burnout time for a dust cloud undergoing kinetic-controlled combustion is equal to the product of the burnout time for a single particle times a modifying factor

$$\tau_{\text{cloud}} = \left(\frac{\rho_p r_0}{k_c C_{\text{ox},0}} \right) f_{\text{kin}}(\mathbb{E}) = \tau_p f_{\text{kin}}(\mathbb{E}) \tag{7.14}$$

There are two limiting cases for the oxygen equivalence ratio: the cloud burnout time is equal to the single particle burnout time when there is an infinite excess of oxygen ($\tau_{\text{cloud}} = \tau_p$ when $\mathbb{E} \to \infty$) and the cloud burnout time becomes infinitely long when the oxygen quantity is exactly stoichiometric ($\tau_{\text{cloud}} \to \infty$ when $\mathbb{E} = 1$). The quantitative effect of these modifying factors for a diminishing oxygen atmosphere can be judged from Table 7.2.

Bryant performed premixed dust flame experiments with graphite (kinetic-controlled combustion) and amorphous boron (diffusion-controlled combustion) and compared the measured particle burning times with the appropriate single particle model. Without consideration of the diminishing oxygen concentration effect, the measured burnout times were 1.5−2 times longer than predicted with single particle model. When he included the effect of the diminishing oxygen concentration, he found significantly better agreement of the observed burning times with model predictions (Bryant, 1971a,b).

While the formulation of this model had produced some very satisfying results with a plausible physical interpretation, you should view this model with some skepticism. The primary assumption upon which this model is built is that there are no spatial gradients in either the particle (fuel) or oxygen concentration. Empirical studies to be described later in this chapter show this assumption to be incorrect. In the next section, we will relax this assumption and discuss flame propagation models that consider a spatial gradient in temperature and, indirectly, a gradient in fuel consumption (or particle diameter).

Table 7.2 Impact of a Diminishing Oxygen Atmosphere on Diffusion-Controlled and Kinetic-Controlled Combustions of a Dust Cloud with Perfect Mixing

Oxygen Equivalence Ratio ($\mathbb{E}$)	$f_{diff}(\mathbb{E})$	$f_{kin}(\mathbb{E})$
1.0	∞	∞
1.1	5.65	3.58
1.2	3.69	2.67
1.3	2.92	2.26
1.4	2.51	2.02
1.5	2.25	1.87
1.6	2.06	1.75
1.7	1.93	1.67
1.8	1.83	1.60
1.9	1.74	1.54
2.0	1.68	1.50
9.0	1.09	1.07
∞	1.00	1.00

Adapted from Bryant, J.T., 1971a. The combustion of premixed laminar graphite dust flames at atmospheric pressure. Combust. Sci. Technol. 2, 389–399.

7.2 SCALING ANALYSIS FOR HETEROGENEOUS FLAME PROPAGATION

We will begin our consideration of dust flame propagation with a simple model originally proposed by Williams for flame propagation through a liquid mist (Williams, 1985, pp. 472–474). Starting with a simple physical picture of how a flame propagates through a cloud of liquid droplets, the model yields an expression for the laminar burning velocity of the flame. The model is readily extended to combustible dusts with the premise that combustible dust is simply less volatile than a liquid fuel.

After developing Williams' model for calculating the laminar burning velocity for a dust flame, we will consider briefly the physical interactions of dispersed multiphase flow, turbulence, and thermal radiation.

7.2.1 WILLIAMS' SIMPLIFIED MODEL FOR THE BURNING VELOCITY OF A HETEROGENEOUS FLAME

Williams observed that mist/spray flames could exhibit two extremes in behavior: homogeneous combustion and heterogeneous combustion. In homogeneous combustion, the flame preheats the droplets fully vaporizing them and forming a fuel

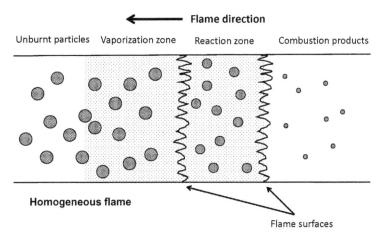

FIGURE 7.2

Homogeneous flame behavior in a dispersed phase fuel cloud. The flame preheats and vaporizes the fuel particles creating a premixed fuel vapor—air mixture.

vapor—air mixture. The flame sweeps through the premixed fuel vapor—air mixture. The cloud of vaporized droplets behaves like a premixed gaseous system. Williams noted that mist flames which burn in the homogeneous limit would have just slightly lower burning velocities than the corresponding premixed gaseous flames due to the slight diminishment of energy due to the heat of vaporization. Fig. 7.2 is a conceptual sketch of a homogeneous flame.

Highly volatile fuel particles of very small diameter will approximate homogeneous combustion. Particle loading also plays a role in the flame behavior. If fuel lean, the particles should burn to completion. If fuel rich, excess fuel—particles with a diameter smaller than their initial diameter—will be found in the combustion products.

At the other extreme, heterogeneous combustion is exemplified by slightly volatile liquids of large particle diameter. In heterogeneous combustion, the flame is not able to preheat and vaporize the droplets sufficiently to cause them to burn as a cloud. Instead, the individual droplets are ignited and burn separately with individual diffusion flames surrounding each droplet. Fig. 7.3 is a sketch of a heterogeneous flame.

Since the fuel volatility is represented by the Spalding transfer number B, the scaling arguments can be applied to combustible dusts as well. Consider a cloud of monodisperse fuel particles undergoing diffusion-controlled combustion. Williams presented the following scaling argument for estimating the heterogeneous burning velocity (Williams, 1985, pp. 472−474; also see Law, 2006, pp. 625−629). Williams began with an expression for the burning velocity (see Section 4.3 for the energy balance model of Annamalai and Puri, 2007). We shall

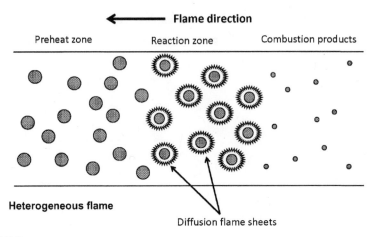

Flame direction

Preheat zone　　　Reaction zone　　　Combustion products

Heterogeneous flame

Diffusion flame sheets

FIGURE 7.3

Heterogeneous flame behavior in a dispersed phase fuel cloud. The individual burning particles preheats and ignites their nearest neighbors with each particle burning essentially independently.

emulate his example, but use the notation from Section 4.3. The burning velocity equation is

$$S_u = \left(\frac{\alpha_T \, \omega'''_{avg}}{\rho_u \, Y_{F,u}} \right)^{1/2} \tag{7.15}$$

$$\omega'''_{avg} = \pi \, d_p^2 \, \rho_s \left(-\dot{d}_p \right) n_{p,0} \tag{7.16}$$

The rate of particle diameter regression is designated as $\dot{d}_p$, the oxidizer to fuel ratio is s, and the initial number concentration of particles is $n_{p,0}$. The rate of particle diameter regression is given by the expression:

$$\dot{d}_p = 2 d_p \frac{d(d_p)}{dt} = \frac{d(d_p)^2}{dt} = -K; \quad K = \frac{8 \left(k_g / C_{p,g} \right)}{\rho_s} \ln(B+1) \tag{7.17}$$

Substituting the expression for the regression rate into the average reaction rate and choosing the initial particle diameter as a characteristic diameter value yield

$$\omega'''_{avg} = \pi \, \rho_s \, d_{p,0} \, K \, n_{p,0} \tag{7.18}$$

An expression for the burning velocity can now be obtained by substituting the expression for the average reaction rate

$$S_u = \alpha_T \left[\left(\frac{2\pi}{Y_{F,u}} \right) d_{p,0} \, n_{p,0} \, \ln(B+1) \right]^{1/2} \tag{7.19}$$

A relation can be derived for the particle number density in terms of the particle diameter. For a monodisperse cloud, the dust concentration ρ_d is related to the number density by the mass balance relation:

$$\rho_d = m_{p,0}\, n_{p,0} = \rho_s \left(\frac{\pi}{6}\, d_{p,0}^3\right) n_{p,0} \;\rightarrow\; n_{p,0} = \frac{6}{\pi}\left(\frac{\rho_d}{\rho_s}\right)\frac{1}{d_{p,0}^3} \tag{7.20}$$

Substituting the expression for the particle number density into the burning velocity equation gives

$$S_u = \alpha_T \left[\left(\frac{12}{Y_{F,u}}\right)\left(\frac{\rho_d}{\rho_s}\right)\frac{1}{d_{p,0}^2}\ln(B+1)\right]^{1/2} \;\rightarrow\; S_u \approx \frac{\alpha_T\, \rho_d^{1/2}}{d_{p,0}} \tag{7.21}$$

For heterogeneous flames, the laminar burning velocity is inversely proportional to the initial particle diameter: smaller particles yield faster burning velocities. It also predicts that the burning velocity increases with the square root of increasing dust concentration. A similar analysis can be applied to the flame thickness. One begins with the following relation for the flame thickness from Section 4.3:

$$\delta = \left(\frac{2\rho_m Y_{F,u}\alpha_T}{\omega_{avg}'''}\right)^{1/2} \tag{7.22}$$

Applying the same reasoning as we did for the burning velocity, we find for the flame thickness

$$\delta = d_{p,0}\left[\frac{Y_{F,u}}{6\ln(B+1)}\left(\frac{\rho_s}{\rho_d}\right)\right]^{1/2} \;\rightarrow\; \delta \approx d_{p,0} \tag{7.23}$$

The thickness of a heterogeneous flame is proportional to the initial particle diameter: smaller particles require smaller reaction zones for complete combustion. Thus, Williams' model establishes both the laminar burning velocity and flame thickness in terms of the fuel volatility (the transfer number B), the fuel concentration, and the particle diameter.

Williams also illustrates how scaling (order of magnitude) arguments can be used to formulate a criterion for distinguishing between homogeneous and heterogeneous combustion during flame propagation through a dispersed phase fuel cloud. The criterion for homogeneous combustion for flame propagation through a mist or dust cloud is essentially the same method used in Section 7.1.4 to evaluate deflagration efficiency. The difference is that in Williams' method the vaporization time is used instead of the combustion time. The premise of Williams' criterion for homogeneous combustion is that the vaporization time of the particle t_{vap} must be less than the deflagration wave time: $t_{vap} \leq t_{wave}$. The deflagration wave time is calculated as before: $t_{wave} = \delta/S_u$.

The vaporization time is calculated in the same manner that the burnout time was calculated for a liquid droplet diffusion flame. The difference between vaporization and combustion lies in the calculation of the transfer number using the enthalpy of vaporization instead of the enthalpy of combustion: $B_{vap} = C_{p,g}(T_s - T_\infty)/\Delta h_{vap}$ (Turns, 2012, pp. 374−383; Law, 2006, pp. 213–217; Kuo, 2005, pp. 569–578). The vaporization time and the minimum particle diameter are calculated for liquid droplets as

$$t_{vap} = \frac{d_{p,0}^2}{K_{vap}} = \frac{\rho_l d_{p,0}^2}{8(k_g/C_{p,g})\ln(B_{vap}+1)} \tag{7.24}$$

$$d_{p,min} = \left[8 \frac{\rho_g}{\rho_l} \left(\frac{\alpha_T}{S_u}\right)^2 \ln(B_{vap}+1) \right]^{1/2} \tag{7.25}$$

Law gives a sample calculation for a liquid droplet assuming properties for an alkane vapor–air mixture. He assigns the following property values: $B_{vap} = 0.5$, $\rho_g/\rho_l = 10^{-3}$, $\alpha_T = 1 \times 10^{-4}$ m^2/s, $S_u = 0.4$ m/s. Substituting these values gives the minimum droplet diameter criterion equal to $d_{p,min} = 10$ μm. Droplets less than or equal to this diameter will give rise to homogeneous combustion. Larger particles will burn heterogeneously. Remember, this calculation can be applied to combustible dusts with the substitution of suitable physical properties. For charring organic solids, the enthalpy of gasification should be used in lieu of the enthalpy of vaporization.

The power of Williams' simplified model for the burning velocity of a heterogeneous flame lies in its simplicity. The models should not be used with the expectation of a high degree of accuracy; they are more appropriately used for developing insight and intuition into combustible dust behavior. More sophisticated analyses yield more complicated formulas—if at all—without really increasing our insight. It is important to appreciate three factors that can confound a simple heterogeneous flame model: they are two-phase flow effects, turbulence, and thermal radiation. We shall discuss each of these first and then proceed to different formulations for dust flame propagation modeling based on closed form analytical solutions.

EXAMPLE 7.2

Using Williams' model for heterogeneous flames, calculate the laminar burning velocity and flame thickness for a monodisperse aluminum dust with a particle diameter of 10 μm. Assume air is the oxidizer. Use the following properties: $\alpha_T = 2.2 \times 10^{-5}$ m^2/s, $Y_{F,u} = 0.2038$, $\rho_d = 300$ g/m^3, $\rho_s = 2700$ kg/m^3, and $B = 0.42$. Compare these results with Ballal's dust flame model (Ballal, 1983, Figs 9 and 15) (reproduced with permission as Fig. 7.5 and Fig. 7.7).

Solution:
The laminar burning velocity is given by Eq. (7.21):

$$S_u = \alpha_T \left[\left(\frac{12}{Y_{F,u}}\right)\left(\frac{\rho_d}{\rho_s}\right)\frac{1}{d_{p,0}^2}\ln(B+1) \right]^{1/2}$$

$$S_u = 2.2 \times 10^{-5} m^2/s \left[\left(\frac{12}{0.2038}\right)\left(\frac{0.3 kg/m^3}{2700 kg/m^3}\right)\frac{1}{(10^{-5}m)^2}\ln(1.42) \right]^{1/2}$$

This gives a laminar burning velocity $S_u = 0.105$ m/s. The flame thickness is given by Eq. (7.23):

$$\delta = d_{p,0} \left[\frac{Y_{F,u}}{12 \ln (B+1)} \left(\frac{\rho_s}{\rho_d} \right) \right]^{1/2}$$

$$\delta = \left(10^{-5} \text{m} \right) \left[\frac{0.2038}{12 \ln (1.42)} \left(\frac{2700 \text{ kg/m}^3}{0.3 \text{ kg/m}^3} \right) \right]^{1/2}$$

This gives a flame thickness of $\delta = 2.09 \times 10^{-4}$ m. For comparison, from inspection of the Figs 9 and 15 in Ballal (1983), the burning velocity is estimated as $S_u = 0.4$ m/s and the flame thickness is $\delta = 5 \times 10^{-4}$ m. The models of Williams and Ballal give comparable (within an order of magnitude) predictions.

7.2.2 MULTIPHASE FLOW EFFECTS ON HETEROGENEOUS FLAME PROPAGATION

In a combustible dust cloud, the discrete nature of the fuel particles introduces a fundamentally different level of complexity into the description of flame propagation behavior compared to premixed gaseous systems. Different levels of description of the multiphase system are available enabling a greater or lesser degree of sophistication for the description of the momentum, energy, and mass transport phenomena in the system (Smoot and Pratt, 1979; Smoot and Smith, 1985; Fan and Zhu, 1998; Crowe et al., 2012). For each physical situation to be considered, you must ask yourself what level of mathematical sophistication must be employed to give you a satisfactory description of the system behavior.

In Chapter 4, we reviewed some simple concepts for multiphase flow. In this section, we illustrate the impact of multiphase flow on the measurement of flammability limits and the laminar burning velocity in dust flame propagation. In particular, the difference in velocity between the fluid and the particles can give rise to differences in combustion characteristics. There are other examples which could be used to illustrate multiphase flow effects, but their influence is less direct and less easily demonstrated. The example to be discussed in this section serves as a concrete illustration of how multiphase flow effects can influence dust flame characteristics.

Several experimental investigations have utilized a vertical flammability tube for measuring the characteristics of a heterogeneous flame (both mists and dust clouds). It was noticed that it was possible to determine the lower flammability limit (LFL, or equivalently, the minimum explosive concentration, MEC) by observing the success or failure of flame propagation in the tube. However, it was soon apparent that the LFL determined by upward flame propagation was different from the LFL determined by downwards flame propagation. With the aid of photography, it was hypothesized that gravitational settling was responsible for the variance in LFL results (Burgoyne, 1963; Berlad, 1981; Sun et al., 2003). In other words, the velocity of the particles was not the same as the velocity of the burnt gas. This difference in velocity is often called *velocity slip*.

Despite investigators' best efforts to achieve a uniform spatial distribution within the flammability tube, terrestrial gravity exerts an unavoidable influence on the dynamics of the particles. In upwards flame propagation experiments, the particles move in the opposite direction as the flame. This means that the local (microscale) particle concentration encountered by the flame is enriched compared to the macroscale average concentration. The opposite effect is at work for downward flame propagation where the flame is chasing the falling particles. This effect leads to a local (microscale) concentration that is lower compared to the macroscale concentration. Therefore, LFL measured in upwards flame propagation tends to be smaller than LFL measured in downwards propagation, ie, $\text{LFL}_{\text{down}} > \text{LFL}_{\text{up}}$. This observation can be used to calculate the burning velocity at the MEC condition. The following derivation is due to Burgoyne (1963). I will subsequently refer to this model as the velocity slip model for the lower flammability limit.

Consider a vertical flammability tube with the necessary apparatus for generating a uniform dust concentration (at least in the macroscale dimension) within the tube. Let C_L denote the local dust concentration and LFL the global dust concentration. Assume that the dust is monodisperse and that LFL concentrations have been measured for both upwards and downwards flame propagation, LFL_{up} and LFL_{down}, respectively. Assume that the settling velocity of the particles can be approximated by the terminal settling velocity for a single particle, $v_{\text{p,TSV}}$ (see Section 4.6.2). It is also assumed that the burning velocity is the same regardless of the direction of flame propagation.

A mass balance on the flammability tube can be written as

$$C_L S_u = \text{LFL}\left(S_u \pm v_{\text{p,TSV}}\right) \tag{7.26}$$

The mass balance adds the settling velocity for upwards propagation and subtracts it for downwards propagation. Since the global dust concentration and the burning velocity are the same for both cases we can write the mass balance as

$$\text{LFL}_{\text{up}}\left(S_u + v_{\text{p,TSV}}\right) = \text{LFL}_{\text{down}}\left(S_u - v_{\text{p,TSV}}\right) \tag{7.27}$$

Rearranging and solving for the burning velocity give the desired result:

$$S_u = v_{\text{p,TSV}}\left(\frac{\text{LFL}_{\text{down}} + \text{LFL}_{\text{up}}}{\text{LFL}_{\text{down}} - \text{LFL}_{\text{up}}}\right) \tag{7.28}$$

This model predicts the value of the burning velocity at the LFL condition as a function of the terminal settling velocity of the particles and the dust concentrations measured at the LFL for the upwards and downwards flame propagation.

Note that the burning velocity at the LFL condition will be less than the burning velocity at the stoichiometric condition, and the maximum burning velocity is usually at a fuel rich concentration. Unfortunately, most of the published flammability tube studies on combustible dusts are not conducted in a manner to collect the data needed to test the validity of this model. Additional, intuitive support can be found by comparing the lower flammability limit for flammable gases or vapors with the minimum explosive concentration for combustible dusts. In Section 1.3.5, it was observed that the lower flammability limit for gaseous fuels was on the order of one-half the stoichiometric concentration ($\Phi \cong 0.5$). Table 7.3 is a summary of these concentration limits, collectively referred to as the lean limits expressed in terms of the equivalence ratio (data from Dobashi, 2007).

The lean limits for flammable gases and vapors in Table 7.3 follow the rule of thumb of $\Phi(\text{LFL}) \cong 0.5$. The lean limits for the combustible dusts are much lower.

To further validate the simple velocity slip model for the LFL requires burning velocity data obtained with monodisperse (or nearly so) powders at a range of dust concentrations in both the upward and downward propagation direction. Data sets of this type are rarely available. So while this model demonstrates the influence of multiphase flow effects on dust flame propagation, it should be considered only as an instructive model.

Table 7.3 Comparison of Lean Limits for Flammable Gases/Vapors and Combustible Dusts (Dobashi, 2007)

Flammable Gas/Vapor	Lean Limit, Φ	Combustible Dust	Lean Limit, Φ
Methane	0.50	Adipic acid	0.28
Ethane	0.52	Lactose	0.22
Ethylene	0.40	Polyethylene	0.17
Benzene	0.49	Sulfur	0.10
Methanol	0.46	Aluminum	0.29
Ethanol	0.49	Iron	0.35

EXAMPLE 7.3

Laminar flame propagation experiments were conducted with a combustible dust in a flammability tube and the minimum explosive concentration was determined for both upwards and downwards propagation. The data are $\text{LFL}_{\text{down}} = 75 \text{ g/m}^3$ and $\text{LFL}_{\text{up}} = 50 \text{ g/m}^3$. Calculate the laminar burning velocity assuming that the dust had a terminal settling velocity $v_{\text{p,TSV}} = 0.50 \text{ m/s}$.

Solution

The burning velocity at the LFL condition is given by Eq. (7.28):

$$S_u = v_{\text{p,TSV}} \left(\frac{\text{LFL}_{\text{down}} - \text{LFL}_{\text{up}}}{\text{LFL}_{\text{down}} + \text{LFL}_{\text{up}}} \right)$$

$$S_u = 0.50 \text{ m/s} \left(\frac{75 - 50}{75 + 50} \right)$$

The burning velocity at the LFL condition is $S_u = 0.10 \text{ m/s}$.

7.2.3 TURBULENCE EFFECTS ON HETEROGENEOUS FLAME PROPAGATION

The investigation of turbulent premixed gaseous flames has been the subject of intensive study and is reflected in the volume of published papers on the subject (Andrews et al., 1975; Lipatnikov, 2013). The modeling of turbulent multiphase combustion is at a sophisticated level of development (Smoot and Pratt, 1979; Smoot and Smith, 1985; Kuo and Acharya, 2012a,b). Turbulent flame propagation in combustible dust clouds has been the subject of a much smaller number of empirical investigations. This is because of the many challenges facing the investigator who wishes to measure turbulence parameters in dense clouds of burning particles (Wolanski, 1991). This brief summary highlights some of the key observations made using flammability studies with either upwards or downwards flame propagation. Many more studies have been published on turbulence measurements in closed vessels and so we will return to this subject in Chapter 8.

In terrestrial gravity, it is necessary to disperse a mass of combustible dust to form a dust cloud. The dispersal process creates a turbulent flow field. In a series of tests with starch dust (20 µm average particle diameter), Veyssière and his colleagues demonstrated the effect of turbulent concentration field on flame propagation (Veyssière, 1992; Rzal et al., 1993). Using a glass flammability tube with a length of 3 m and a square cross-section of 0.2 m, they were able to observe upward flame propagation through the dust dispersion. They observed concentration gradients in the dispersion with regions of fuel rich and fuel lean composition. The characteristic length scale of the fuel lean regions correlated with the integral scale of the turbulent flow field, calculated to be on the order of 0.01 m in the smaller tube and 0.02 m in the larger tube. The flame traveled preferentially through the regions of greater dust concentration and bypassing the regions of lean concentration with flame velocities ranging from 0.2 to 1.0 m/s. Their experiments were conducted at the approximate stoichiometric dust concentration range of 200–250 g/m^3 with a maximum laminar burning velocity of 0.35 m/s. By varying the dust concentration and dispersion characteristics, maximum turbulent burning velocities of 1.2 m/s were observed. Tests conducted in a flammability tube of similar design but having a smaller cross-section (0.1 m × 0.1 m) gave smaller burning velocities, around 0.22 m/s, for the same starch material. Turbulent flame structures were recorded using high-speed photography, but turbulence parameters were not measured.

Krause and Kasch conducted flame propagation studies with lycopodium, cornstarch, and wheat flour using flammability tubes with a length of 2 m and a cylindrical cross-section with a diameter of 0.06 m for the smaller tube and 0.1 m for the larger tube (Krause and Kasch, 2000). They observed maximum laminar burning velocities for lycopodium of 0.28 m/s (smaller tube) and 0.50 m/s (larger tube); these maxima were measured at approximately twice the stoichiometric concentration. They were not able to disperse the wheat flour satisfactorily to obtain laminar burning velocity measurements. Turbulent intensity measurements were performed using hot wire anemometers in the tubes during dust-free dispersion experiments. Thus, they assumed that the presence of dust did not alter the flow field.

Wang and colleagues performed flammability tube experiments and measured turbulence parameters (Wang et al., 2006a, 2006b). They used cornstarch with a mean particle diameter of 14 μm. Turbulence measurements were performed using particle image velocimetry (PIV) and high-speed photography. They observed that the vertical turbulent intensity measurements were consistently 20–50% greater than the horizontal intensities. They attributed this enhancement of turbulence in the vertical direction to the effect of the gravitational settling of the particles. Their studies offer proof that dust-free turbulent flows may offer insight into the dust dispersion process, but they will not offer an accurate assessment of the effect of particle motion on the turbulent flow field.

Proust and his colleagues also performed flammability experiments in a vertical tube with starch dust and measured the turbulent flow field parameters with a specially modified pitot tube (Hamberger et al., 2007; Schneider and Proust, 2007). They used different dispersion pressures to create varying levels of turbulent intensity. They were able to correlate the turbulent burning velocity with the turbulent intensity.

The measurement of turbulence in multiphase combustion systems is fraught with difficulty. The measurement results published thus far are apparatus and scale dependent. There has been some modest success with correlating these results with scaling relations developed for premixed turbulent combustion in gases, but it is too early to determine their validity or their general applicability. A good overview of both the computational techniques and experimental methods can be found in the book by Crowe et al. (2012).

7.2.4 THERMAL RADIATION EFFECTS ON HETEROGENEOUS FLAME PROPAGATION

The contribution of thermal radiation to mist flame propagation is generally considered to be negligible because the mist cloud is transparent (Ballal and Lefebvre, 1981; Williams, 1985, p. 473). Dust clouds, on the other hand, tend to exhibit a broad range optical thickness depending on particle density (dust concentration) and particle size. Good reviews on the modeling of thermal radiation in dust flames can be found in the books by Smoot and his colleagues (Smoot and Pratt, 1979; Smoot and Smith, 1985). The role of thermal radiation in dust flame propagation is controversial. This seems to be largely due to the absence of empirical dust flame studies where fuel composition, dust concentration, and particle size are carefully controlled and manipulated over a broad range of values to achieve the widest variation possible in optical thickness. The problem is further compounded by the length scale dependence of radiation heat transfer.

Some investigators have claimed that radiation is a significant contributor to laminar flame propagation. The proponents of this position cite the following categories of evidence:

- Comparison of laminar burning velocity measurements obtained from laboratory burners with simple flame propagation models which incorporate radiation heat transfer (Cassel et al., 1948; Cassel et al., 1957; Cassel, 1964; Ballal, 1983b).

- Insights derived from simple flame propagation models addressing various aspects of thermal radiation modeling (Essenhigh and Csaba, 1963; Arpaci and Tabaczynski, 1982; Ogle et al., 1984).

Other investigators have claimed that radiation is not a significant contributor to laminar flame propagation. These proponents tend to rely on

- Flame propagation in flammability tubes supported by conceptual interpretation of the test results (Proust and Veyssière, 1988; Proust, 2006a, 2006b)
- Flame quenching measurements in flammability tubes supported by conceptual interpretation or flame modeling of the test results (Goroshin et al., 1996a, 1996b).

The investigations into the role of radiation in dust flame propagation are too few, too limited, and their data sets are too sparse to form any definitive generalizations. It seems probable that the ultimate answer to this question will be, "it depends." For a given combustible dust, fuel lean concentration of larger particles may be more conduction controlled while fuel rich concentrations of smaller particles may be radiation controlled. To settle this debate, more investigations are needed which comprehensively manipulate chemical composition, dust concentration, and particle size.

7.2.5 SUMMARY

Heterogeneous flame behavior is far more complex than premixed gaseous flames. Williams' model is a simple yet insightful tool for understanding the role that fuel volatility, fuel concentration, and particle size play in laminar flame propagation. But the simplicity of Williams' model is also its weakness when trying to evaluate the relative importance of other transport processes. A simple model was introduced to evaluate multiphase flow effects for a specific scenario. The significance of turbulence has been demonstrated empirically, but there remains much work to be done before concise generalizations can be formulated and applied with confidence. A similar problem exists with the evaluation of the contribution of thermal radiation to laminar flame propagation.

We will now explore the development of more detailed flame propagation models progressing from the simpler to the more complex.

7.3 THERMAL THEORIES OF LAMINAR DUST FLAME PROPAGATION

A thermal theory of laminar flame propagation is based on a one-dimensional form of the thermal energy equation. With thermal theories we seek models for the laminar burning velocity based on the flame temperature, physical properties of the dust cloud, and consideration of the relevant heat transfer mechanisms. Two thermal theories are presented here not because they are "right" but because they illustrate the utility of simple models. The models presented are a hybrid version of the Mallard−Le Chatelier flame propagation model that considers the relative importance of conduction, convection, and radiation heat transfer (Cassel et al., 1948). A drawback of this model is that it relies on the use of an ignition temperature, a somewhat artificial device.

The second model uses the energy balance technique described in Chapter 4 to derive a flame propagation model in which the dust cloud is assumed to be optically thick (Ogle et al., 1984). This model does not require the introduction of an ignition temperature, but it does require the specification of the flame temperature (the introduction of radiation heat losses prevents the use of the adiabatic flame temperature).

Both models rely on the homogeneous mixture approximation. It is assumed that the physical properties of the dust cloud can be estimated by the mixture properties, and it is assumed that the particles and gas are in both thermal equilibrium and velocity equilibrium. Furthermore, it is assumed that the temperature distribution inside the particles is uniform (particles are isothermal $Bi \ll 1$).

In both thermal theories, the end result is an expression for the burning velocity of the dust cloud. This result can be combined with the single particle models for either diffusion-controlled or kinetic-controlled combustion to infer the effect of particle size on the burning velocity and other deflagration parameters.

7.3.1 THERMAL THEORY OF CASSEL ET AL.

The thermal theory of Cassel et al. is based on a modification of Mallard−Le Chatelier model for laminar flame propagation (Cassel et al., 1948). The Mallard−Le Chatelier model is similar to the energy balance models introduced in Chapter 4. It is based on a steady one-dimensional flame and divides the flame into two regions, a reaction zone and a preheat zone. The physical picture is illustrated in Fig. 7.4.

The derivation of the model begins by writing an energy balance across the flame. The essential argument is that the heat required to raise the temperature of the mixture in the preheat zone from the ambient temperature to the ignition temperature is released in the reaction zone and transported to the preheat zone by conduction. The temperature profile is assumed to be linear. The energy balance is written as

$$\rho_{m,0} S_u C_{p,m}(T_i - T_u) = \frac{k_m(T_f - T_i)}{\delta} + \varepsilon \sigma F_{f-0}\left(T_f^4 - T_u^4\right) \tag{7.29}$$

On the left-hand side of the equation, the subscript m designates the unburnt dust−air mixture properties, S_u is the laminar burning velocity, and the subscripts i, u designate the ignition and unburnt mixture temperatures. On the right-hand side δ is the flame thickness, k_m is the unburnt mixture thermal conductivity, ε is the emissivity of the flame, σ is the Stefan−Boltzmann constant, and F_{f-0} is the radiation view factor from the flame to the ambient environment and has a value of 1 for planar flames.

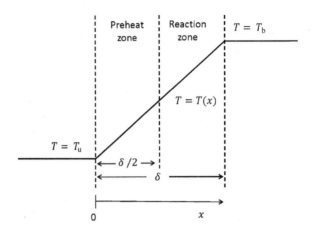

FIGURE 7.4

Mallard–Le Chatelier model for a dust flame (Cassel et al., 1948).

Invoking the flame relation $\delta = S_u\, t_b$, the flame thickness is eliminated from the energy balance. There are two cases to consider. The first is the case where there is *no radiation*. This is the classic Mallard–Le Chatelier model. Solving for the burning velocity gives us

$$S_u = \left[\left(\frac{k_m}{\rho_{m,0}\, C_{p,m}}\right)\left(\frac{T_f - T_i}{T_i - T_u}\right)\frac{1}{t_b}\right]^{1/2} \tag{7.30}$$

Similar to our premixed gas flame models, the laminar burning velocity is proportional to the square root of the thermal diffusivity and a chemical reaction time scale (see Chapter 4). For dust flames, it is convenient to select the single particle burnout time for the chemical time scale.

The second case to consider is flame propagation *with radiation*. Substituting the burning velocity relation $\delta = S_u\, t_b$ and rearranging Eq. (7.15) yields a quadratic equation. Solving for the burning velocity and rejecting the nonphysical root gives the result:

$$S_u = K_2 + \left(K_2^2 + K_1\right)^{1/2} \tag{7.31}$$

$$K_1 = \left(\frac{k_m}{\rho_{m,0}\, C_{p,m}}\right)\left(\frac{T_f - T_i}{T_i - T_u}\right)\frac{1}{t_b}; \quad K_2 = \frac{\varepsilon\sigma F_{f-0}\left(T_f^4 - T_u^4\right)}{2C_{p,m}(T_i - T_u)} \tag{7.32}$$

The first term is simply the Mallard–Le Chatelier model for the burning velocity and the second term can be considered to be the radiation contribution. To calculate a burning velocity for a dust flame using the radiation-modified Mallard–Le Chatelier model, one must estimate the ignition temperature, the burnout time, and the emissivity of the flame. Of these parameters, the ignition temperature is the one that carries with it the most uncertainty. Ogle et al. give a sample calculation for the burning velocity of a coal flame, and the agreement is within 10% when comparing with predictions from a comprehensive two-phase flow model (Ogle et al., 1984).

EXAMPLE 7.4

Cassel and his colleagues experimented with premixed aluminum dust flames and measured a laminar burning velocity $S_u = 0.20$ m/s (Cassel et al., 1948). Compare this experimental measurement with their flame propagation model. Use the following values for the physical properties: $k_m = 2.6 \times 10^{-2}$ W/m · K, $\rho_{m,0} = 1.48$ kg/m³, $C_{p,m} = 1110$ J/kg · K, $T_f = 3000$ K, $T_i = 2300$ K, $T_u = 300$ K, $\varepsilon = 0.5$, $F_{f-0} = 1$, and $t_{wave} = 30$ ms.

Solution

The burning velocity is calculated from Eqs. (7.31) and (7.32).

$$K_1 = \left(\frac{k_m}{\rho_{m,0} C_{p,m}} \right) \left(\frac{T_f - T_i}{T_i - T_u} \right) \frac{1}{t_{wave}}$$

$$K_1 = \left(\frac{2.6 \times 10^{-2} \text{ W/m} \cdot \text{K}}{(1.48 \text{ kg/m}^3) 1110 \text{ J/kg} \cdot \text{K}} \right) \left(\frac{3000 - 2300}{2300 - 300} \right) \frac{1}{0.030 \text{ s}} = 1.85 \times 10^{-4}$$

$$K_2 = \frac{\varepsilon \sigma F_{f-0} (T_f^4 - T_u^4)}{2 C_{p,m} (T_i - T_u)}$$

$$K_2 = \frac{0.5 (5.67 \times 10^{-8} \text{ W/m}^2 \cdot \text{K}^4)(1)(3000^4 - 300^4)}{2 (1110 \text{ J/kg} \cdot \text{K})(2300 - 300)} = 0.517$$

$$S_u = K_2 + \left(K_2^2 + K_1 \right)^{1/2} = 1.03 \text{ m/s}$$

With an estimated emissivity of $\varepsilon = 0.5$, the calculated laminar burning velocity is $S_u = 1.03$ m/s, a value that does not compare well to the measured value of 0.20 m/s. Adjusting the estimated emissivity to $\varepsilon = 0.1$ brings the calculated burning velocity to $S_u = 0.208$ m/s, a value that compares well with the measurement. This example illustrates the importance of seeking the best estimates for physical properties so that one can avoid the temptation to deliberately tune parameters to get the "right" answer.

7.3.2 THERMAL THEORY OF OGLE ET AL.

The greatest weakness of the modified Mallard−Le Chatelier model is its dependence on specifying an ignition temperature. Ogle et al. derived an expression for the burning velocity of a dust flame without invoking an ignition temperature (Ogle et al., 1984). The model assumes that the dust cloud is optically thick. This assumption is better satisfied for very fuel rich dust clouds. The model is based on a mixture formulation for the two-phase mixture with constant properties. The continuity equation steady, one-dimensional planar flame can be readily integrated

$$\frac{d}{dx}(\rho_m v_x) = 0 \rightarrow \rho_m v_x = \rho_{m,0} \, S_u = \text{constant} \tag{7.33}$$

The energy equation for a steady, one-dimensional planar flame takes the following form:

$$\rho_m \, C_{p,m} \, v_x \frac{dT}{dx} = k_m \frac{d^2T}{dx^2} + 2 \, \sigma \, a \, T_f^4 \, E_2(a \, x) \tag{7.34}$$

Most of the symbols in the energy equation have been previously defined, but there are two new ones: a is the absorption coefficient and E_2 is the exponential integral of order 2. The exponential integral term represents the absorption of radiant energy in an optically thick dust cloud from a flame sheet at temperature T_f. The boundary conditions for this flame model are

$$T(x = 0) = T_u; \quad \frac{dT}{dx}(x \rightarrow \infty) = 0; \quad T(x \rightarrow \infty) = T_f \tag{7.35}$$

Three boundary conditions are needed to solve the energy equation: two for determining the temperature profile and one for solving for the burning velocity (which appears as an eigenvalue in this model). Using the exponential kernel approximation (Özisik, 1973, pp. 332−333) for the exponential integral, the temperature profile and the burning velocity are given by the expressions:

$$\frac{(T - T_u)}{(T_f - T_u)} = 1 - \exp(-1.5ax) \tag{7.36}$$

$$S_u = \frac{1}{\rho_{m,0} C_{p,m}(T_f - T_u)} \left[\sigma T_f^4 - \frac{3}{2} k_m a(T_f - T_u) \right] \tag{7.37}$$

In the limit where heat conduction within the flame can be neglected, the temperature profile is unchanged but the expression for the burning velocity simplifies to

$$S_u = \frac{\sigma T_f^4}{\rho_{m,0} \, C_{p,m}(T_f - T_u)} \tag{7.38}$$

Comparing the two expressions for the burning velocity, it can be shown that for typical property values for dust clouds, the contribution of heat conduction compared to radiation is negligible. This is likely due to the assumption that the transport of thermal radiation within the dust cloud can be modeled as an optically thick medium. In other words, the relative magnitude of heat conduction is negligible only because we have assumed it to be so.

Weber suggested an alternative solution to the problem specified by Ogle et al. based on a more general formulation of the solution to the differential equation (Weber, 1989). His solution is based on the argument that the ignition temperature can be defined as the inflection point on the temperature profile (he refers to this as the vanishing second derivative). Weber's alternative model for the burning velocity becomes

$$S_u = \frac{1}{\rho_{m,0}\,C_{p,m}(T_i - T_u)}\left[\frac{1}{2}\sigma T_f^4 + \left(\left(\frac{1}{2}\sigma T_f^4\right)^2 + 2\,k_m(T_i - T_u)\rho_{m,0}a\,T_f^4\right)^{1/2}\right] \quad (7.39)$$

Weber identified the following improvements with this model: the square root dependence on thermal conductivity in Weber's model is similar to the result for premixed gases and the burning velocity in Weber's model increases with increasing thermal conductivity. The disadvantage of Weber's model is that it requires the specification of an ignition temperature.

EXAMPLE 7.5

Cassel and his colleagues experimented with premixed aluminum dust flames and measured a laminar burning velocity $S_u = 0.20$ m/s (Cassel et al., 1948). Compare this experimental measurement with their flame propagation model. Compare the burning velocity models of Ogle et al. (1984) and Weber (1989). Use the following values for the physical properties: $k_m = 2.6 \times 10^{-2}$ W/m·K, $\rho_{m,0} = 1.48$ kg/m^3, $C_{p,m} = 1110$ J/kg·K, $T_f = 3000$ K, $T_i = 2300$ K, $T_u = 300$ K, $\varepsilon = 0.5$.

Solution

Compare the burning velocity calculations from Eqs. (7.38) and (7.39).

$$S_u = \frac{\sigma T_f^4}{\rho_{m,0}C_{p,m}(T_f - T_u)}$$

$$S_u = \frac{(5.67 \times 10^{-8}\,\text{W/m}^2\cdot\text{k}^4)(3000)^4}{(1.48\,\text{kg/m}^3)(1110\,\text{J/kg}\cdot\text{K})(3000 - 300)} = 1.04\,\text{m/s}$$

$$S_u = \frac{1}{\rho_{m,0}C_{p,m}(T_i - T_u)}\left[\frac{1}{2}\sigma T_f^4 + \left(\left(\frac{1}{2}\sigma T_f^4\right)^2 + 2\,k_m\,(T_i - T_u)\rho_{m,0}aT_f^4\right)^{1/2}\right]$$

$$S_u = \frac{1}{(1.48\,\text{kg/m}^3)(1110\,\text{J/kg}\cdot\text{K})(2300 - 300)}$$

$$\left[\frac{1}{2}(5.67\times10^{-8}\,\text{W/m}^2\cdot\text{k}^4)(3000)^4 + \left(\left(\frac{1}{2}(5.67\times10^{-8}\,\text{W/m}^2\cdot\text{K}^4)(3000)^4\right)^2\right.\right.$$

$$\left.\left.+ 2(2.6\times10^{-2}\,\text{W/m}\cdot\text{K})(2300-300)(1.48\,\text{kg/m}^3)(10\,\text{m}^{-1})(3000)^4\right)^{1/2}\right]$$

$$= 108\,\text{m/s}$$

The burning velocity model of Ogle et al. differs from the experimental value by a factor of 5, while Weber's model is off by a factor of 540.

7.3.3 SUMMARY OF OTHER THERMAL THEORIES OF LAMINAR DUST FLAME PROPAGATION

The advantage of thermal theories of flame propagation is their simplicity. This is also the source of their weakness. Smoot and Horton surveyed the progress in the characterization of coal dust flames in terms of both experimental measurements and modeling (Smoot and Horton, 1977). They noted that the radiation model of Cassel and his colleagues was never formally tested against experimental measurements. Essenhigh and Csaba explored the use of thermal theories of flame propagation (also in the context of coal dust) with an emphasis on the effects of unequal gas and particle temperatures (Essenhigh and Csaba, 1963).

Arpaci and Tabaczinski presented an insightful analysis on the combined effects of emission, absorption, and scattering on flame propagation and quenching (Arpaci and Tabaczynski, 1982, 1984). Using scaling analysis, they identified a family of useful dimensionless parameters for evaluating the relative importance of radiation versus conduction and convection. In brief, their analysis confirmed that as the magnitude of radiation increased, the burning velocity and flame thickness increased, the quenching distance increased, and the flame temperature decreased.

The utility of thermal theory for deriving scaling parameters has been described in relation to the many contributions of Russian investigators (Goltsiker et al., 1994). However, these simple theories do not seem capable of accurate prediction of the burning velocity. An improvement on the thermal theory is due to Ballal and his colleagues in which they combine a time scale analysis to an energy balance on the flame.

7.4 BALLAL'S THEORY OF DUST FLAME PROPAGATION

Ballal published a dust flame propagation model based on time scale arguments (Ballal, 1983b). This work was an extension of work he had published with Lefebvre on flame propagation in liquid mists (Ballal and Lefebvre, 1981). In this section, we will review Ballal's theory for predicting the burning velocity and flame thickness in dust flames. We will see that an important advantage of Ballal's theory is that it interprets dust flame behavior in terms of the single particle combustion model. In a later section within this chapter, we will study Ballal's use of these ideas to predict ignition and quenching behavior of dust clouds.

Ballal identified two classes of solid fuels: Type A, fuels which burn with heterogeneous reactions (carbon and low volatiles coal), and Type B, fuels which burn in the vapor phase. A premixed flame of Type A fuel requires sufficient evolution of vapor phase intermediates like carbon monoxide to support gas phase combustion. I will also refer to Type A fuels as nonvolatile fuels. For Type B fuels, sufficient fuel vapor must be released to support a flame. I will refer to Type B fuels as volatile fuels. The time scale for evolution (or vaporization) of gas phase fuel is designated t_e. Once the fuel vapor has been formed in the oxidant atmosphere, it must then react (oxidize). The chemical reaction time scale is designated t_c.

Table 7.4 Dust Flame Dynamics Based on Time Scales

Time Scale Relation	Dust Flame Behavior
$t_q < t_e + t_c$	Flame is quenched
$t_q = t_e + t_c$	Steady flame propagation
$t_q > t_e + t_c$	Flame accelerates

The heat released by chemical reaction is dissipated or quenched by conduction to the fresh unburnt mixture and radiation losses to the environment. The time scale for quenching is designated t_q. Ballal asserted that the fundamental constraint for steady flame propagation was based on these three time scales in the following manner:

$$t_q = t_e + t_c \qquad (7.40)$$

This additive relation is similar to the reasoning we employed in single particle kinetics (see Section 6.2). Ballal identified three types of flame dynamics based on these three time scales (Table 7.4).

The assumption that the sum of the evolution and chemical time are additive implies that these processes are independent of each other and occur in sequence. A further assumption of the analysis is that each particle is completely consumed within the flame thickness. Ballal justified this assumption based on his analysis of flame photographs which indicated that the flame thickness was typically 10–100 times the diameter of a particle. He argued that a particle traveling a flame thickness of this magnitude should have adequate residence time in the flame to completely react. An additional constraint that he imposed on this analysis is that he restricted his attention to fuel lean flames.

We shall see that Ballal's model has both strengths and weaknesses. Its strength lies in the simplicity of its formulation and its reliance on single particle combustion models. Its primary weakness is that it requires the same input parameters required by the thermal theories, plus many more. Perhaps its greatest weakness is that it requires the specification of an ignition temperature. Thus, while many of the input parameters can be calculated a priori as physical properties, some, like ignition temperature, require empirical measurement.

EXAMPLE 7.6

The reasoning employed by Ballal involves the use of the time-averaged particle diameter relation $\overline{d}_p = (2/3)d_0$. Derive this relation. Assume the particle combustion behavior is described by the d-squared law, $d_p^2(t) = d_{p,0}^2 - Kt$.

Solution

Since we seek to calculate an averaged quantity, we will invoke the mean value theorem of calculus. First write the d-squared law in dimensionless terms by defining $\hat{d} = d_p/d_{p,0}$ and $\hat{t} = t/\tau$, and recalling that $\tau = d_{p,0}^2/K$. Substituting into the d-squared law,

$$\hat{d}^2(t) = 1 - \hat{t} \;\rightarrow\; \hat{d}(t) = \sqrt{1 - \hat{t}}$$

The mean value of calculus provides us the definition of the average value of particle diameter, $\overline{d}_p$.

$$\overline{d}_p = \int_0^1 \hat{d}(t)\, d\hat{t} = \int_0^1 \sqrt{1 - \hat{t}}\, d\hat{t}$$

Consulting a standard table on indefinite integrals, one finds

$$\int \sqrt{ax + b}\, dx = \left(\frac{2b}{3a} + \frac{2x}{3}\right)\sqrt{ax + b}$$

Employing this solution and evaluating between the limits of $\hat{t} = 0$ and 1 gives the result $\overline{d} = 2/3$ or $\overline{d}_p = (2/3)d_{p,0}$.

7.4.1 BURNING VELOCITY AND FLAME THICKNESS

The analysis for dust flame propagation follows his previous work on mist flame propagation with suitable modifications (Ballal and Lefebvre, 1981; Ballal, 1983b). One such modification is the inclusion of a radiation loss term. The radiation loss term is based on the assumption of an optically thin flame (a suitable approximation for fuel lean flames). For a single particle the radiant heat loss rate is

$$Q_{rad} = \pi d_p^2\, \varepsilon \sigma T_p^4 \tag{7.41}$$

Ballal generalizes his analysis for polydisperse particles. Therefore, he substitutes the Sauter mean diameter d_{32} into this expression: $d_p^2 \rightarrow (C_1 d_{32})^2$. The factor C_1 is a conversion factor derived from the frequency distribution of the particle diameters; it is defined as $C_1 = d_{20}/d_{32}$, where d_{20} is a surface area mean diameter (refer to the discussion on particle size statistics in Chapter 2). The number density of particles is calculated by the expression:

$$n_p = \frac{(F/A)\rho_g \delta_r A_{flame}}{\rho_s\left(\frac{\pi}{6}\right)(C_3 d_{32})^3} \tag{7.42}$$

The factor C_3 is another conversion factor derived from the frequency distribution of the particle diameters; it is defined as $C_3 = d_{30}/d_{32}$, where d_{30} is defined as the volume mean diameter.

The radiation heat loss depends on the size of the particles, but in combustion the particles change size as the reaction progresses. Ballal states that the average particle diameter during its lifetime is $\overline{d}_p = (2/3)d_0$. This relation is derived in the example that preceded this section. The radiation heat loss term for a cloud of polydisperse particles becomes

$$Q_{rad} = 9\,(F/A)\,\delta_r A_{flame} \left(\frac{\rho_g}{\rho_s}\right)\left(\frac{C_1^2}{C_3^3}\right)\left(\frac{1}{d_{32}}\right)\varepsilon\sigma T_p^4 \tag{7.43}$$

The quench time t_q is defined as the ratio of the excess enthalpy of the reaction zone to the rate of heat loss by conduction to the fresh unburnt mixture and radiation loss to the environment. To find an expression for the quench time, we begin with an energy balance across the flame

Energy released by combustion reaction
= Energy required to preheat the unburnt mixture by conduction   (7.44)
+ energy lost to the environment by radiation

$$\rho_g\,C_{p,g}\Delta T_r\,\delta_r\,A_{flame} = \left[k_g\left(\frac{\Delta T_r}{\delta_r}\right)A_{flame} + 9\,(F/A)\,\delta_r A_{flame}\left(\frac{\rho_g}{\rho_s}\right)\left(\frac{C_1^2}{C_3^3}\right)\left(\frac{1}{d_{32}}\right)\varepsilon\sigma T_p^4\right]t_q \tag{7.45}$$

Solving for the quench time and simplifying the equation results in the expression:

$$t_q = \left[\left(\frac{\alpha_g}{\delta_r^2}\right) + \left(\frac{9\,(F/A)}{\rho_g\,C_{p,g}}\right)\left(\frac{C_1^2}{C_3^3}\right)\left(\frac{1}{d_{32}}\right)\frac{\varepsilon\sigma T_p^4}{\Delta T_r}\right]^{-1} \tag{7.46}$$

The evolution (or vaporization) time t_e is defined as the time it takes for sufficient fuel to be deposited into the gas phase to promote flame propagation. Ballal developed two different ways to calculate the evolution time, one for Type B fuels and one for Type A. For a Type B fuel, the evolution time is defined as the mass of fuel in the reaction zone divided by the average rate of fuel vapor evolution. It is calculated by performing a mass balance across the flame:

Mass of fuel in combustion zone = (average rate of fuel volatilized)t_e (7.47)

$$(F/A)\rho_g\delta_r A_{flame} = \left[8\left(\frac{C_1}{C_3^3}\right)\left(\frac{k_g}{\rho_g\,C_{p,g}}\right)\left(\frac{F/A}{\rho_s}\right)\frac{\delta_r A_{flame}}{d_{32}^2}\ln(1+B)\right]t_e \tag{7.48}$$

Solving for t_e, the final expression for the evolution time for Type B (volatile) fuels is

$$t_e(Type\ B) = \frac{C_3^3\rho_s d_{32}^2}{8C_1\left(\frac{k_g}{C_{p,g}}\right)\ln(1+B)} \tag{7.49}$$

The Type A fuels are basically solid carbon (eg, graphite or low volatile coal). For carbon, recall that heterogeneous reactions must first create CO gas which then, in turn, is oxidized into CO_2. The oxidation of CO_2 is the primary exothermic reaction for the combustion of carbon. Ballal recommended an energy balance approach for this calculation.

$$\text{Energy released in reaction zone by surface reaction} = (\text{average heat release by CO evolution})t_e \tag{7.50}$$

$$\rho_g C_{p,g} \Delta T \delta_r A_{\text{flame}} = \left[\dot{m}_s C_{p,g} \Delta T (F/A)^{-1} \right] t_e \tag{7.51}$$

Solving for t_e for Type A fuels, the final expression for the evolution time is

$$t_e(\text{Type A}) = \frac{C_3^3 \rho_s d_{32}^2}{8 C_1 \left(\frac{k_g}{C_{p,g}} \right) \phi \ln(1 + B)} \tag{7.52}$$

It is important to recognize that the expression for the evolution time for Type A fuels differs from that for Type B fuels in two ways: the equivalence ratio appears in the denominator of Eq. (7.52), and the manner in which the transfer number B is calculated is different for either scenario.

The time scale for chemical reaction t_c is defined explicitly as the usual deflagration time scale, $t_c = \delta/S_L$. For Type A fuels (essentially solid carbon), the laminar burning velocity S_L refers to the burning velocity for a carbon monoxide flame. The thickness of the reaction zone is defined as

$$\delta_L = \frac{\alpha_g}{S_L} \left(\frac{\Delta T_r}{\Delta T_{pr}} \right) \tag{7.53}$$

Substituting the expressions for the chemical time t_c, the quench time t_q, and the evolution time t_e into Eq. (7.40), and then solving for the flame thickness in the dust cloud, δ_r, you obtain the following expression:

$$\delta_r = \alpha_g^{0.5} \left\{ \left[\frac{C_3^3 \rho_s d_{32}^2}{8 C_1 \left(\frac{k_g}{C_{p,g}} \right) \phi \ln(1+B)} + \frac{\alpha_g}{S_L^2} \left(\frac{\Delta T_r}{\Delta T_{pr}} \right) \right]^{-1} - \left(\frac{9(F/A)}{\rho_s C_{p,g}} \right) \left(\frac{C_1^2}{C_3^3} \right) \left(\frac{1}{d_{32}} \right) \frac{\varepsilon \sigma T_p^4}{\Delta T_r} \right\}^{-0.5} \tag{7.54}$$

The physical interpretation of Eq. (7.54) is

$$\delta_r = \{\text{diffusion term} + \text{reaction term} - \text{radiation term}\} \tag{7.55}$$

The burning velocity of the dust flame, S_D, is calculated by this expression:

$$S_D = \frac{\alpha_g}{\delta_r} \left(\frac{\Delta T_r}{\Delta T_{pr}} \right) \tag{7.56}$$

Aluminum and magnesium metals are representative of Type B (volatile) fuels. Fig. 7.5 is a plot of burning velocity as a function of dust concentration for aluminum and magnesium powders. For a given particle diameter, the burning velocity increases with dust concentration for fuel lean mixtures. For a given dust concentration, the burning velocity decreases as the particle size increases.

The value of B for aluminum and magnesium is 0.42 and 1.14, respectively.

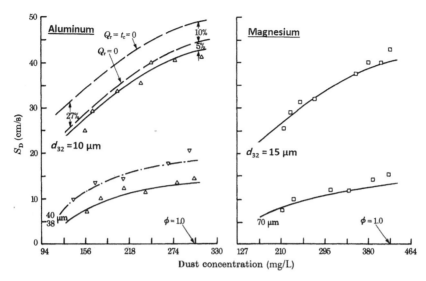

FIGURE 7.5

Burning velocity of dust flame for aluminum (left side) and magnesium (right side) as a function of dust concentration. Families of curves represent different mean particle diameters. Symbols are measured velocities and solid lines are the model predictions.

Reprinted with permission from Ballal, D.R., 1983b. Flame propagation through dust clouds of carbon, coal, aluminium and magnesium in an environment of zero gravity. Proc. Royal Soc. London, A385, 21–51.

As an example of the use of Ballal's model for a Type A (nonvolatile) fuel, Ballal considered three types of bituminous coal designated Bersham (volatile matter 39.2%, $d_{32} = 12\ \mu m$), Beynon (volatile matter 27.2%, $d_{32} = 11\ \mu m$), and Annesley (volatile matter 37.2%, $d_{32} = 47\ \mu m$). Fig. 7.6 is a presentation of this data.

A similar trend of burning velocity increasing with dust concentration is observed (for fuel lean mixtures). For the same dust concentration, increasing particle size reduces the burning velocity. Finally, the burning velocity increases with increasing volatile matter.

7.4.2 SIGNIFICANCE OF THE FUEL VOLATILITY

Ballal's experiments show that for a given dust concentration (or equivalence ratio), the flame thickness *decreases* and the burning velocity *increases* with increasing fuel volatility. Fuel volatility can be judged from actual volatile matter measurements or by the value of the transfer number *B*. Fig. 7.7 is a plot of the dimensionless flame thickness (flame thickness divided by the mean particle diameter) as a function of the equivalence ratio for different fuels.

For the fuels included in the plot, *iso*-octane vapor is the most volatile and solid carbon is the least volatile fuel. For a given fuel, the flame thickness grows as the equivalence ratio approaches the lean limit.

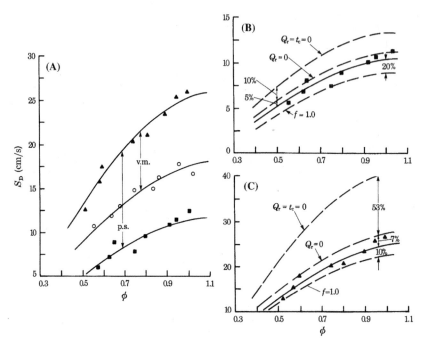

FIGURE 7.6

Burning velocity of coal dust flames as a function of equivalence factor (dust concentration). (A) Comparison of Bersham (▲), Beynon (○), and Annesley (■) coals. Influence of volatile matter (v.m.) and particle size (p.s.) indicated by arrows. (B) Annesley coal: Theory versus experiment. (C) Bersham coal: Theory versus experiment.

Reprinted with permission from Ballal, D.R., 1983b. Flame propagation through dust clouds of carbon, coal, aluminium and magnesium in an environment of zero gravity. Proc. Royal Soc. London, A385, 21–51.

One of the more important features of Ballal's model is that it is capable of predicting burning velocities for fuels with a wide range of volatilities (Fig. 7.8).

Ballal's model for dust flame propagation is significant because it incorporates the important factors that govern the magnitude of the burning velocity for a dust flame: the equivalence ratio (dust concentration), the volatility of the fuel (the Spalding transfer number), the mean particle diameter, the polydispersity of the particle size distribution, and the radiant heat loss. These are features that are not easily incorporated into simple thermal theories of dust flame propagation.

However, there are some significant disadvantages to Ballal's flame propagation model, namely its dependence on the specification of an ignition temperature and a laminar flame velocity for the vapor component of the fuel particle. Furthermore, an incorrect assignment of the particle emissivity or particle temperature can introduce a significant error in the burning velocity calculation.

To free ourselves of the approximations introduced by thermal theories or Ballal's time scale analysis, we must find a way to solve the equations of change for a cloud of burning fuel particles. The direct solution of the equations of change is the subject of

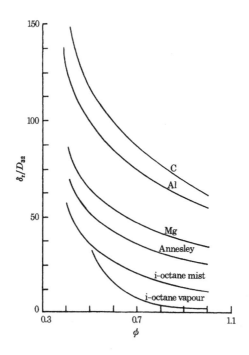

FIGURE 7.7

Dimensionless flame thickness for different fuels as a function of the equivalence ratio.

Reprinted with permission from Ballal, D.R., 1983b. Flame propagation through dust clouds of carbon, coal, aluminium and magnesium in an environment of zero gravity. Proc. Royal Soc. London, A385, 21–51.

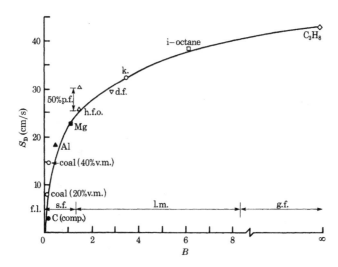

FIGURE 7.8

Burning velocity versus the Spalding transfer number B. The types of fuel depicted in this figure range from propane vapor (g.f., gaseous fuel) to *iso*-octane liquid mist (l.m.), and from volatile solid fuels (s.f.) to nonvolatile solids. Calculations based on the stoichiometric concentration of fuel, a particle diameter of 30 μm, atmospheric pressure, and ambient temperature. Fuel abbreviations: k, kerosene; d.f., diesel fuel; h.f.o., heavy fuel oil; p.f., prevaporized fuel; and f.l., flammability limit.

Reprinted with permission from Ballal, D.R., 1983b. Flame propagation through dust clouds of carbon, coal, aluminium and magnesium in an environment of zero gravity. Proc. Royal Soc. London, A385, 21–51.

Chapter 10. In the next section, we will briefly discuss a powerful mathematical technique that allows the investigator to seek analytical solutions—or at least simplified numerical solutions—to the equations of change. This technique is based on the perturbation method of solving differential equations called activation energy asymptotics.

7.5 MODELS BASED ON ACTIVATION ENERGY ASYMPTOTICS

The method of activation energy asymptotics (AEA) offers a means of analyzing and solving the equations of change for dust flame propagation. The procedure relies a systematic consideration of order of magnitude arguments (sometimes called scaling arguments) that permit simplification of the governing differential equations (Krantz, 2007). Another aspect of the analysis relies on exploiting the special mathematical properties of the Arrhenius temperature dependence of the reaction rate terms. We will restrict our attention to steady, one-dimensional, laminar flame propagation and consider premixed gas flames, spray flames, and dust flames. Because the mathematical technique is somewhat complex, we will only describe the basic concepts and illustrate the type results that can be generated with this type of analysis.

The governing equations for combustion are nonlinear differential equations. A number of mathematical techniques are available to the investigator who wishes to solve *linear* differential equations. The presence of nonlinearity complicates the search for solutions. One technique that has proven to be especially helpful in the solution of nonlinear differential equations is the method of perturbations (Varma and Morbidelli, 1997, Chapter 9; Aziz and Na, 1984). The essential idea behind perturbation methods is to postulate for a differential equation a solution of the form

$$y(x) = y_0 + \varepsilon y_1 + \varepsilon^2 y_2 + \cdots \tag{7.57}$$

where ε is a parameter that is small ($\varepsilon \ll 1$). By first assigning $\varepsilon = 0$, one obtains the zeroth order "unperturbed" solution. Allowing $\varepsilon \ll 1$, one can obtain a first-order "perturbed" solution by retaining only the linear term. Higher order solutions are possible as well. Following well accepted guidelines for rendering the governing equations dimensionless, the definition of the parameter ε falls out naturally such that it will have a physical interpretation.

In the nonisothermal chemically reacting systems encountered in combustion, the nonlinear Arrhenius temperature dependence of the chemical reactions is particularly troublesome. A whole set of mathematical techniques specifically tailored to combustion theory has developed by exploiting the mathematical behavior of the Arrhenius term. This set of mathematical tools is called activation energy asymptotics. In AEA, the perturbation parameter most often selected is the reciprocal of the Zeldovich number, $\varepsilon = \mathrm{Ze}^{-1}$. The Zeldovich number Ze is defined as

$$\mathrm{Ze} = \frac{E_\mathrm{a}(T_\mathrm{f} - T_0)}{R_g T_\mathrm{f}^2} \tag{7.58}$$

The flame temperature and ambient temperature are denoted as T_f and T_0, respectively, E_a is the Arrhenius activation energy, and R_u is the universal gas constant. Values of Ze for hydrocarbon−air mixtures are typically in the range of 5−15; this range is sufficiently large to make the perturbation technique effective (Williams, 1985, p. 155).

There are a number of different perturbation techniques that have been developed for combustion analysis (Buckmaster and Ludford, 1982). These various techniques have a common strategy to formulate the governing equations in dimensionless form and then to apply order of magnitude arguments to simplify the equations. Then key variables are expanded as a power series and substituted into the simplified equations. In many combustion problems, it makes sense to divide the combustion problem into discrete zones based on the dominant physical or chemical phenomena occurring in that zone. For example, in a steady, one-dimensional planar flame it is convenient to divide a premixed flame into three zones: an upstream preheat zone, a reaction zone, and a downstream equilibrium zone. The three regions are then "joined" to each other through matching conditions. While many thermal theories have been formulated using this convenient formulation, they have done so in an intuitive or ad hoc manner. In AEA, this formulation has a more rigorous mathematical basis. Good brief discussions of the application of asymptotic methods to premixed gas flames can be found in the books by Williams and Law (Williams, 1985, pp. 154−165; Law, 2006, pp. 255−263). I refer the reader to these references for a detailed discussion of the solution technique.

The two key attributes of a laminar premixed flame are the burning velocity and the flame thickness. As an example, Law describes the analysis of a premixed flame of a prototypical hydrocarbon−air mixture using a global kinetic expression for the combustion reaction (Law, 2006, pp. 255−263). Physical properties are assumed to be constant. To simplify the nomenclature, I have assumed a zeroth order reaction. The original derivation is for a more general nth order kinetics in both fuel and oxygen concentration. For the details of the mathematical solution, I refer the interested reader to the book by Law with the warning that AEA is not for the faint hearted!

The burning velocity is given by the following dimensionless expression in terms of the Lewis number Le, Damkohler number Da, and Zeldovich number Ze:

$$\frac{2\,\text{Le}\,\text{Da}}{\text{Ze}^2} = 1,\, \text{Le} = \frac{\alpha}{\mathcal{D}},\; \text{Da} = \left[\frac{(\lambda/C_p)A}{(\rho_u v_u)^2}\right]\exp\left(\frac{-E_a}{R_g T_b}\right),\; \text{Ze} = \left[1 - \left(\frac{T_u}{T_b}\right)\right]\frac{E_a}{R_g T_b} \quad (7.59)$$

In dimensional terms, the burning velocity becomes

$$v_u = \left[\frac{1}{\rho_u}\left(\frac{\alpha^2}{\mathcal{D}}\right)A\exp\left(\frac{-E_a}{R_g T_b}\right)\right]^{1/2}\left(1 - \frac{T_u}{T_b}\right)\left(\frac{E_a}{R_g T_b}\right) \quad (7.60)$$

The functional form of the AEA burning velocity is similar to the expressions derived from thermal theories (see Chapter 4) but there is now an additional contribution due to the nonunity value of the Lewis number and a contribution due to the temperature effect on the reaction rate. The important point here is that AEA provides a rigorous basis for exploring the frontiers of combustion theory, and as a body of knowledge it complements the capabilities of computational fluid dynamics.

There is a significant body of literature on applications of AEA in gaseous systems. We turn our attention now to heterogeneous combustion systems (liquid mists and dust clouds).

Premixed heterogeneous flames (two-phase systems consisting of fuel mists or dust clouds) present a more complex challenge for analysis by AEA. The two-phase mixture characteristics, the single particle reaction kinetics, unequal gas and particle temperatures and velocities, and particle concentration nonuniformities all conspire to complicate the analysis of flame propagation. I give a brief survey of some of the literature on steady, one-dimensional, laminar flame propagation, first in liquid mists and then in dust clouds. Liquid fuels are considered first since they will typically represent simpler combustion behavior with evaporation—rather than gasification/pyrolysis—being the rule.

One of the earliest papers on laminar flame propagation in a fuel mist was by Williams (1960). In this paper, he considered a monodisperse fuel spray and derived a criterion for determining which mode would prevail as the dominant mechanism: homogenous combustion (droplets fully vaporized prior to passage of flame) or heterogeneous combustion (droplet vaporization and combustion occurs sequentially within the flame). He derived burning velocity expressions for the two limits of homogeneous and heterogeneous combustion. The expression for homogeneous combustion gave burning velocity predictions very similar to but slightly less than premixed gaseous flames due to the energy decrement consumed by vaporization. The formula for heterogeneous combustion was compared with experimental data and found to give results in reasonable agreement. The equations derived in this study were sufficiently complex that they required an iterative numerical solution.

Several investigations have been directed at a deeper understanding of spray or mist flames using AEA. This paragraph will cite just a small sample of the peer-reviewed literature on spray flames. Lin et al. investigated the fuel lean and fuel rich behaviors of dilute monodisperse mist flames (Lin et al., 1988). They derived an expression for the burning velocity that predicted a decreasing flame speed with increasing liquid fuel loading and increasing droplet size. Silverman et al. considered the influence of polydispersity on the burning velocity for homogeneous combustion (Silverman et al., 1993). The same investigators later extended this work to heterogeneous combustion of polydisperse sprays (Greenberg et al., 1996). While the previous papers refer to one-dimensional planar flames, Han and Chen were able to apply AEA to the problem of one-dimensional spherical flame propagation, but the resulting analysis requires a numerical solution for the burning velocity (Han and Chen, 2015).

The main lesson to learn from these investigations is that the burning velocity exhibits a dependence on both fuel concentration and particle size. The effect that fuel concentration exerts on burning velocity is not simple; increasing vapor concentration increases the velocity while increasing liquid concentration tends to dampen it. Similarly, the impact of particle size is not simple even in monodisperse sprays.

One would expect that smaller particles would give rise to faster velocities. But larger particles also give rise to wrinkling of the flame front which increases the surface area, and therefore the mass burning rate, of the premixed flame. Experimental studies which will be described later in this chapter confirm that there is in monodisperse sprays an optimum particle diameter that yields the maximum burning velocity. With polydisperse sprays, it is far more difficult to generalize the complex range of combustion behavior. This last statement should serve as a caution that liquid mists are supposed to demonstrate simpler combustion behavior than dust clouds.

There has been some notable work using AEA and other analytical methods to study dust flame propagation. The reaction kinetics of solid particle combustion introduces additional complexity to the flame propagation process. A further complication is the significance of thermal radiation with increasing fuel concentrations and decreasing particle diameter. For the most part, the formulas for burning velocity of a dust flame are so convoluted that they defy simple interpretation. Therefore, I will not cite these results but instead will only discuss the investigations in a general sense.

Separately, Mitani and Joulin and his colleagues investigated the inhibition or quenching effect of adding inert particles to the propagation of a premixed flame (Mitani, 1981; Joulin, 1981; Joulin and Deshaies, 1986, Blouquin and Joulin, 1996). These studies served as a basis for then considering the behavior of reactive (solid fuel) particles and the relative importance of thermal radiation (Deshaies and Joulin, 1986). Berlad and his colleagues formulated a premixed dust flame model based on AEA (Seshadri et al., 1992). Goroshin and his colleagues focused on aluminum dust flames and published an analytical model (based on AEA) for fuel lean and fuel rich combustions and compared their results with quenching diameter experiments (Goroshin et al., 1996a). The analytical model was partly the basis for Bidabadi's PhD thesis (Bidabadi, 1995). Bidabadi subsequently has become a strong proponent for the use of AEA and other analytical methods for studying dust flame ignition, propagation, and quenching (a small selection of his work: Bidabadi et al., 2010; Bidabadi and Mafi, 2012; Jadidi et al., 2010).

7.6 IGNITION AND QUENCHING OF DUST FLAMES

In Chapter 4 I defined ignition as the initiation of combustion in a fuel-oxidizer system. The same definition applies whether it is a premixed gaseous mixture or a combustible dust mixture. There are several different energy sources that can provide an ignition stimulus to a dust cloud (Palmer, 1973, Chapter 7; Eckhoff, 2003, Chapter 5; Amyotte, 2013, Chapters 8 and 9):

- Flames (eg, welding, torch cutting)
- Electrical sparks
- Electrostatic discharge (static electricity)
- Hot surfaces
- Spontaneous heating/smoldering nests
- Friction/mechanical sparks (eg, grinding, impact of dissimilar metals, tramp metal in rotary equipment).

Generally, combustible dust clouds are more resistant to ignition than flammable gases and vapors. This is because of the additional heat and mass transfer resistances inherent to the chemical reaction of solid particles. Heterogeneous ignition models tend to be more complex than those for heterogeneous flame propagation. Many of the transport processes neglected in flame propagation are important in ignition phenomena. The body of literature on dust cloud ignition is not as well developed as the ignition theory for premixed gaseous systems. Quenching, on the other hand, has received even less attention than ignition and is not as well developed theoretically as the topic of dust cloud ignition.

There are two parameters used to characterize dust cloud ignition: the *minimum autoignition temperature* (MAIT) and the *minimum ignition energy* (MIE). The autoignition temperature is the temperature of the dust cloud that results in a prompt thermal runaway reaction resulting in the appearance of flame. The MIE is the minimum amount of thermal energy that must be deposited into the center of a dust cloud (under standardized conditions) that results in flame propagation. Dust flame quenching is characterized by a characteristic dimension—the spacing of two parallel plates or the diameter of a cylindrical tube—that results in the extinction of a propagating premixed dust flame. This characteristic dimension is called the *quench diameter* (d_q). To be meaningful, all three of these parameters require the specification of the system geometry, the boundary conditions, and the initial conditions.

We will first briefly review the experimental methods used to determine ignition properties of dust clouds. Then I will present an overview of the theory of dust flame ignition and quenching.

7.6.1 SUMMARY OF EXPERIMENTAL METHODS

Several organizations have developed test standards for evaluating the hazard characteristics for combustible dusts (Britton et al., 2005). There are standardized ignition tests for the autoignition of a dust cloud by a hot surface (ASTM E1491, 2012), the MIE for a dust cloud from an electrical spark (ASTM E2019, 2007), and the hot surface ignition temperature for a dust layer (ASTM E2021, 2009). The self-heating and ignition of dust layers were covered in Chapter 5, Smoldering Phenomena, and will not be discussed further here. The intent of the standardized tests is to provide characteristic data—an autoignition temperature and a minimum ignition energy—that one may use to evaluate a variety of combustible dust hazard scenarios.

The operating principle behind the standardized test for the MAIT of a dust cloud is to inject a dust sample into a furnace preheated to a specified temperature and ambient pressure. The test is conducted with air at its ambient composition. Ignition is determined by the appearance of flame discharged from the furnace. The test procedure and the design of the furnace are described in an ASTM standard (ASTM E1491, 2012). The standard furnace design is a 0.27-L volume called the Godbert Greenwald furnace, but a total of four different designs are endorsed as permissible with the standard procedure. Work performed by the U.S. Bureau of Mines compared the performance of the 1.2-L furnace with a 6.8-L furnace. They determined that based on tests using a 6.8-L furnace, the test results from the smaller 1.2-L volume furnace were similar (Conti et al., 1993). As noted in the standard, the MAIT measurements obtained from the Godbert–Greenwald furnace tend to be higher than those obtained from the other three furnaces (the third furnace being a 0.27-L furnace designed by the German standards organization BAM).

The operating principle behind the standardized test for the MIE is to discharge an electrical arc of specified energy in the center of a dust cloud (ASTM E2019, 2007). The test is conducted with air at its ambient composition, temperature, and pressure. Following the procedure of the test, one obtains as the output the minimum energy spark that can ignite a given combustible dust cloud. Correlations are available to convert the measured result into an equivalent ignition stimulus (eg, mechanical sparks) (Siwek and Cesana, 1995).

There is no standardized test procedure for performing dust flame quenching measurements. There is a standardized procedure for determining the quenching diameter of premixed gaseous mixtures, but this procedure is for a laboratory apparatus better suited for gases than for combustible dusts (ASTM 582, 2013). Dust flame quenching measurements have been reported in the scientific literature (cited below), but currently there is no consensus on the laboratory apparatus or procedure to be used.

As a final word on the matter, I caution the reader that standardized tests such as those described above are intended for comparative purposes. The test measurements obtained do not have a fundamental meaning, but rather, the results are meaningful only in comparison with a reference standard. Such standardized tests are extremely useful for promoting industrial safety, but the performance of these test methods is sometimes extremely difficult to simulate with mathematical models.

7.6.2 IGNITION

As with our discussion of heterogeneous flame propagation, there is value in examining single particle behavior first, and as an example of completely volatile fuel particles, we first consider the ignition of single liquid fuel droplets before advancing to solid particles. Whether it is liquid fuel droplets or solid fuel particles, an effective ignition stimulus liberates enough fuel vapor into a volume sufficiently large—the ignition kernel—that the subsequent combustion of the fuel–air mixture generates enough heat to ignite the adjacent fuel unburnt particles. While the qualitative features of heterogeneous ignition are well-established, there is no single model capable of rationalizing the published ignition data. The preferred dust cloud ignition model would be capable of explaining the trends observed in autoignition temperature or minimum ignition energy with respect to changes in dust concentration, particle size, dust cloud size, and oxygen concentration.

The subject of single liquid fuel droplet ignition has been reviewed by Aggarwal (2014). Citing over 100 references, it is apparent that this subject is well developed, and provided that one is able to obtain or measure the necessary physical property and kinetic data, one is able to predict the ignition delay time for an individual fuel droplet.

Generalizing from individual droplets to sprays is more complicated. The ignition behavior of sprays has been the subject of two especially good reviews (Annamalai and Ryan, 1992; Aggarwal, 1998). In comparing the ignition behavior of a spray with a single droplet, one key difference is that a single droplet can only be ignited by an external ignition source. A spray, on the other hand, can be ignited by an external ignition source or by spontaneous (self-heating and thermal runaway) ignition. A second key difference is the range of potential interactions between droplets in a spray during ignition: a spray can be ignited as individual droplets, as a cluster of droplets, or as the entire spray volume. Although some investigators have postulated criteria for describing different types of interaction during ignition, these methods do not lend themselves to direct comparison with solid particles or dust clouds. Further information can be found in the references already cited and in Sirignano's book (Sirignano, 2010).

There are few comprehensive surveys of single fuel particle ignition. Essenhigh et al. have surveyed the literature for coal particle ignition and Annamalai and Ryan have updated this review (Essenhigh et al., 1989; Annamalai and Ryan, 1993, pp. 523−567). With some caution, the findings presented in these papers can be applied to other charring organic solid fuels. Two ignition mechanisms are identified: a homogeneous mechanism operative for larger particles ($d_p > 100$ μm) heated slowly (heating rate $< 100°C/s$) and a heterogeneous mechanism operative for smaller particles ($d_p < 100$ μm) (Essenhigh et al., 1989, p. 4). Homogeneous ignition occurs in two steps. First, the heat source drives volatiles out of the particle and ignites them. The burning volatiles shield the char particle from ignition until the volatiles are mostly consumed. Then the second stage begins which is the ignition and burning of the char. Heterogeneous ignition is a three stage process. The oxidizer gas first begins to heterogeneously react with the unburnt particle. As the heterogeneous reaction proceeds, it eventually transitions to the homogeneous combustion of volatile material followed by the third stage, the combustion of the solid char. An accurate calculation of the ignition properties of a given particle will require the accurate specification of the physical and kinetic properties of the solid.

With a dust cloud, ignition is again complicated by the interaction of burning particles with each other. Cassel and Liebman dubbed this a cooperative effect for dust cloud ignition because the ignition temperature of the cloud was less than the ignition temperature of a single particle (Cassel and Liebman, 1959). The primary form of interaction between particles is through the heat transfer processes dominant within the cloud. At sufficiently high dust concentrations, the cloud will become optically thick. Thus, radiant heat losses from the ignition kernel will be reduced and the average temperature within the kernel will increase. Furthermore, as indicated in the previous paragraph, there is a particle size effect that will determine if the ignition process is controlled by heterogeneous or homogeneous kinetics. In a polydisperse dust cloud, it is probable that both types of processes will be in force.

The discrete nature of the fuel particles has inspired theoretical approaches to the problem of ignition that are quite different from the discussion presented in Chapter 4 on the ignition of premixed gaseous mixtures. Instead, many of the dust cloud ignition models have been based on the thermal explosion theory described in Chapter 5 regarding the initiation of smoldering combustion. Several models have been developed over the last 50 years, but these models have rarely been challenged with empirical data for a variety of combustible dusts under a variety of conditions (Cassel, 1964; Mitsui and Tanaka, 1973; Krishna and Berlad, 1980; Essenhigh et al., 1989; Proust, 2006a). Hence, these models must be regarded somewhat cautiously; they may be an excellent source of insight into the dust cloud ignition process, but their strengths and limitations are not fully known.

Krishna and Berlad presented a model for dust cloud ignition assuming an absence of spatial gradients in the dust cloud (Krishna and Berlad, 1980). The model was based on energy equations written for both the particle and gas phases and the reaction was based on heterogeneous kinetics. Their analysis yielded an expression for the autoignition temperature of the dust cloud as a function of dust concentration and particle size. Their model was in qualitative agreement with the data on metal particle ignition presented in Cassel and Liebman's (1959) paper, but they did not make a quantitative comparison of their model predictions with the experimental data. An interesting feature of their model is that it predicted for *high* dust concentrations the autoignition temperature decreased with *decreasing* particle diameters, but for *dilute* dust concentrations it predicted that the autoignition temperature would decrease for *increasing* particle diameters.

Zhang and Wall extended the Krishna—Berlad model by including both heterogeneous and homogeneous kinetics expressions for coal combustion (Zhang and Wall, 1993). Their model enabled consideration of other effects on dust cloud ignition such as the influence of oxygen concentration, but they made only qualitative comparisons.

Higuera et al. formulated a model for predicting the ignition temperature of a coal dust cloud undergoing heterogeneous kinetics (Higuera et al., 1989). This model was based on an AEA analysis of the governing equations for a heterogeneous mixture of particles in the gas phase. Their paper explored the mathematical behavior of their model, but did not make any comparisons with experimental data. Baek et al. also investigated the behavior of a model based on a heterogeneous description of the particles and gas phase with a particular emphasis on the effect of the temperature distribution within the particles (Baek et al., 1994). Again, this analysis was not compared with experimental data.

Mittal and Guha critiqued several dust cloud autoignition models and concluded that they were better suited for evaluating dusts that burn by heterogeneous kinetics (Mittal and Guha, 1997b). These same investigators performed autoignition experiments with polyethylene dust using the Godbert—Greenwald furnace and developed a model based on thermal explosion theory and homogeneous kinetics to explain the data (Mittal and Guha, 1996; Mittal and Guha, 1997a). They concluded that their model would be more successful in predicting autoignition temperatures for volatile organic solids. It does not appear, however, that this model has been challenged with experimental data for other materials. A model with a more sophisticated kinetics submodel has been tested against the polyethylene data and gives further support for the success of the use of homogeneous combustion kinetics for volatile organic solids (Di Benedetto et al., 2010).

The work of Ballal and Lefebvre stands out as one of the few concerted investigations into the development of models to predict liquid mist and dust cloud ignition energies (Ballal and Lefebvre, 1978, 1979; Ballal, 1980; Ballal and Lefebvre, 1981; Ballal, 1983a). Their approach based on the analysis of time scales has been discussed earlier in this chapter with regards to the prediction of dust flame thickness and laminar burning velocity. Their work stands out in particular because it was based on substantial experimental efforts with a variety of different fuels (first liquid fuels and then solid fuels). With regards to the development of a model for the MIE, they also developed a model for predicting quenching diameters. Thus, while their efforts do not yield an in-depth perspective of the physical and chemical phenomena occurring in the microscale, their efforts do result in one of the more comprehensive models with a practical output that has been tested empirically. The primary disadvantage to note is, like the flame propagation model discussed earlier, the ignition model has been tested only in the dust concentration range from fuel lean to just slightly fuel rich.

As a final note on dust cloud ignition, some investigators have noticed the effect of the chemical composition on the particle surface on dust cloud ignition characteristics. Baudry and colleagues took aluminum powders and anodized them to create different thicknesses of aluminum oxide (Al_2O_3) coating on the particle surfaces humidity (Baudry et al., 2007). They found that the ignition energy increased with increasing oxide layer thickness. In a later study, Bernard et al. confirmed the effect of aluminum oxide thickness on the MIE, but they also noted that the presence of moisture could confound the trend (Bernard et al., 2012). Their interpretation was that water vapor trapped between the oxide layer and base metal could enhance the ignition and combustion process. There are few studies that explore the possible role of surface contaminants on dust particle ignition or combustion, but the potential for this type of interaction should be borne in mind.

7.6.3 QUENCHING

Flame quenching tests are usually conducted with a flame propagating either upwards or downwards in a vertical direction. There are several very good descriptive surveys of premixed gas flame propagation and quenching in tubes (Guenoche, 1964; Lewis and von Elbe, 1987, pp. 226–301; Bjerketvedt et al., 1997). Interestingly, there is a difference in the quenching mechanism for upward versus downwards flame propagation with downwards quenching being caused by heat losses to the wall and upwards quenching being caused by flame stretching (Jarosinski et al., 1982).

The mechanism for quenching dust flames has received less attention than premixed gases. There are only a few published studies on the subject and they are mostly of an experimental nature. Furthermore, they are based on different materials, equipment, and procedures. This makes it difficult to compare the work of different investigators.

Ballal performed a series of experiments with a variety of dusts (three types of bituminous coal, graphite, electrode carbon, aluminum, magnesium, titanium) and developed a model to predict the quenching distance (Ballal, 1980, 1983a). His experiments were conducted in the dust concentration range from fuel lean to stoichiometric (equivalence ratio from 0.4 to 1.0). The asymptotic minimum quenching distance was determined to be approximately 2 mm for particles with a Sauter mean diameter of 40 μm.

Jarosinksi and his colleagues published experimental studies using aluminum powders (Jarosinski et al., 1986). They used a flammability tube and measured the quench distance during downward flame propagation. Their experiments were conducted in the dust concentration range of fuel rich (equivalence ratio from 1.1 to 3.7). They determined the asymptotic minimum quenching distance was approximately 9 mm for particles with a Sauter mean diameter of 9.5 μm.

Goroshin et al. conducted quenching experiments in a similar flammability tube by observing downward flame propagation (Goroshin et al., 1996b). Using aluminum powder with a Sauter mean diameter of 5.4 μm, they explored a range of dust concentrations with an equivalence ratio from 0.5 to 1.9. They determined the asymptotic minimum quenching distance to be 5 mm.

Habibzadeh and Keyhani performed quenching experiments using the downward flame propagation technique and a flammability tube similar to that used by Goroshin et al. (Habibzadeh and Keyhani, 2008). Using an aluminum powder with a Sauter mean diameter of 18 μm, and over a dust concentration range corresponding to an equivalence ratio of 0.5−4.7, they determined the asymptotic minimum quenching distance to be 3 mm.

The conclusion to be drawn from these four studies is that there are too few dust flame quenching studies to permit generalizations about dust chemical composition, dust concentration, or particle size. This is a field in its infancy, and investigators wishing to use published quench distance data on dust flames must be thoughtful and selective of which data sets to use.

The determination of the quenching distance is not only of theoretical importance, but it also has a practical significance. Quenching is related to but not the same as the maximum experimental safe gap (MESG). The MESG is the maximum gap that will not permit a gas or dust deflagration in an enclosure to vent through and successfully ignite a flammable mixture of the same type that exists outside the enclosure (Eckhoff, 2003, pp. 346−351). The MESG should normally be smaller than the quench distance. The MESG is useful in the evaluation of the performance of rotary air locks and flameless venting devices (Siwek, 1989; Siwek and Cesana, 1995; Going and Chatrathi, 2003; Snoeys et al., 2012).

7.7 SURVEY OF HETEROGENEOUS FLAME PROPAGATION BEHAVIOR

Heterogeneous flame, consisting of either liquid mists or dust clouds, propagates differently than premixed gaseous flames. In this section, we will derive insights drawn from experiments using both stationary flames on burners and propagating dust flame experiments in flammability tubes. The objective of this section is to discuss how the laminar burning velocity, lower flammability limit, flame temperature, and flame structure are influenced by fuel volatility, equivalence ratio, mean particle size, and particle size distribution (monodisperse, bimodal, polydisperse). As with single particle combustion, I have organized the discussion of combustible dusts into two basic classes, organic solids and metallic (inorganic) solids. Organic solids are further divided into three groups based on volatility: noncharring solids (completely volatile), charring solids (partially volatile), and nonvolatile solids. Metallic solids are divided into volatile and nonvolatile solids.

Unconfined heterogeneous flame propagation experiments are difficult to perform and so a wide variety of experimental apparatuses have been developed. Some investigations have been particularly systematic and thorough while others have been strongly curtailed due to the limitations of the apparatus or other practical difficulties. Therefore, this survey of flame behavior will seem less comprehensive than the discussion on single particle combustion. A more comprehensive survey on the literature of dust flame propagation investigations can be found in Eckhoff's book (Eckhoff, 2003, Chapters 4 and 9).

The reader is cautioned to bear in mind that some of the investigations discussed below measured the *burning velocity* and others measured the *flame speed*. The *burning velocity* is the velocity of the unburnt mixture relative to the flame. The *flame speed* is the speed of a moving flame front as viewed from a stationary frame of reference (laboratory coordinates).

7.7.1 ORGANIC SPRAY OR MIST FLAMES

Deflagrations in liquid fuel aerosol clouds are a demonstrated hazard (Santon, 2009). As an accident scenario, they are less common than dust explosions. For example, in Santon's review article he identified a 35 flash fires or explosions associated with the ignition of an accidental release of an aerosol cloud over a period of 50 years (1959–2009) suggesting an incidence rate of less than 1 per year. According to the study by the U.S. Chemical Safety Board, the incidence rate for combustible dust accidents (281 combustible dust accidents in a 25 year period, 1980–2005) may be more than 10 per year (CSB, 2006). Since liquid fuels can be expected to be more volatile than solid fuels, the smaller number of accidental deflagrations with mists or sprays is likely to be more a reflection of the lower incidence rate of aerosol cloud formation compared to combustible dust clouds. We will briefly review some of the key experimental studies of spray or mist flames because of their relevance by analogy to dust deflagrations.

There have been two basic experimental designs for generating aerosol clouds of liquid droplets: cloud chambers (Burgoyne and Cohen, 1954) and atomization (Ballal and Lefebvre, 1981). There have been many variations on these techniques primarily inspired by the desire to create droplets of a certain size range or droplet concentrations within a particular set of values. These experiments are difficult because of the tendency of liquid droplets to deposit onto the apparatus walls or to coagulate in the air stream. Bowen and Cameron reviewed the state of the art for both experimental and theoretical studies of aerosol deflagrations (Bowen and Cameron, 1999). They noted that at larger droplet sizes ($d_p > 30$ μm) the heterogeneous burning velocity cannot exceed the burning velocity of the premixed vapor.

Early investigations into mist flames demonstrated that there is a critical particle size range below which the flame burns homogeneously and above which the flame burns heterogeneously (Burgoyne and Cohen, 1954). Working with monodisperse mists of tetralin (a combustible liquid with the chemical formula $C_{10}H_{12}$), they found that the critical size for homogeneous burning was 10 and 40 μm for heterogeneous burning. The homogeneous flame was described as a continuous blue flame front with a parabolic profile. The heterogeneous flame was described as shrinking in thickness as the droplet diameter increased until at 40 μm it became a cluster of isolated droplets burning with yellow flames. They conducted experiments over a range of droplet sizes from 8 to 38 μm, the burning velocity varied from 0.28 to 0.68 m/s. Due to the limitations of their experimental apparatus, they were not able to independently vary droplet size and fuel concentration. The range of equivalence ratios was from approximately 0.5 to 1.0; thus their experiments were conducted in the fuel lean to stoichiometric range. Chan and Jou later expanded this work also working with monosized tetralin mists and confirmed that at an equivalence ratio of 0.5 and the droplet size range of 10−40 μm, the burning velocity matched the premixed mixture value (at 10 μm) and then rose to a maximum before falling as it approached the droplet size of 40 μm (Chan and Jou, 1988).

Several flame propagation studies have been conducted since then with the goal of establishing the relationship between the burning velocity and droplet size. Mizutani and Nakajima studied the impact of the overall equivalence ratio defined as the sum of the equivalence ratio of fuel vapor plus the equivalence ratio of droplets (Mizutani and Nakajima, 1973). They observed that the addition of kerosene droplets to a propane−air flame initially increased the burning velocity, but as the concentration of droplets increase the burning velocity went through a maximum. They investigated the impact of turbulence and demonstrated that turbulence caused an enhancement of the burning velocity by wrinkling the flame front. Due to the limitations of the experimental apparatus, Mizutani and Nakajima were not able to systematically investigate the effect of droplet diameter and equivalence ratio on the burning velocity.

Ballal and Lefebvre conducted experiments that systematically varied the droplet diameter and equivalence ratio in the fuel lean range (Ballal and Lefebvre, 1981). They demonstrated a decrease in the burning velocity with increasing droplet diameter. They also demonstrated that at constant overall equivalence ratio, as the fraction of fuel vapor increased (designated as Ω in their paper), so did the burning velocity. Experimental results for *iso*-octane at three Sauter mean particle diameters are shown in Fig. 7.9.

The inverse trend of burning velocity with particle diameter is apparent as is the direct relationship of the burning velocity with the vapor fraction. They developed a flame propagation model (described earlier in this chapter) that gave reasonable agreement with the experimental results as shown in the figure.

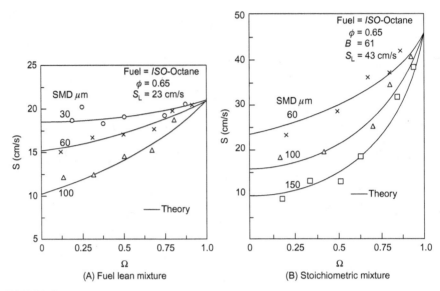

FIGURE 7.9

Laminar burning velocity for *iso*-octane mist flames at different Sauter mean diameters (SMD) and fuel vapor fractions (Ω) for two different equivalence ratios.

Reprinted with permission from Ballal, D.R., Lefebvre, A.H., 1981. Flame propagation in heterogeneous mixtures of fuel droplets, fuel vapour and air. In: Eighteenth Symposium (International) on Combustion, The Combustion Institute. Elsevier, pp. 321–328.

Myers and Lefebvre investigated different fuel chemical compositions (all combustible liquids) and noted that above a certain droplet size, evaporation rates controlled flame propagation and the magnitude of the burning velocity varied inversely with droplet diameter (Myers and Lefebvre, 1986). They observed a difference in the flame structure as a function of the droplet size with small droplets burning homogeneously and larger drops presenting a cloud of individually burning droplets. They also observed a dramatic increase in the amount of soot and smoke with increasing droplet size. Their results indicated that for evaporation-controlled combustion, the burning velocity was inversely related to the mean droplet size, and the burning velocity increased with increasing equivalence ratio up to the stoichiometric condition. The burning velocity was directly proportional to the fuel volatility as evaluated by the fuel transfer number B. They also found an interesting influence of fuel chemical structure on burning velocity that transcended fuel volatility. Fuels with a higher aromatic content (more benzene rings or related structures) tended to burn with a greater luminosity, ie, they burned more brightly. The work of Myers and Lefebvre teaches us that even liquid fuels can display complex combustion behavior.

Richards and Lefebvre examined turbulent burning velocities for kerosene, decalin, and toluene fuel–air mixtures over a droplet size range from 20 to

110 μm and an equivalence ratio from 0.37 to 1.84 (Richards and Lefebvre, 1989). At equivalence ratios less than 1.1, they found the expected trend that the turbulent burning velocity increased with decreasing droplet size but showed little variation with equivalence ratio. But at fuel rich concentrations, they found that the burning velocity *increased* with increasing droplet size. They observed in their paper that the burning velocity trend in the fuel rich condition was likely due to the fact that the equivalence ratio was calculated based on the total fuel quantity, both liquid and vapor, and did not indicate the local vapor concentration that entered the flame. Since larger fuel droplets have smaller surface-to-volume ratios, it should be expected that the burning velocity maximum will occur at higher fuel concentrations.

Several studies have reported on the ignition characteristics of aerosol clouds (Burgoyne, 1963; Ballal and Lefebvre, 1978, 1979, 1981; Puttick, 2008; Gant et al., 2012). The general trends are the MIE decreases with decreasing droplet size and decreases with increasing fuel concentration in the fuel lean condition reaching a minimum value near the stoichiometric value. The minimum explosive concentration for aerosols is highly dependent on droplet size and has thus far eluded a simple theoretical explanation.

7.7.2 NONCHARRING ORGANIC DUST FLAMES

There is a close chemical similarity between noncharring organic solids and combustible organic liquids. Butlin studied upward flame propagation in polyethylene dust clouds in a vertical flammability tube (Butlin, 1971). The glass flammability tube had a diameter of 75 mm and a height of 2 m. His powder had a mean particle diameter of 200 μm and a nominal dust concentration of 20 g/m^3. Flame motion was recorded by high-speed photography. He observed in his experiments a reversal in the flow direction of the unburnt particles. Initially they descended at their settling velocity, on the order of 1.5 m/s, but would reverse direction and flow upwards as they were accelerated ahead of the flame. In these experiments, the upward flame velocity attained values of 1.5 m/s. He observed a range of combustion behavior, from bluish flames surrounding individual particles to the formation of coherent yellow flames that were composed of clusters of particles. Butlin observed at this fuel lean condition that the polyethylene particles tended to burn to completion.

EXAMPLE 7.7

Verify that the flammability tube experiments conducted by Butlin were performed in fuel lean conditions (Butlin, 1971).

Solution

Polyethylene is an olefin polymer with the repeating unit $(-CH_2-)$. The molar mass of this repeating unit is $\mathcal{M}_d = 14.03$ g/mol. The stoichiometric concentration for polyethylene is calculated from the formula derived in Chapter 3:

$$\rho_{d,st} = \frac{8.59 \, \mathcal{M}_d}{\left(x + \frac{y}{4} - \frac{z}{2}\right)} = \frac{8.59(14.03 \text{ g/mol})}{1.5} = 80.4 \text{ g/m}^3$$

The equivalence ratio is calculated as $\Phi = \rho_{d,actual}/\rho_{d,st} = (20 \text{ g/m}^3)/(80.4 \text{ g/m}^3) = 0.249$. The equivalence ratio less than 1 indicates that the mixture was fuel lean.

Panagiotou and Levendis observed the burning behavior of four polymers using a laminar flow drop tube furnace with a length of 250 mm and a diameter of 35 mm (Panagiotou and Levendis, 1998). Their apparatus and experimental procedures were developed specifically to observe differences in particle group (dust cloud) combustion and single particle combustion. They tested three noncharring polymers: polyethylene, polymethyl methacrylate, and polystyrene. The fourth polymer, polyvinyl chloride, was a charring polymer. All samples were sieved to a particle size range of 125—212 μm. They recorded flame motion with high-speed photography and measured flame temperature with a three-color pyrometer (a description of the pyrometer can be found in Panagiotou et al., 1996). They controlled the dust concentration entering the furnace, but do to the limitations of the experimental apparatus, they were able to only estimate the concentration.

They observed that polyethylene particle had a strong tendency to form group flames. The group flames formed a coherent yellow flame while individual particles were observed to burn with a faint bluish flame. Polystyrene particles also exhibited group combustion behavior with yellow flames but considerably more soot production than the polyethylene flames. The polymethyl methacrylate particles burned with behavior intermediate to polyethylene and polystyrene. The polyvinyl chloride particles did not exhibit group combustion in their experiments; they strictly burned as individual particles. They also noted that the flame diameters were smaller in the dust cloud than the flame diameter of a single burning particle. Overall, they observed that the flame temperatures of the dilute dust clouds were approximately 200 K lower than the flame temperatures measured for single particles, and dense cloud temperatures were approximately 200 K lower than dilute dust clouds.

Investigators at the University of Tokyo in Tokyo, Japan, have published more than 10 papers just on unconfined flame propagation in dust clouds of noncharring organic solids. The organic materials used in their experiments fall into two classes: fatty acids and long chain alcohols. Using essentially the same flammability tube arrangement, they have studied flame behavior of these organic solids with a variety of instrumentation technologies.

Using 1-octadecanol ($C_{18}H_{38}O$) and stearic acid ($C_{18}H_{36}O_2$), the investigators observed that the flame structure and flammability limits were dependent on the particle diameter (Chen et al., 1996; Ju et al., 1998a,b; Dobashi and Senda, 2002). Small diameter particles (10−20 μm) vaporized with the fuel vapor forming a yellow flame front while larger particles ($d_p > 80$ μm) burned with individual blue diffusion flames surrounding each particle. Similar features had been reported with mist flames (Burgoyne and Cohen, 1954). These experiments were performed with a transparent flammability tube with central ignition. Their optical measurements were performed during the spherical expansion of the flame. Their instrumentation included direct and Schlieren photography to record flame front motion, laser light scattering for measuring particle size, a thermocouple for temperature measurement, and an ionization probe to detect the flame reaction zone.

The particle diameter was also a significant determinant in the flammability limits for stearic acid. The lower flammability limit was found to be influenced primarily by the concentration of smaller particles ($d_p < 60$ μm) and was determined to be a concentration of smaller particles equal to 30 g/m^3 or an equivalence ratio of 0.32. The upper flammability limit was determined by the total concentration of particles and was found to be 340 g/m^3 or an equivalence ratio of 3.6. The flame thickness was found to be generally in the range of 14 mm.

Subsequent studies expanded the types of instrumentation used and varied the organic solids tested in order to investigate the effect of fuel volatility. With the addition of ultraviolet (UV) band filters, the investigators were able to make additional observations of the flame structure (Dobashi and Senda, 2006). The UV band filters reduced the luminosity of the flame which tended to overwhelm the resolution of ordinary photography. This technique allowed them to confirm that the leading edge of the flame is a continuous zone consistent with a premixed flame structure (due to the vaporization and combustion of the smaller particles). These observations were extended to the fatty acids, myristic acid ($C_{14}H_{28}O_2$) and behenic acid ($C_{22}H_{44}O_2$) (Anezaki and Dobashi, 2007; Dobashi, 2007).

Gao and his colleagues investigated the effects of particle characteristics using three fatty alcohols: 1-hexadecanol ($C_{16}H_{34}O$), 1-octadecanol ($C_{18}H_{38}O$), and 1-eisosanol ($C_{20}H_{42}O$) (Gao et al., 2012). Performing flame propagation tests at a dust concentration of 250 g/m^3 ($\Phi \cong 3.90$ for all three alcohols), they found the most volatile alcohol, 1-hexadecanol, gave the highest flame speeds and emitted more light. Interestingly, the highest flame temperatures were associated with the lowest volatility alcohol, 1-eisosanol. Fig. 7.10 shows plots of the maximum temperature measured in the flame zone and the "mean propagation velocity," ie, the *flame speed* (not the burning velocity), as a function of dust concentration.

I leave it to the reader to verify that the stoichiometric dust concentration for these three fatty alcohols is approximately 64 g/m^3. Thus the data in Fig. 7.10 was obtained exclusively in the fuel rich region.

In another investigation, Gao et al. compared the flame structure of mist flames and dust flames of homologous compounds (Gao et al., 2013a). The liquid fuels were methanol (CH_4O), propanol (C_3H_8O), hexanol ($C_6H_{14}O$), and octanol ($C_8H_{18}O$). The solid fuels were the same fatty alcohols used in the prior study (1-hexadecanol, 1-octadecanol, and 1-eisosanol). The flame propagation experiments

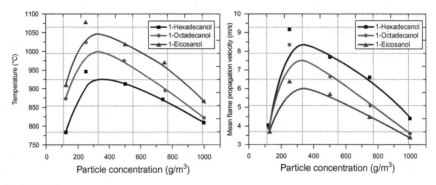

FIGURE 7.10

Plot of maximum flame temperature and mean flame speed as a function of dust concentration for three fatty alcohols.

Reprinted with permission from Gao, W., Dobashi, R., Mogi, T., Sun, J., Shen, X., 2012. Effects of particle characteristics on flame propagation behavior during organic dust explosions in a half-closed chamber. J. Loss Prev. Process Ind. 25, 993–999.

demonstrated clearly that more volatile fuels form continuous flame zones and less volatile fuels form flame zones of discrete burning particles. Images of the mist and dust flames from direct photography and UV band images are shown in Fig. 7.11.

The more volatile fuels are at the top of Fig. 7.11, and the less volatile fuels are toward the bottom. Additional evidence of the role of fuel volatility on flame propagation is illustrated in the flame speed measurements shown in Table 7.5.

It was also observed that flame thicknesses increased with decreasing fuel volatility.

The effect of particle size distribution on flame propagation behavior was investigated by creating three different particle size distributions of octadecanol dust flames (Gao et al., 2013b). Their interpretation is that in polydisperse dust clouds the combustion of small particles creates a continuous flame structure (yellow flame) that provides the heat source that volatilizes and ignites the larger particles (discrete blue flames). The three particle size distributions are depicted in Fig. 7.12 with characteristic size parameters presented in the adjacent table.

In their analysis of the experimental data, Gao and his colleagues determined that particles with a diameter less than 61 μm contributed to the continuous zone at the leading edge of the dust flame and larger particles burned individually in the latter portion of the flame. The Type A powder had the highest fraction of particles smaller than 61 μm, Type C had the smallest fraction, and Type B was intermediate. The quantitative effect of the contribution of the small particles is apparent in the photographs of the flame development for the three powders (Fig. 7.13).

The maximum flame speed attained in these tests was 0.50 m/s (Type A), 0.29 m/s (Type B), and 0.23 m/s (Type C). The importance of small particles in

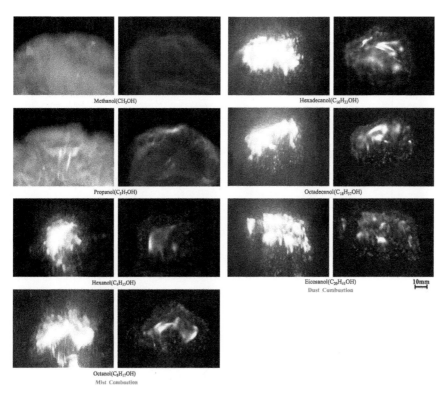

Methanol(CH₃OH)

Hexadecanol(C₁₆H₃₃OH)

Propanol(C₃H₇OH)

Octadecanol(C₁₈H₃₇OH)

Hexanol(C₆H₁₃OH)

Eicosanol(C₂₀H₄₁OH)   10mm
Dust Combustion

Octanol(C₈H₁₇OH)
Mist Combustion

FIGURE 7.11

Images of alcohol mist flames and dust flames using direct photography (left-hand side columns) and UV band filter photography (right-hand side columns).

Reprinted with permission from Gao, W., Mogi, T., Sun, J., Dobashi, R., 2013a. Effects of particle thermal characteristics on flame structures during dust explosions of three long-chain monobasic alcohols in an open-space chamber. Fuel 113, 86–96.

Table 7.5 Flame Speeds Measured for Alcohol Mist and Dust Flames

Fuel	Phase	Flame Speed (m/s)
Methanol	Liquid	2.44
Propanol	Liquid	1.40
Hexanol	Liquid	0.32
Octanol	Liquid	0.31
Hexadeconal	Solid	0.52
Octadecanol	Solid	0.48
Eicosanol	Solid	0.42

Data from Gao, W., Mogi, T., Sun, J., Dobashi, R., 2013a. Effects of particle thermal characteristics on flame structures during dust explosions of three long-chain monobasic alcohols in an open-space chamber. Fuel 113, 86–96.

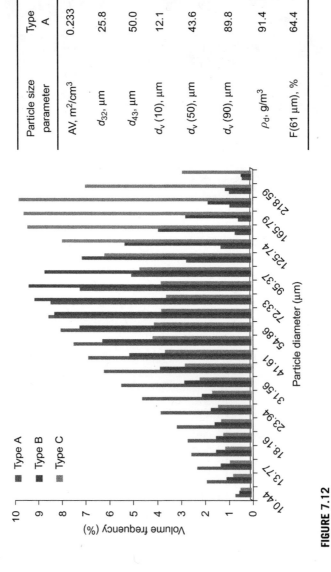

Particle size parameter	Type A	Type B	Type C
AV, m²/cm³	0.233	0.197	0.177
d_{32}, μm	25.8	30.4	33.9
d_{43}, μm	50.0	68.9	102.4
d_v (10), μm	12.1	16.6	18.5
d_v (50), μm	43.6	62.9	102.0
d_v (90), μm	89.8	126.7	190.1
ρ_d, g/m³	91.4	64.1	42.3
F(61 μm), %	64.4	45.1	29.8

FIGURE 7.12

Particle size distributions of three octadecanol powders with their size parameters indicated.

Reprinted with permission from Gao, W., Mogi, T., Sun, J., Yu, J., Dobashi, R., 2013b. Effects of particle size distributions on flame propagation mechanism during octadecanol dust explosions. Powder Technol. 249, 168–174.

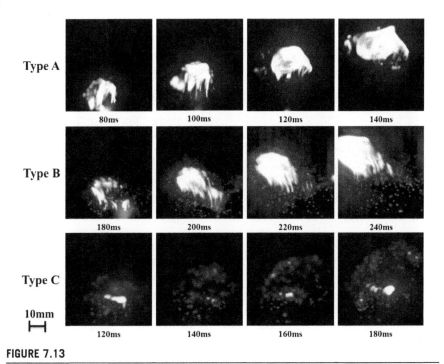

FIGURE 7.13

Photographs of flame propagation for three 1-octadecanol powders.

Reprinted with permission from Gao, W., Mogi, T., Sun, J., Yu, J., Dobashi, R., 2013b. Effects of particle size distributions on flame propagation mechanism during octadecanol dust explosions. Powder Technol. 249, 168–174.

driving flame propagation was further corroborated with the other two fatty alcohols, 1-hexadecanol and 1-eisosanol (Gao et al., 2015a). Additionally, this last study demonstrated the contribution of fuel volatility.

7.7.3 COAL DUST FLAMES

Coal dust flames have been the subject of extensive study for many years due to the practical importance of pulverized coal combustion for power generation as well as for underground coal mine safety. Studies on industrial combustion are usually not relevant to the combustion conditions encountered in accidental dust explosions and flash fires. For example, most of the studies discussed in this section involve stationary burners. Therefore, this review is focused more on experimental studies that are relevant to combustible dust hazard scenarios. Excellent books covering most aspects of coal combustion and gasification are available (Smoot and Pratt, 1979; Smoot and Smith, 1985; Borman and Ragland, 1998). A good review of the challenges in coal dust flame propagation studies can be found in the paper by Essenhigh (1977).

Maguire et al. investigated the minimum and maximum concentration limits of coal dust mixtures in upward flame propagation in a flammability tube (Maguire et al., 1962). The flammability tube had a length of 4.68 m and a diameter of 0.142 m. The bituminous coal dust used in this study had a mean particle diameter of 30 μm and a volatiles content of 36% on a dry ash-free basis. They performed 238 experiments with dust concentrations ranging from 0 to 2600 g/m^3. Flame motion was recorded with a drum camera. They defined the concentration limit as a dust concentration that resulted in a 50% probability of flame propagation. One difficulty with this definition is that the length of flame travel varied considerably as they approached the concentration limits. Table 7.6 is a summary of their results.

Each concentration value in this table is a mean value derived from several trials for each condition. The coefficient of variation (the standard deviation divided by the mean value) for each of the measured values was on the order of 10%. Based on a 50% probability of successful ignition, the recommended values for the lower flammability limit was determined to be 310 g/m^3 and the upper flammability limit was 1020 g/m^3. Although they acknowledged the effect of velocity slip on the determination of concentration limits, they did not attempt to correct for this effect.

Smoot and his colleagues performed measurements of premixed coal dust flames in a downward flowing flat flame burner with a diameter of 100 mm (Smoot et al., 1977). They were able to measure the burning velocity, the axial temperature profile, and gaseous and solid combustion product concentrations as a function of dust concentration, particle size, and volatiles content. The used coal dust samples with two mean particle diameters, 10 and 33 μm. They also tested two levels of volatiles content: a high volatiles content with 38% volatiles and 6% ash and a low volatiles content with 18% volatiles and 5% ash. Stoichiometric dust concentrations were estimated by the authors to be between 100 and 200 g/m^3.

They found that burning velocities ranged from 0.20 to 0.35 m/s in the fuel rich dust concentration range of 300–600 g/m^3. The dust flame thickness, determined by temperature measurement, was typically on the order of 0.01 m.

Table 7.6 Lower and Upper Flammability Limits of Coal Dust Based on a 50% Probability of Ignition

Flame Propagation (m)	Lower Flammability Limit (g/m^3)	Upper Flammability Limit (g/m^3)
0.4	230	2070
0.5	240	1730
1.0	270	1220
1.5	310	1020

Data from Maguire, B.A., Slack, C., Williams, A.J., 1962. The concentration limits for coal dust–air mixtures for upward propagation of flame in a vertical tube. Combust. Flame 6, 287–294.

The composition of gaseous combustion products was determined by gas chromatography. Solid combustion products were evaluated by proximate and ultimate analyses. It was estimated that over the range of dust concentrations, approximately 25−50% of the volatile content was consumed to feed the flame. Although the passage of the coal particles through the flame changed their shape, it did not significantly change their size.

Using the same experimental setup, Horton et al. reported additional data on coal dust flames (Horton et al., 1977). Their conclusion was that particle size had the dominant impact on the magnitude of the burning velocity and volatiles content was a secondary factor. They also noted that two coals having the same proximate analyses but coming from different mine produced the same combustion results.

Milne and Beachey performed coal dust flame experiments using an upright flat flame burner with a diameter of 63 mm (Milne and Beachey, 1977a,b). They tested coal with a size range from 10 to 20 μm. They were able to establish stable flames with fuel rich dust concentrations ranging from 140 to 300 g/m^3 or more (compared to a reported stoichiometric concentration of 120 g/m^3). Their measured burning velocities were 0.10−0.15 m/s, smaller than those measured by Smoot and his colleagues. Milne and Beachey attributed this difference to heat losses. Their detailed measurements of the composition profiles of the flame indicated that in rich flames almost all of the oxygen was consumed.

Smoot and Horton reviewed the literature on pulverized coal dust flame propagation (Smoot and Horton, 1977). They surveyed a number of different studies and found laminar burning velocities for coal dust reported in the range from 0.05 to 0.35 m/s. They commented that there were no published reports of stable burner flames with volatiles content less than 18%. It was also noted that the observed burning velocities were dependent on the burner size. They also confirmed that flame propagation was driven by volatiles combustion and that only a portion of the volatiles content participated in the flame. Finally, they confirmed that the microstructural changes in burnt coal particles observed in burner studies were consistent with observations drawn from single particle studies. Krazinski et al. reviewed the coal flame modeling literature and noted the significant difference in burning velocity measurements obtained from unconfined burner studies versus confined burner studies (Krazinski et al., 1979).

Bradley and his colleagues performed an interesting study where they compared methane−graphite−air flames and coal dust flames (Bradley et al., 1986). Their idea was to suggest that methane−graphite−air mixtures could be used to simulate coal combustion. The advantage of this approach is that the chemical kinetics of methane−graphite−air combustion is simpler and more deterministic than coal−air mixtures. They measured burning velocities in the range of 0.30−0.40 m/s. in fuel lean conditions. This work is of more theoretical importance and is probably of less relevance to combustible dust hazard investigations. On the other hand, this work does present a nice example of a hybrid fuel system (flammable gas plus combustible dust) which will be a topic in Chapter 8.

A more recent review of coal combustion in enriched oxygen mixtures and hybrid systems can be found in the paper by Chen et al. (2012).

Cao and his colleagues investigated fireball growth and thermal radiation heat transfer in unconfined coal dust flames (Cao et al., 2014). They used two types of bituminous coal dust, one with a volatiles content of 42% and the other with 35%. The coals had median particle diameters in the $32-34\ \mu m$ range. They used a vertical flammability tube with a height of 600 mm and a diameter of 68 mm. They tested the coal dusts at concentrations of 250, 500, and 750 g/m^3. They recorded the flame motion with high-speed photography and infrared imaging. The maximum flame speeds for the two dust ranged from 10 to 12 m/s and coincided with a dust concentration of 500 g/m^3.

Graphite provides an interesting point of reference for evaluating the rate of coal combustion in a deflagration. Char oxidation does not play an important role in accidental dust deflagrations. Experiments with graphite are a very approximate substitute for coal char, but it can be viewed as a model compound. Bryant, in work described earlier in this chapter, measured the burning velocity of graphite flames on an upward flowing Bunsen burner (Bryant, 1971). He used a graphite powder with an ash content of 0.52% and a mean particle diameter of $0.91\ \mu m$. It was not possible to establish a steady flame with graphite in air, but in a pure oxygen atmosphere, he measured a burning velocity of 0.086 m/s at a dust concentration of 210 g/m^3. This is a remarkably small burning velocity when one considers the particle size and oxidizer concentration.

One conclusion to draw from these coal dust flame propagation studies is that only a portion of the volatile content actually contributes to the deflagration. The results of the graphite flame study lend additional evidence to the inference that the contribution of char oxidation to coal dust flame propagation is minimal compared to the contribution of volatiles combustion. It is sobering to realize that for all of the damage potential of coal dust flames, only a fraction of the total chemical energy is unleashed.

7.7.4 BIOMASS DUST FLAMES

In this section, I will discuss biomass flame studies with biomass including starch, grain flour, cork, and lycopodium. This somewhat random collection of combustible dusts is a consequence of the availability of published studies.

One of the primary differences between charring and noncharring organic dust flames is that charring organic dusts have a lower volatiles content. To a certain degree, it might be expected that charring organic solids would yield slower flames than noncharring organics. But as we saw with coal dust flames, only a fraction of the volatile matter actually contributes fuel to the flame. A more direct determinant of burning velocity or flame speed is the mean particle diameter of the dust.

Essenhigh and Woodhead measured the flame speed of cork dust using two glass flammability tubes, each with a length of 5.18 m; the diameter of one tube was 50 mm and the other was 75 mm (Essenhigh and Woodhead, 1958). The cork dust was sieved into three powders of different particle size distributions. The average proximate analysis was 3.5% moisture and ash. The ultimate analysis gave an empirical formula of $C_5H_7O_2$. They estimated the stoichiometric dust concentration to be 150 g/m^3. Flame motion was recorded with a high-speed drum camera system. They performed tests with ignition at an open end, which gave steady propagation flames, and with a closed end, resulting in accelerating flames. The maximum flame speed of an accelerating flame was 20 times larger than the comparable steady propagation flame. They also observed the velocity slip in downward flame propagation with the terminal settling velocity of the dust cloud exceeding the flame speed. In the flame propagation experiments, Essenhigh and Woodhead observed a coherent flame front followed by incandescent (burning) particles. The lower limit for all three of the cork dust samples was 50 g/m^3 and the upper limit (established with one sample) was 1800 g/m^3. The maximum steady propagation flame speeds ranged from 0.60 to 0.95 m/s.

The Fire Research Station in England developed a vertical flammability tube for combustible dusts with a length of 5.2 m and a diameter of 0.254 m (Palmer and Tonkin, 1965a). Palmer and Tonkin tested six industrial dusts with the vertical flammability tube as a sort of go/no-go test (Palmer and Tonkin, 1965b). The dusts tested were methyl cellulose, manioc, sodium carboxy methyl cellulose, processed starch, polyvinylidene chloride, and calcium citrate. Dust concentrations ranged from 20 to 4000 g/m^3. Observed flame speeds ranged from 1.0 to 9.4 m/s. The results of these tests were compared and correlated with small-scale laboratory tests. They also performed some experiments with inert materials to test explosion suppression characteristics (Palmer and Tonkin, 1968).

Van Wingerden and Stavseng described flammability tube tests with four charring organic dusts: lignite, cornstarch, and maize starch (van Wingerden and Stavseng, 1996). The dimensions of the polycarbonate flammability tube were 1.6 m in length and a diameter of 0.128 m. They observed agglomeration of the unburnt dusts which affected the gravitational settling of the dust cloud during the test. They observed flame thicknesses on the order of several centimeters in some cases. In some of their tests, they observed afterburning following the passage of the flame through the dust cloud. They calculated the burning velocity from flame speed measurements. They observed a discernable trend of increasing burning velocity with increasing dust concentration for three of the four dusts (lignite being the exception). The measured burning velocities ranged from 0.25 to 0.60 m/s.

Proust and Veyssiere have described a number of investigations with starch. In one of their earlier studies, Proust and Veyssiere investigated the propagation of starch flames in a vertical tube apparatus (Proust and Veyssière, 1988). The flammability tube had a total length of 3 m and a diameter of 0.2 m. Working with a starch with a 20 μm mean diameter, they tested dust concentrations ranging from 50 to 500 g/m^3 (they estimated the stoichiometric dust concentration to be

236 g/m^3). Flame motion was recorded using high-speed photography. They used ionization probes to track the flame front, thermocouples to measure the maximum flame temperature, and photodiodes to record the light emitted by the flame. In upward propagation, the flame speed ranged from 0.46 to 0.63 m/s, with the maximum value corresponding to the approximate stoichiometric dust concentration of 250 g/m^3. The laminar burning velocity was calculated from the flame speed data and determined to be 0.20–0.25 m/s (using the method described by Andrews and Bradley, 1972a or see Lewis and von Elbe, 1987, pp. 305–309). They also reported a lean flammability limit of approximately 70 g/m^3.

Veyssière described upward and downward flame propagation behaviors in starch–air mixtures and indicated that the measured burning velocity is a function of the diameter of the flammability tube (Veyssière, 1992). Smaller tubes give lower burning velocities due to heat losses at the wall. Proust summarized a number of flame propagation studies performed in his laboratory with an emphasis on flame structure observations (Proust, 2006a,b). Proust indicates that there can be considerable uncertainty in laminar burning velocity measurements (on the order of $\pm 25\%$ to 50%). Proust indicates that typical results obtained in his laboratory using a flammability tube for the maximum burning velocities are 0.20 m/s for starch, 0.23 m/s for sulfur, and 0.47 m/s for lycopodium (Proust, 2006a).

Although it is a natural material, lycopodium (the spores of club moss) has been a popular reference material for dust flame studies because of its fairly uniform composition ($CH_{1.58}O_{0.71}$) and its nearly monodisperse particle size distribution with a particle diameter of approximately 30 μm (Amyotte and Pegg, 1989; Amyotte et al., 1990; Slatter et al., 2015). The stoichiometric dust concentration is 118 g/m^3. Lycopodium has been the subject of several flame propagation studies.

Mason and Wilson studied laminar flames of lycopodium dust–air mixtures on a stationary burner with a diameter of 10.9 mm (Mason and Wilson, 1967). They were able to obtain stable flames with dust concentrations in the range of $125–190 \text{ g/m}^3$. They recorded particle motion through the flame with photography and observed the velocity slip phenomenon. A bluish leading edge to the flame was noted along with a conical reddish yellow flame region above it. The average laminar burning velocity of their flames was 0.11–0.14 m/s.

Berlad described lycopodium dust flame propagation studies conducted in both terrestrial and microgravity (Berlad, 1981). The motivation for performing flame propagation experiments in two different levels of gravity was to isolate and eliminate the velocity slip phenomenon and its effect on flame speed measurements. A Perspex flammability tube with a length of 0.76 m and a diameter of 50 mm was used for the terrestrial gravity experiments. At a dust concentration of 130 g/m^3, a flame speed of 0.17–0.19 m/s was obtained. In microgravity, the same dust concentration gave a flame speed of 0.11 m/s.

These studies were later augmented with additional measurements to discern the flame structure of lycopodium dust flames (Joshi and Berlad, 1986). In this paper, they emphasize that the velocity slip effect is likely to be significant in the particle size range of interest for evaluating combustible dust hazards. They used a quartz tube for their flame propagation tests. The dimensions of the tube was 1.54 m in length and 96 mm in diameter. They employed a water-cooled flame holder to enable calorimetric measurements of the flame.

Han et al. also investigated vertical flame propagation in lycopodium dust clouds (Han et al., 2000, 2001). They were able to observe and record the fine structure of the dust flame. Their flammability tube had a length of 1.8 m and the square cross-section had a width of 0.15 m. Flame motion was recorded by high-speed photography. They performed flame propagation tests over the range of dust concentrations from 47 to 592 g/m^3. The maximum upward flame speed attained was 0.50 m/s at a dust concentration of 170 g/m^3. Fig. 7.14 is a plot of the laminar burning velocity as a function of dust concentration.

Upward flame propagation was accompanied by smaller flame clusters that propagated downwards. An example of the flame appearance is shown in Fig. 7.15 for the dust concentration of 592 g/m^3. The flame front assumed the characteristic parabolic shape (sometimes called "tulip" shape) with a cluster of burning particles traveling just behind the flame front. Below the flame front, a less-defined cloud of burning particles trails drifts downwards to the bottom of the flammability tube.

The authors described this as a double flame structure.

Based on an analysis of the Schlieren images and the ionization probe signals, the flame thickness of the upward flame front was estimated to be 20 mm. The maximum flame temperature measured by thermocouple spanned from 950°C to 1100°C for the dust concentration range tested. This is lower than the calculated adiabatic flame temperature of 1960°C. The investigators suggested that the

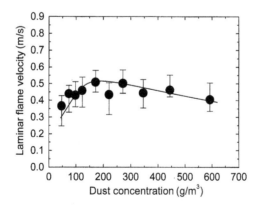

FIGURE 7.14

Laminar burning velocity for lycopodium powder as a function of dust concentration.

Reprinted with permission from Han, O.-S., Yashima, M., Matsuda, T., Matusi, H., Miyake, A., Ogawa, T., 2000. Behavior of flames propagating through lycopodium dust clouds in a vertical duct. J. Loss Prev. Process Ind. 13, 449–457.

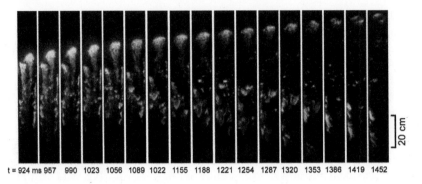

t = 924 ms 957 990 1023 1056 1089 1022 1155 1188 1221 1254 1287 1320 1353 1386 1419 1452

FIGURE 7.15

The sequence of laminar flame propagation through a lycopodium dust cloud with a concentration of 592 g/m^3.

Reprinted with permission from Han, O.-S., Yashima, M., Matsuda, T., Matusi, H., Miyake, A., Ogawa, T., 2000. Behavior of flames propagating through lycopodium dust clouds in a vertical duct. J. Loss Prev. Process Ind. 13, 449–457.

difference in flame temperature was the combined result of radiant heat loss to the tube walls and incomplete combustion.

Following this initial work, Han et al. employed a PIV system to better probe the dust flame structure. In this study, they confirmed the presence of particle clusters caused by agglomeration. The suggested mechanism for agglomeration was the formation of electrostatic charge by the collision of lycopodium particles in the fluidization chamber. In their experimental apparatus, the dust cloud was formed by elutriation of dust particles from a fluidized bed located beneath the flammability tube. The initial dust cloud in the tube was created by opening a shutter that separated the fluidized bed from the tube. Once the tube was filled with the desired dust cloud, the shutter would be closed thereby isolating the fluidized bed from the flame propagation experiment.

Han et al. argue in their paper that the particle agglomerates are present in both the unburnt and burnt mixtures. During the flame propagation process, they observed luminous flames, some of which were consistent with agglomerates and some whose size was more consistent with individual particles. They interpreted the images as indicating that some, but not all, of the agglomerates disintegrate into individual particles in the preheat zone of the flame. Their explanation is that the preheating of the particles leads to pyrolysis which, in turn, causes an outward flow of pyrolysis gases from the surfaces of the individual particles and a shrinking of the particle size. For the agglomerates that do not disintegrate, a diffusion flame surrounds the agglomerated cluster.

With the PIV instrumentation, they were able to examine more closely the motion of the lycopodium particles as they descend in the flammability tube and encounter the upward propagating flame. Far from the flame, the unburnt particles traveled downwards due to gravitational settling. As the particles in the centerline of the tube approach the flame, they are slowed by the upward flow of gas just in front of the flame. This deceleration of the unburnt particles causes an enrichment of the local concentration of fuel just in front of the flame. As the particles enter the flame front and travel through the reaction zone, the particles reverse flow direction and move upwards with the flame (Fig. 7.16).

They documented the upward propagating flame carried burning particles with it. As the flame approached an unburnt particles not located at the centerline of the tube, the flame caused a displacement of the unburnt particles toward the tube walls. These particles follow a reverse flow pattern that entrains them into the rear portion of the flame. The trajectories of individual lycopodium particles are depicted in Fig. 7.17.

An unburnt particle initially travels downwards due to gravitational settling. As it approaches the flame front, the particle decelerates and reverses direction. This reversal of particle motion as the flame approaches has also been observed in hexadecanol flames by Gao et al. (2015b). As the particle accelerates, it is heated in the flame and begins to pyrolyze, thus contributing fuel vapor to the flame. Once the pyrolyzing particle is heated to a high enough temperature, it ignites and burns.

A detailed conceptual picture of the particle—flame interactions is shown in Fig. 7.18. Note that particle agglomerates and single particles both contribute to the flame propagation process.

In this detailed conceptual mode of flame propagation, it is inferred that individual particles will heat faster than agglomerates. Individual particles ignite first and form single particle flames. The single particles form the leading edge of the

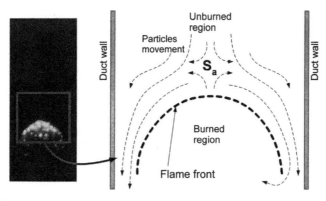

FIGURE 7.16

Dust particle motion near the flame front.

Reprinted with permission from Han, O.-U., Yashima, M., Matsuda, T., Matusi, H., Miyake, A., Ogawa, T., 2001. A study of flame propagation mechanisms in lycopodium dust clouds based on dust particles' behavior. J. Loss Prev. Process Ind. 14, 153–160.

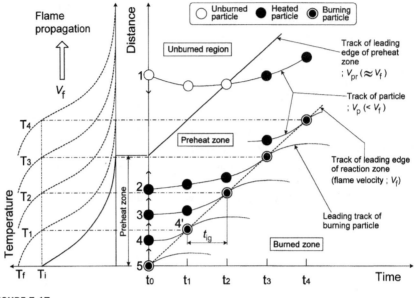

FIGURE 7.17

Lycopodium particle trajectory diagram.

Reprinted with permission from Han, O.-U., Yashima, M., Matsuda, T., Matusi, H., Miyake, A., Ogawa, T.,
2001. A study of flame propagation mechanisms in lycopodium dust clouds based on dust particles'
behavior. J. Loss Prev. Process Ind. 14, 153–160.

flame front. The heat released at the leading edge preheats and pyrolyzes the agglomerates. In the photographic record, the agglomerate exhibits a bluish flame. As the agglomerate flame progresses, some of the agglomerates disintegrate and those particles burn individually with a luminous yellow flame.

7.7.5 ALUMINUM DUST FLAMES

Aluminum is a volatile metal. It burns with a bright flame that is much hotter than typical hydrocarbon flames. Early experiments conducted by the U.S. Bureau of Mines demonstrated the feasibility of creating stable aluminum dust flames using a Bunsen burner configuration (Cassel et al., 1948). In their earliest published study, Cassel and his colleagues achieved burning velocities between 0.19 and 0.25 m/s with fuel lean flames. The burner diameter for these measurements was 25 mm. In subsequent studies, they performed additional experiments with a flat flame burner and determined that thermal radiation accounted for approximately 35% of the heat transfer required for preheating the unburnt fuel lean mixture (Cassel et al., 1957). Cassel later summarized their work on single particle studies and compared these results with burner flames (Cassel, 1964). He discussed the trend of faster burning velocities with smaller particles, faster burning velocities with increasing dust

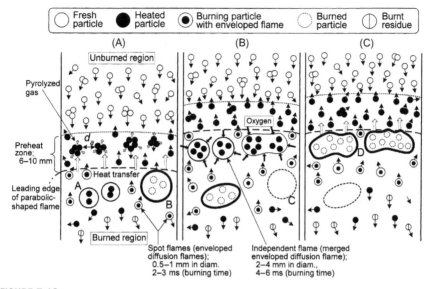

FIGURE 7.18

Detailed conceptual model of lycopodium dust flame propagation: (A) heating and ignition of single particles; (B) heating and ignition of particle agglomerates; and (C) final stage of pyrolysis and combustion of particles.

Reprinted with permission from Han, O.-U., Yashima, M., Matsuda, T., Matusi, H., Miyake, A., Ogawa, T., 2001. A study of flame propagation mechanisms in lycopodium dust clouds based on dust particles' behavior. J. Loss Prev. Process Ind. 14, 153–160.

concentration (fuel lean mixtures), and the direct proportionality of the burning velocity on the square root of the oxygen concentration.

Alekseev and Sudakova conducted spherical flame propagation tests with a number of metal powders including aluminum (Alekseev and Sudakova, 1983). They tested dust concentrations up to 2000 g/m^3 which for aluminum corresponds to an equivalence ratio of approximately 6.5 (stoichiometric dust concentration = 307 g/m^3). Over this range of aluminum dust concentrations, they measured flame speeds as large as 2.5 m/s. The flame speed was initially a very strong function of dust concentration up to the stoichiometric limit, followed by a weaker dependence at fuel rich conditions. They also noted that the completeness of combustion was concentration dependent: combustion was 90% complete at a dust concentration of 100 g/m^3, but only 25% complete at 1000 g/m^3.

Goroshin and his colleagues performed a number of flame propagation experiments with aluminum dusts (Goroshin et al., 1996a,b, 2000). They performed their flame propagation experiments in a 50-mm-diameter Bunsen burner. They measured the lean limit, quenching distance, and laminar burning velocity of aluminum powder with a Sauter mean diameter of approximately 5 μm. The lean limit was 150 g/m^3 and the quenching distance was approximately 5 mm over the concentration range from 150 to 600 g/m³. Over the same dust concentration range, they found the laminar burning velocity to be approximately 0.20 m/s. This lack of dependence of the burning velocity on the dust concentration is contrary to the results of Cassel and Ballal (Cassel, 1964; Ballal, 1983b). This discrepancy is yet to be resolved.

An additional experimental technique used by these same investigators was to vary the oxidizer mixture to determine its effect on burning velocity. Increasing the oxygen content in a nitrogen—oxygen mixture led to increasing values of the burning velocity (Fig. 7.19).

They also investigated the effect of oxygen diffusivity on the burning velocity by changing the carrier gas from nitrogen to argon or helium. The measurements are summarized in Fig. 7.20.

In theory, the burning velocity in helium should be 3.9 times larger than the burning velocity in argon. The experimental result gave a ratio of 3.2, the difference being attributed by the authors to the effect of dissimilar Lewis numbers. From Fig. 7.21, it can be seen that the difference between nitrogen and helium as the carrier gas is evident in the shape of the premixed flame.

Goroshin et al. have also examined the structure of aluminum dust flames using emission spectroscopy (Mamen et al., 2005; Goroshin et al., 2007). This technique has the potential to probe the interior of the dust flame to measure the temperature profile of the gas and particle phases. Detailed measurements of this sort can be a valuable means of testing more detailed models of dust flame kinetics. But the optical thickness of the flame at higher (fuel rich) dust concentrations makes the application of the technique difficult. Flame propagation tests have also been

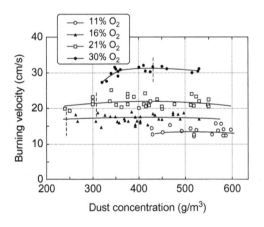

FIGURE 7.19

Laminar burning velocity of aluminum dust flames at different dust concentrations and with different oxygen concentrations and nitrogen as the carrier gas.

Reprinted with permission from Goroshin, S., Fomenko, I., Lee, J.H.S., 1996b. Burning velocities in fuel-rich aluminium dust clouds. Twenty-Sixth Symposium (International) on Combustion. The Combustion Institute, Elsevier, pp. 1961–1967.

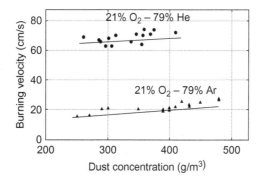

FIGURE 7.20

Laminar burning velocity of aluminum in 21% oxygen and two different carrier gases, helium and argon.

Reprinted with permission from Goroshin, S., Fomenko, I., Lee, J.H.S., 1996b. Burning velocities in fuel-rich aluminium dust clouds. Twenty-Sixth Symposium (International) on Combustion. The Combustion Institute, Elsevier, pp. 1961–1967.

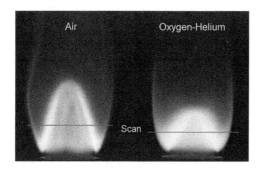

FIGURE 7.21

Premixed aluminum dust flames stabilized on a Bunsen burner apparatus with different oxidizing mixtures: air (oxygen—nitrogen) and oxygen—helium.

Reprinted with permission from Goroshin et al., 2007.

performed at different oxygen concentrations as a test of the flame propagation mechanism (Wright et al., 2015). The essential aspect of the test was to discern the effect of changing the oxygen concentration on the observed flame speed. With a continuous flame zone, the flame speed should be proportional to the square root of the oxygen concentration, and with a discrete flame zone the flame speed should be independent of the oxygen concentration. Due to the scatter in the flame speed measurements, the results reported in this study were inconclusive.

Risha et al. investigated the flame propagation behavior of micron-sized and nano-sized aluminum dust clouds (Risha et al., 2005). Using a 17.3-mm-diameter Bunsen burner arrangement, they measured laminar flame speeds of aluminum dust flames with a range of equivalence ratios from 0.81 to 1.62. The particle size was reported as 5−8 μm and the measured flame speeds ranged from roughly 0.10−0.15 m/s. These values are lower than those reported by others. They then tested blends of this aluminum powder with nano-aluminum powder (particle diameter 100 nm) and found that increasing the percentage of nano-aluminum increased the flame speed. The maximum effect was observed with 30% nano-aluminum blend yielding a flame speed of roughly 0.33 m/s.

Bocanegra and his colleagues also examined the flame propagation behavior of aluminum dust cloud composed of particles in the micro-size range (6 μm) or nano-size range (250 nm) (Bocanegra et al., 2009). They employed a quartz flammability tube with an diameter of 14 mm and a length of 180 mm. They measured the laminar flame speed of upward propagating flames in the dust concentration range of 200−2000 g/m^3. The micro-powder attained a maximum flame speed of approximately 0.30 m/s at a dust concentration of 600 g/m^3 and the nano-powder attained a maximum flame speed of approximately 45 m/s at a dust concentration of 1500 g/m^3.

Dreizin and his colleagues have conducted several investigations aimed at understanding the role of phase equilibrium in the formation of combustion products in aluminum dust flames (Trunov et al., 2005a,b; Corcoran et al., 2013). In particular, they have examined the role of the oxide layer on the unburnt aluminum powder to better understand its effect on dust cloud ignition. They have found that the ignition temperature of aluminum particles depends on the particle size distribution and the thermal history of the particles. At the current time, it is difficult to see how to incorporate their results into flame propagation studies, but their work is significant because it reveals some of the difficulties one encounters in trying to characterize the combustion characteristics of real combustible dusts.

Sun et al. investigated the structure of aluminum dust flames using a flammability chamber (Sun et al., 2006b). They recorded flame motion with high-speed photography and Schlieren photography. They tested fuel rich dust concentrations of 420, 490, and 600 g/m^3. They estimated that the flame front had a preheating zone width of 3 mm and a reaction zone of 5−7 mm. The burning dust cloud had a granular appearance suggesting that the diffusion flames surrounding individual burning particles were not completely merged into a continuous flame zone. Fig. 7.22 shows the flame motion as recorded by high-speed photography and Schlieren photography.

There was also evidence of asymmetric diffusion flames surrounding some particles suggesting the jetting and fragmenting behaviors observed in single particle combustion.

7.7.6 IRON DUST FLAMES

Iron is a nonvolatile metallic solid. Iron dust burns exclusively as a heterogeneous flame and thus, it can be expected that it will not form a continuous flame zone. This is because the adiabatic flame temperature of iron combustion in air is less than the boiling point of iron (2285 K flame temperature vs 3023 K boiling point).

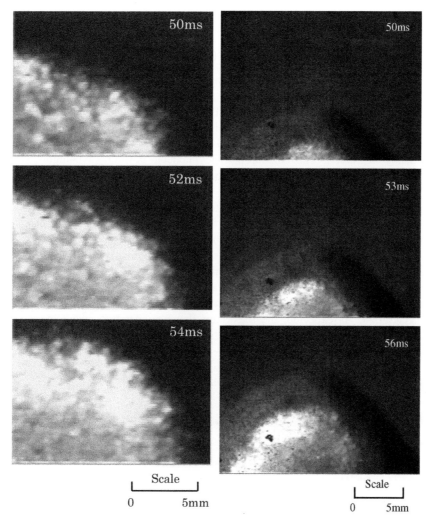

FIGURE 7.22

Aluminum dust flame propagation recorded by high-speed photography (dust concentration of 420 g/m^3) and Schlieren photography (dust concentration 490 g/m^3). Times indicated in images is the elapsed time from ignition and length scale indicated below image.

Reprinted with permission from Sun, J.-H., Dobashi, R., Hirano, T., 2006b. Structure of flames propagating through aluminum particles cloud and combustion process of particles. J. Loss Prev. Process Ind. 19, 769–773.

Thus the iron oxide combustion product grows as a layer surrounding the shrinking iron core (see Chapter 6).

Researchers at the University of Tokyo have conducted a number of studies on iron powders to better understand the multiphase flow effects in dust flame propagation (Sun et al., 1998, 2000, 2001, 2003, 2006a). They have worked with two kinds of iron powders, one with a size distribution between 1 and 3 μm and another with a distribution between 2 and 4.5 μm. They used a flammability chamber to create spherically expanding flames. They observed a luminous zone of 3−5 mm in width with a granular appearance which they interpret to be the heterogeneous flame zone. Fig. 7.23 illustrates the motion of the iron dust flame.

Testing flame propagation over a dust concentration range from approximately 50 to 2000 g/m^3, they found that the flame speed increased from the lean limit to a maximum that was beyond the stoichiometric limit (about 652 g/m^3) and then gradually decreased. The smaller particle diameter powder gave higher flame speeds. Fig. 7.24 is a plot of the flame speed data.

The experimental observations were found to be in alignment with a single particle combustion model based on iron oxidation kinetics (Sun et al., 2000).

The flame temperature was found to follow the same trend as the flame speed with respect to its dependence on the dust concentration (Sun et al., 2001). The axial temperature profile in the gas phase was measured with a 20-μm-diameter thermocouple (platinum/13% platinum−rhodium junction). The temperature measurements were corrected for the temperature lag due to finite heat transfer rates. The maximum temperature in the flame was observed to be a function of the dust concentration. As expected, the maximum flame temperatures were lower than

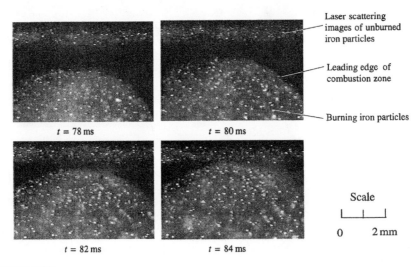

FIGURE 7.23

High-speed photographs of flame propagation in an iron dust cloud with a dust concentration of 1050 g/m^3. Time is elapsed time from ignition.

Reprinted with permission from Sun, J.-H., Dobashi, R., Hirano, T., 1998. Structure of flames propagating through metal particle clouds and behavior of particles. In: Twenty-Seventh Symposium (International) on Combustion. The Combustion Institute, Elsevier, pp. 2405−2411.

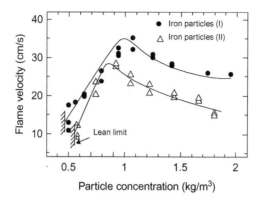

FIGURE 7.24

Flame speed for two different iron powders as a function of dust concentration.

Reprinted with permission from Sun, J.-H., Dobashi, R., Hirano, T., 1998. Structure of flames propagating through metal particle clouds and behavior of particles. In: Twenty-Seventh Symposium (International) on Combustion. The Combustion Institute, Elsevier, pp. 2405–2411.

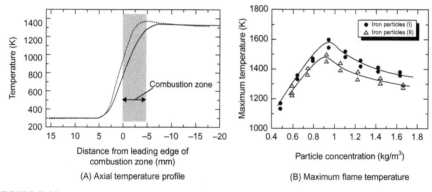

FIGURE 7.25

Iron dust flame temperature measurements. The figure on the left is the axial temperature profile measurement (solid line) and the corrected temperature measurement (dashed line). The figure on the right is the maximum flame temperature.

Reprinted with permission from Sun, J.-H., Dobashi, R., Hirano, T., 2001. Temperature profile across the combustion zone propagating through an iron particle cloud. J. Loss Prev. Process Ind. 14, 463–467.

the calculated adiabatic flame temperature. The temperature data are plotted in Fig. 7.25.

Sun and his colleagues also observed the velocity slip effect which gave rise to a similar family of particle trajectories as described for the lycopodium studies described earlier in this chapter (Sun et al., 2003, 2006a). A significant advantage with iron dust flames is that the particles can be observed as they pass through the flame zone and beyond into the postflame region. The particles never disappear: as they travel through the reaction zone, the iron particles are transformed into iron oxide particles by the heterogeneous oxidation reaction at the surface of the particle. This feature of iron particle combustion permitted the measurement of both particle velocity and concentration in a way that was not possible with particles that burn homogeneously.

Fig. 7.26 is a plot of the gas and particle velocity axial profiles as the flame advances horizontally from left to right. Since the flame propagates horizontally, the effect of gravitational settling is neglected and, therefore, this figure shows only the effect of velocity slip. The velocity profile for a premixed gaseous flame is also shown for comparison.

Consider the gas velocity in the premixed flame on the right-hand side of Fig. 7.26. As the flame advances from left to right, it pushes the unburnt gas mixture in front of it and the gas accelerates in the same direction as flame travel. The unburnt gas velocity attains its maximum value when the flame overtakes it. After passing through the flame, the burnt gas is accelerated backwards away from the flame. The same sequence of events occurs for the dust flame, but the changing gas velocity is accompanied by a changing particle velocity.

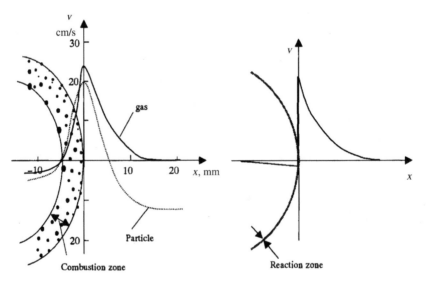

FIGURE 7.26

Gas and particle velocity profiles in an iron dust flame (left-hand side) and a premixed gaseous flame (right-hand side).

Reprinted with permission from Sun, J.-H., Dobashi, R., Hirano, T., 1998. Structure of flames propagating through metal particle clouds and behavior of particles. In: Twenty-Seventh Symposium (International) on Combustion. The Combustion Institute, Elsevier, pp. 2405–2411.

By continuity, if there are changes in the gas and particle velocity axial profiles during the passage of the flame, there should also be changes in the dust concentration. Using laser light scattering, this variation in dust concentration was indeed observed. Fig. 7.27 shows the axial profiles of the particle concentration and velocity in an iron dust flame. Dust flame propagation experiments were performed with an iron dust concentration of 1050 g/m^3. The particle size distribution ranged from 1 to 5 µm. However, upon sampling of the dust cloud and inspection by scanning electron microscope, the investigators observed agglomeration of the iron particles with the characteristic size of the agglomerates on the order of 10 µm or larger.

The measurements show an enrichment of the unburnt particle concentration as the flame approaches. The ratio of the particle concentration at the entrance into the combustion zone compared to the free stream concentration is approximately 2.6. This enrichment factor has a very practical consequence: in the determination of the lower flammability limit for dust flame propagation, the nominal dust concentration is not the actual dust concentration that the flame "sees."

Sun et al. also investigated the effect of gravitational settling and velocity slip on iron dust flame propagation (Sun et al., 2006a). Fig. 7.28 compares the particle velocity profile for upward and downward flame propagations. The experiments were conducted at an iron dust concentration of 1050 g/m^3. The particle size distribution ranged from 1.5 to 4.5 µm, but there were indications of agglomeration in the unburnt mixture. The terminal settling velocity was measured as approximately 0.10 m/s. In upward flame propagation, the particles were observed to fall toward the approaching flame at a steady velocity and then, at a distance of approximately 10 mm from the flame, the particles decelerated, stopped, and then

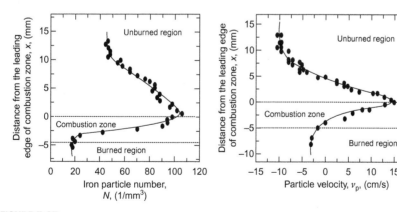

FIGURE 7.27

Comparison of the axial profile measurements of the iron particle concentration and particle velocity. The iron dust concentration was 1050 g/m^3.

Reprinted with permission from Sun, J.-H., Dobashi, R., Hirano, T., 2003. Concentration profile of particles across a flame propagating through an iron particle cloud. Combust. Flame 134, 381–387.

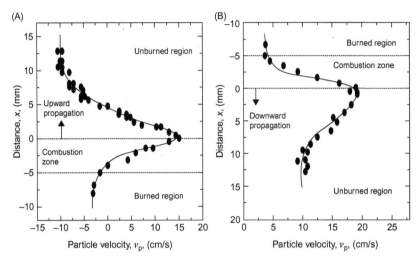

FIGURE 7.28

Axial particle velocity measurements for upward flame propagation (left-hand side) and downward flame propagation (right-hand side).

Reprinted with permission from Sun, J.-H., Dobashi, R., Hirano, T., 2006a. Velocity and number density profiles of particles across upward and downward flame propagating through iron particle clouds. J. Loss Prev. Process Ind. 19, 135–141.

accelerated in the same direction as the flame propagation. The change in particle motion occurred at approximately 4.5 mm from the flame. The particles attained their maximum velocity at the leading edge of the flame. The particles then decelerated as they passed through the flame and again reversed their direction of travel and drifted downwards.

In downwards flame propagation, the particles moved in the same direction as the flame throughout the flame propagation event. The unburnt particles settled at their terminal velocity and then accelerated with the approach of the flame. The particles attained their maximum velocity upon entering the leading edge of the flame. The particles then decelerated as they passed through the combustion zone of the flame and drifted downwards.

The axial particle velocity behavior caused an enrichment of the particle concentration as the unburnt dust cloud approached the flame. Fig. 7.29 shows the particle concentration behavior for both upward and downward flame propagations. The particle concentrations were normalized by the free stream (unburnt) dust concentration. For upward flame propagation, the maximum normalized particle concentration was approximately 3.5 and for downward flame propagation it was approximately 2.3.

The enrichment of the local dust concentration means that the actual dust concentration entering the flame was higher than the nominal dust concentration. Furthermore, the enrichment factor was greater for upward propagation than it was

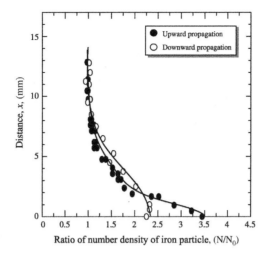

FIGURE 7.29

Axial profile of the particle concentration enrichment for upward and downward flame propagations. Concentrations are normalized by the free stream (unburnt) particle concentration.

Reprinted with permission from Sun, J.-H., Dobashi, R., Hirano, T., 2006a. Velocity and number density profiles of particles across upward and downward flame propagating through iron particle clouds. J. Loss Prev. Process Ind. 19, 135–141.

for downward propagation. These experiments were conducted at fuel rich conditions (equivalence ratio $\Phi \cong 1.6$), but the trend in concentration enrichment should hold as well for fuel lean conditions including the lean limit. These observations provide empirical support for the simple velocity slip model for the lower flammability limit derived earlier in this chapter. Given the degrees of enrichment observed in these studies, one would expect that a lean limit determined by upward flame propagation would be *lower* than a lean limit determined by downward propagation.

Goroshin and his colleagues have also conducted a number of studies on iron dust flame propagation in a microgravity environment (Tang et al., 2009, 2011; Goroshin et al., 2011). They have tested flame propagation in the dust concentration range from 900 to 1200 g/m³ or an equivalence ratio from 1.43 to 1.90. The increasing granularity of the flame with increasing particle size is evident from Fig. 7.30.

They tested a range of particle size distributions and demonstrated, like Sun et al., that the flame speed decreases with increasing particle diameter. Furthermore, as in their investigations with aluminum dust, these investigators explored the effect of different oxidizing mixtures by testing with air and then substituting argon and helium for the carrier gas. The flame speed measurements for different carrier gases are shown in Fig. 7.31.

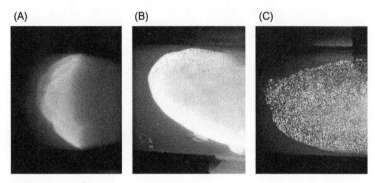

FIGURE 7.30

Iron dust flames propagating from right to left. Average particle diameter increases from
left to right (A < B < C).

*Reprinted with permission from Tang, F.-D., Goroshin, S., Higgins, A., Lee, J., 2009. Flame propagation and
quenching in iron dust clouds. Proc. Combust. Inst. 32, 1905–1912.*

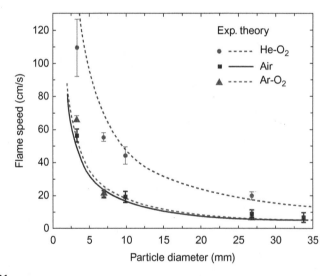

FIGURE 7.31

Flame speed as a function of particle diameter for different carrier gases.

*Reprinted with permission from Goroshin, S., Tang, F.-D., Higgins, A.J., Lee, J.H.S., 2011 Laminar dust
flames in a reduced-gravity environment. Acta. Astronaut. 68, 656–666.*

Since a heterogeneous dust flame depends entirely on the diffusion of oxygen
to the surface of the burning particle to promote combustion, the trend of increasing flame speed with increasing oxygen diffusivity is reasonable.

7.7.7 SUMMARY ON FLAME PROPAGATION STUDIES

It should be apparent that dust flame propagation experiments are not easy to perform. Each type of apparatus has its merits and its deficiencies, and each dust has its own individual peculiarities. Stationary burners offer the most control over the development of a premixed dust flame, but do not replicate a practical combustible dust hazard scenario. Vertical flammability tubes have the potential to simulate a hazard scenario, but suffer from velocity slip due to gravitational settling. An additional problem is that flame propagation results may be scale dependent, eg, small flames can experience greater heat loss than large flames.

Despite these practical challenges, some generalizations are possible. Laminar flame characteristics are dependent on the fuel volatility, equivalence ratio (dust concentration), and mean particle diameter. For a given dust concentration and monodisperse particle size, the laminar burning velocity is directly related to the fuel volatility. For fixed fuel volatility and particle size, the burning velocity increases with dust concentration from the lean limit to a maximum value in the fuel rich region that may then either decrease or taper off asymptotically. For fixed fuel volatility and dust concentration, the burning velocity increases with decreasing particle diameter.

A polydisperse particle population exhibits more complex behavior than a monodisperse population. In a qualitative sense, the role of the particle size distribution can be summarized by dividing the distribution into fine particles and larger particles. The size criterion for fine particles cannot be defined universally, but as fuel volatility increases, the size limit for fine particles increases.

If the fuel particle burns by homogeneous chemical reactions, then the flame structure tends to be a continuous flame zone at the flame front followed by a heterogeneous zone burning particles. The fine particles contribute the fuel vapor for the continuous flame zone with the larger particles lagging behind the continuous zone and burning individually or in clusters.

In vertical flammability tube experiments, the effect of gravitational settling on dust flame propagation has been demonstrated. The dust flame has a tendency to accelerate the fuel particles and enrich the dust concentration at its leading edge. Thus, the actual lean limit concentration may be less than the nominal dust concentration. This is also one possible explanation for why lean limits expressed as an equivalence ratio are lower for combustible dusts than they are for premixed flammable gases.

7.8 COMBUSTIBLE DUST FLASH FIRES

In Chapter 1 I introduced the hazard scenario for a combustible dust flash fire. A flash fire is a transient event with a flame sweeping through a combustible dust cloud without the generation of an overpressure. A number of industrial activities that could lead to a flash fire are described in the CCPS book on safe handling of powders and bulk solids (CCPS/AIChE 2005, pp. 122—124). Flash fires are especially heinous because of the nature of the injuries they cause. In this section, I will first discuss the human injury potential of flash fires illustrates these through two published accident case studies. The characteristics of a flash fire are the next topic followed by a survey of flash fire studies from the peer-reviewed scientific literature.

7.8.1 HUMAN INJURY POTENTIAL AND ACCIDENT CASE STUDIES

Flash fires cause burn injuries. A burn injury is defined as an area of tissue damage caused by the effects of heat (Cooper, 2006, Chapter 9). In a combustible dust flash fire, the source of heat can be flame contact, contact with burning particles, or thermal radiation (Grimard and Potter, 2011). The extent or depth of tissue damage is divided into four categories: first and second degree burns which describe partial skin thickness damage, third degree burns which are full skin thickness damage, and fourth degree burns which are defined by heat damage or charring of underlying tissue (NFPA 921, 2014, p. 250).

Medical case studies on the treatment of dust explosion burn injuries have appeared in the medical literature (Russell et al., 1980; Beausang and Herbert, 1994; Still, Jr. et al., 1996). As reported in these medical references, the treatment is typical of other types of burn injuries incurred by fire exposure.

Burn injuries can be caused by thermal conduction, convection, or radiation (NFPA 921, 2014, p. 245). As a point of reference, direct skin contact with a brass block heated to 60°C caused pain in 1 s, a partial thickness burn in 10 s, and a full thickness burn in 100 s. When exposed to convection (hot air), the onset of pain and burn injury occurred at a temperature above 120°C. Radiant heat flux can cause injuries at the following levels: 1 kW/m^2 is the threshold for pain, 4 kW/m^2 can cause partial thickness burns, and full thickness burns can be caused with only a brief exposure to 20 kW/m^2. The degree of injury depends on the thermal exposure which is usually characterized as either a temperature—time or heat flux—time correlation (Purser, 2002). Engineering correlations are available to assist the safety professional in the prediction of burn injuries from thermal radiation (Prugh, 1994; Purser, 2002).

Two accident investigation reports published by the U.S. Chemical Safety Board (CSB) highlight the hazards of flash fires. The first report is actually about three separate incidents that occurred at a manufacturing plant which produced atomized steel and iron powders (CSB, 2011). Two of these incidents specifically involved flash fires with iron dust. In one of these incidents, a worker was performing maintenance work on a hydrogen furnace while standing on a ladder. The worker was using a hammer and the action of the hammer dislodged fugitive iron dust that had accumulated on the upper surfaces of I-beams. The iron dust engulfed the worker and ignited causing a flash fire. The worker received first and second degree burns from the fire, but recovered from his injuries. The second iron dust flash fire incident in the report resulted in two fatalities.

The second accident investigation report concerned a newspaper ink manufacturing facility (CSB, 2013). Depending on the specific ink recipe being produced, the facility handled carbon black, gilsonite (naturally occurring hydrocarbon mineral), and petroleum distillate oils (combustible liquids). The incident occurred shortly after installing a new air pollution control system. During production, a fire was discovered and several workers went to investigate. A deflagration occurred within the exhaust vents and a flash fire/fireball vented from a feed hopper located at ground level. Several workers were burned by the flash fire. It was not possible to determine the exact chemical composition of the fuel mixture (carbon black, gilsonite, oil) involved in the flash fire. As an aside, this case study reminds us that it is important to recognize that the installation of air pollution control equipment often introduces new fire and explosion hazards that must be managed (Ogle et al., 2005).

Both reports address many safety issues and is worth a careful reading by anyone engaged in the prevention and control of combustible dust hazards. An important point to recognize is that fact finding in accident investigations is rarely easy and it often happens that more than one opinion about the cause of the accidents will emerge. Thus, any accident report, including those produced by government agencies, must be read with the understanding that other scientifically defensible opinions may exist. However, what can be learned without question from these two reports is this: (1) despite the twice-demonstrated flash fire hazard of the iron dust described in the first CSB report, upon standard explosibility testing, the iron dust presented as only a marginal dust explosion hazard and (2) in a real industrial setting, it may be difficult to determine the exact chemical composition of the combustible dust hazards.

7.8.2 SURVEY OF FLASH FIRE STUDIES

There is at present no standardized laboratory test that characterizes the hazards of a combustible dust flash fire. One objective of developing a standardized test would be to be able to rank combustible dusts in terms of the magnitude of flash fire hazard. Criteria for ranking flash fire hazards would include the maximum fireball size, the duration of the fireball, the peak (spatially averaged) flame temperature, and the magnitude of the peak radiant heat flux.

Stern and his colleagues have demonstrated that dust explosibility parameters do not accurately characterize flash fire hazards (Stern et al., 2015a). They selected three organic fuels: a food additive (84% volatiles, 22% less than 75 μm by sieving), a concrete additive (4.4% volatiles, 27% less than 75 μm by sieving), and a flame retardant material (a chlorinated hydrocarbon with 82% volatiles, 38% less than 75 μm by sieving). The materials were tested for the standard dust explosibility parameters in a 20-L sphere (K_{st} and P_{max}) and the K_{st} values were found to be quite similar, ranging from 70 to 85 bar-m/s. They designed a constant pressure deflagration apparatus with a volume of 20 L for simulating a flash fire with a combustible dusts. The dust was dispersed vertically and ignited. The materials were tested at fuel rich nominal dust concentrations corresponding to the maximum K_{st} values measured by the 20-L sphere. The food additive and concrete additive were both tested in the nominal dust concentration range of $125-1000$ g/m^3. The fire retardant material was tested at a dust concentration range of $1000-2000$ g/m^3. They recorded the flame motion with both still and video photography. They found that under the same test conditions, the materials gave radically different fireball volumes despite the similar K_{st} values. They found that the fireball volume correlated better with the volatiles content of the materials.

They then added additional instrumentation to the test setup: a heat flux gauge, Type K thermocouples (36 gauge or 127 μm), and an infrared video camera (Stern et al., 2015b). They performed fireball tests with aluminum powder (100% less than 75 μm) and nondairy creamer (largely composed of sucrose sugar and 22% less than 75 μm). They also performed calibration runs with methane. The dusts were tested at fuel rich nominal concentrations of 500 and 1000 g/m^3. The heat flux gauge and infrared video camera gave better indications of the fireball properties than the thermocouples. The primary conclusion from this work was that the instrumentation was able to demonstrate that the peak heat flux and temperatures increased directly with dust loading. The authors identified a number of potential improvements for future work.

Skjold et al. devised an apparatus for conducting constant pressure deflagration tests with balloons (Skjold et al., 2013). Their concept was inspired by the soap bubble method for measuring burning velocity of a premixed gas (Strehlow and Stuart, 1953; Lewis and von Elbe, 1987, pp. 216−218; Strehlow, 1984, pp. 256−258). Their procedure was to preinflate the balloon, inject the combustible dust, and ignite the mixture at the center of the balloon. The initial size of the filled balloon was estimated as 0.5 m^3 (corresponding to a diameter of approximately 0.98 m). This gave them a spherical dust cloud with a constant pressure flame. The flame motion was tracked by a high-speed video camera. Tests were conducted at fuel rich dust concentrations with lycopodium (140, 210, and 285 g/m^3) and maize starch (425 and 500 g/m^3). From their video data they were able to estimate the burning velocity and flame speed for the cloud.

Julien and his colleagues at McGill University devised a similar balloon test and compared their flame speed measurements with those obtained from free field dispersion tests (Julien et al., 2015a,b). Their balloon tests followed a similar procedure as Skjold et al. tests but using balloons with a volume of 14 L (0.3 m diameter). They used aluminum powder for all of their tests. For the laboratory tests, the Sauter mean diameter was equal to 5.6 μm. They tested fuel rich dust concentrations (approximate equivalence ratios from 1.5 to 2.0) and measured flame speeds of 2.0−2.5 m/s, from which they calculated burning velocities ranging from 0.20 to 0.23 m/s.

The field tests involved the vertical dispersal of dust from a canister. The apparatus could disperse up to 1 kg of dust. For the field tests they used two aluminum powders, one with a median diameter of 8.0 μm and the other with a median diameter of 12.0 μm. The dispersal of the dust formed a turbulent jet which was then ignited. They measured the flame speed by the vertical advancement of the flame and found a range of values from 10 to 14 m/s. They also conducted tests with varying concentrations of oxygen and different oxygen–carrier gas blends using helium and argon. With these various oxidizer mixtures, they observed various types of flame instability that go beyond the range of our interest in combustible dust hazards.

As discussed in Section 7.7.3, Cao et al. performed a series of flash fire tests with coal using a vertical flammability tube (Cao et al., 2014). The surface temperature of the propagating coal dust flames was measured using an infrared camera and the diameter of the fireball was measured by the high-speed video camera. However, the results thus obtained are specific to their apparatus. It is unclear at this time how to generalize these results.

A series of tests performed by investigators at the Health and Safety Executive in the United Kingdom examined the thermal radiation emitted by fireballs from dust explosion venting (Holbrow et al., 2000). This is a particularly useful scenario to consider as it relates directly to evaluating the potential hazard to personnel in an industrial setting. They tested six different combustible dusts: coal, cornflour, toner, polyethylene, anthraquinone, and aluminum. In addition to the thermal radiation measurements obtained by infrared video, they also examined the thermal response of 15 types of target materials including garment fabrics. Their findings, however, reinforce the theme of this section that the magnitude of the fireball hazard is very dependent on how the fireball is created. Holbrow et al. conclude that their thermal radiation measurements do not correlate well with the K_{st} parameter. They observed that larger explosion vents tended to create larger fireballs. They also noted that the fireball duration was typically too brief to ignite fabric samples unless the target fabric was very close to the fireball. This test program was quite extensive and the interested reader is encouraged to consult the original paper for more details.

7.8.3 A KINEMATIC MODEL FOR THE PROPAGATION OF A FLASH FIRE

In Section 3.8, I presented a thermodynamic model to differentiate a flash fire from a confined deflagration. The essential conclusion is that the volume of the dispersed dust cloud must be a small fraction of the enclosure volume (in the absence of any venting capability in the enclosure, perhaps $\alpha_{fill} < 1\%$). To relate the behavior of a flash fire to the burning velocity, we must consider the kinematics of the flame motion. In an uncontrolled setting, a combustible dust cloud could take any arbitrary shape. The size of the dust cloud will depend on the mass of combustible dust dispersed. The propagation behavior of the flame will depend on where ignition occurs. It should be clear that this problem is not easily addressed in all its generality. Therefore, we will consider a simple model that relates the flame speed to the burning velocity through a thermodynamic factor: the ratio of the burnt gas density to the unburnt gas density. This is the balloon model for a constant pressure deflagration (Skjold et al., 2013; Julien et al., 2015a).

The objective is to derive a relation that connects the flame speed to the laminar burning velocity. Assume a spherical dust cloud with an initial radius $r_{c,0}$. A flame is ignited at the center of the cloud at time zero and propagates radially outwards. The flame radius is denoted r_f. The flame propagates with a constant laminar burning velocity S_u and a constant flame speed S_b. The flame is idealized as a surface of discontinuity. The physical situation is depicted in Fig. 7.32.

The derivation begins with the geometric relation that partitions the dust cloud into two regions: the burnt gas and the unburnt mixture (Strehlow and Stuart, 1953; Skjold et al., 2013). It is anticipated that as the flame advances the boundary of the dust cloud r_c will expand to maintain the constant pressure condition. Thus, the cloud boundary is subject to the condition $r_c \geq r_{c,0}$.

$$V_b = \frac{4\pi}{3}\left[r_f^3 - \left(r_c^3 - r_{c,0}^3 \right) \right] \tag{7.61}$$

The flame speed is the time rate of change of the flame position with respect to stationary coordinates. The assumption that the pressure is constant means that there is no flame acceleration. Therefore, it is inferred that the flame speed is

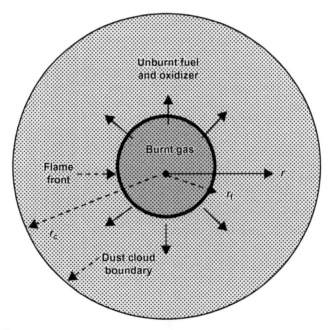

FIGURE 7.32

Constant pressure deflagration model for a combustible dust flash fire.

constant. The flame speed can be expressed as a differential equation that is readily integrated with the initial condition that $r_f(t = 0) = 0$.

$$S_b = \frac{dr_f}{dt} \to \frac{r_f}{t} \tag{7.62}$$

The jump condition across the flame gives the relation between the burning
velocity and the flame speed:

$$\rho_b S_b = \rho_u S_u \to S_b = S_u \left(\frac{\rho_u}{\rho_b}\right) = S_u \left(\frac{T_b}{T_u}\right)\left(\frac{\mathcal{M}_u}{\mathcal{M}_b}\right) \tag{7.63}$$

where it is assumed that the ideal gas equation of state applies. Theoretically, the
burnt gas temperature is the adiabatic isobaric flame temperature. In practice, the
actual burnt gas temperature is substantially less than the adiabatic value. Recall
from Chapter 3 that practical values for the density ratio in an isobaric flame are
in the range $(\rho_u/\rho_b) \cong 5 - 6$.

The volume of the burnt mixture can be calculated from the burning velocity.

$$V_b = S_u \int_0^t 4\pi r_f^2 dt = 4\pi S_u \int_0^t S_b^2 \, t^2 dt = \left(\frac{4\pi}{3}\right) S_u \, S_b^2 \, t^3 \tag{7.64}$$

Eliminating the burnt gas volume V_b in Eqs. (7.61) and (7.64), and making
use of Eq. (7.62) yields the expression linking the burning velocity and the flame
speed with the radial positions of the flame and the cloud boundary:

$$S_u = S_b \left[1 - \left(\frac{r_c^3 - r_{c,0}^3}{r_f^3}\right)\right] \tag{7.65}$$

Thus, by tracking the radial position of the cloud boundary and the flame, and
determining the flame speed from the slope of the flame position data, one can calculate the burning velocity. The application of this model to real data is not trivial.
One must be on guard to prevent numerical errors in the radial measurements from
creeping into the calculations as the value is cubed, thereby magnifying the error.

Skjold and his colleagues presented data in their paper that corroborates the
basic physics of this model (Skjold et al., 2013). Fig. 7.33 is a sequence of photographs taken during a constant pressure deflagration of lycopodium dust in a balloon. Two features are immediately apparent. First, the balloon expanded during
the course of the deflagration which is consistent with the assumption of constant
pressure within the balloon. Second, the flame expanded at a faster rate than the
balloon boundary which allowed the flame to intercept the balloon boundary
causing the balloon failure.

Fig. 7.34 is a plot of the radius versus time data.

A comparison of the slopes of the flame radius and the balloon radius in
Fig. 7.34 clearly indicates that the flame travels at a faster speed. The linearity of
the balloon and flame radii data is a further corroboration of the assumptions of
the model.

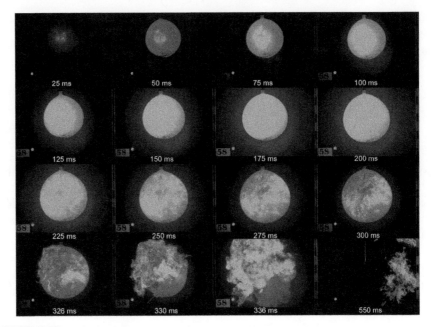

FIGURE 7.33

Time sequence of photographs of constant pressure deflagration of lycopodium dust in a balloon. Dust concentration 285 g/m³.

Reprinted with permission from Skjold, T., Olsen, K.L., Castellanos, D., 2013. A constant pressure dust explosion experiment. J. Loss Prev. Process Ind. 26, 562–570.

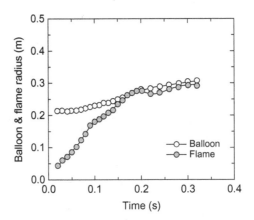

FIGURE 7.34

Plot of balloon radius and flame radius measurements versus time.

Reprinted with permission from Skjold, T., Olsen, K.L., Castellanos, D., 2013. A constant pressure dust explosion experiment. J. Loss Prev. Process Ind. 26, 562–570.

EXAMPLE 7.8

Use the kinematic model for a constant pressure dust deflagration and the data from Fig. 7.34 to estimate the burning velocity for lycopodium at a dust concentration of 285 g/m³. *Hint*: Use the data corresponding to the time at which the flame just contacts the balloon surface. Designate the time of contact as t_c.

Solution

Begin by writing down the final form of the kinematic model that relates the burning velocity to the flame speed.

$$S_u = S_b \left[1 - \left(\frac{r_c^3 - r_{c,0}^3}{r_f^3} \right) \right]$$

At the time of flame contact, $r_f = r_c$. Substitute this condition and use the relation that $S_b = r_f / t$.

$$S_u = \frac{r_f}{t_c} \left[1 - \left(\frac{r_f^3 - r_{c,0}^3}{r_f^3} \right) \right] = \frac{r_f}{t_c} \left(\frac{r_{c,0}}{r_f} \right)^3$$

From Fig. 7.34, $t_c \cong 0.17$ s, $r_f(t = t_c) \cong 0.25$ m, and $r_{c,0} \cong 0.21$ m. The value of S_u is calculated to be

$$S_u = \frac{r_f}{t_c} \left(\frac{r_{c,0}}{r_f} \right)^3 = \frac{0.25 \text{ m}}{0.17 \text{ s}} \left(\frac{0.21 \text{ m}}{0.25 \text{ m}} \right)^3 = 0.87 \text{ m/s}$$

This value is close but a little high compared to the values reported by Han et al. (2000), which are approximately 0.5 m/s. An important source of error in this method is the precision and accuracy of reading the values off of the figure.

Further empirical validation of the balloon model comes from the experiments of Julien et al. (2015a). Fig. 7.35 shows data from a constant pressure deflagration with aluminum powder.

The investigators then used the flame speed data and the ratio of the unburnt to burnt gas densities to calculate burning velocities. The data are shown in Fig. 7.36.

The burning velocity calculations are compared with measurements obtained from their Bunsen burner apparatus and show reasonable agreement (Goroshin et al., 1996b).

There are obvious limitations to this simple model. It neglects the turbulence of the dispersion process and turbulent flame propagation. It also precludes air entrainment into the fireball which will cool its temperature. This model does not lend itself to modeling the fireball temperature or flame radiation. The majority of the studies

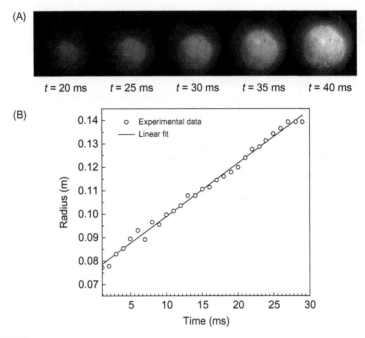

FIGURE 7.35

Constant pressure deflagration test with aluminum powder at a dust concentration of 500 g/m³: (A) sequence of photographs showing flame growth in the balloon and (B) flame radius as a function of time.

Reprinted with permission from Julien, P., Vickery, J., Whiteley, S., Wright, A., Goorshin, S., Bergthorson, J.M., et al., 2015a. Effect of scale on freely propagating flames in aluminum dust clouds. J. Loss Prev. Process Ind. 36, 230–236.

that have examined these issues with flash fires have been conducted with flammable gases or vapors and in much larger quantities (from 1 to 1000 tonnes fuel) than are practical in combustible dust hazard scenarios (Prugh, 1994).

7.9 CONTROLLING FLASH FIRE HAZARDS

A flash fire can be prevented or controlled. Fire prevention is the preferred strategy as it eliminates the potential consequences of human injury, property damage, or environmental impact. Fire control depends on a number of factors such as building construction, inventory control, fire alarm systems, fire protection systems, and administrative controls like evacuation plans and drills. Fire control takes us too far afield of from our chosen subject and so the reader is referred to standard reference materials for more information (Cote, 2008).

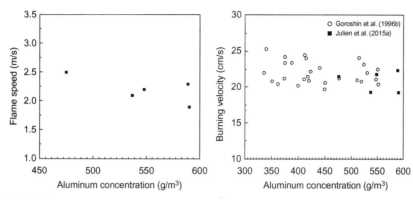

FIGURE 7.36

Flame speed measurements and burning velocity calculations for aluminum powder.

Reprinted with permission from Julien, P., Vickery, J., Whiteley, S., Wright, A., Goorshin, S., Bergthorson, J.M., et al., 2015a. Effect of scale on freely propagating flames in aluminum dust clouds. J. Loss Prev. Process Ind. 36, 230–236.

A systematic strategy for controlling flash fire hazards can be developed with reference to the flash fire square. The flash fire square is composed of four elements: combustible dust (fuel), oxidizer, ignition, and dust dispersal. The strategy to prevent a combustible dust flash fire is to remove one of these factors from the workplace. Two cautionary notes: (1) this is sometimes far more difficult than it sounds and (2) the probability of success in removing any given factor depends on many considerations. The four elements for the prevention of combustible dust flash fires are fuel control, ignition control, oxidant control, and dust dispersal control.

The two primary means of fuel control are housekeeping and the addition of inert solid additives (Amyotte, 2013, Chapter 4; Eckhoff, 2003, pp. 112−113). Housekeeping refers to the disciplined and systematic control of fugitive dust accumulation. The first order of business is to minimize leakage of combustible dusts to the greatest extent possible. Then on a routine basis, remove dust accumulations and monitor for spills and other unexpected accumulations.

Solid inert materials are applied where dust accumulations cannot be avoided. A typical example is in working underground coal mines. Coal dust is everywhere and cannot be eliminated. So inert materials like limestone powder are deployed to coat the floor, walls, and ceiling of the coal mine. This strategy is called rock dusting. With sufficiently large quantities of rock dust, a coal dust flame will be quenched before it can propagate. Obviously, in an industrial setting, the use of inert materials in this manner is an exceptional circumstance.

Another tactic to prevent a flash fire is through oxidant control. This is done by reducing or eliminating oxygen through the use of inert gases (Crowl, 2003, Chapter 3 and Appendix A). Examples of inert gases are nitrogen and carbon dioxide. The immediate problem with this tactic is that inert gases are physical asphyxiants, ie, an oxygen-deficient atmosphere is fatal to humans. Thus, this tactic is unlikely to be a feasible alternative.

The next tactic is ignition control. Ignition control covers a large and diverse array of potential energy sources including smoking materials, cutting or welding torches, static electricity, and electrical arcs. The NFPA handbook is an excellent entry point into the safety literature on ignition control (Cote, 2008).

Dust dispersal control is the final tactic for flash fire prevention. Dust dispersal can be caused by a steady ventilation flow by fans or blowers or by a jet of air such as a compressed air line. Perhaps the best approach to preventing dust dispersal is fuel control (housekeeping). In the event this is not practical, then a liquid additive such as water or oil (petroleum-derivative or vegetable) may be an option to reduce the ease of dispersion of the dust. However, care must be taken to avoid using a liquid that is chemically reactive with the dust. For example, water is a potential oxidizing agent with certain metals and linseed oil generates heat and is prone to spontaneous ignition.

Finally, it should be mentioned that it may be appropriate to have workers wear flame-resistant garments (Frank and Rodgers, 2012, Chapter 4). Personal protection equipment (PPE) should be considered as a last resort for flash fire protection. Stated differently, PPE should be considered an extra layer of protection above and beyond the existing safeguards implemented to prevent a flash fire. After all, the hazard scenario for flame-resistant garments is fire engulfment or contact. It is unlikely that a comfortable flame-resistant garment will cover 100% of the worker's surface area, and so it is not a guarantee that the worker will not receive a burn injury; it simply reduces the surface area of the injury. Finally, like any engineered product, flame-resistant garments have a design basis (Camenzind et al., 2007). If the design basis is exceeded—the fire is too large or its duration is too long—then the garment may not provide the desired level of protection.

7.10 SUMMARY

Under the right set of conditions, a flame can sweep through a cloud of combustible dust particles. This chapter has considered a range of topics relevant to unconfined dust flame propagation. We began with a description of steady one-dimensional flame propagation and broadly identified the experimental techniques used to produce these flames and described how to calculate the burning velocity. This led to a discussion of the similarities and differences between gas, mist, and dust flames. At the microscale level, heterogeneous flames sometimes burn as individual particles and sometimes as a cluster or group of particles. Criteria were presented to help predict when a given dust cloud will exhibit single particle or group combustion.

A simple model due to F.A. Williams was derived for the burning velocity and flame thickness of heterogeneous flames.

For particles undergoing diffusion controlled combustion, the burning velocity was found to be inversely proportional to the particle diameter. The Williams model predicts that the burning velocity increases with the square root of the dust concentration. The burning velocity was also predicted to increase with fuel volatility, but this dependence was weaker than the particle size or dust concentration effects. A final insight derived from the Williams model is that the flame thickness was proportional to the particle diameter.

Then some of the complicating factors in heterogeneous flame behavior were examined including velocity slip, turbulence, and thermal radiation.

Next we explored three types of theories of laminar dust flame propagation were presented. The first type of model discussed was the thermal theory, a model based on the continuity and thermal energy equations. Two variations on this theme were used to derive expressions for the burning velocity. The advantage of thermal theories is their simplicity; their disadvantage is it is difficult to relate these models to the properties of the individual particles.

The next dust flame model presented was Ballal's flame propagation theory which is based on an analysis of characteristic time scales. One important advantage of Ballal's model is that it can be directly related to the expression for diffusion-controlled combustion of single particles. Both the burning velocity and the flame thickness were derived from Ballal's model. Next, more comprehensive models of flame propagation were described that use the mathematical technique of activation energy asymptotics.

To complement the discussion on steady flame propagation, the transient phenomena of ignition and quenching in dust clouds were introduced using Ballal's model as the theoretical framework. Some characteristic results were discussed and a critique of this approach was summarized based on the work of Jarosinski and his colleagues.

From there we explored a wide range of heterogeneous flame propagation behavior based primarily on laboratory investigations under carefully controlled conditions. The survey included a review of flame propagation in organic mists, organic dusts (noncharring and charring), and metallic dusts (volatile and nonvolatile). And from these various studies, we examined the effect of the equivalence ratio and the mean particle diameter on flame propagation.

After exploring these various topics related more to combustion science, we turned our attention to the hazards of unconfined dust flames. The relationship between unconfined dust flame propagation and flash fire phenomena was explored. Finally, an overview of flash fire hazard control completed the chapter.

This investigation of unconfined dust flame propagation has set the stage for advancing to the next level of complexity: confined flame propagation, the subject of Chapter 8. We shall see that imposing the additional constraint of confinement on dust flame propagation can lead to further complications.

REFERENCES

Aggarwal, S.K., 1998. A review of spray ignition phenomena: present status and future research. Prog. Energy Combust. Sci. 24, 565−600.

Aggarwal, S.K., 2014. Single droplet ignition: theoretical analyses and experimental findings. Prog. Energy Combust. Sci. 45, 79−107.

Alekseev, A.G., Sudakova, I.V., 1983. Flame propagation rate in air suspension of metal powders. Combust. Explos. Shock Waves 19, 564−566.

Amyotte, P., 2013. An Introduction to Dust Explosions. Butterworth Heinemann, Elsevier, New York.

Amyotte, P., Pegg, M.J., 1989. Lycopodium dust explosions in a Hartmann bomb; effects of turbulence. J. Loss Prev. Process Ind. 2, 87−94.

Amyotte, P., Baxter, B.K., Pegg, M.J., 1990. Influence of initial pressure on spark-ignited dust explosions. J. Loss Prev. Process Ind. 3, 261−263.

Andrews, G.E., Bradley, D., 1972a. Determination of burning velocities: a critical review. Combust. Flame 18, 133−153.

Andrews, G.E., Bradley, D., 1972b. The burning velocity of methane−air mixtures. Combust. Flame 19, 275−288.

Andrews, G.E., Bradley, D., Lwakabamba, S.B., 1975. Turbulence and turbulent flame propagation—a critical appraisal. Combust. Flame 24, 285−304.

Anezaki, T. and Dobashi, R., 2007. Effects of particle materials on flame propagation during dust explosions. In: Proceedings of the Fifth International Seminar on Fire and Explosion Hazards, Edinburgh, UK, April 23−27, pp. 247−253.

Annamalai, K., Puri, I.K., 2007. Combustion Science and Engineering. CRC Press, Boca Raton, FL.

Annamalai, K., Ryan, W., 1992. Interactive processes in gasification and combustion. Part I: Liquid drop arrays and clouds. Prog. Energy Combust. Sci. 18, 221−295.

Annamalai, K., Ryan, W., 1993. Interactive processes in gasification and combustion—II. Isolated carbon, coal and porous char particles. Prog. Energy Combust. Sci. 19, 383−446.

Annamalai, K., Ryan, W., Dhanapalan, S., 1994. Interactive processes in gasification and combustion—Part III: Coal/char particle arrays, streams and clouds. Prog. Energy Combust. Sci. 20, 487−618.

Arpaci, V.S., Tabaczynski, R.J., 1982. Radiation-affected laminar flame propagation. Combust. Flame 46, 315−322.

Arpaci, V.S., Tabaczynski, R.J., 1984. Radiation-affected laminar flame quenching. Combust. Flame 57, 169−178.

ASTM E582, 2013. Standard Test Method for Minimum Ignition Energy and Quenching Distance in Gaseous Mixtures. ASTM International, West Conshohocken, PA.

ASTM E1491, 2012. Standard Test Method for Minimum Autoignition Temperature of Dust Clouds. ASTM International, West Conshohocken, PA.

ASTM E2019, 2007. Standard Test Method for Minimum Ignition Energy of a Dust Cloud in Air. ASTM International, West Conshohocken, PA.

ASTM E2021, 2009. Standard Test Method for Hot-Surface Ignition Temperature of Dust Layers. ASTM International, West Conshohocken, PA.

Aziz, A., Na, T.Y., 1984. Perturbation Methods in Heat Transfer. Hemisphere Publishing Corporation, New York.

Baek, S.W., Ahn, K.Y., Kim, J.U., 1994. Ignition and explosion of carbon particle clouds in a confined geometry. Combust. Flame 96, 121−129.

Ballal, D.R., 1980. Ignition and flame quenching of quiescent dust clouds of solid fuels. Proc. Royal Soc. Lond. A369, 479−500.

Ballal, D.R., 1983a. Further studies on the ignition and flame quenching of quiescent dust clouds. Proc. Royal Soc. Lond. A385, 1−21.

Ballal, D.R., 1983b. Flame propagation through dust clouds of carbon, coal, aluminium and magnesium in an environment of zero gravity. Proc. Royal Soc. Lond. A385, 21−51.

Ballal, D.R., Lefebvre, A.H., 1978. Ignition and flame quenching of quiescent fuel mists. Proc. Royal Soc. Lond. A364, 277−294.

Ballal, D.R., Lefebvre, A.H., 1979. Ignition and flame quenching of flowing heterogeneous fuel−air mixtures. Combust. Flame 35, 155−168.

Ballal, D.R., Lefebvre, A.H., 1981. Flame propagation in heterogeneous mixtures of fuel droplets, fuel vapour and air. Eighteenth Symposium (International) on Combustion. The Combustion Institute, Elsevier, New York, pp. 321−328.

Baudry, G., Bernard, S., Gillard, P., 2007. Influence of the oxide content on the ignition energies of aluminum powders. J. Loss Prev. Process Ind. 20, 330−336.

Beausang, E., Herbert, K., 1994. Burns from a dust explosion. Burns 20, 551−552.

Berlad, A., 1981. Combustion of particle clouds. Combustion experiments in a zero-gravity laboratory. Progress in Astronautics and Aeronautics, Volume 73, pp. 91−127.

Bernard, S., Gillard, P., Foucher, F., Mounaim-Rousselle, C., 2012. MIE and flame velocity of partially oxidised aluminium dust. J. Loss Prev. Process Ind. 25, 460−466.

Bidabadi, M., 1995. An Experimental and Analytical Study of Laminar Dust Flame Propagation, Ph.D. thesis. McGill University, Montreal, Quebec.

Bidabadi, M., Mafi, M., 2012. Analytical modeling of combustion of a single iron particle burning in the gaseous oxidizing medium. J. Mech. Eng. Sci. 227, 1006−1021.

Bidabadi, M., Shahbabaki, A.S., Jadidi, M., Montazerinejad, S., 2010. An analytical study of radiation effects on the premixed laminar flames of aluminium dust clouds. Proc. Inst. Mech. Eng. C J. Mech. Eng. Sci. 224, 1679−1695.

Bjerketvedt, D., Bakke, J.R., van Wingerden, K., 1997. Gas explosion handbook. J. Hazard. Mater. 52, 1−150.

Blouquin, R., Joulin, G., 1996. On the influence of gradient transports upon the burning velocities of dust flames. Combust. Sci. Technol. 115, 355−367.

Bocanegra, P.E., Sarou-Danian, V., Davidenko, D., Chauveau, C., Gokalp, I., 2009. Studies on the burning of micro- and nanoaluminum particle clouds in air. Prog. Propul. Phys. 1, 47−62.

Borman, G.L., Ragland, K.W., 1998. Combustion Engineering. McGraw-Hill, New York.

Bowen, P.J., Cameron, L.R.J., 1999. Hydrocarbon aerosol explosion hazards: a review. Trans. Inst. Chem. Eng. 77, 22−30.

Bradley, D., Habik, S.E.-D., Swithenbank, J.R., 1986. Laminar burning velocities of CH_4−air−graphite mixtures and coal dust. Twenty-first Symposium (International) on Combustion. The Combustion Institute, Elsevier, New York, pp. 249−256.

Britton, L.G., Cashdollar, K.L., Fenlon, W., Frurip, D., Going, J., Harrison, B.K., et al., 2005. The role of ASTM E27 methods in hazard assessment. Part II: flammability and ignitibility. Process Saf. Prog. 24, 12−28.

Bryant, J.T., 1971a. The combustion of premixed laminar graphite dust flames at atmospheric pressure. Combust. Sci. Technol. 2, 389−399.

Bryant, J.T., 1971b. Amorphous boron dust flames. Combust. Sci. Technol. 3, 145–152.

Buckmaster, J.D., Ludford, G.S.S., 1982. Theory of Laminar Flames. Cambridge University Press, Cambridge.

Burgoyne, J.H., 1963. The flammability of mists and sprays. Second Symposium on Chemical Process Hazards. Institution of Chemical Engineers, pp. 1–5.

Burgoyne, J.H., Cohen, L., 1954. The effect of drop size on flame propagation in liquid aerosols. Proc. Royal Soc. Lond. A Math. Phys. Eng. Sci. 225, 375–392.

Butlin, R.N., 1971. Polyethylene dust-air flames. Combust. Flame 17, 446–448.

Camenzind, M.A., Dale, D.J., Rossi, R.M., 2007. Manikin test for flame engulfment evaluation of protective clothing: historical review and development of a new ISO standard. Fire Mater. 31, 285–295.

Cao, W., Gao, W., Liang, J., Xu, S., Pan, F., 2014. Flame-propagation behavior and a dynamic model for the thermal-radiation effect in coal-dust explosions. J. Loss Prev. Process Ind. 29, 65–71.

Cassel, H.M., 1964. Report of Investigations 6551. Some Fundamental Aspects of Dust Flames. U.S. Department of the Interior, Bureau of Mines.

Cassel, H.M., Liebman, I., 1959. The cooperative mechanism in the ignition of dust dispersions. Combust. Flame 3, 467–475.

Cassel, H.M., das Gupta, A.K., Guruswamy, S., 1948. Factors affecting flame propagation through dust clouds. Third Symposium on Combustion, Flames and Explosions. William Wilkins and Co., Elsevier, New York, pp. 185–190.

Cassel, H.M., Liebman, I., Mock, W.K., 1957. Radiative transfer in dust flames. Sixth Symposium (International) on Combustion. The Combustion Institute, Elsevier, New York, pp. 602–605.

CCPS/AIChE, 2005. Guidelines for Safe Handling of Powders and Bulk Solids. American Institute of Chemical Engineers. John Wiley & Sons, New York.

Chan, K.-K., Jou, C.-S., 1988. An experimental and theoretical investigation of the transition phenomenon in fuel spray deflagration: 1. The experiment. Fuel 67, 1223–1227.

Chen, J.-L., Dobashi, R., Hirano, T., 1996. Mechanisms of flame propagation through combustible particle clouds. J. Loss Prev. Process Ind. 9, 225–229.

Chen, L., Yong, S.Z., Ghoniem, A.F., 2012. Oxy-fuel combustion of pulverized coal: characterization, fundamentals, stabilization and CFD modeling. Prog. Energy Combust. Sci. 38, 156–214.

Conti, R.S., Cashdollar, K.L., Thomas, R.A., 1993. Report of Investigations 9467. Improved 6.8-L Furnace for Measuring the Autoignition Temperatures of Dust Clouds. U.S. Bureau of Mines, U.S. Department of the Interior.

Cooper, P.N., 2006. Chapter 9. Burn injury. Essentials of Autopsy Practice. Springer, Berlin, pp. 215–232.

Corcoran, A.L., Hoffmann, V.K., Dreizin, E.L., 2013. Aluminum particle combustion in turbulent flames. Combust. Flame 160, 718–724.

Cote, A.E. (Editor in Chief). Fire Protection Handbook, twentieth ed. National Fire Protection Association, Quincy, MA, 2008.

Crowe, C.T., Schwarzkopf, J.D., Sommerfield, M., Tsuji, Y., 2012. Multiphase Flows with Droplets and Particles, second ed. CRC Press, Boca Raton, FL.

Crowl, D., 2003. Understanding Explosions. Center for Chemical Process Safety, American Institute of Chemical Engineers. John Wiley & Sons, New York.

CSB, November 2006. Investigation Report: Combustible Dust Hazard Study. U.S. Chemical Safety and Hazard Investigation Board, Washington, DC.

CSB, December 2011. Case Study: Metal dust flash fires and hydrogen explosion— Hoeganaes Corporation, Gallatin, TN. U.S. Chemical Safety and Hazard Investigation Board, Washington, DC.

CSB, January 2013. Case Study: Ink Dust Explosion and Flash Fires in East Rutherford, New Jersey. U.S. Chemical Safety and Hazard Investigation Board, Washington, DC.

Dahoe, A.E., Hanjalic, K., Scarlett, B., 2002. Determination of the laminar burning velocity and the Markstein length of powder–air flames. Powder Technol. 122, 222–238.

Deshaies, B., Joulin, G., 1986. Radiative transfer as a propagation mechanism for rich flames of reactive suspensions. SIAM J. Appl. Math. 46, 561–581.

Di Benedetto, A., Di Sarli, V., Russo, P., 2010. On the determination of the minimum ignition temperature for dust–air mixtures. Chem. Eng. Trans. 19, 189–194.

Dobashi, R., Senda, K., 2002. Mechanisms of flame propagation through suspended combustible particles. J. Phys. IV France 12, 459–465.

Dobashi, R., Senda, K., 2006. Detailed analysis of flame propagation during dust explosions by UV band observations. J. Loss Prev. Process Ind. 19, 149–153.

Dobashi, R., 2007. Flame propagation during dust explosions. In: Proceedings of the Fifth International Seminar on Fire and Explosion Hazards, Edinburgh, UK, April 23–27, pp. 16–27.

Eckhoff, R., 2003. Dust Explosions in the Process Industries, third ed. Gulf Professional Publishing, Elsevier, New York.

Essenhigh, R.H., 1977. Combustion and flame propagation in coal systems: a review. Sixteenth Symposium (International) on Combustion. The Combustion Institute, Elsevier, pp. 353–374.

Essenhigh, R.H., Csaba, J., 1963. The thermal radiation theory for plane flame propagation in coal dust clouds. Ninth Symposium (International) on Combustion. The Combustion Institute, Elsevier, pp. 111–125.

Essenhigh, R.H., Woodhead, D.W., 1958. Speed of flame in slowly moving clouds of cork dust. Combust. Flame 2, 365–382.

Essenhigh, R.H., Misra, M.K., Shaw, D.W., 1989. Ignition of coal particles: a review. Combust. Flame 77, 3–30.

Fan, L.-S., Zhu, C., 1998. Principles of Gas–Solid Flows. Cambridge University Press, Cambridge.

Frank, W.L., Rodgers, S.A., 2012. NFPA Guide to Combustible Dusts. National Fire Protection Association, Quincy, MA.

Friedman, R., Maček, A., 1963. Combustion studies of single aluminum particles. Ninth Symposium (International) on Combustion. The Combustion Institute, Elsevier, New York, pp. 703–712.

Gant, S., Bettis, R., Santon, R., Buckland, I., Bowen, P., Kay, P., 2012. Generation of flammable mists form high flashpoint fluids: literature review. Hazards XXIII. Institution of Chemical Engineers, pp. 327–339.

Gao, W., Dobashi, R., Mogi, T., Sun, J., Shen, X., 2012. Effects of particle characteristics on flame propagation behavior during organic dust explosions in a half-closed chamber. J. Loss Prev. Process Ind. 25, 993–999.

Gao, W., Mogi, T., Sun, J., Dobashi, R., 2013a. Effects of particle thermal characteristics on flame structures during dust explosions of three long-chain monobasic alcohols in an open-space chamber. Fuel 113, 86—96.

Gao, W., Mogi, T., Sun, J., Yu, J., Dobashi, R., 2013b. Effects of particle size distributions on flame propagation mechanism during octadecanol dust explosions. Powder Technol. 249, 168—174.

Gao, W., Mogi, T., Yu, J., Yan, X., Sun, J., Dobashi, R., 2015a. Flame propagation mechanisms in dust explosions. J. Loss Prev. Process Ind. 36, 186—194.

Gao, W., Mogi, T., Rong, J., Yu, J., Yan, X., Dobashi, R., 2015b. Motion behaviors of the unburned particles ahead of flame front in hexadecanol dust explosion. Powder Technol. 271, 125—133.

Going, J.E., Chatrathi, K., 2003. Efficiency of flameless venting devices. Process Saf. Prog. 22, 33—42.

Goltsiker, A., Chivilikhin, S., Belikov, A., 1994. Nonsteady heterogeneous flame propagation: a development of Todes and Zel'dovich scaling ideas in 1969—1994. In: Proceedings of the Zel'dovich Memorial International Conference on Combustion, Moscow, September 12—17.

Goroshin, S., Bidabadi, M., Lee, J.H.S., 1996a. Quenching distance of laminar flame in aluminum dust clouds. Combust. Flame 105, 147—160.

Goroshin, S., Fomenko, I., Lee, J.H.S., 1996b. Burning velocities in fuel-rich aluminum dust clouds. Twenty-Sixth Symposium (International) on Combustion. The Combustion Institute, Elsevier, pp. 1961—1967.

Goroshin, S., Kolbe, M., Lee, J.H.S., 2000. Flame speed in a binary suspension of solid fuel particles. Proc. Combust. Inst. 28, 2811—2817.

Goroshin, S., Mamen, J., Higgins, A., Bazyn, T., Glumac, N., Krier, H., 2007. Emission spectroscopy of flame fronts in aluminum suspensions. Proc. Combust. Inst. 31, 2011—2019.

Goroshin, S., Tang, F.-D., Higgins, A.J., Lee, J.H.S., 2011. Laminar dust flames in a reduced-gravity environment. Acta. Astronaut. 68, 656—666.

Greenberg, J.B., Silverman, I., Tambour, Y., 1996. A new heterogeneous burning velocity formula for the propagation of a laminar flame front through a polydisperse spray of droplets. Combust. Flame 104, 358—368.

Grimard, J.K., Potter, K., 2011. Effect of Dust Deflagrations on Human Skin. Major qualifying project submitted in partial fulfillment of B.S. degree at Worcester Polytechnic Institute.

Guenoche, H., 1964. Flame propagation in tubes and in closed vessels. In: Markstein, G.H. (Ed.), Nonsteady Flame Propagation. Pergamon Press, Oxford.

Habibzadeh, M.R., Keyhani, M.H., 2008. Experimental Investigation on Quenching Distance for Aluminum Dust Flames. ASME 2008 Fluids Engineering Division Summer Meeting collocated with the Heat Transfer, Energy Sustainability, and 3rd Energy Nanotechnology Conferences. American Society of Mechanical Engineers.

Hamberger, P., Schneider, H., Jamois, D., Proust, C., 2007. Correlation of turbulent burning velocity and turbulence intensity for starch dust air mixtures. In: Third European Combustion Meeting.

Han, O.-S., Yashima, M., Matsuda, T., Matusi, H., Miyake, A., Ogawa, T., 2000. Behavior of flames propagating through lycopodium dust clouds in a vertical duct. J. Loss Prev. Process Ind. 13, 449—457.

Han, O.-U., Yashima, M., Matsuda, T., Matusi, H., Miyake, A., Ogawa, T., 2001. A study of flame propagation mechanisms in lycopodium dust clouds based on dust particles' behavior. J. Loss Prev. Process Ind. 14, 153–160.

Han, W., Chen, Z., 2015. Effects of finite-rate droplet evaporation on the ignition and propagation of premixed spherical spray flame. Combust. Flame 162, 2128–2139.

Higuera, F.J., Linan, A., Trevino, C., 1989. Heterogeneous ignition of coal dust clouds. Combust. Flame 75, 325–342.

Holbrow, P., Hawksworth, S.J., Tyldelsey, A., 2000. Thermal radiation from vented dust explosions. J. Loss Prev. Process Ind. 13, 467–476.

Horton, M.D., Goodson, F.P., Smoot, L.D., 1977. Characteristics of flat, laminar coal-dust flames. Combust. Flame 28, 187–195.

Jadidi, M., Bidabadi, M., Shahrbabaki, A.Sh, 2010. Quenching distance and laminar flame speed in a binary suspension of solid fuel particles. Latin Amer. Appl. Res. 40, 39–45.

Jarosinski, J., Strehlow, R.A., Azarbarzin, A., 1982. The mechanism of lean limit extinguishment of an upward and downward propagating flame in a standard flammability tube. Nineteenth Symposium (International) on Combustion. The Combustion Institute, Elsevier, New York, pp. 1549–1557.

Jarosinski, J., Lee, J.H.S., Knystautas, R., Crowley, J.D., 1986. Quenching of dust-air flames. Twenty-first Symposium (International) on Combustion. The Combustion Institute, Elsevier, New York, pp. 1917–1924.

Joshi, N.D., Berlad, A.L., 1986. Gravitational effects on stabilized, premixed, lycopodium-air flames. Combust. Sci. Technol. 47, 55–68.

Joulin, G., 1981. Asymptotic analysis of non-adiabatic flames: heat losses towards small inert particles. Eighteenth Symposium (International) on Combustion. The Combustion Institute, Elsevier, New York, pp. 1395–1404.

Joulin, G., Deshaies, B., 1986. On radiation-affected flame propagation in gaseous mixtures seeded with inert particles. Combust. Sci. Technol. 47, 299–315.

Ju, W.-J., Dobashi, R., Hirano, T., 1998a. Dependence of flammability limits of a combustible particle cloud on particle diameter distribution. J. Loss Prev. Process Ind. 11, 177–185.

Ju, W.-J., Dobashi, R., Hirano, T., 1998b. Reaction zone structures and propagation mechanisms of flames in stearic acid particle clouds. J. Loss Prev. Process Ind. 11, 423–430.

Julien, P., Vickery, J., Whiteley, S., Wright, A., Goorshin, S., Bergthorson, J.M., et al., 2015a. Effect of scale on freely propagating flames in aluminum dust clouds. J. Loss Prev. Process Ind. 36, 230–236.

Julien, P., Vickery, J., Whiteley, S., Wright, A., Goroshin, S., Frost, D.L., et al., 2015b. Freely-propagating flames in aluminum dust clouds. Combust. Flame 162, 4241–4253.

Krantz, W.B., 2007. Scaling Analysis in Modeling Transport and Reaction Processes: a Systematic Approach to Model Building and the Art of Approximation. John Wiley & Sons, New York.

Krause, U., Kasch, T., 2000. The influence of flow and turbulence on flame propagation through dust–air mixtures. J. Loss Prev. Process Ind. 13, 291–298.

Krazinski, J.L., Buckius, R.O., Krier, H., 1979. Coal dust flames: a review and development of a model for flame propagation. Prog. Energy Combust. Sci. 5, 31–71.

Krishna, C.R., Berlad, A.L., 1980. A model for dust cloud autoignition. Combust. Flame 37, 207–210.

Kuo, K., 2005. Principles of Combustion, second ed. John Wiley & Sons, New York.

Kuo, K., Acharya, R., 2012a. Applications of Turbulent and Multiphase Combustion. John Wiley & Sons, New York.

Kuo, K., Acharya, R., 2012b. Fundamentals of Turbulent and Multiphase Combustion. John Wiley & Sons, New York.

Law, C.K., 2006. Combustion Physics. Cambridge University Press, Cambridge.

Law, C.K., Faeth, G.M., 1994. Opportunities and challenges of combustion in microgravity. Prog. Energy Combust. Sci. 20, 65−113.

Lewis, B., von Elbe, G., 1987. Combustion, Flames, and Explosions of Gases, third ed. Academic Press, New York.

Lin, T.H., Law, C.K., Chung, S.H., 1988. Theory of laminar flame propagation in off-stoichiometric dilute sprays. Int. J. Heat Mass Transfer 31, 1023−1034.

Lipatnikov, A., 2013. Fundamentals of Premixed Turbulent Combustion. CRC Press, Boca Raton, FL.

Maguire, B.A., Slack, C., Williams, A.J., 1962. The concentration limits for coal dust−air mixtures for upward propagation of flame in a vertical tube. Combust. Flame 6, 287−294.

Mamen, J., Goroshin, S., Higgins, A., 2005. Spectral structure of the aluminum dust flame. In: 20th International Colloquium on the Dynamics of Explosions and Reactive Systems.

Mason, W.E., Wilson, M.J.G., 1967. Laminar flames of lycopodium dust in air. Combust. Flame 11, 195−200.

Milne, T.A., Beachey, J.E., 1977a. The microstructure of pulverized coal−air flames. I. Stabilization on small burners and direct sampling techniques. Combust. Sci. Technol. 16, 123−138.

Milne, T.A., Beachey, J.E., 1977b. The microstructure of pulverized coal−air flames. II. Gaseous species, particulate and temperature profiles. Combust. Sci. Technol. 16, 139−152.

Mitani, T., 1981. A flame inhibition theory by inert dust and spray. Combust. Flame 43, 243−253.

Mitsui, R., Tanaka, T., 1973. Simple models of dust explosion. Predicting ignition temperature and minimum explosive limit in terms of particle size. Ind. Eng. Chem. Process Des. Dev. 12, 384−389.

Mittal, M., Guha, B.K., 1996. Study of ignition temperature of a polyethylene dust cloud. Fire Mater. 20, 97−105.

Mittal, M., Guha, B.K., 1997a. Minimum ignition temperature of polyethylene dust: a theoretical model. Fire Mater. 21, 169−177.

Mittal, M., Guha, B.K., 1997b. Models for minimum ignition temperature of organic dust clouds. Chem. Eng. Technol. 20, 53−62.

Mizutani, Y., Nakajima, A., 1973. Combustion of fuel−vapor−drop-air systems: part I—open burner flames. Combust. Flame 21, 343−350.

Myers, G.D., Lefebvre, A.H., 1986. Flame propagation in heterogeneous mixtures of fuel drops and air. Combust. Flame 66, 193−210.

NFPA 921, 2014. Guide for Fire and Explosion Investigations. National Fire Protection Association, Quincy, MA.

Ogle, R.A., Beddow, J.K., Vetter, A.F., 1984. A thermal theory of laminar premixed dust flame propagation. Combust. Flame 58, 77−79.

Ogle, R.A., Carpenter, A.R., Morrison III, D.R., 2005. Lessons learned from fires and explosions involving air pollution control systems. Process Saf. Prog. 24, 120−125.

Özisik, M.N., 1973. Radiative Transfer and Interactions with Conduction and Convection. Wiley-Interscience, New York.

Palmer, K.N., 1973. Dust Explosions and Fires. Chapman and Hall, London.

Palmer, K.N., Tonkin, P.S., 1965a. Fire Research Note No. 605. Explosions of Marginally Explosible Dust Mixtures Dispersed in a Large Scale Vertical Tube. Fire Research Station

Palmer, K.N., Tonkin, P.S., 1965b. Fire Research Note No. 607. The Explosibility of Some Industrial Dusts in a Large Scale Vertical Tube Apparatus. Fire Research Station

Palmer, K.N., Tonkin, P.S., 1968. The explosibility of dusts in small-scale tests and large-scale industrial plant. I. Chem. E. Symposium Series, No. 25. Institution of Chemical Engineers, London, pp. 66–75.

Panagiotou, T., Levendis, Y., 1998. Observations on the combustion of polymers (plastics): from single particles to groups of particles. Combust. Sci. Technol. 137, 121–147.

Panagiotou, T., Levendis, Y., Deichatsios, M., 1996. Measurements of particle flame temperatures using three-color optical pyrometry. Combust. Flame 104, 272–287.

Proust, C., 2006a. Flame propagation and combustion in some dust–air mixtures. J. Loss Prev. Process Ind. 19, 89–100.

Proust, C., 2006b. A few fundamental aspects about ignition and flame propagation in dust clouds. J. Loss Prev. 19, 104–120.

Proust, C., Veyssière, B., 1988. Fundamental properties of flames propagating in starch dust–air mixtures. Combust. Sci. Technol. 62, 149–172.

Prugh, R.W., 1994. Quantitative evaluation of fireball hazards. Process Saf. Prog. 13, 83–91.

Purser, D.A., 2002. Toxicity assessment of combustion products, Section 2, Chapter 6, pp. 2-83 to 2-171. In: DiNenno, P.J., et al., (Eds.), SFPE Handbook of Fire Protection Engineering, third ed. National Fire Protection Association, Quincy, MA.

Puttick, S., 2008. Liquid mists and sprays flammable below the flash point—the problem of preventative bases of safety. Hazards XX. Institution of Chemical Engineers, pp. 1–13.

Rallis, C.J., Garforth, A.M., 1980. The determination of laminar burning velocity. Prog. Energy Combust. Sci. 6, 303–329.

Richards, G.A., Lefebvre, A.H., 1989. Turbulent flame speeds of hydrocarbon fuel droplets in air. Combust. Flame 78, 299–307.

Risha, G.A., Huang, Y., Yetter, R.A., Yang, V., 2005. Experimental investigation of aluminum particle dust cloud combustion. In: 43rd Aerospace Sciences Meeting and Exhibit, Reno, Nevada, January 10–13.

Ronney, P.D., 1998. Understanding combustion processes through microgravity research. Twenty-Seventh Symposium (International) on Combustion. The Combustion Institute, Elsevier, New York, pp. 2485–2506.

Ross, H.D. (Ed.), 2001. Microgravity Combustion: Fire in Free Fall. Academic Press, New York.

Russell, R.C., Baldwin, J.R., Law, E.J., 1980. Burns due to grain dust explosions. J. Trauma 20, 767–771.

Rzal, F., Veyssière, B., Mouilleau, Y., Proust, C., 1993. Experiments on turbulent flame propagation in dust–air mixtures. In: Progress in Astronautics and Aeronautics, vol. 152, pp. 211–231.

Santon, R.C., 2009. Mist fires and explosions—an incident survey. Hazards XXI. Institution of Chemical Engineers, pp. 370–374.

Schneider, H., Proust, C., 2007. Determination of turbulent burning velocities of dust air mixtures with the open tube method. J. Loss Prev. Process Ind. 20, 470–476.

Seshadri, K., Berlad, A.L., Tangirala, V., 1992. The structure of premixed particle-cloud flames. Combust. Flame 89, 333–342.

Shoshin, Y., Dreizin, E., 2002. Production of well-controlled laminar aerosol jets and their application for studying aerosol combustion processes. Aerosol Sci. Technol. 36, 953–962.

Silverman, I., Greenberg, J.B., Tambour, Y., 1993. Stoichiometry and polydisperse effects in premixed spray flames. Combust. Flame 93, 97–118.

Sirignano, W.A., 2010. Fluid Dynamics and Transport of Droplets and Sprays, second ed. Cambridge University Press, Cambridge.

Sirignano, W.A., 2014. Advances in droplet array combustion theory and modeling. Prog. Energy Combust. Sci. 42, 54–86.

Siwek, R., 1989. New knowledge about rotary air locks in preventing dust ignition breakthrough. Plant Oper. Prog. (later changed to Process Saf. Prog.) 8, 165–176.

Siwek, R., Cesana, C., 1995. Ignition behaviour of dusts: meaning and interpretation. Process Saf. Prog. 14, 107–119.

Skjold, T., Olsen, K.L., Castellanos, D., 2013. A constant pressure dust explosion experiment. J. Loss Prev. Process Ind. 26, 562–570.

Slatter, D.J.F., Sattar, H., Medina, C.H., Andrews, G.E., Phylaktou, H.N., Gibbs, B.M., 2015. Biomass explosion testing: accounting for the post-test residue and implications on the results. J. Loss Prev. Process Ind. 36, 318–325.

Smoot, L.D., Horton, M.D., 1977. Propagation of laminar pulverized coal–air flames. Prog. Energy Combust. Sci. 3, 235–258.

Smoot, L.D., Horton, M.D., Williams, G.A., 1977. Propagation of laminar pulverized coal–air flames. Sixteenth (International) Symposium on Combustion. The Combustion Institute. Elsevier, New York, pp. 375–387.

Smoot, L.D., Pratt, D.T. (Eds.), 1979. Pulverized Coal Combustion and Gasification: Theory and Applications for Continuous Flow Processes. Springer, Berlin.

Smoot, L.D., Smith, P.J., 1985. Coal Combustion and Gasification. Plenum Press, New York.

Snoeys, J., Going, J.E., Taveau, J.R., 2012. Advances in dust explosion protection techniques: flameless venting. Proc. Eng. 45, 403–413.

Stern, M.C., Rosen, J.S., Ibarreta, A.F., Myers, T.J., Ogle, R.A., 2015a. Unconfined deflagration testing for the assessment of combustible dust flash fire hazards. In: 11th Global Congress on Process Safety. American Institute of Chemical Engineers, Spring 2015 Meeting, Austin, TX, April 27–29.

Stern, M.C., Rosen, J.S., Ibarreta, A.F., Ogle, R.A., Myers, T.J., 2015b. Quantification of the thermal hazard from metallic and organic dust flash fires, October 27–29. 2015 International Symposium. Mary Kay O'Connor Process Safety Center, Texas A&M University

Still Jr., J.M., Law, E.J., Pickens Jr., H.C., 1996. Burn due to a sawdust explosion. Burns 22, 164–165.

Strehlow, R.A., 1984. Combustion Fundamentals. McGraw-Hill, New York.

Strehlow, R.A., Stuart, J.G., 1953. An improved soap bubble method of measuring flame velocities. Fourth Symposium (International) on Combustion. The Combustion Institute, Elsevier, New York, pp. 329–336.

Sun, J.-H., Dobashi, R., Hirano, T., 1998. Structure of flames propagating through metal particle clouds and behavior of particles. Twenty-Seventh Symposium (International) on Combustion. The Combustion Institute, Elsevier, New York, pp. 2405–2411.

Sun, J.-H., Dobashi, R., Hirano, T., 2000. Combustion behavior of iron particles suspended in air. Combust. Sci. Technol. 150, 99–114.

Sun, J.-H., Dobashi, R., Hirano, T., 2001. Temperature profile across the combustion zone propagating through an iron particle cloud. J. Loss Prev. Process Ind. 14, 463−467.

Sun, J.-H., Dobashi, R., Hirano, T., 2003. Concentration profile of particles across a flame propagating through an iron particle cloud. Combust. Flame 134, 381−387.

Sun, J.-H., Dobashi, R., Hirano, T., 2006a. Velocity and number density profiles of particles across upward and downward flame propagating through iron particle clouds. J. Loss Prev. Process Ind. 19, 135−141.

Sun, J.-H., Dobashi, R., Hirano, T., 2006b. Structure of flames propagating through aluminum particles cloud and combustion process of particles. J. Loss Prev. Process Ind. 19, 769−773.

Tang, F.-D., Goroshin, S., Higgins, A., Lee, J., 2009. Flame propagation and quenching in iron dust clouds. Proc. Combust. Inst. 32, 1905−1912.

Tang, F.-D., Goroshin, S., Higgins, A., 2011. Modes of particle combustion in iron dust flames. Proc. Combust. Inst. 33, 1975−1982.

Trunov, M.A., Schoenitz, M., Dreizin, E.L., 2005a. Ignition of aluminum powders under different experimental conditions. Propel. Explos. Pyrotech. 30, 36−43.

Trunov, M.A., Schoenitz, M., Zhu, X., Dreizin, E.L., 2005b. Effect of polymorphic phase transformation in Al_2O_3 film on oxidation kinetics of aluminum powders. Combust. Flame 140, 310−318.

Tseng, L.-K., Ismail, M.A., Faeth, G.M., 1993. Laminar burning velocities and Markstein numbers of hydrocarbon/air flames. Combust. Flame 95, 410−426.

Turns, S.R., 2012. An Introduction to Combustion, third ed. McGraw-Hill, New York.

van Wingerden, K., Stavseng, L., 1996. Measurements of the laminar burning velocities in dust−air mixtures. VDI. Ber. 1272, 553−564.

Varma, A., Morbidelli, M., 1997. Mathematical Methods in Chemical Engineering. Oxford University Press, Oxford.

Veyssière, B., 1992. Development and propagation regimes of dust explosions. Powder Technol. 71, 171−180.

Wang, S., Pu, Y., Jia, F., Gutowski, A., 2006a. Effect of turbulence on flame propagation in cornstarch dust−air mixtures. J. Thermal Sci. 15, 186−192.

Wang, S., Pu, Y., Jia, F., Gutkowski, A., Jarosinski, J., 2006b. An experimental study on flame propagation in cornstarch dust clouds. Combust. Sci. Technol. 178, 1957−1975.

Weber, R.O., 1989. Thermal theory for determining the burning velocity of a laminar flame, using the inflection point in the temperature profile. Combust. Sci. Technol. 64, 135−139.

Williams, F.A., 1960. Monodisperse spray deflagration. Prog. Astronaut. Rocket. 2, 229−264.

Williams, F.A., 1985. Combustion Theory, second ed. Benjamin/Cummings Publishing Company, Menlo Park, CA.

Wolanski, P., 1991. Deflagration and detonation combustion of dust mixtures. Dyn. Deflag. React. Syst. Heterogen. Combust. Prog. Astronaut. Aeronaut. 132, 3−31.

Wright, A., Goroshin, S., Higgins, A., 2015. An attempt to observe the discrete flame propagation regime in aluminum dust clouds. In: 25th International Colloquium on the Dynamics of Explosions and Reactive Systems, Leeds, UK.

Zhang, D.-K., Wall, T.F., 1993. An analysis of the ignition of coal dust clouds. Combust. Flame 92, 475−480.

Confined unsteady dust flame propagation

A dust deflagration is most commonly associated with the confined unsteady propagation of a dust flame within an enclosure. Confining a propagating dust flame introduces new complexities to the analysis, chief among them being the progressive combustion of the fuel and the accompanying pressure rise in the enclosure. The confined deflagration scenario serves two important analytical roles. First, it is the basis for the laboratory measurement of combustible dust properties in either the 20-L sphere or the ISO cubic meter vessel. Second, it represents the simplest conceptual model for a dust explosion (flame acceleration, to be discussed in Chapter 9, adds additional complexity). This chapter is largely restricted to integral balance models for flame propagation in a closed vessel. More sophisticated mathematical models of confined deflagrations will be discussed in Chapter 10.

This chapter begins with a description of one-dimensional flame propagation in a spherical vessel. After a brief survey of combustion studies in closed vessels, we will consider the flame propagation process in a spherical vessel in some detail. The specification of a spherical geometry is less restrictive than one might think. Dust clouds ignited at the center of the cloud will propagate a spherical flame. Two factors can cause a departure from a spherical geometry. One is flame contact with an obstacle or confining wall. The second is ignition closer to the edge of the dust cloud rather than its center. Both of these complicating factors will be addressed briefly as this chapter unfolds. This section closes with a classification of mathematical models used to study flame propagation in closed vessels.

Next, two important characteristic features of unsteady confined deflagrations are described. The first characteristic feature is the formation of a radial temperature gradient with the maximum temperature at the center of the sphere. This is called the Flamm−Mache temperature gradient. The second feature is the direct relationship between the fractional pressure rise and the fractional conversion of fuel. These two features, observed only in confined deflagrations, are important considerations in the development of mathematical models for the combustion process.

The next topic to be considered is the relative importance, in terms of the order of magnitude, of the rate processes associated with dust combustion: chemical kinetics, velocity slip due to gravitational settling, turbulence, and gravitational settling. Due to the presence of the confining walls, we will need to consider one additional rate process: the speed of sound and how this relates to the existence or absence of a pressure gradient within the enclosure. These considerations will focus primarily on the calculation of characteristic time scales relative to the 20-L and 1-m^3 test vessels. This analysis provides a basis for developing intuition regarding the physical and chemical processes involved in confined dust deflagrations.

In fluid dynamics, integral models are a statement of the conservation principles defined for a control volume. The primary objective for developing integral models for confined deflagrations is to provide insight into the calculation of dust explosion parameters such as explosion severity (the maximum rate of pressure rise), explosion pressure (the maximum pressure), and the burning velocity. In the published literature, there are many variations on this theme of the integral model. We will examine these models in terms of how the control volume is partitioned into zones and by the type of thermodynamic process assumed for the transition from the unburnt to the burnt gas state. From the relationship between the flame radius and pressure histories, it is possible to extract an estimate of the burning velocity. All of these fluid dynamic models have limitations that introduce uncertainty into the determination of the burning velocity. A further difficulty is that the burning velocity is sensitive to small variations in the thermodynamic properties of the combustible dust—air mixture. In sum, the burning velocity is not a universal material property of a combustible dust; it is, instead, a property of the system.

This presentation then leads us naturally to a discussion of two cubic relations: the cubic pressure rise and the cube root scaling law. The cubic pressure rise describes the early behavior of the deflagration. The cube root scaling law is related to the late behavior of the deflagration. Thus, these two cubic relations are not related to each other. The cube root scaling law is perhaps the single most important relation in dust explosion dynamics. It provides the theoretical justification for applying laboratory scale measurements to full-scale systems using the deflagration index, better known as the K_{st} parameter. The K_{st} parameter plays a central role in the design of dust explosion protection technologies. It is important to understand the limitations of the cube root scaling law as it is a first-order approximation to deflagration behavior.

We will then review the influence of experimental conditions on the laboratory measurement of the explosibility parameters K_{st} and P_{max}. Our primary focus in that review will be to understand the effects of dust dispersion and igniter strength and how these must be manipulated for proper comparison with laboratory vessels of different size. These factors relate to precombustion turbulence. Following ignition, the flame propagates and generates more turbulence. The turbulent burning velocity is faster than its laminar counterpart and increases the explosion severity. Some of the empirical work on the measurement and prediction of turbulent burning velocities for combustible dust will be reviewed.

Next, we will explore how different fuel types influence dust explosion severity. We will survey noncharring organic solids, charring organic solids, and both volatile and nonvolatile metallic solids. Our goal will be to understand how fuel volatility, fuel concentration, the mean particle size, and the particle size distribution affect the dust explosion parameters K_{st} and P_{max}.

Following this discussion of deflagration behavior, the concentration limits of explosion are discussed. We will begin the discussion with an observation of a key feature of confined dust deflagrations: incomplete combustion. Incomplete combustion is observable at both the microscale (partially burnt particles) and at the macroscale (partially unburnt powder). The relative importance of this feature will depend on the objectives of your investigation. Perhaps the greatest significance of incomplete combustion is that it is observable not only in the laboratory but also in the field following an accidental dust deflagration. With confined dust deflagrations, incomplete combustion is the rule rather than the exception. The next topic will be a review of four explosion concentration limits: the minimum explosive concentration, the maximum explosive concentration, the limiting oxygen concentration, and the minimum inerting solids concentration. The nature of these limits is quite different for combustible dusts compared to flammable gases. It will be shown that the flame temperature and the discrete nature of the dust particles are significant factors in establishing the values of explosion limits for combustible dusts.

Section 8.9 is a brief discussion of how to control dust explosion hazards from the perspective of the dust explosion pentagon. The final section is a summary of the key ideas developed in this chapter.

8.1 FLAME PROPAGATION IN CLOSED VESSELS

Combustion of a fuel in a closed vessel causes a rise in gas temperature and pressure. This simple observation has become the basis for much of the combustible dust characterization work that has occurred over the last 100 years (see, for example, the experimental work on flammable gases in Hopkinson, 1906 and combustible dusts in Brown, 1917). This section begins with a brief historical survey that explains how the spherical test chamber became the standard combustion test device. Some of the important features of flame propagation in a closed vessel is described next. This is followed by a brief taxonomy of mathematical models.

8.1.1 HISTORICAL PERSPECTIVE

Early investigators realized that igniting a flammable gas mixture in the center of a closed vessel resulted in the propagation of a spherical flame (Ellis and Wheeler, 1927; Lewis and von Elbe, 1934; Fiock and Marvin, 1937). They further recognized that information about the combustion process could be extracted by recording the pressure history and photographing the flame motion. Flame propagation models were developed to predict the pressure history and flame motion, and from these data to extract the laminar burning velocity (Ellis and Wheeler, 1927; Lewis and von Elbe, 1934, 1987, pp. 381–395; Fiock and Marvin, 1937; Fiock et al., 1940; Jost, 1946, pp. 136–159). These techniques were used to study a variety of flammable gas-oxidizer systems.

In the 1960s the Hartmann bomb, developed by Hartmann and his colleagues at the U.S. Bureau of Mines, became one of the more popular laboratory test vessels for measuring explosion severity (Dorsett et al., 1960). Hartmann bomb data was published by investigators from around the world, but there was uncertainty on how to use the data to evaluate dust explosion severity. The U.S. Bureau of Mines had developed explosion severity and ignition sensitivity indexes for organizing and ranking their dust explosion data in comparison with Pittsburgh seam bituminous coal dust (Hertzberg, 1987). But it was not clear how to apply this data to the design of engineering safeguards for dust explosion protection. The Bureau of Mines created an extensive body of test data on dust explosion venting, but the work was largely empirical (Hartmann and Nagy, 1957).

In addition to the Hartmann bomb, investigators tried a variety of test chambers of different sizes and shapes and, as a result, found it difficult to compare results generated by different laboratories (Nagy et al., 1971; Nagy and Verakis, 1983, pp. 91−95; Eckhoff, 2003, pp. 534−535). Investigators experimented with a variety of test vessels, dust dispersal methods, igniters, and test procedures in a search for reproducible and scientifically defensible dust explosion measurements. By the 1980s, the scientific community had converged on a preferred test vessel geometry: the spherical vessel with a minimum volume of 20 L (Bartknecht, 1981, pp. 27−44; Field, 1989, pp. 76−81; Bartknecht, 1989, pp. 56−80). The primary benefit of the spherical test vessel with a minimum volume of 20 L is that it yields an explosion severity parameter that is scalable to larger vessel sizes (assuming that the vessel is properly calibrated). This explosion severity parameter, the K_{st} value, provides a sound basis for the design of dust explosion protection technologies like venting and suppression.

8.1.2 FLAME TRAVEL AND PRESSURE RISE

Consider a combustible dust uniformly distributed inside a spherical vessel. If a combustible dust cloud is ignited at its center, the flame will propagate in the outward radial direction and exhibit spherical symmetry. Buoyancy or an asymmetric dust concentration distribution can distort the spherical shape, but the spherical dust flame is an important and useful conceptual model for a confined deflagration. As the flame advances into the unburnt mixture, it behaves like a leaky piston: it pushes the unburnt mixture toward the confining wall but allows some of the unburnt mixture to enter the flame. The flame converts the unburnt mixture into burnt mixture and drives the flow of the burnt mixture toward the center of the sphere. Thus, from the perspective of an observer in the laboratory, as the flame front advances, there is a flow of unburnt mixture toward the wall and a flow of burnt mixture toward the center. The mixture velocity is zero at both the center of the vessel and at the wall. Fig. 8.1 illustrates the propagation of a spherical flame inside a spherical vessel.

Because the volume grows as the radius cubed, the majority of the fuel is burned near the end of the flame travel. Although this seems a simple observation, the cubic behavior of the burnt gas volume growth is a very important driver

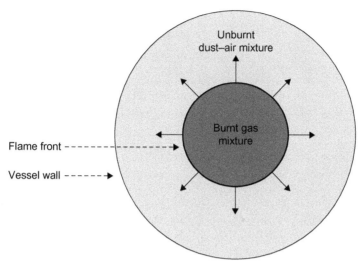

FIGURE 8.1

Spherical dust flame propagation in a spherical vessel.

Table 8.1 Tabulation of Dimensionless Burnt Gas Volume as a Function of Flame Radius for a Spherical Combustion Vessel

r_b/r_{sphere}	0	0.1	0.2	0.3	0.4	0.5	0.6	0.7	0.8	0.9	0.95	1
V_b/V_{sphere}	0	0.001	0.008	0.027	0.064	0.125	0.216	0.343	0.512	0.729	0.857	1

behind confined deflagration behavior. Table 8.1 is a calculation of the dimensionless burnt gas volume, V_b/V_{sphere}, expressed as a function of the dimensionless flame radius, r_b/r_{sphere}, assuming central ignition.

Table 8.1 shows that when the flame has traveled to the midpoint of the vessel, the corresponding burnt gas volume occupies only 1/8 of the vessel volume; at 90% of the vessel radius, less than 75% of the vessel volume is occupied by burnt gas.

This cubic relationship between flame travel and burnt gas volume results in a slow pressure rise for most of the flame travel distance. As the flame approaches the vessel wall, the pressure suddenly rises quickly reaching its final value when the flame first touches the vessel wall (or nearly so). This behavior is illustrated in Fig. 8.2, a plot of explosion data from a carbon monoxide–oxygen mixture in a 7.6-L spherical bomb (Fiock and Marvin, 1937; Fiock et al., 1940).

The shapes of the flame position and pressure history curves are a distinctive feature of confined deflagrations. The rather large explosion pressure of this system arises because of the lack of diluent gas (64.87% CO, 32.44% O_2, and 2.69% H_2O).

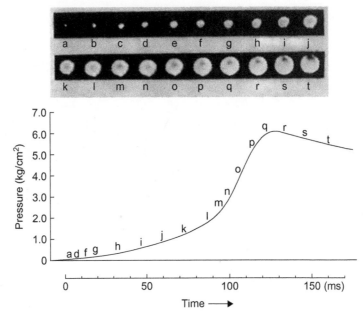

FIGURE 8.2

Plot of pressure history measurements and flame motion photographs for a carbon monoxide–oxygen mixture in a closed vessel. The letters on the flame photographs coincide with the timing indicated on the pressure history.

Reprinted with permission from Fiock, E.F., Marvin, Jr., C.F., 1937. The measurement of flame speeds. Chem. Rev. 21, 367–387.

Since the flame travels from the center of the vessel outwards, it should be apparent that the combustion process will not create a uniform distribution of temperature or species concentrations. Instead, the combustion process can be described as progressive burning, meaning that the combustion process follows a one-dimensional trajectory and creates one-dimensional temperature and concentration profiles.

The pressure changes as a function of time but is spatially uniform. The radial growth of the spherical flame correlates directly with the increasing pressure. This is illustrated in Fig. 8.3 which depicts the pressure history of an aluminum dust (particle size −400 mesh) explosion in a 10-L cylindrical vessel (Ishihama and Enomoto, 1973). Ishihama and Enomoto tracked the flame motion with high-speed photography and correlated these photographs with the pressure history. The deflagration was initiated at the center of the vessel and the flame propagated radially outwards. They observed that the peak pressure coincided with the flame front contacting the bottom vessel wall.

The rising pressure compresses the burnt and unburnt gas mixture thereby increasing the local temperature. In conjunction with the progressive combustion process, this compression process results in a significant temperature gradient with the highest temperature at the center (one of the first reports of this phenomenon was by Hopkinson, 1906).

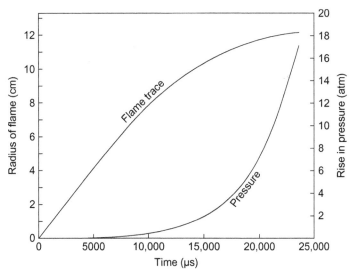

FIGURE 8.3

Pressure rise history plotted with a timed sequence of photographs showing the growth of an aluminum dust flame in a 10-L test vessel. The dust concentration was 1000 g/m^3 and the shape of the vessel was a right circular cylinder. The conversion for the pressure units is 1 kg/cm^2 = 0.9807 bar.

Reproduced with permission from Ishihama, W., Enomoto, H., 1973. New experimental method for studies of dust explosions. Combust. Flame 21, 177–186.

The characteristic features of the pressure history have become the primary measurements for analyzing dust deflagration behavior. It is relatively straightforward to record the pressure history over the course of the deflagration. At a minimum, the dust explosion parameters $(dP/dt)_{max}$ and P_{max} can be extracted from the pressure history (see Chapter 1). The explosion test vessel has evolved into a convenient experimental technology that is well-accepted for characterizing combustible dust hazards and results in test data that are appropriate for process safety design. Additionally, with an appropriate formulation of an integral model, it is possible to extract estimates of the burning velocity from the pressure history.

8.1.3 MODEL TAXONOMY

There is a large number of papers published on the modeling of flame propagation in closed vessels. To help make sense of this diverse literature, it is useful to create a taxonomy or classification scheme to determine the similarities and differences between the various models. For example, Eckhoff organized his discussion of dust flame propagation by degree of increasing mass combustion rate: single particle burning, laminar flames, turbulent flames, and detonation (Eckhoff, 2003, Chapter 4). Continillo differentiates models by the mathematical character of the model: lumped parameter models (consisting of a set of ordinary differential equations) and distributed parameter models (consisting of a set of partial differential equations) (Continillo, 1989). Cloney and his colleagues discuss the importance of length and time scale resolution in helping one to decide on the appropriate level (Cloney et al., 2014).

I will present in total four classes of models for confined flame propagation. In this chapter, I will introduce the first class of models: integral models. The second class, based on gas dynamics, will be discussed in Chapter 9. The third and fourth classes, based on the equations of change, will be discussed in Chapter 10. The third class is called hydrodynamic models. Hydrodynamic models are models of the flame propagation process that depend on the unsteady, one-dimensional, inviscid equations of change to describe the gas motion. The flame itself is treated as an infinitely thin discontinuity. The advantage of the hydrodynamic models is that they provide a means of probing some of the assumptions underlying the integral models.

The fourth class will be based on computational fluid dynamics (CFD). CFD models use numerical methods to solve the equations of change. With CFD models, it is possible to incorporate multiphase flow effects, turbulence, and thermal radiation.

In Chapter 3 we applied simple thermodynamic models to extract the basic information about the constant volume flame temperature and explosion pressure. The simplified thermodynamic models assumed uniform combustion within a closed volume. The uniform combustion model is but a first approximation for describing the progressive motion of the flame in a confined deflagration. To examine the nature of the progressive combustion process that is characteristic of confined flame propagation, we first turn to integral models. Typically, integral models are models based on mass and energy balances written for control volumes. The physical and chemical properties within the control volume are assumed to be spatially uniform. Since flame propagation in closed vessels is an inherently non-uniform process, investigators over the years have employed a number of clever, insightful mathematical tricks to coax gradients out of integral models. We shall briefly survey this field before examining in detail two specific models, an adiabatic model by Bradley and Mitcheson (1976) and an isothermal model due to Nagy et al. (1969).

8.1.4 SUMMARY

The study of flame propagation in closed vessels serves two important functions: it serves as a convenient, rational method for the laboratory evaluation of combustible dusts, and it is the fundamental conceptual model for a primary dust explosion. At the beginning of the 20th century, investigators realized that a spherical explosion vessel with central ignition could give them the best quality data. By the 1980s, a consensus emerged that the minimum size vessel that would yield scalable dust explosion test results was the 20-L sphere. Meanwhile many investigators around the world sought to discover and model the relationship between the flame motion and the pressure history that evolved during a confined deflagration test. One of the more striking differences between unconfined and confined flame propagations is the temperature gradient that forms within the confinement due to adiabatic compression.

8.2 THE FLAMM—MACHE TEMPERATURE GRADIENT

Consider a combustible dust cloud dispersed in a rigid, impermeable, and adiabatic vessel. If the mixture is ignited and it burns at a uniform rate throughout the mixture, the final temperature within the vessel will be spatially uniform and equal the isochoric flame temperature (see Chapter 3). But in combustion vessels with central ignition, the temperature is not spatially uniform. Instead experimental observations have shown that the temperature is highest at the center of the vessel and decreases monotonically toward its lowest value at the vessel wall (Hopkinson, 1906; Lewis and von Elbe, 1987, pp. 381–395, 616–621). The temperature gradient, called the Flamm—Mache temperature gradient, is created by the progressive combustion process as the deflagration wave sweeps through the unburnt mixture (Zeldovich et al., 1985, pp. 472–478). Fig. 8.4 illustrates the Flamm—Mache temperature gradient in relation to the isochoric flame temperature T_{ex}.

The Flamm—Mache temperature gradient can be analyzed with a thermodynamic model which considers the processes of combustion and adiabatic compression. We will consider the two endpoints of the radial temperature profile: the

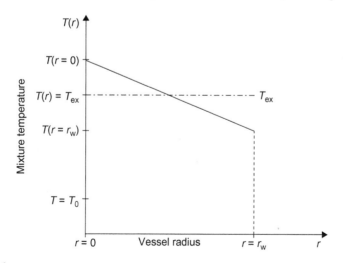

FIGURE 8.4

Depiction of the Flamm—Mache temperature profile in a spherical vessel.

vessel center and the wall. The dust mixture is assumed to obey the ideal gas equation of state. All properties are assumed constant and we neglect dissociation effects. The temperature at the center can be calculated by assuming that the first increment of fuel burned occurs at constant pressure (the initial pressure P_0), corresponding to the adiabatic flame temperature, T_{ad}. The burnt mixture at that point is then compressed to the final explosion pressure, P_{ex}. We assume that $T_0 = 300$ K, $\Delta T_{ad} = 2000$ K, $P_{ex} = 10$ atm (absolute pressure), $P_0 = 1$ atm, and $\gamma = 1.2$. Invoking the equation for the adiabatic compression of an ideal gas, $(T_2/T_1) = (P_2/P_1)^{(\gamma-1)/\gamma}$, the temperature of the burned mixture at the center of the vessel is given by the equation (Smith et al., 1996, pp. 65–66):

$$T(r=0) = (T_0 + \Delta T_{ad})\left(\frac{P_{ex}}{P_0}\right)^{(\gamma-1)/\gamma} \tag{8.1}$$

Substituting the values given, the center temperature is $T(r=0) = 3380$ K.

The temperature at the vessel wall can be calculated by assuming that the unburnt mixture temperature is compressed adiabatically from the initial pressure to the explosion pressure. The unburnt mixture then reacts and further increases its temperature by the amount ΔT_{ad}.

$$T(r=r_w) = T_0\left(\frac{P_{ex}}{P_0}\right)^{(\gamma-1)/\gamma} + \Delta T_{ad} \tag{8.2}$$

Substituting the values given, the temperature at the wall is $T(r=r_w) = 2440$ K. This value is substantially less than the center temperature. It is evident from these two equations that the temperature at the vessel center will always be hotter than the temperature at the wall.

In Fig. 8.4, the isochoric flame temperature is shown as an intermediate temperature between the center and wall temperatures. The Flamm–Mache temperature gradient does not invalidate the constant volume combustion analysis of Chapter 3. The mass-averaged temperature of the burnt mixture is the isochoric flame temperature. This assertion will be proven in Section 8.5 when we discuss integral models. From the point of view of hazard evaluation, the isochoric flame temperature and the explosion pressure are sufficient as thermodynamic metrics for combustible dust hazards.

8.3 THE FRACTIONAL PRESSURE RISE RELATION

In Section 3.7, we derived a thermodynamic expression which related the fractional pressure rise with the mass fraction of the mixture burnt.

$$f = \frac{P - P_0}{P_{ex} - P_0} \quad \text{and} \quad f = \frac{m_{mix,b}}{m_{mix,0}} \tag{8.3}$$

The mass fraction f ranges in value from 0 at the start of combustion (zero quantity of fuel consumed) to 1 when combustion is complete (all fuel consumed in fuel lean systems or all oxygen consumed in fuel rich systems). This relation was derived for a system undergoing uniform combustion (ie, a well-mixed system) (Guenoche, 1964, pp. 160−165; Zeldovich et al., 1985, pp. 472−478). It was not explicitly derived for systems undergoing progressive combustion.

This relation has often been applied to progressive combustion systems. It plays a central role in the derivation of integral models for simulating confined dust deflagrations and is often invoked as a simplifying relation. It is usually attributed to the combustion investigators Lewis and von Elbe (Jost, 1946, pp. 145−146; Lewis and von Elbe, 1987, pp. 386−389). The primary assumptions in their derivation were that the heat capacities and molar masses were constant and equal in the unburnt and burnt mixtures. It was implied that the heat capacity ratio was also a constant. They further assumed that the ideal gas equation of state was obeyed by the mixture and that the system was adiabatic. This simple relation greatly simplifies the analysis of progressive combustion in a closed vessel. I shall refer to Eq. (8.3) as the LvE (Lewis and von Elbe) fractional pressure rise relation.

There have been a number of critiques of the fractional pressure rise relation involving both theoretical and empirical combustion studies (Oppenheim and Maxson, 1993; Saeed and Stone, 2004a,b; van den Bulck, 2005; Luijten and de Goey, 2007; Luijten et al., 2009a,b; Saeed, 2009). The general consensus of these studies is that the fractional pressure rise relation is only approximate and has the potential to introduce errors on the order of 10% into burning velocity determinations. Whether this is a significant source of error in modeling a confined dust deflagration will depend on your specific application.

Luijten and his colleagues have given a very careful analysis of the fractional pressure rise relation (Luijten et al., 2009b). Their derivation is quite informative and will be summarized here. Although their derivation is for a flammable gas mixture, it is readily generalized to combustible dust mixtures by substituting multiphase mixture properties for the gas properties. Consider a combustible dust mixture undergoing progressive combustion in a spherical vessel. We will assume that the mixture was ignited at the center of the vessel and that a spherical flame is propagating outward toward the wall. The flame is assumed to be infinitely thin. The vessel is rigid, adiabatic, and impermeable. The gas is assumed to obey the ideal gas equation of state. As shown in Fig. 8.5, the flame divides the mixture into two regions, burnt and unburnt.

The total specific volume v_t is the mass of dust−air mixture divided by the volume of the vessel. The total specific internal energy u_t at the initial condition is the internal energy of combustion. The total specific volume and internal energy can be written in terms of the burnt and unburnt mass contributions:

$$v_t = f v_b + (1 - f) v_u \tag{8.4}$$

$$u_t = f u_b + (1 - f) u_u \tag{8.5}$$

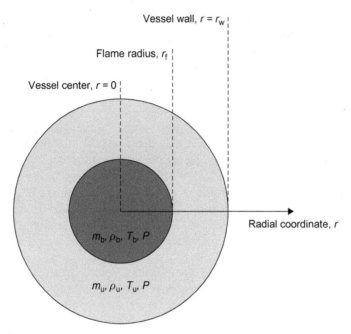

FIGURE 8.5

Coordinate frame for dust deflagration in a spherical vessel.

Substituting the ideal gas equation of state into the specific volume equation, we obtain

$$\frac{R_g T_0}{P_0 \mathcal{M}_0} = f\left(\frac{R_g T_b}{P \mathcal{M}_b}\right) + (1-f)\left(\frac{R_g T_u}{P \mathcal{M}_u}\right) \tag{8.6}$$

As before, the symbol R_g denotes the universal gas constant. It is important to emphasize that the following symbols are constants: $P_0, P_{ex}, T_0, \mathcal{M}_0, C_{v,u}, C_{v,b}$. These symbols are variables which vary with time: T_b, T_u, P, f. Finally, $\mathcal{M}_0 = \mathcal{M}_u$.

I have exercised two changes in the derivation presented in Luijten et al. paper. First, I will assign the reference temperature for the internal energy to be equal to the initial temperature of the vessel contents, $T_{reference} = T_0$. Second, I have written out specifically the ratio $R_g/\mathcal{M}$ whereas they combine the ratio into a single symbol. The specific internal energy equation can be written in terms of the thermal and chemical (combustion) energy components

$$u_u = C_{v,u}(T_u - T_0) + \Delta u_c \quad \text{and} \quad u_b = C_{v,b}(T_b - T_0) \tag{8.7}$$

Substituting Eqs. (8.6) and (8.7) into (8.5) gives the following expression of the burnt gas temperature:

$$T_b = T_0 + \frac{1}{f C_{v,b}}\left\{f\Delta u_c - (1-f)\left[C_{v,u}(T_u - T_0)\right]\right\} \tag{8.8}$$

The burnt gas temperature variable T_b can be eliminated in Eq. (8.6) by substituting Eqs. (8.7) and (8.8). Solving for the pressure ratio yields

$$\frac{P}{P_0} = (1-f)\frac{T_u}{T_0} + f\left(\frac{\mathcal{M}_u}{\mathcal{M}_b}\right) + \frac{\mathcal{M}_u}{\mathcal{M}_b C_{v,b} T_0}\left[f\Delta u_c - (1-f)C_{v,u}(T_u - T_0)\right] \qquad (8.9)$$

To simplify this expression, we solve Eq. (8.9) for the explosion pressure P_{ex} by setting $f = 1$. This gives us

$$\frac{P_{ex}}{P_0} = \left(\frac{\mathcal{M}_u}{\mathcal{M}_b}\right)\left[\frac{C_{v,b}T_0 + \Delta u_c}{C_{v,b}T_0}\right] \qquad (8.10)$$

Substituting this into Eq. (8.9) simplifies to this equation:

$$\frac{P}{P_0} = (1-f)\frac{T_u}{T_0} + f\left(\frac{P_{ex}}{P_0}\right) - (1-f)\frac{\mathcal{M}_u}{\mathcal{M}_b}\left[\frac{C_{v,u}(T_u - T_0)}{C_{v,b}T_0}\right] \qquad (8.11)$$

Employing the definition for the specific heat capacity ratio, $\gamma_u = C_{p,u}/C_{v,u}$ and $\gamma_b = C_{p,b}/C_{v,b}$, the last term on the right-hand side of Eq. (8.11) can be written as

$$\frac{\mathcal{M}_u}{\mathcal{M}_b}\left[\frac{C_{v,u}(T_u - T_0)}{C_{v,b}T_0}\right] = \frac{\gamma_b - 1}{\gamma_u - 1}\left(1 - \frac{T_u}{T_0}\right) \qquad (8.12)$$

Substituting Eq. (8.12) into Eq. (8.11) and solving for the mass fraction burnt f

$$f = \frac{P - P_0\left[\left(\frac{\gamma_b - 1}{\gamma_u - 1}\right) + \frac{T_u}{T_0}\left(\frac{\gamma_u - \gamma_b}{\gamma_u - 1}\right)\right]}{P_{ex} - P_0\left[\left(\frac{\gamma_b - 1}{\gamma_u - 1}\right) + \frac{T_u}{T_0}\left(\frac{\gamma_u - \gamma_b}{\gamma_u - 1}\right)\right]} \qquad (8.13)$$

Finally, we assume adiabatic compression of the ideal gas mixture $(T_u/T_0) = (P/P_0)^{(\gamma_u - 1)/\gamma_u}$, and upon substitution we get an expression entirely in pressures and specific heat capacities:

$$f = \frac{P - P_0 F(P)}{P_{ex} - P_0 F(P)} \quad \text{where} \quad F(P) = \left(\frac{\gamma_b - 1}{\gamma_u - 1}\right) + \left(\frac{\gamma_u - \gamma_b}{\gamma_u - 1}\right)\left(\frac{P}{P_0}\right)^{(\gamma_u - 1)/\gamma_u} \qquad (8.14)$$

The final result, Eq. (8.14) (the LDdG formula), reduces to the Lewis–von Elbe (LvE) expression, Eq. 8.3, when $\gamma_u = \gamma_b$. Realistically, $\gamma_u = \gamma_b$ will be the exception, not the rule, but one can estimate a mean value such that $\gamma_{mean} = \gamma_u = \gamma_b$ (Nagy and Verakis, 1983, p. 83). The magnitude of error introduced by this assumption depends on the functional form of the equation. Luijten et al. examined the constant volume combustion of methane–air mixtures and determined that the error in determining the laminar burning velocity introduced by using the LvE expression was on the order of 1.5% (Luijten et al., 2009b). Studies by other authors have essentially concurred with this assessment (Saeed and Stone, 2004a,b; van den Bulck, 2005; Luijten and de Goey, 2007; Luijten et al., 2009a; Saeed, 2009).

EXAMPLE 8.1

Compare the performance of the LvE model with the LDdG model. Consider a combustible dust–air mixture contained in a spherical combustion vessel with rigid, adiabatic, and impermeable walls. Assume central ignition of the mixture. The properties of the mixture are $P_0 = 1.00$ bar, $P_{ex} = 9$ bar, $\gamma_u = 1.4$, and $\gamma_b = 1.2$. Compute the burn fraction f as a function of the pressure ratio P/P_0 in increments of 1.0 bar. Calculate the error in f as defined by

$$\text{Error}(f)(\%) = \left[\frac{f(\text{LvE}) - f(\text{LDdG})}{f(\text{LDdG})}\right] \times 100$$

Solution

The burn mass fraction for the two models is given by the following equations:

$$f(\text{LvE}) = \frac{(P/P_0) - 1}{(P_{ex}/P_0) - 1}$$

$$f(\text{LDdG}) = \frac{(P/P_0) - F(P)}{(P_{ex}/P_0) - F(P)} \quad \text{where} \quad F(P) = \left(\frac{\gamma_b - 1}{\gamma_u - 1}\right) + \left(\frac{\gamma_u - \gamma_b}{\gamma_u - 1}\right)\left(\frac{P}{P_0}\right)^{(\gamma_u - 1)/\gamma_u}$$

The comparison is tabulated below.

P/P_0	$f(\text{LvE})$	$f(\text{LDdG})$	Error(%)
1	0	0	0
2	0.125	0.113	10.6
3	0.250	0.232	7.8
4	0.375	0.355	5.6
5	0.500	0.481	4.0
6	0.625	0.609	2.6
7	0.750	0.738	1.6
8	0.875	0.868	0.8
9	1	1	0

For the numerical values given for this example, the greatest error in the calculation of the burn fraction introduced by the LvE model is at low pressure ratios. The magnitude of the error decreases as the pressure ratio increases.

In many applications, the linear LvE approximation will be satisfactory. For applications demanding the utmost in numerical accuracy, the LDdG formula is recommended.

8.4 TIME SCALE ANALYSIS FOR CONFINED UNSTEADY DUST FLAME PROPAGATION

In our investigation of unconfined dust flame propagation, I introduced the concept of an efficient deflagration wave wherein the time scale of the deflagration wave (the flame thickness divided by the burning velocity, $t_{wave} = \delta/S_u$) is equal to the burnout time of a single particle, $t_{wave} = \tau$. This was a simple example of time scale analysis. In a confined deflagration, two additional macroscale parameters are apparent: the length scale of the vessel, its smallest half-dimension, and the time scale for complete flame travel, which can be approximated by the time to maximum pressure. Although several investigators have referenced the importance of time and length scale analyses to understand the relative magnitude of the physical and chemical processes occurring during a deflagration, the work by van der Wel and his colleagues stands out in this regard (van der Wel et al., 1992). Their experimental program involved testing three combustible dusts with different levels of volatility (lycopodium, potato starch, and activated carbon) and varied the igniter strength and ignition delay time. They specifically compared the performance of the 20-L and 1-m^3 vessels in terms of three time scales and demonstrated their value in providing insight into the dust deflagration process.

In the analysis below, I not only address some of the same time scale ratios but also will discuss some additional considerations. It should be noted that none of these time scale ratios rise to the level of a critical dimensionless parameter like the Reynolds number where a specific value can signify a transition to a new transport regime. Instead, these time scale ratios are primarily of value as a comparison in the sense of being greater than or less than 1 to indicate the relative importance of competing physical processes.

It is important to identify characteristic time and length scale ratios in a physical problem because they offer insight into the underlying physics and they offer a means of testing assumptions. This insight can help guide the development of mathematical models and it is also a way to better understand the limitations of model assumptions. The system of interest to be analyzed will be the laboratory test vessel. This is because it is easier to evaluate the effect of rate processes under the controlled conditions of the laboratory. The same principles will apply in accident settings, but the presence of greater uncertainty may hinder their application. The analysis of this section will be based on one of the more common laboratory test vessels, the 20-L sphere ($r_w = 0.168$ m). It is assumed that the combustible dust is a population of monodisperse spheres with a particle diameter d_p.

The reference time scale for a confined deflagration in an explosion test vessel is the time to maximum pressure, $t_{P_{max}}$, which is the time required for the flame to travel the distance equal to the radius of the test vessel (see Chapter 1). Unfortunately, the time to maximum pressure is not routinely reported with dust explosibility data. Working with a 20-L sphere, Ogle et al. tested aluminum powders at dust concentrations from 300 to 900 g/m^3 ($\Phi \cong 1-3$); they reported time to maximum pressure data in the range $5 \leq t_{P_{max}} \leq 40$ ms (Ogle et al., 1988). Cashdollar presented example results from a 20-L sphere for eight dusts (featuring a wide range of volatilities) all tested at a dust concentration of 600 g/m^3; he reported time to maximum pressure data from 25 to 250 ms (see Figures 1−3, reproduced with permission from Cashdollar, 2000). It is proposed that the order of magnitude of the time to maximum pressure in a 20-L sphere for most combustible dusts will range between 10 and 300 ms. This rather broad range of values arises from the range in dust concentrations, fuel volatility, and particle sizes encountered in combustible dust testing.

We will examine five characteristic time scale ratios based on the following dynamic variables:

- The particle burnout time, τ, is the time for complete combustion of a single particle of initial diameter d_{p0}
- The particle momentum relaxation time, t_{mom}, is the time for a particle at rest to accelerate to 67% of the free stream gas velocity
- The turbulent integral time scale, t_{int}, is the characteristic time scale of the turbulent velocity fluctuations
- Gravitational settling time, t_{TSV}, is the time required for a particle of initial diameter d_{p0} to fall through a distance equal to the radius of the test vessel
- Acoustic wave travel time, t_{ac}, is the time for a sound wave to travel the distance equal to the radius of the test vessel.

Each time ratio will be a dimensionless number and will be designated by the symbol Π_i.

The first time scale ratio is the combustion time ratio, $\Pi_{combustion} = \tau/t_{P_{max}}$. If this ratio is less than 1 then a single particle has the potential to burn to completion before the flame has reached the vessel wall. An alternative interpretation is that it indicates that the flame thickness is thin compared to the radius of the vessel. A thin flame thickness, $\delta \ll r_w$, is one of the prerequisites for the cubic scaling law employing the K_{st} value. But there is insufficient data to provide further quantitative resolution on this question of flame thickness and the cubic scaling law. It is well recognized that dust flames are thicker than premixed gaseous flames and this may be a source of departure from the cubic scaling law, but there are as yet no easily implemented quantitative criteria for establishing the degree of departure (see, for example, the discussion in Dahoe et al., 1996a). It is likely that the time to maximum pressure is proportional to the particle diameter of the dust tested, but this relationship has not yet been systematically investigated.

A further note of caution is in order regarding the burnout time τ. The particle burnout time is a theoretical prediction of combustion behavior for a single particle. As a flame sweeps through a suspended dust cloud, the combustion behavior of an individual particle may differ considerably from the single particle combustion ideal. Even if the combustion time ratio is much smaller than one, it is possible that the particles will not be completely consumed. The fact that so little is known about this seemingly obvious time scale ratio is an indication that much work remains to be done to better understand combustible dusts.

The next time scale ratio is the momentum time ratio, defined as $\Pi_{\mathrm{mom}} = t_{\mathrm{mom}}/t_{P_{\max}}$. If this ratio is less than 1 then the velocity slip will tend to be negligible. A quantitative example illustrating a range of magnitudes relevant to this analysis was presented in Section 4.6.2 and Table 4.5. From Section 4.6.2, the equation for the momentum time scale is given by the following equation:

$$\tau_{\mathrm{mom}} = \frac{\rho_p d_p^2}{18\mu} \tag{8.15}$$

For most organics and metals, velocity slip will be negligible for particle diameters $d_p \leq 30$ μm. This means that these smaller particles will follow the fluctuations of the turbulent flow field.

As described in Section 4.8, there are different length scales that can help describe the spectrum of turbulent velocity fluctuations. Of these, the integral time scale is useful as it correlates with the largest eddies of the flow field. The turbulent integral time scale ratio is $\Pi_{\mathrm{int}} = t_{\mathrm{int}}/t_{P_{\max}}$. If this time scale ratio is larger than 1 then the flame front will become wrinkled and the mass combustion rate of the flame will be increased due to the increased flame surface area (van der Wel et al., 1992). If this time scale is less than 1, then the local interphase (interparticle) heat and mass transfer rates will be accelerated by the fine scale turbulence. The dust explosibility parameters are sensitive to the level of turbulence created by the dust dispersion process in the test vessel (data to support this assertion will be discussed later in this chapter). Test vessels of different sizes can be calibrated by adjusting the igniter strength and ignition delay time so that the explosibility parameters better conform to the cubic scaling law. In addition to the effects of the turbulent dust dispersion process, explosibility parameters can be affected by the magnitude of the turbulent burning velocity. The characteristics of the turbulent burning velocity and its impact on the behavior of a dust deflagration are sufficiently complex that I will devote an entire section of this chapter to this topic.

The gravitational settling time ratio is defined as $\Pi_{\mathrm{TSV}} = t_{\mathrm{TSV}}/t_{P_{\max}}$. This metric is a measure of the relative importance of particles settling in comparison with the flame travel time. This is essentially the same issue analyzed for vertical flame propagation in Chapter 7. The particle settling time is calculated from the terminal settling velocity of the particle (see Section 4.6.2): $t_{\mathrm{TSV}} = r_w/v_{p,\mathrm{TSV}}$. It can be expected that the effect of particle settling will be negligible if the ratio is greater than 1. This problem was examined in Example 4.9. Referring to that example, particle settling should be unimportant for the particle size range $d_p \leq 100$ μm. For larger particle sizes, it is likely that gravity settling will distort the dust concentration distribution.

The final time scale ratio to consider is the acoustic time ratio. This ratio, designated by $\Pi_{ac} = t_{ac}/t_{P_{max}}$, is important because it offers a means for evaluating the assumption of spatial uniformity of the pressure field. The justification for neglecting the pressure gradient is an important simplification for the modeling of confined deflagrations. The relaxation of this assumption will be addressed in Chapter 9. The acoustic time scale is simply the ratio of the vessel radius to the sound speed: $t_{ac} = r_w/a_g$. At a temperature of 300 K (27°C), the sound speed in air is 347 m/s (Zucrow and Hoffman, 1976, p. 701). The acoustic time scale is 0.48 ms, which makes the acoustic time ratio much smaller than one which, in turn, implies that the pressure gradient can be neglected. As an aside, it has been assumed that the sound speed within the dust cloud can be approximated by the sound speed of air. This assumption can be relaxed, but for the dust concentrations of interest, it is a reasonable assumption (Wallis, 1969, pp. 143−144, 207−209).

The reasoning behind this inference can be explained thus. As the flame grows with each infinitesimal advance, the heat released by combustion causes an infinitesimal increase in pressure. The infinitesimal increase in pressure propagates as a wave pulse from the flame toward the vessel wall. This wave pulse is the beginning of a pressure gradient. But the pressure pulse is reflected at the wall and propagates back toward the flame. The reflected pulse "smooths" the spatial gradient in pressure (but does not diminish it) and propagates toward the center of the vessel. Each incremental advancement of the flame is accompanied by additional heat release and generation of pressure pulses. In the time scale of the time to maximum pressure, the generation, reflection, and interaction of pressure pulses ensure a steady rise in pressure that is spatially uniform. This analysis has been subjected to rigorous analysis and validated by investigators at the U.S. Bureau of Mines (Perlee et al., 1974; Kansa and Perlee, 1976). In these studies, it was concluded that pressure gradients within the test vessel could be neglected if the flame Mach number did not exceed the criterion, $Ma = S_u/a_g < 0.2$.

In summary, time scale analysis in some cases can place bounds on some of the common assumptions employed to analyze confined deflagrations of combustible dust. In other cases, it reveals how little we know about the fundamental behavior of combustible dust. The affirmative conclusions we can draw from this section is that for fine particle sizes velocity slip and gravitational settling can be neglected. Turbulence will enhance the burning velocity of the deflagration, but the exact mechanism depends on the particle size. The spatial gradient of pressure can be neglected and this is not affected by particle size or dust concentration.

8.5 INTEGRAL MODELS

Integral models are mathematical statements of mass and energy conservation in a control volume. The accumulation of mass or energy within the control volume is expressed as the difference of input and output streams. By now it should be expected that the models that we formulate for flame propagation in a spherical vessel must somehow accommodate the motion of the flame within the vessel. This is accomplished by dividing the contents of the spherical vessel into two regions, a burnt gas region and an unburnt gas region. The two regions are divided by the flame which is treated as an infinitely thin surface of discontinuity. The essential ingredients of an integral model for flame propagation in a closed vessel are mass balances for both the burnt and unburnt gas regions, energy balances for both regions, the jump condition for mass balance across the flame, an equation of state, and just a little bit of geometry. Of course one must make the appropriate arrangements for estimating the thermodynamic properties of the fuel−air mixture and for specifying the initial and boundary conditions.

I will begin this section with a discussion of the elementary relations that are fundamental to flame propagation in a spherical vessel. Next I survey the range of integral models emphasizing the history of development and giving a very brief overview of the many contributions from the international combustion community. I then consider the derivation of two specific flame propagation models, the adiabatic model of Bradley and Mitcheson and the isothermal model of Nagy, Conn, and Verakis. The adiabatic model yields a very good simulation of flammable gas and combustible dust flame propagation behavior in a closed vessel. Unfortunately, it is a nonlinear differential equation that resists analytical integration. The isothermal model is less accurate, but it is simpler to evaluate analytically and it provides a direct path toward the cubic pressure rise relation. Throughout this section, I will be drawing heavily from the gas combustion literature. The models used for gas combustion are easily adapted to dust combustion by using mixture properties. The potential limitations to this approach will be addressed as we encounter them.

8.5.1 ELEMENTARY RELATIONS FOR FLAME PROPAGATION IN A SPHERICAL VESSEL

Early in the history of gas explosion investigations, investigators realized that spherical test chambers would be the most convenient shape with which to work because centrally ignited flammable mixtures propagate spherical flames. Nonspherical explosion vessels permit asymmetric flame contact with the cold walls of the vessel, thereby leading to lower explosion pressures and confounding a reliable measure of the maximum rate of pressure rise. At first glance, it might seem that the mathematical modeling of a long cylinder closed on both ends and ignited at one end would be an easier system to analyze. But such an experimental arrangement leads to acoustic oscillations and turbulence, effects that complicate a simple control volume analysis (Guenoche, 1964). Thus, the spherical vessel has largely prevailed as the preferred configuration.

To facilitate a control volume analysis, it is convenient to invoke the following assumptions.

- Spherical vessel with rigid, adiabatic, and impermeable walls
- Constant thermodynamic properties
- Constant molar masses
- Ideal gas equation of state
- Adiabatic compression of the gas as the pressure rises
- Linear fractional pressure rise relation.

The experimental validation of these assumptions will be examined later.

8.5.1.1 Mass balances and geometrical relations for a spherical vessel

Most integral models are derived using the same basic formulation of mass balances and geometrical relations (Bradley and Mitcheson, 1976; Nagy and Verakis, 1983, Chapter 5; Harris, 1989; Mannan, 2005, pp. 17/37 to 17/47). We begin with some definitions for the unburnt and burnt mass of reacting mixture.

$$m_i = m_u + m_b; \quad m_i = \rho_i V_{tot}; \quad m_u = \rho_u V_u; \quad m_b = \rho_b V_b \qquad (8.16)$$

The subscripts i, u, b refer to initial, unburnt, and burnt quantities. The following geometrical relations are evident from Fig. 8.5.

$$V_{tot} = V_u + V_b; \quad V_{tot} = \frac{4\pi}{3} r_w^3; \quad V_u = \frac{4\pi}{3}(r_w^3 - r_f^3); \quad V_b = \frac{4\pi}{3} r_f^3 \qquad (8.17)$$

The mass balance equations for the unburnt and burnt mixture regions and the jump condition across the flame front are as follows:

$$\frac{dm_u}{dt} = -\rho_u S_u A_f; \quad \frac{dm_b}{dt} = \rho_b S_b A_f; \quad \rho_u S_u = \rho_b S_b \qquad (8.18)$$

Given the equality of pressure throughout the explosion vessel, the ideal gas equation of state for the unburnt and burnt gas mixtures can be written as follows:

$$P = P_u = \frac{m_u R_g T_u}{\mathcal{M}_{mix,u} V_u} = P_b = \frac{m_b R_g T_b}{\mathcal{M}_{mix,b} V_b} \qquad (8.19)$$

We now turn our attention to deriving an expression for the average burnt gas temperature as an example of the utility of the relations just presented.

8.5.1.2 Average burnt gas temperature

In Section 8.2, I introduced the Flamm—Mache temperature gradient and stated that upon completion of the deflagration the average temperature of the burnt mixture was the isochoric flame temperature. In this section, I prove that assertion with the aid of an expression for the average burnt gas temperature derived by

Luijten et al. for any value of the burn fraction f (Luijten et al., 2009). While this exercise provides insight into the confined deflagration process, it is likely to be somewhat in error in its quantitative predictions due to the assumption of constant thermodynamic properties. However, it is not difficult to modify this analysis and allow for chemical equilibrium with variable properties. The modified algorithm will then require solution by numerical analysis.

Consider a flame propagating through a flammable gas mixture in a spherical vessel. An expression for the pressure within the vessel can be written in terms of the flame position r_f and the burn mass fraction f. Substitute the expressions for the unburnt and burnt gas volumes into Eq. (8.19), and write the mass variables in terms of the burn mass fraction:

$$\frac{m_i(1-f)R_gT_u}{\frac{4\pi}{3}(r_w^3 - r_b^3)\mathcal{M}_{mix,u}} = \frac{m_i f R_g \bar{\bar{T}}_b}{\frac{4\pi}{3}(r_b^3)\mathcal{M}_{mix,b}} \tag{8.20}$$

The burnt gas temperature $\bar{\bar{T}}_b$ is written with the double overbar to signify that it is a mass-averaged temperature. Both the burnt gas temperature and the unburnt gas temperature are functions of the burn mass fraction; they are not constants. Solving Eq. (8.20) for the burnt gas temperature gives

$$\bar{\bar{T}}_b = T_u\left(\frac{1-f}{f}\right)\left(\frac{r_b^3}{r_w^3 - r_b^3}\right)\frac{\mathcal{M}_{mix,b}}{\mathcal{M}_{mix,u}} \tag{8.21}$$

The next step is to eliminate the radii from Eq. (8.21) since they are unknown. We can derive relations for the radii in terms of pressure and temperature. We make use of the following definitions:

$$m_u = \rho_i V_{tot}(1-f); \quad P_i = \frac{m_i R_g T_i}{V_{tot}\mathcal{M}_{mix,i}}; \quad \mathcal{M}_{mix,i} = \mathcal{M}_{mix,u} \tag{8.22}$$

Solving for the unburnt mixture volume,

$$V_u = \frac{m_u R_g T_u}{P\mathcal{M}_{mix,u}} = \frac{P_i}{P}\left(\frac{T_u}{T_i}\right)(1-f)V_{tot} \tag{8.23}$$

Invoking the geometrical relations for the volume of a sphere in terms of its radius, we get for the radius of the burnt region

$$r_b = r_w\left[1-(1-f)\frac{P_i}{P}\frac{T_u}{T_i}\right]^{1/3} \tag{8.24}$$

Eq. (8.24) can be used to express the ratio of radii that appears in Eq. (8.21)

$$\left(\frac{r_b^3}{r_w^3 - r_b^3}\right) = \frac{r_w^3\left[1-(1-f)\left(\frac{P_i}{P}\right)\left(\frac{T_u}{T_i}\right)\right]}{r_w^3 - r_w^3\left[1-(1-f)\left(\frac{P_i}{P}\right)\left(\frac{T_u}{T_i}\right)\right]} \tag{8.25}$$

This expression is now used in Eq. (8.21) to give

$$\bar{\bar{T}}_b = T_i \left(\frac{1}{f}\right)\left(\frac{M_{mix,b}}{M_{mix,u}}\right)\left(\frac{P}{P_i}\right)\left[1-(1-f)\left(\frac{P_i}{P}\right)\left(\frac{T_u}{T_i}\right)\right] \qquad (8.26)$$

Finally, we use the adiabatic relation for an ideal gas, $T_u/T_i = (P/P_i)^{(\gamma_u-1)/\gamma_u}$, to eliminate the temperature ratio. This gives us the desired result:

$$\bar{\bar{T}}_b = T_i \left(\frac{1}{f}\right)\left(\frac{M_{mix,b}}{M_{mix,u}}\right)\left(\frac{P}{P_i}\right)\left[1-(1-f)\left(\frac{P}{P_i}\right)^{-1/\gamma_u}\right] \qquad (8.27)$$

P_{ex} is the constant volume explosion pressure. This expression gives us the desired limits for the temperature of the mixture. Prior to ignition, the mass averaged temperature is equal to the initial temperature: $P(f=0)=P_i \to \bar{\bar{T}}_b = T_i$. Upon complete combustion, the mass averaged temperature is equal to the constant volume explosion temperature: $\bar{\bar{T}}_b(f=1)=T_i(P_{ex}/P_i)(M_{mix,b}/M_{mix,u})=T_{ex}$. If the Lewis–von Elbe linear fractional pressure rise relation can be used, then it is possible to calculate the unburnt gas temperature and the average burnt gas temperature as a function of the mass fraction burnt. This is illustrated in the next example.

EXAMPLE 8.2

Here is an example for calculating the average burnt gas and unburnt gas temperatures generated by a confined deflagration. Assume that the Lewis–von Elbe fractional pressure relation is valid, that the unburnt and burnt mixtures are ideal gases with constant properties, and that the unburnt and burnt mixture molar masses are constant and equal.

$$f = \frac{P-P_0}{P_{ex}-P_0}$$

$$\frac{T_u}{T_i} = \left(\frac{P_i}{P}\right)^{(\gamma_u-1)/\gamma_u}$$

$$\bar{\bar{T}}_b = T_i \left(\frac{1}{f}\right)\left(\frac{M_{mix,b}}{M_{mix,u}}\right)\left(\frac{P}{P_i}\right)\left[1-(1-f)\left(\frac{P}{P_i}\right)^{-1/\gamma_u}\right]$$

Given: $T_i = 300$ K, $P_i = 1$bar, $P_{ex} = 9$bar, and $\gamma = 1.4$.

Solution

Advance the pressure by a convenient increment from $P = P_i$ to $P = P_{ex}$. Use the Lewis–von Elbe relation to calculate the burnt mass fraction f. The calculation of the unburnt and burnt gas temperatures follows directly.

P/P_i	$f(LvE)$	$\bar{\bar{T}}_b/T_i$	T_u/T_i
1	0	1	1
2	0.125	7.46	1.22
3	0.250	7.89	1.37
4	0.375	8.18	1.49
5	0.500	8.42	1.58
6	0.625	8.60	1.67
7	0.750	8.75	1.74
8	0.875	8.88	1.81
9	1	9	1.87

Notice that this calculation predicts a very large Flamm–Mache temperature gradient. This is likely overestimated. To get a more reasonable result will require a more rigorous evaluation of the thermodynamic properties.

8.5.2 SURVEY OF INTEGRAL MODELS

This literature survey of integral models for flame propagation loosely follows a chronological progression and, to a lesser extent, is organized by groups of investigators. The description of the early studies is more focused on key experimental findings and validation of some of the assumptions used in formulating integral models. With flammable gases, investigators emphasized the development of methods for extracting laminar burning velocities. This effort has achieved a high level of sophistication and success in part because the starting mixture can be made well-mixed yet quiescent. The extraction of laminar burning velocities for combustible dust is far more difficult because the dust dispersion process is by necessity turbulent (in terrestrial gravity) if one hopes to achieve a well-mixed state. For combustible dusts, a simpler approach based on scaling called the cube root scaling law has become the more favored means of assessing the explosion hazard.

The physical and chemical phenomena involved in a confined deflagration are complex. In order to simulate this complex combustion system with an integral model requires simplification of some features and neglect of others. The key features of integral models developed by different investigators are based on certain specific assumptions invoked to achieve that simplification. Some of these efforts have been more successful than others. I will largely concentrate on the structural aspects of these models. For a more complete description of any one of these models, I direct the reader to the original publications.

The structural features most commonly addressed by investigators are the following:

- Spatial structure of the unburnt and burnt gas regions: two-zone models assume a thin flame, three-zone models assume a flame of finite thickness, and multi-zone models represent a coarse discretization of the unburnt and burnt regions.
- Thermodynamic model for gas compression: adiabatic, polytropic (adiabatic model with the heat capacity ratio adjusted for heat losses), and isothermal models. Usually based on the ideal gas equation of state.
- Combustion progression model: typically either the Lewis−von Elbe fractional pressure relation or an approach based on the expansion ratio $(E = \mathcal{M}_u T_b / \mathcal{M}_b T_u)$ and a kinematic approximation for the flame speed.

Again, this survey of integral models covers both models developed for flammable gases as well as combustible dusts. A distinction will be drawn only in those cases where the model is applicable only to combustible dusts.

8.5.2.1 Early investigations (pre-1970)

Many of the early deflagration studies (pre-1970) concentrated on the reporting of empirical observations and the development of instrumentation for measuring pressure, temperature, and flame motion. Methods were developed for the direct calculation of burning velocity from pressure and flame position history data. These studies were conducted with flammable gases (Hopkinson, 1906; Ellis and Wheeler, 1927; Ellis and Morgan, 1934). The methods of data analysis were based on integral model concepts but were not solved explicitly for the laminar burning velocity. Instead, the burning velocity was extracted by direct measurement of the slope of the flame trace obtained photographically.

The analysis of deflagration experimental data took on a new level of sophistication with the work of Lewis and von Elbe, scientists at the U.S. Bureau of Mines (Lewis and von Elbe, 1934). The subject of this paper was to present deflagration test data for ozone explosions $(O_3 \rightarrow \frac{3}{2} O_2)$ in a 150-mm-diameter spherical vessel. They reported burning velocities on the order of 0.5−7.8 m/s. Lewis and von Elbe developed an integral model based on the geometric relations for the spherical bomb, the adiabatic compression of the gas, and an algebraic expression for the burnt volume fraction. They demonstrated the calculation of the laminar burning velocity, the temperature gradient in the vessel, and the particle path lines from the pressure history data.

The linear fractional pressure rise relation was first published nearly simultaneously in their book and in a paper presented at a meeting of The Combustion Institute (the book, Lewis and von Elbe, 1987, pp. 381−395, the first edition was published in 1951; Manton et al., 1953). In these and subsequent publications, well summarized in their book, they emphasized the importance of accurate measurement of thermodynamic properties and calculation of chemical equilibria in determining the laminar burning velocity. In the paper by Manton et al., they compared burning velocity determinations from the spherical vessel with stationary flame measurements and found good agreement between the two. They also indicated that they obtained good agreement between the predicted flame motion and the flame radius history data obtained by photography.

The integral model of Lewis and von Elbe was quickly adopted and tested by others (Fiock and Marvin, 1937; Fiock et al., 1940). Much of this earlier work was motivated by internal combustion engine research. Fiock and his colleagues tested the carbon monoxide–oxygen system and a variety of flammable gases and vapors using a 250-mm spherical vessel. They examined the effect of the initial pressure on flame propagation and compared their burning velocity determinations with those obtained by the constant pressure soap bubble technique.

Eschenbach and Agnew investigated the potential role of compressibility effects on flame propagation in the methane–oxygen system (Eschenbach and Agnew, 1958). They derived a correction to the burning velocity calculation to account for compressibility effects (ie, burning velocities on the order of 10% or greater of the unburnt mixture speed of sound). Typical burning velocities were determined to be on the order of 3–4 m/s. They tracked flame motion using ionization probes and measured the pressure history. They derived an expression for the burnt mass fraction different from the Lewis–von Elbe expression:

$$f = \frac{P - P_0}{P_0 \gamma_u (E - 1)} \qquad (8.28)$$

The variable $E = T_b \mathcal{M}_u / T_u \mathcal{M}_b$ is the expansion ratio. They noted close agreement with their burning velocity determinations versus stationary burner experiments. The test data indicated that they were achieving complete combustion if the initial pressure was 1 atmosphere. However, they noted that at subatmospheric initial pressures the spherical propagating flame did not appear to result in complete combustion.

O'Donovan and Rallis expressed the concern that the Lewis–von Elbe fractional pressure rise relation may suffer from serious errors and derived an alternative expression (O'Donovan and Rallis, 1959). Luijten et al. observed that this expression can be derived from their result, Eq. 8.14, by assuming that $\gamma_b = 1$ (Luitjen et al., 2009).

$$f = \frac{P - P_0 G(P)}{P_{ex} - P_0 G(P)}; \ G(P) = \left(\frac{P}{P_0}\right)^{(\gamma_u - 1)/\gamma_u} \qquad (8.29)$$

O'Donovan and Rallis describe how to employ their model equations to extract a burning velocity from either pressure–time data or flame radius–time data. However, they did not compare their model with any experimental data.

Rallis and Tremeer took up the challenge to compare existing integral models with experimental data to discover the strengths and limitations of these models (Rallis and Tremeer, 1963). They compared the burning velocity predictions of eight flame propagation models (including the Lewis–von Elbe model) with experimental data obtained with stoichiometric acetylene–air deflagration tests in a 160-mm-diameter sphere. Most of these models used the same mass balance and geometric relations, but differed in how they approximated the thermodynamic processes and the burnt mass fraction history. These models tended to be valid only at low pressures (small burnt mass fractions). Rallis and Tremeer derived a new flame propagation model based on the gas density ratio which did a better job of predicting the measured flame radii. Their analysis also indicated that the burnt gas mixture was not in chemical equilibrium.

8.5.2.2 U.S. Bureau of Mines contributions

As indicated in Chapter 1, the U.S Bureau of Mines had embarked on studies of combustible dust (coal dust in particular) as early as 1920. In the 1960s, some of these investigators turned their attention to the use of integral models to investigate and correlate their dust explosion data. One of the experimental difficulties encountered with dust deflagrations is that while pressure history data can be collected, flame radius measurements cannot. Thus, the models had to be tested by their ability to reproduce pressure histories. Nagy et al. derived two flame propagation models, one based on adiabatic gas compression and one based on isothermal compression (the burnt gas was assigned a constant temperature and the unburnt gas temperature was assigned the initial temperature of the mixture) (Nagy et al., 1969). They made explicit comparisons of model predictions versus measurement data to determine the goodness of fit and concluded that the adiabatic model gave a better match to the measured pressure histories than the simpler isothermal model. The greatest errors occurred at the end of the deflagration process. They determined that prediction errors were caused by heat losses and incomplete combustion. Although less accurate than the adiabatic model, the isothermal model did compare reasonably well with the empirical data. Plus, the isothermal model offered two advantages. First, the isothermal model could be solved analytically for the complete pressure history of the deflagration. Second, it offered a path to the derivation of the cubic pressure rise equation, thus, providing a theoretical foundation for what had been previously offered as an empirical observation (Zabetakis, 1965).

In 1971, Nagy and his colleagues published an extensive analysis of confined deflagration data obtained from 17 different explosion vessels varying in both size and shape (Nagy et al., 1971). The results included 50 gas explosion tests and 149 dust explosion tests. The gas mixtures tested were methane−air and propane−air mixtures. The combustible dusts tested were cornstarch, cellulose acetate, and Pittsburgh coal. The volume of the vessels ranged from 0.51 to 25,600 L, and the shapes tested included spherical, cylindrical, and rectilinear. They reported that for a fixed set of conditions, the maximum rate of pressure rise decreased with increasing volume. It also decreased as the surface-to-volume ratio of the vessel decreased. They further stated that neither vessel size nor shape affected the maximum pressure (neglecting heat losses to the vessel walls). The test results indicated that the explosion development of combustible dusts was essentially the same as that observed with flammable gases.

Perlee et al. derived a flame propagation model using an approach based on the expansion ratio E and an approximation for the burnt gas temperature (Perlee et al., 1974). Assuming an adiabatic compression of the gas mixture, they derived a slightly different form of the cubic pressure rise equation. The model was tested with explosion data for a stoichiometric methane−air mixture. They also estimated the expansion ratio and heat capacity ratio using chemical equilibrium calculations instead of assuming average values. They evaluated the sensitivity of the flame propagation model as a function of the input parameters. For example, their calculations indicated that a 10% error in the heat capacity ratio, burning velocity, or expansion ratio resulted in a 20−40% error in the pressure history.

Finally, the work of Kansa and Perlee deserves a brief summary (Kansa and Perlee, 1976). Kansa and Perlee investigated the effects of compressibility (finite sound speed) on flame propagation in a spherical vessel. They solved the governing unsteady one-dimensional equations of change using the method of characteristics. They noted that at small Mach numbers, Ma < 0.2, the pressure gradient was negligible. Their finite difference model displayed good agreement with an integral model calculation. At higher Mach numbers, the flame motion generated finite pressure waves (pressure oscillations) that affected the development of the pressure history.

8.5.2.3 Post-1970s contributions

Since the 1970s, a great deal of modeling and experimental work on confined deflagrations has occurred worldwide. Much of this work has been with flammable gases (Bradley and Mitcheson, 1976; Rallis and Garforth, 1980; Fairweather and Vasey, 1982; Singh, 1988; Harris, 1989; Evans, 1994; van den Bulck, 2005; Lautkaski, 2005). The modeling investigation has involved a variety of variations on spatial structure, thermodynamic equilibrium computations, and flame progression submodels. Most of these studies used flame propagation models that were embellishments of those models already discussed with the goal of achieving a better fit of model predictions to empirical data. Some of these will be discussed later in terms of model validation.

Fewer models have been specifically developed for combustible dusts. Nomura and Tanaka published a model based on a microscale description of the dust cloud envisioned as a lattice model (Nomura and Tanaka, 1980, 1992). The model assumed that spherical particles were distributed uniformly in air. The particle combustion was assumed to be diffusion limited. A macroscale mass balance tracked the flame motion and a macroscale energy balance was used to calculate the temperature and pressure of the gas. The model is an interesting contribution because of its incorporation of single particle combustion kinetics into an integral model formulation. However, the assumptions used in its derivation are severe and not easily subjected to verification.

A different approach was taken by Dahoe and his colleagues (Dahoe et al., 1996a,b). They developed a model which divided the dust cloud into three zones: the burnt mixture, a flame of finite thickness, and the unburnt mixture. The model assumed adiabatic compression, constant heat capacity ratios, and a single step irreversible combustion reaction. Instead of a single expression for the burnt mass fraction, the model requires the calculation of the burnt mass fraction as a function of the radius of the vessel. The Dahoe model is probably one of the more innovative approaches to simulate dust flame propagation with an integral model. Its solution, however, is far more complex than the simpler integral models described earlier.

8.5.3 THE ADIABATIC MODEL OF BRADLEY AND MITCHESON

In this section, I derive the integral model formulated by Bradley and Mitcheson for gaseous deflagrations (Bradley and Mitcheson, 1976). I choose this model because it typifies the adiabatic integral models and can easily accommodate refinements like variable thermodynamic properties. The Bradley–Mitcheson (BM) model is based on the following assumptions:

- A combustible dust is dispersed with a uniform concentration in a spherical vessel
- Ignition occurs at the center of the vessel; the ignition event is time zero
- The vessel walls are rigid, adiabatic, and impermeable
- The mixture heat capacity ratios are constant
- The molar masses of the unburnt and burnt mixture are equal
- The ideal gas equation of state applies to both the unburnt and burnt mixtures
- The flame propagation process can be described as a thin flame with a laminar burning velocity
- The LvE linear fractional pressure rise relation relates the burnt mass fraction f to the instantaneous pressure P.

The objective of this analysis is to derive an equation for the pressure history in terms of measureable parameters. Recall that in dust deflagrations it is difficult to monitor the flame motion within the closed vessel. The easiest parameter to measure is the instantaneous pressure. Thus, our final result must permit the calculation of the pressure history of the deflagration in terms of the initial and final (explosion) pressures, the initial mixture density, the radius of the vessel, and the mixture thermodynamic properties.

The derivation begins with the mass balance on the unburnt mixture (see Fig. 8.5):

$$\frac{dm_u}{dt} = -4\pi r_b^2 \rho_u S_u \tag{8.30}$$

The symbols have been previously defined. This expression indicates that the mass of unburnt mixture is depleted by the flow of unburnt mixture through the flame surface. We seek an expression for the unburnt mass in terms of the pressure. This requires expressions that will link the unburnt and burnt mass quantities, the mixture pressure, and the radius of the spherical vessel. These variables can be obtained from the total mass balance on the dust cloud, the LvE fractional pressure relation, and the specification of the mass of the initial combustible dust mixture:

$$\frac{m_u}{m_i} = 1 - \frac{m_b}{m_i} = 1 - f; \quad f = \frac{m_b}{m_i} = \frac{P - P_0}{P_{ex} - P_0}; \quad m_i = \frac{4\pi}{3} r_w^3 \rho_i \tag{8.31}$$

These equations are combined to give an expression for the unburnt mass in the vessel for any value of the burnt mass fraction f:

$$m_u = \frac{4\pi}{3} r_w^3 \rho_i \left[1 - \left(\frac{P - P_0}{P_{ex} - P_0} \right) \right]$$ (8.32)

Substituting the expression for the unburnt mass into the mass balance and solving for the pressure derivative yield the following differential equation:

$$\frac{dP}{dt} = \left(\frac{3S_u}{r_w} \right) \left(\frac{\rho_u}{\rho_i} \right) (P_{ex} - P_0) \left(\frac{r_b}{r_w} \right)^2$$ (8.33)

We now require an expression for the squared ratio of radii. Using the geometrical relations for a sphere, we can write the following equation:

$$\frac{m_u}{m_i} = 1 - f = \frac{\rho_u}{\rho_i} \left[1 - \left(\frac{r_b}{r_w} \right)^3 \right]$$ (8.34)

This expression is solved for the ratio of radii. Then we make two substitutions: invoke the adiabatic compression relation for the density ratio, $P\rho_u^{-\gamma_u} = \text{constant}$, and the LvE fractional pressure relation:

$$\left(\frac{r_b}{r_w} \right)^3 = 1 - \left(\frac{\rho_i}{\rho_u} \right)(1 - f) = 1 - \left(\frac{P_0}{P} \right)^{1/\gamma_u} \left[1 - \left(\frac{P - P_0}{P_{ex} - P_0} \right) \right]$$ (8.35)

Eq. (8.35) is substituted into Eq. (8.33) to give us the desired result:

$$\frac{dP}{dt} = \left(\frac{3S_u}{r_w} \right)(P_{ex} - P_0) \left(\frac{P}{P_0} \right)^{1/\gamma_u} \left[1 - \left(\frac{P_0}{P} \right)^{1/\gamma_u} \left(\frac{P_{ex} - P}{P_{ex} - P_0} \right) \right]^{2/3}$$ (8.36)

This is a nonlinear ordinary differential equation which must satisfy both the initial condition, $P(t = 0) = P_0$, and the final condition, $P(t = t_{P_{ex}}) = P_{ex}$, where the symbol for the time to maximum pressure is t_{MP}. To solve it directly, one must resort to numerical methods. The specific algorithm chosen will depend on the information available to the investigator. For example, if the pressure history is given then the burning velocity can be estimated. If the burning velocity is known, the time to maximum pressure can be found. Additional details on the solution of Eq. (8.36) can be found in the original paper (Bradley and Mitcheson, 1976).

Bradley and Mitcheson chose methane−air mixtures as their test case for comparing model predictions with experimental data. They solved the flame propagation model using numerical methods. They incrementally advanced the burnt mass fraction, computed the thermodynamic properties of the mixture with each step, and assumed a variable burning velocity. They compared their full numerical solution with approximate solutions where they assumed that the burning velocity was constant (0.45 m/s). Fig. 8.6 shows the flame radius history, three fluid particle trajectories (path lines), and the final burnt gas temperature radial profile in the spherical vessel.

The fluid particle trajectories represent locations at 0.25, 0.5, and 0.75 of the vessel radius. The trajectories indicate that the unburnt mixture at a given location is first accelerated away from the flame and toward the wall, captured and ingested by the flame, and then accelerated behind the flame toward the vessel center. The net displacement is zero: individual fluid particles are first displaced from and then returned to their original position. Compare this description of flame kinematics with the discussion of unconfined flame propagation in Section 7.7.

Fig. 8.7 illustrates the calculated results for the explosion pressure versus equivalence ratio and a typical pressure history.

The BM model exhibited a good fit to the experimental data for the methane−air system. A final important feature of this model is that the pressure rises without bound until $P(t = t_{P_{ex}}) = P_{ex}$, and the maximum rate of pressure rise occurs at $t = t_{P_{ex}}$. This is not in accord with typical empirical data where the rate of pressure rise goes through a maximum and then decreases as the pressure approaches P_{ex}.

An expression of the K_{st} value can be derived from the adiabatic model. Recognizing that the maximum rate of pressure rise occurs when $P = P_{ex}$, Eq. (8.36) gives

$$\left(\frac{dP}{dt}\right)_{max} = \left(\frac{3S_u}{r_w}\right)(P_{ex} - P_0)\left(\frac{P_{ex}}{P_0}\right)^{1/\gamma_u} \qquad (8.37)$$

Substituting the expression for the volume of a sphere,

$$V^{1/3} = \left(\frac{4\pi}{3}\right)^{1/3} r_w \qquad (8.38)$$

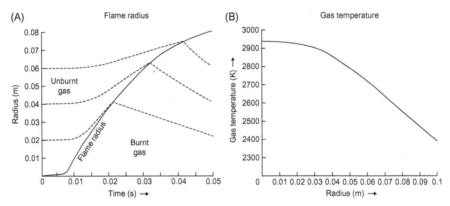

(A) Flame radius (B) Gas temperature

FIGURE 8.6

Flame propagation model predictions: (A) flame radius and particle path histories and (B) burnt gas temperature profile.

Reproduced with permission from Bradley, D., Mitcheson, A., 1976. Mathematical solutions for explosions in spherical vessels. Combust. Flame 26, 201−217.

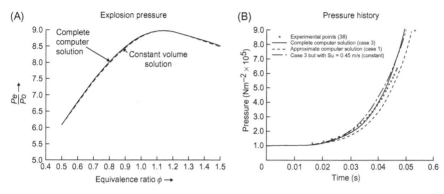

FIGURE 8.7

Explosion pressures for methane–air mixtures: (A) explosion pressure versus equivalence ratio (computer solution compared with isochoric explosion pressure) and (B) pressure history showing performance of numerical solution with approximate solutions and experimental data.

Reproduced with permission from Bradley, D., Mitcheson, A., 1976. Mathematical solutions for explosions in spherical vessels. Combust. Flame 26, 201–217.

One obtains the K_{st} value for the BM model:

$$K_{st}(BM) = \left(\frac{dP}{dt}\right)_{max} V^{1/3} = (36\pi)^{1/3}(P_{ex} - P_0)\left(\frac{P_{ex}}{P_0}\right)^{1/\gamma_u} S_u \qquad (8.39)$$

Recall that in this derivation, the laminar burning velocity is assumed to be a constant.

EXAMPLE 8.3

Using the Bradley–Mitcheson adiabatic model, calculate the K_{st} values for the following confined deflagration test data. $P_{ex} = 9.7$ bar, $P_0 = 1.01$ bar, $\gamma_u = 1.4$, and $S_u = 0.5$ m/s

Solution

The formula for the Bradley–Mitcheson K_{st} value is

$$K_{st}(BM) = (36\pi)^{1/3}(P_{ex} - P_0)\left(\frac{P_{ex}}{P_0}\right)^{1/\gamma_u} S_u$$

Substituting the input parameters, the result is

$$K_{st}(BM) = (36\pi)^{1/3}(9.7 - 1.01 \text{ bar})\left(\frac{9.7 \text{ bar}}{1.01 \text{ bar}}\right)^{1/1.4}(0.5 \text{ m/s}) = 106 \text{ bar} \cdot \text{m/s}$$

8.5.4 ISOTHERMAL MODEL OF NAGY, CONN, AND VERAKIS

With the BM adiabatic model, the adiabatic compression constraint introduces a nonlinear term that makes the integration of the pressure history very difficult. Nagy, Conn, and Verakis derived a similar adiabatic model and encountered the same problem (Nagy et al., 1969; Nagy and Verakis, 1983, pp. 75−84). They discovered that if they assumed that the compression was isothermal instead of adiabatic, a closed-form solution was possible. In particular, the isothermal compression approximation assumes that the unburnt gas temperature remains constant at its initial mixture temperature and the burnt gas temperature equals the flame temperature. Although not specifically stated in their derivation, it is clear from the context that the burnt gas temperature should be assigned the value of the isochoric combustion temperature.

The NCV model as presented here assumes a spherical geometry, the ideal gas equation of state, and equal molar masses for the unburnt and burnt gas mixtures. A second distinguishing feature of the Nagy−Conn−Verakis (NCV) isothermal model is that it does not require the LvE mathematical relation linking the fractional pressure rise to the burnt mass fraction. Instead, the isothermal assumption leads naturally to an algebraic expression linking pressure changes to changes in the burnt volume fraction.

We first derive the burnt volume fraction relation. Consider a mole number balance on the mixture and substitute the ideal gas equation of state

$$n_i = n_u + n_b \rightarrow \frac{P_0 V_w}{R_g T_u} = \frac{P V_u}{R_g T_u} + \frac{P V_b}{R_g T_b} \tag{8.40}$$

Collecting like terms, recognizing that $T_b/T_u = P_{ex}/P_0$, and solving for the burnt volume fraction, V_b/V_w, we get the desired result:

$$\frac{V_b}{V_w} = \left(\frac{r_b}{r_w}\right)^3 = \frac{1 - (P_0/P)}{1 - (P_0/P_{ex})} \tag{8.41}$$

The next step is to derive an expression for the time rate of change in pressure. We begin with mass balances for the unburnt and burnt mixtures.

$$\frac{dm_u}{dt} = \frac{M}{R_g T_u} \cdot \frac{d(PV_u)}{dt} \rightarrow \frac{R_g T_u}{M} \cdot \frac{dm_u}{dt} = V_u \frac{dP}{dt} + P \frac{dV_u}{dt} \tag{8.42}$$

$$\frac{dm_b}{dt} = \frac{M}{R_g T_b} \cdot \frac{d(PV_b)}{dt} \rightarrow \frac{R_g T_b}{M} \cdot \frac{dm_b}{dt} = V_b \frac{dP}{dt} + P \frac{dV_b}{dt} \tag{8.43}$$

Observing that $dm_u/dt = - dm_b/dt$ and $dV_u/dt = - dV_b/dt$, Eqs. (8.42) and (8.43) can be added together to obtain an expression for the time rate of change in pressure:

$$V_w \frac{dP}{dt} = \frac{R_g}{M}(T_b - T_u)\frac{dm_b}{dt} = \rho_u S_u A_b \left(\frac{R_g}{M}\right)(T_b - T_u) \tag{8.44}$$

Applying the ideal gas equation of state for the unburnt density ρ_u, substituting the burnt volume fraction relation (Eq. 8.41), and simplifying the geometric identities give the following result:

$$\frac{dP}{dt} = \beta P^{1/3}(P - P_0)^{2/3}; \quad \beta = 3 \left(\frac{P_{ex}}{P_0}\right)^{2/3} \left[\left(\frac{P_{ex}}{P_0}\right) - 1\right]^{1/3} \left(\frac{S_u}{r_w}\right) \tag{8.45}$$

The mathematical behavior of the pressure rise is similar to the behavior of the BM model: the pressure rate is predicted to increase monotonically with the maximum rate at $P = P_{ex}$. The value of K_{st} is given by the expression:

$$K_{st}(NCV) = \left(\frac{dP}{dt}\right)_{max} V^{1/3} = (36\pi)^{1/3} \left(\frac{P_{ex}}{P_0}\right)(P_{ex} - P_0)S_u \tag{8.46}$$

EXAMPLE 8.4

Using the Nagy–Conn–Verakis isothermal model, calculate the K_{st} values for the following confined deflagration test data. $P_{ex} = 9.7$ bar, $P_0 = 1.01$ bar, and $S_u = 0.5$ m/s

Solution

The formula for the Nagy–Conn–Verakis K_{st} value is

$$K_{st}(NCV) = (36\pi)^{1/3} \left(\frac{P_{ex}}{P_0}\right)(P_{ex} - P_0)S_u$$

Substituting the input parameters, the result is

$$K_{st}(NCV) = (36\pi)^{1/3} \left(\frac{9.7 \text{ bar}}{1.01 \text{ bar}}\right)(9.7 - 1.01 \text{ bar}) \ (0.5 \text{ m/s}) = 41.7 \text{ bar·m/s}$$

Using the same deflagration input data as in Example 8.3, the isothermal value is much smaller than the adiabatic value: $K_{st}(BM) \cong 2.5 \ K_{st}(NCV)$.

To obtain the complete pressure history, Eq. (8.45) must be integrated from the initial condition $P(t = 0) = P_0$ to the indefinite endpoint $P(t)$.

$$\int_{P_0}^{P} \frac{dP}{P^{1/3}(P - P_0)^{2/3}} = \beta t \tag{8.47}$$

This integral can be solved directly if it is transformed into a standard form. The paper by Lautkaski considers this problem in some detail and is highly recommended to the reader (Lautkaski, 2005). To do so, first define a new variable x

$$x = \left(1 - \frac{P_0}{P}\right)^{1/3} \to P = \frac{P_0}{1 - x^3}; \quad dP = \frac{3P_0 x^2}{(1 - x^3)^2} dx \tag{8.48}$$

Substituting the change in variable into the integral and its limits gives this modified form:

$$\int_0^{(1-P_0/p)^{1/3}} \frac{dx}{1-x^3} = \beta t \tag{8.49}$$

The solution for this integration can be found in standard tables of integrals (Beyer, 1976, p. 344)

$$\int \frac{dx}{1-x^3} = \frac{1}{6c^2}\ln\left[\frac{(x-c)^2}{x^2+cx+c^2}\right] - \frac{1}{c^2\sqrt{3}}\tan^{-1}\left(\frac{2x+c}{c\sqrt{3}}\right) \tag{8.50}$$

Substituting $c = 1$ and the definition for x gives the pressure history for the NCV isothermal model:

$$\beta t = \frac{-1}{6}\ln\left\{\frac{\left[(1-P_0/P)^{1/3}-1\right]^2}{(1-P_0/P)^{2/3}+(1-P_0/P)^{1/3}+1}\right\}$$
$$+ \frac{1}{\sqrt{3}}\tan^{-1}\left[\frac{2(1-P_0/P)^{1/3}+1}{\sqrt{3}}\right] - \frac{1}{\sqrt{3}}\tan^{-1}\left(\frac{1}{\sqrt{3}}\right) \tag{8.51}$$

While integration does lead to a closed-form solution, the functional structure of Eq. (8.51) does not easily lend itself to developing physical insight. Thus, Nagy and his colleagues suggested a clever device for simplifying the integration of Eq. (8.47): expand the integrand with a power series, retain only the first term of the expansion, and integrate this simpler functional form.

Specifically, the term in parentheses can be expanded using the binomial expansion (Beyer, 1976, p. 396). It is helpful to simplify the integrand with the transformation

$$z = \frac{P}{P_0} - 1; \ dz = \frac{dP}{P_0} \tag{8.52}$$

The integral now takes the form

$$\int_{P_0}^{P} \frac{dP}{P^{1/3}(P-P_0)^{2/3}} = \int_0^{\frac{P}{P_0}-1} \frac{dz}{z^{2/3}(1+z)^{1/3}} \tag{8.53}$$

The following expansion is valid when n is not a positive integer and the domain is $-1 < x < 1$.

$$(1+x)^{-n} = 1 + nx + \frac{1}{2}\cdot\frac{n(n-1)}{2!}x^2 + \frac{n(n-1)(n-2)}{3!}x^3 + \cdots \tag{8.54}$$

The expansion of the parenthetical term in the integrand becomes

$$(1+z)^{-1/3} = 1 - \frac{1}{3}z + \frac{2}{9}z^2 - \frac{14}{81}z^3 + \cdots \tag{8.55}$$

The complete integrand becomes

$$z^{-2/3}(1+z)^{-1/3} = z^{-2/3} - \frac{1}{3}z^{1/3} + \frac{2}{9}z^{4/3} - \frac{14}{81}z^{7/3} + \cdots \tag{8.56}$$

Substituting this power series into the integrand, retaining only the first term, and integrating give the result:

$$\int_0^{\frac{P}{P_0}-1} \frac{dz}{z^{2/3}(1+z)^{1/3}} = \int_0^{\frac{P}{P_0}-1} z^{-2/3}\,dz = 3\left(\frac{P-P_0}{P_0}\right)^{1/3} = \beta t \tag{8.57}$$

Inserting the definition of β and simplifying we obtain the final result:

$$\frac{P - P_0}{P_{\text{ex}} - P_0} = \left(\frac{P_{\text{ex}}}{P_0}\right)^2 \left(\frac{S_u t}{r_w}\right)^3 \tag{8.58}$$

The limitation on this cubic equation, based on the use of the binomial expansion, is that $P < 2P_0$. It will be shown in the next section that this pressure limitation does not seriously limit the utility of the model. There are other routes to deriving this cubic relation which differ only in the coefficient preceding the cubic term (see Perlee et al., 1974, for an approach assuming adiabatic compression or Harris, 1989, for another approach assuming isothermal compression). Luitjen et al. have correctly indicated that the isothermal compression model violates the first law of thermodynamics (Luijten et al., 2009). However, in comparisons of the model with experimental data, the error in pressure history is usually smaller than the experimental error ($<20\%$) and is therefore tolerable. This will be discussed in greater detail in the following section.

8.5.5 CUBIC PRESSURE RISE VERSUS THE CUBE ROOT SCALING LAW

Evans warns of the potential for "cubic confusion" in the study of confined deflagrations (Evans, 1991). Specifically, the isothermal compression model predicts that the early pressure history of a confined deflagration is proportional to time cubed. This observation of a *cubic pressure rise relation* is limited (at least mathematically) to small pressure values $P < 2P_0$. Alternatively, the K_{st} parameter is associated with the *cube root scaling law*, a relation which predicts that the maximum rate of pressure rise for a given combustible dust and fixed conditions is inversely proportional to the volume of the enclosure. The maximum rate of pressure rise tends to occur at high pressure values $P \approx P_{\text{max}} \approx P_{\text{ex}}$. Although both of these empirical laws involve a cubic function, they are clearly *not* related: the cubic pressure rise relation applies at the beginning of the confined deflagration and the cube root scaling law applies near the end. To make matters worse, some authors refer to the cube root scaling law as the cubic law (Cross and Farrer, 1982, pp. 203−206; Frank and Rodgers, 2012, pp. 58−63).

Both of these relations are first-order approximations, ie, they capture the essential physics but gloss over the details. Nevertheless, both relations have played a critical role in the design of explosion safeguards. The cubic pressure rise model is relevant to active explosion protection technologies such as explosion venting, isolation, and suppression. Most process equipment enclosures and most buildings are structurally weak in comparison with the isochoric explosion pressure a combustible dust. To successfully protect a weak enclosure, the protection technology must actuate as early in the deflagration process as possible before the explosion overpressure causes structural failure. This design constraint for early actuation can be easily understood with the cubic pressure rise model, and in fact some early investigations were based on this model (Mannan, 2005, pp. 17/268 to 17/281).

The cubic pressure rise model was originally based on empirical observations of deflagration tests with flammable gases (Zabetakis, 1965, p. 15). Later it was placed on a theoretical foundation when Nagy et al. derived it from an integral flame propagation model by assuming isothermal compression restricting the model to the small overpressure limit (Nagy et al., 1969, 1971; Nagy and Verakis, 1983, pp. 81–83).

Knapton et al. published a derivation of the model that was also based on the isothermal approximation (Knapton et al., 1973). They performed a number of deflagration tests on jet fuel (JP4)–air mixtures at high initial pressures in a cylindrical pressure cell with a volume of 56.3 cm³. Fig. 8.8 is a plot of the pressure histories from five test conditions.

They analyzed the flammable gas deflagration data by plotting pressure history data versus time cubed and found good agreement in the early portion of the deflagration travel time.

Perlee et al. derived a cubic pressure rise model assuming adiabatic compression (Perlee et al., 1974). They then analyzed experimental pressure histories and tested a power law fit for stoichiometric methane–air deflagrations

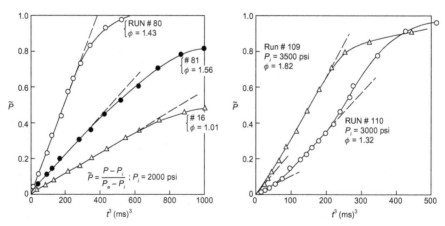

FIGURE 8.8

Deflagration pressure histories plotted as dimensionless pressure rise versus time cubed.

Reproduced with permission from Knapton, J.D., Stobie, I.C., Krier, H., 1973. Burning rate studies of fuel air mixtures at high pressures. Combust. Flame 21, 211–220.

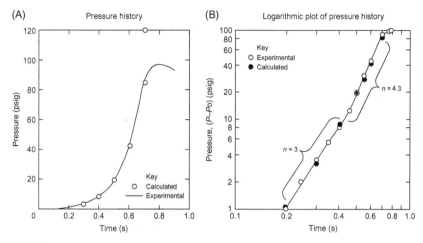

FIGURE 8.9

Pressure history of stoichiometric methane−air deflagration in a 12-ft-diameter sphere:
(A) comparison of computed pressure history and experimental data and (B) cubic
pressure rise in early portion (public domain, Perlee et al., 1974).

in 12-ft-diameter sphere and found the data matched an exponent of three at low
overpressures (Perlee et al., 1974). Fig. 8.9 is a plot of their pressure history data.

Sapko et al. published additional data on confined deflagration tests with metha-
ne−air mixtures in the 12-ft-diameter sphere (Sapko et al., 1976). Again, at early
times, the pressure data yielded a reasonable fit to the cubic pressure rise equation.
Finally, in a series of papers by Romanian investigators, a modified form of the cubic
pressure rise relation has been tested with flammable gas mixtures including propa-
ne−air and propylene−oxygen over a range of fuel concentrations (Razus et al.,
2006b; Brinzea et al., 2010; Brinzea et al., 2011). Interestingly, they used both spherical and cylindrical test vessels. Again, the cubic pressure rise model exhibited reasonable agreement with the empirical data in the early stage of the deflagrations.

EXAMPLE 8.5

Recall that the cubic pressure rise model is limited to the pressure range
$P \leq 2P_0$. Use the NCV cubic pressure rise relation and compute the time
when $P = 2P_0$. Assume the following data: $P_{ex} = 9.7$ bar, $P_0 = 1.01$ bar,
$r_w = 0.168$ m, and $S_u = 0.5$ m/s.

Solution

The NCV cubic pressure rise model is

$$\frac{P - P_0}{P_{ex} - P_0} = \left(\frac{P_{ex}}{P_0}\right)^2 \left(\frac{S_u t}{r_w}\right)^3$$

Substituting the input parameters gives

$$t = \sqrt[3]{\left(\frac{P - P_0}{P_{ex} - P_0}\right)\left(\frac{P_0}{P_{ex}}\right)^2\left(\frac{r_w}{S_u}\right)^3} = \sqrt[3]{\left(\frac{2.02 - 1.01 \text{ bar}}{9.7 - 1.01 \text{ bar}}\right)\left(\frac{1.01 \text{ bar}}{9.7 \text{ bar}}\right)^2\left(\frac{0.168 \text{ m}}{0.5 \text{ m/s}}\right)^3}$$

The result is $t_{2P_0} = 36.3$ ms.

In the time period from the 1960s to the 1980s, the cylindrical Hartmann bomb had become the standard test vessel in many countries, but it was recognized that it had its limitations. Its utility was the ease with which it was possible to generate explosion pressure data. The data could be used to make comparisons against a reference material like Pittsburgh seam coal. But by the 1980s, it became apparent that there was a need for dust deflagration measurements that were reproducible and which conveyed physically meaningful information (Lee, 1987; Kauffman et al., 1987). The maximum explosion pressure could be measured with some confidence, but not the "violence" of the explosion. It was eventually realized that the maximum rate of pressure rise was the best parameter for measuring the explosion severity. But it was also recognized that different results were obtained with test vessels of different sizes and shapes. Investigators realized that many of the integral models proposed for deflagration modeling could be simplified into the cube root law and the K_{st} parameter. This provided a means for scaling the maximum rate of pressure rise with the size of the vessel. There was a need to develop a standardized test vessel with a standardized test procedure that could yield consistent and reproducible K_{st} parameters. This need led to the adoption of the 20-L sphere and the 1-m^3 sphere and standardized test procedures (Eckhoff, 1987; Bartknecht, 1989; ASTM E1226, 2012). These test vessels are depicted in Fig. 8.10.

The cube root law and the K_{st} parameter have attained great significance for the design of deflagration venting, suppression, and isolation technologies (Bartknecht, 1981, 1989; Eckhoff, 1987; Barton, 2002, Chapters 6–12; Frank and Rodgers, 2012, pp. 201–220). The K_{st} parameter is used as a proxy for the "speed" of a dust deflagration wave. Many explosion protection design methods use the K_{st} parameter as an input parameter to an empirical correlation. Design methods are developed by technical committees through standards setting organizations like the British Standards Institute or the National Fire Protection Association, and the methods should be followed as presented in the standards.

Flame propagation models provide a theoretical basis for calculating the K_{st} parameter. The greatest challenge in using a flame propagation model to calculate a K_{st} parameter is to find a suitable estimate for the burning velocity as a function of dust concentration, oxidizing atmosphere, and the degree of turbulence. In practice, the strategy is usually the reverse: the K_{st} parameter is measured for a range of dust concentrations under standardized conditions (fixed oxidizing atmosphere and degree of turbulence). Many design methods use the K_{st} parameter

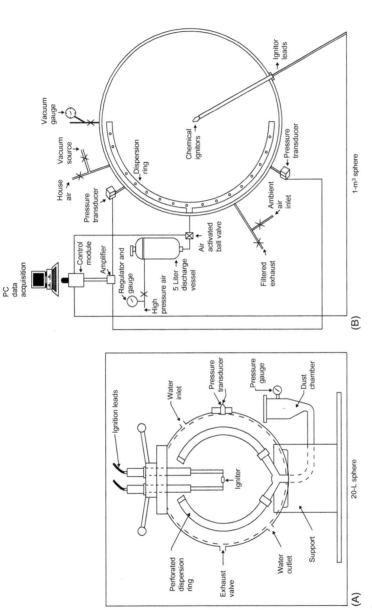

FIGURE 8.10

Drawings of the Siwek 20-L (A) and Fike 1-m^3 (B) spherical test vessels.

Reproduced with permission from (A) Pu, Y.K., Jarosinski, J., Johnson, V.G., Kauffman, C.W. 1990. Turbulence effects on dust explosions in the 20-liter spherical vessel.

In: Twenty-Third Symposium (International) on Combustion. The Combustion Institute, Elsevier, pp. 843–849 and (B) Chatrathi, K., 1994. Dust and hybrid explosibility

in a 1 m3 spherical chamber. Process Saf. Prog. 13, 183–189.

directly. For those investigators with a more fundamental interest, and a willingness to deal with turbulence, the burning velocity can be estimated from the explosibility data.

EXAMPLE 8.6

Using the LvE fractional pressure rise relation, calculate the conversion of fuel based on the maximum pressure measurement for coal with a coal dust concentration of 0.4 kg/m^3 using the data of Amyotte et al. (1991a). Assume that the initial conditions were $P_0 = 1$bar and $T_0 = 300$K. Further assume that the constant pressure adiabatic flame temperature for coal is 2505 K (Sami et al., 2001, p. 186, stoichiometric value).

Solution

The solution to this problem is a straightforward application of the LvE fractional pressure rise relation:

$$f = \frac{P - P_0}{P_{ex} - P_0} = \frac{P - 1}{P_{ex} - 1}$$

The problem is, we do not have a value for the constant volume explosion pressure. Returning to Chapter 3, Thermodynamics of Dust Combustion, P_{ex} can be estimated using Eq. (3.58):

$$P_{ex} = \gamma P_0 \left(\frac{T_{ad}}{T_0}\right) = 1.2 \ (1 \text{ bar}) \left(\frac{2505 \text{ K}}{300 \text{ K}}\right) = 10.0 \text{ bar}$$

We have further assumed that the heat capacity ratio is 1.2 (burnt gas value). At a coal dust concentration of 0.4 kg/m^3, the measured explosion pressure is 6.5 bar. Substituting into the LvE relation, we get our estimated burn fraction:

$$f = \frac{6.5 - 1}{10.0 - 1} = 0.611$$

This result must be treated cautiously as it is based on three approximations. First, neither the composition nor the properties of the coal were specified. Second, the adiabatic flame temperature at the stoichiometric condition for coal dust was used to estimate the thermodynamic value for the explosion pressure using a simplified thermodynamic relation. Third, the specified coal dust concentration was 400 g/m^3, whereas the stoichiometric concentration for coal dust can be expected to range between 100 and 150 g/m^3. A more accurate analysis could be performed using the specific coal composition and property data with a chemical equilibrium software package.

8.5.6 EXPERIMENTAL VALIDATION OF THE INTEGRAL MODEL APPROACH

This section discusses the experimental testing and validation of integral flame propagation models. The basic question to consider is how well do integral models describe confined dust deflagrations? The topics to be discussed include the basic comparison of flame propagation in combustible dust versus flammable gas mixtures, the uncertainty analysis in evaluating the merits of different forms of flame propagation models, and the various ways in which investigators have used chemical kinetics to yield improved predictions of laminar burning velocities. Turbulence, which is so important in both combustible dust testing and accidental dust explosions, will be addressed in a separate section.

It will appear that much of this discussion is centered on flammable gases and less of it on combustible dusts. There are two reasons for this relative emphasis. First, the volume of scientific literature on the confined deflagration behavior of flammable gases is much greater than that for combustible dusts. Second, combustible dust deflagrations present far more experimental challenges than flammable gases. Multiphase flow effects, dispersion turbulence, optical thickness, and complex reaction kinetics confound efforts to use the more sophisticated combustion diagnostic tools developed for gases. So we will take advantage of the flammable gas literature where we can and underscore the similarities and differences with the deflagration behavior of combustible dusts.

8.5.6.1 Flame kinematics and dynamics

The first step to consider in the validation of integral models is to evaluate whether they are successful in describing the motion of the flame (kinematics) and linking it with the driving forces that cause that motion (dynamics). There is a long history of observing the motion of a confined flame by high-speed photography or flame ionization sensors. These measurements provide a record of the flame radius history. The measurement of the pressure history gives one insight into the dynamics of the flame. The observation of the burnt gas temperature or gas density provides additional information on the deflagration process. All of these data sources can be used to test the validity of an integral model for flame propagation. The coverage of this subject will not be exhaustive. Instead, I will concentrate on the highlights that demonstrate the validity of integral models and will leave some of the more intricate details to the interested reader.

In an effort to demonstrate the validity of their integral models, U.S. Bureau of Mines investigators made explicit comparisons of flammable gas deflagration behavior with both adiabatic and isothermal NCV integral models (Nagy et al., 1969). They compared the performance of their models against two types of data: flame motion and pressure histories. Fig. 8.11 shows the flame radius histories for two flammable gas deflagrations performed by different investigators.

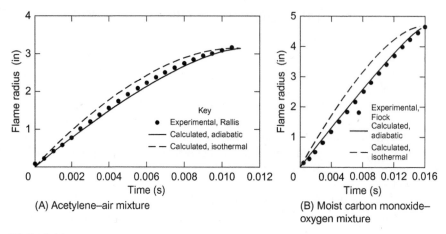

FIGURE 8.11

Comparison of predicted versus measured flame radius histories using the NCV adiabatic and isothermal integral models (public domain, Nagy et al., 1969).

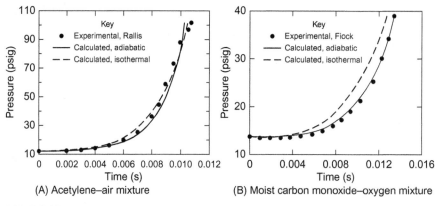

FIGURE 8.12

Comparison of predicted versus measured pressure histories using the NCV adiabatic and isothermal integral models (public domain, Nagy et al., 1969).

It can be seen that the adiabatic model gives a better fit to the data than the isothermal model, but the fit of the isothermal model is still adequate. Fig. 8.12 compares the pressure histories for the same two data sets.

Again, the adiabatic model fits the experimental data better than the isothermal NCV model. The adequacy of the fit will depend on the accuracy and precision with which one calculates the thermodynamic properties and the burning velocity.

Garforth developed an optical technique for measuring the unburnt gas density during a gaseous deflagration (Garforth, 1976). The technique involved the use of

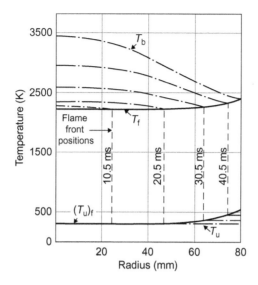

FIGURE 8.13

Temperature profile at distinct times during flame propagation.

Reproduced with permission from Garforth, A.M., Rallis, C.J., 1978. Laminar burning velocity of stoichiometric methane–air: pressure and temperature dependence. Combust. Flame 31, 53–68.

a laser interferometer setup and was determined to be capable of density measurements with an error of less than 1%. He then used this technique to determine the nature of the unburnt gas compression and concluded that the compression was indeed adiabatic. This is one of the few studies—and perhaps the only study—that directly confirmed that the unburnt gas region in a confined deflagration test is subjected to an adiabatic compression process. Garforth and Rallis derived a method for calculating the motion of the gas within a combustion vessel using the unburnt gas density measurements (Garforth and Rallis, 1976). They compared their calculated path lines (gas particle trajectories) with particle track measurements with methane–air deflagration tests and found good agreement. They then applied this experimental technique to a series of methane–air deflagration tests and concluded that the improved measurements of the unburnt gas velocity permitted them to calculate the laminar burning velocity over a range of conditions with an error of less than 5% (Garforth and Rallis, 1978).

Applying their integral model, Garforth and Rallis computed the temperature profile at distinct times for a methane–air deflagration in a spherical chamber (Fig. 8.13).

The temperature profile exhibits some interesting features. First, the adiabatic compression of the unburnt and burnt gases is evident. Second, the acceleration of the flame can be inferred from the flame positions at what are essentially equal time increments.

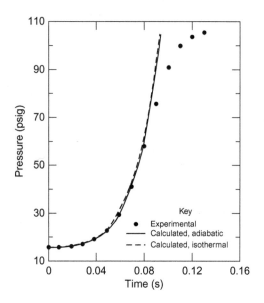

FIGURE 8.14

Comparison of predicted versus measured pressure histories using the NCV adiabatic and isothermal integral models (public domain, Nagy et al., 1969).

Finally, as a demonstration that combustible dust deflagrations are described well by integral models developed for flammable gases, Fig. 8.14 presents a comparison of the pressure history from a cornstarch dust deflagration test with the adiabatic and isothermal NCV models.

The data shown in Fig. 8.14 was obtained from a deflagration test with cornstarch at a concentration of 800 g/m³ in a rectilinear vessel with a volume of 3113 L (110 ft³). The adiabatic and isothermal model predictions fit the data well and, in fact, seem almost indistinguishable. The model predictions significantly depart from the data near the end of the deflagration due to cooling at the vessel walls.

A very important feature to note about the performance of the NCV models is that the initial state of the gas mixture is quiescent. There is no turbulence imposed on the mixture prior to ignition. In normal terrestrial gravity, this is not possible with a combustible dust. The measurement of explosibility parameters for a combustible dust will normally be accompanied by turbulence. Thus, unlike the case for gases, it is not generally possible to obtain a laminar burning velocity from a stationary burner measurement and apply it to a confined deflagration investigation. The burning velocity measurement from a stationary burner will most likely be laminar, and the environment inside the test vessel will be turbulent. As an aside, the effect of dispersion turbulence can be nearly eliminated in microgravity (Pu et al., 1998). But explosibility parameters obtained from microgravity are not a practical answer to managing terrestrial combustible dust hazards. So, if it is desired to extract a burning velocity value from a deflagration test, one must generally estimate the burning velocity from the pressure history data by trial and error to obtain a best fit.

8.5.6.2 *Chemical kinetics and laminar burning velocity*

A number of studies have been performed with the objective of improving the accuracy of laminar burning velocity predictions of hydrocarbon−air mixtures by incorporating better chemical equilibrium data and chemical kinetics models (Metghalchi and Keck, 1980, 1982; Tufano et al., 1983; Dahoe and de Goey, 2003; Saeed and Stone, 2004a, 2004b; Farrell et al., 2004; Razus et al., 2006a; Razus et al., 2012; Dahoe et al., 2013). These refinements tend to lead to more complex, less transparent mathematical models. Even one of the simplest flammable gas systems, hydrogen−air mixtures, requires careful calculation of thermodynamic properties and chemical equilibria to achieve satisfactory accuracy in the calculation of burning velocities (Dahoe, 2005; Lautkaski, 2005; Jo and Crowl, 2009, 2010).

The quality of information required for detailed chemical equilibria and kinetics calculations is not routinely available for combustible dust studies. Di Benedetto and Russo reported on a modeling effort involving cornstarch, polyethylene, and cellulose dusts (Di Benedetto and Russo, 2007). In their investigation, they combined three elements to calculate K_{st} values: a chemical equilibrium calculation for P_{max}, a laminar flame propagation model (the commercially available code CHEMKIN), and an expression for $(dP/dt)_{max}$ based on the integral model of Dahoe and de Goey (2003). They obtained reasonable agreement in their prediction of K_{st} as a function of dust concentration. Their work demonstrates that it is possible to incorporate detailed chemical equilibria and kinetics submodels into the integral model framework. However, CHEMKIN is a very sophisticated chemical kinetics software package that is intended for the specialist. Thus, the use of such tools may appeal to a smaller audience than the traditional, less accurate, integral models. The incorporation of such sophisticated approaches may be better suited for comprehensive dust flame modeling, the subject of Chapter 10.

8.5.7 SUMMARY OF INTEGRAL MODELS FOR FLAME PROPAGATION

This section on integral models began with an introduction to the mass balances and geometrical relations needed for modeling flame propagation in a spherical vessel. The calculation of the average unburnt and burnt gas temperatures during the combustion process was also presented. This was followed by a survey of integral models that have been developed over the past 50 years. Two examples of integral models were derived in detail. The first, the BM model, assumed an adiabatic compression of an ideal gas mixture and employed the LvE linear pressure rise model. The second model was the Bureau of Mines NCV model which was based on an isothermal compression of an ideal gas and did not employ the LvE linear pressure rise model. The calculation of the K_{st} parameter, the basis for the cube root scaling law, was demonstrated with both models. Furthermore, it was shown how the cubic pressure rise model could be derived from the isothermal model at small overpressures. Finally, evidence for the experimental validation of integral models was presented. This evidence demonstrates that the integral model is a satisfactory description of the behavior of a confined dust deflagration.

8.6 DUST EXPLOSION TESTING: CALIBRATION, IGNITER STRENGTH, AND TURBULENCE

Finally, it is time to address the subject of turbulence and its effect on confined deflagrations. To yield meaningful results, the explosion test apparatus must create a uniform dust cloud within the vessel. Most explosion test apparatuses rely on a short burst of compressed air to disperse the dust in the vessel. The dispersal action results in an unsteady turbulent flow field which exhibits a maximum level of turbulent intensity (velocity fluctuations) at the start of the dispersal process. Following the dispersal event, the turbulent intensity decays with time. To ensure comparability of test results, tests conducted in different explosion vessels must be conducted under the same test conditions. The cube root scaling law can be used to compare test results from vessels of different sizes, but the turbulence level in the vessels must be the same (Eckhoff, 1987).

Turbulence has a significant effect on the magnitude of the dust explosion parameters P_{max} and $(dP/dt)_{max}$ (Nagy et al., 1971; Nagy and Verakis, 1983, Chapter 5; Amyotte et al., 1989; Eckhoff, 2003, pp. 325–367; Amyotte, 2013, Chapters 14 and 19). In an effort to ensure comparable test conditions, the dust dispersion hardware and test procedures have become standardized, and competent laboratories involved in combustible dust measurements have implemented quality assurance programs. As one example of the calibration requirements, the ASTM standard for dust explosibility requires that a 20-L sphere must be calibrated with a minimum of five different dust samples over three K_{st} ranges: 1–200, 201–300, and >300 bar-m/s. The 20-L test results must agree with the 1-m^3 test results to within $\pm 10\%$ for P_{max} and $\pm 20\%$ for K_{st} (ASTM E1226, 2012; Myers and Ibarreta, 2013).

8.6.1 CALIBRATION OF TEST VESSELS

In addition to dust dispersion turbulence, the nature and strength (energy release) of the igniter also play an important role in the measurement of dust explosion parameters (van der Wel et al., 1992). Investigators have tested a variety of ignition sources for dust explosion testing. The more popular methods are pyrotechnic matches and electric spark discharges. Since different combustible dusts have different minimum ignition energies, it is necessary to design an ignition source which will ignite most dusts. But if the magnitude of energy is too great, the igniter can mask the energy release of the dust deflagration. This situation reveals itself by distorting, ie, artificially increasing the explosion severity (maximum pressure and maximum rate of pressure rise) of the test. When the igniter energy is too great, the resulting deflagration is said to be *overdriven* (Cashdollar et al., 1992; Mintz, 1995; Zhen and Leuckel, 1997). Due to its smaller size, the 20-L sphere is more susceptible to causing overdriven deflagrations. Empirical evidence suggests the following criterion to avoid overdriven deflagrations: for combustible dusts with $K_{st} \leq 50$ bar-m/s, the deflagration test should be repeated in a 1-m^3 vessel (Myers and Ibarreta, 2013).

The impact of igniter type and strength on the measured dust explosibility parameters has been documented by several investigators (Nagy et al., 1971; Cashdollar et al., 1992; Cashdollar and Chatrathi, 1992; Mintz, 1995; Zhen and Leuckel, 1997). The work of Zhen and Leuckel is especially relevant. They observed the effect of igniter performance on the deflagration behavior of stoichiometric methane—air mixtures using Schlieren photography (Schlieren photography was not possible with the cornstarch tests because the clouds were optically thick). The ignition sources tested were a 1 J electric spark and four different pyrotechnic igniters (75 J, 500 J, 5 kJ, and 10 kJ). They observed that while the electrical spark created a small ignition volume, the pyrotechnic igniters created larger initial flame kernels and ejected glowing sparks. These glowing sparks could be responsible for ignition at multiple points within the vessel volume.

Their tests spanned a range of both ignition energies and delays. Fig. 8.15 shows the effect of igniter energy on deflagration behavior: as the igniter energy increased, the rate of pressure rise increased, and the time to maximum pressure decreased.

The disparity in the shape of the pressure history for the 10 kJ igniter suggests that it caused an overdriven deflagration.

They then repeated the experiments under the similar test conditions with cornstarch. By adjusting the ignition delay time, they were able to vary the level of turbulence at the time of ignition. Longer ignition delays allow more time for the turbulence to decay which reduces the rate of pressure rise (Fig. 8.16).

They stated in their paper that, at higher turbulence levels, they were rarely able to initiate a cornstarch deflagration with a 75 J igniter but were usually successful with a 500 J igniter.

As illustrated in the work by Zhen and Leuckel, ignition and dispersion behavior are coupled. Calibration of a test vessel requires specification of the dust dispersion apparatus, the type and energy of the igniter, and the timing of the ignition event (van der Wel et al., 1992; Zhen and Leuckel, 1997). The calibration of different sized vessels has been demonstrated with the K_{st} parameter and the cube root scaling law (Cashdollar et al., 1992; Kumar et al., 1992; Cashdollar and Chatrathi, 1992; Going et al., 2000; Castellanos et al., 2011; Sattar et al., 2013). In this sense, investigators can have confidence that a sample of combustible dust tested at one laboratory will yield similar results at another laboratory.

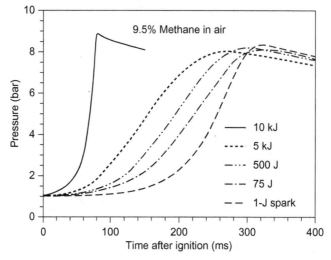

FIGURE 8.15

Ignition of a stoichiometric methane—air mixtures with a range of ignition energies.

Reproduced with permission from Zhen, G., Leuckel, W., 1997. Effects of ignitors and turbulence on dust explosions. J. Loss Prev. Process Ind. 10, 317–324.

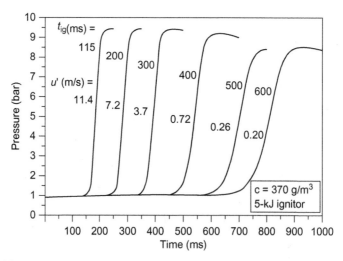

FIGURE 8.16

Variation of ignition delay time for cornstarch dust deflagrations at a fixed dust concentration and igniter energy.

Reproduced with permission from Zhen, G., Leuckel, W., 1997. Effects of ignitors and turbulence on dust explosions. J. Loss Prev. Process Ind. 10, 317–324.

Microgravity experiments are one way to overcome the influence of turbulence and gravitational settling on the dust explosion parameters. Lee et al. demonstrated the effect of gravity by performing aluminum dust explosibility testing at both terrestrial and microgravity (Lee et al., 1993). Under terrestrial gravity, the pressure history was sensitive to the ignition delay following dispersion. Testing under microgravity, the pressure history was relatively insensitive to changes in the delay time. Pu et al. tested both aluminum and cornstarch and found similar trends with changes in the ignition delay (Pu et al., 1998). Bozier and Veyssière characterized the velocity field inside their dust test vessel using particle image velocimetry (PIV) and determined that the dust concentration did indeed tend to remain uniform in microgravity while in terrestrial gravity the concentration in the same vessel decreases with time due to gravitational settling (Bozier and Veyssière, 2010). Collectively, these microgravity experiments indicate the importance of the uniformity of the dust cloud dispersion and its sensitivity to change due to gravitational settling.

A final issue to consider regarding the calibration of deflagration test vessels and procedures is the problem of marginally explosible dusts (defined approximately as dusts with $K_{st} \leq 50$ bar-m/s). The current state of the art in the design of explosion safeguards rests on the foundation of dust explosibility measurements obtained from either 20-L or 1-m^3 vessels. By virtue of cube root scaling, one should be able to use the data from one test vessel to predict the results for the other. There is increasing evidence that this is not the case (Proust et al., 2007; Myers et al., 2013). The discrepancy is especially acute for marginally explosible dusts where a dust tests positive in a 20-L sphere but fails to ignite in a 1-m^3 vessel. The source of the problem in the 20-L sphere is twofold: turbulence and the igniter strength. The turbulent flow field in the 20-L sphere is not equivalent to that in the 1-m^3 vessel. The strong ignition source used in the 20-L sphere causes preheating of the dust mixture (some investigators call it preconditioning). The energy release of the same ignition source is diluted by a dust mixture volume that is 50 times greater in the 1-m^3 vessel.

This places us in a bit of a quandary in trying to decide which test vessel to use. Despite this disparity between the two vessels, the 20-L vessel is still considered to be useful as long as one realizes that it is likely overestimating the K_{st} value of the dust if the dust is marginally explosible. In many design applications, this may be acceptable. As a practical criterion, the combustible dust test methods stipulate that measurements from the 20-L sphere should match results from the 1-m^3 vessel to within $\pm 20\%$. This gives the investigator an idea of the potential uncertainty in using 20-L sphere test data.

8.6.2 DUST DISPERSION TURBULENCE MEASUREMENTS WITHOUT COMBUSTION

Thus far this discussion has indicated that it is possible to calibrate different test vessels to yield comparable dust explosion parameters. However, some investigators have questioned the robustness of such calibrations. The source of their concern is derived from more fundamental efforts to understand the turbulent dust dispersal process. These investigations were performed without combustion. Either the tests were conducted only with air (in the combustion community these are called "cold flow" experiments) or a combustible dust was injected into the test vessel without ignition (using an inert gas such as nitrogen). The measurements were performed using modern combustion diagnostic techniques such as laser Doppler anemometry (LDA) or PIV. The test vessels were fabricated from plastic with optical windows to facilitate flow visualization and turbulence measurements.

Dahoe and his colleagues tested three different dispersion devices in a 20-L sphere: the deflector plate (rebound nozzle), the perforated annular ring, and a circular nozzle (the "Dahoe" nozzle). Fig. 8.17 shows the outline of the three nozzles.

They measured the turbulent intensity as a function of radial position along two perpendicular axes (Dahoe et al., 2001a,b; Mercer et al., 2001). They found that the turbulent intensity decayed with time with the maximum value occurring at the time of injection. From these measurements, they were able to deduce that all three dispersion nozzles produced a turbulent flow field that was homogeneous (the same value at each radial location along one axis) and isotropic (the same value at the same location on different perpendicular axes). Each dispersion nozzle achieved homogeneous, isotropic turbulence within 60 ms (the recommended time to fire the igniter according to the test method) and maintained that quality throughout the remainder of the decay period. The mean turbulent intensity at 60 ms was different for each dispersion device: 2.68 m/s for the perforated dispersion ring, 2.79 m/s for the Dahoe nozzle, and 3.75 m/s for the rebound nozzle.

Fig. 8.18 is an example of the turbulent intensity decay behavior following dispersal within the 20-L sphere.

Referencing work done by other investigators, Dahoe et al. claimed that the turbulent flow field in the 20-L sphere is different from that in the 1-m^3 vessel and warns that these differences could invalidate the cube root scaling law. They recommend using an integral balance calculation to predict dust explosion behavior for safety design purposes. Perhaps a way to rephrase the findings of Dahoe and his colleagues is that the cube root scaling law is equivalent to a simple first-order approximation of scaling behavior, while the integral balance method is a better, second-order approximation.

The work of Dahoe et al. was cold flow tests with no dust; they measured the gas velocity and turbulent intensity. Using an optical dust probe developed at the U.S. Bureau of Mines (now called the Pittsburgh Research Laboratory or PRL), Kalejaiye et al. investigated the uniformity of the dust concentration during the dust dispersal process (Kalejaiye et al., 2010; Conti et al., 1982). The light transmission measurements indicate the local dust concentration. The Lambert–Beer law (see Chapter 4) relates the reduction in light intensity to the dust concentration. They performed their work using a Siwek 20-L sphere with two standard dispersion nozzles, the rebound nozzle and the perforated annular ring. They

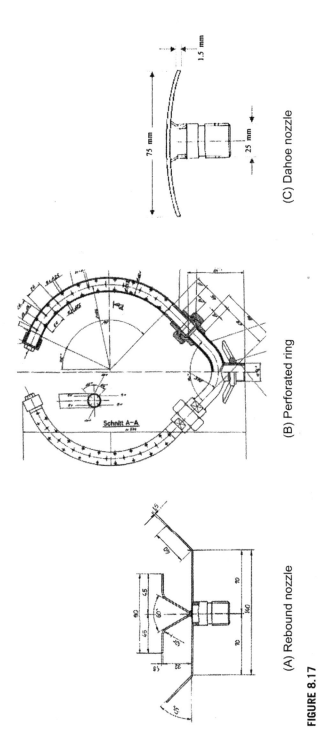

(A) Rebound nozzle

(B) Perforated ring

(C) Dahoe nozzle

1.5 mm

75 mm

25 mm

FIGURE 8.17

Three different dispersion nozzles designed for the 20-L test vessel.

Reproduced with permission form Mercer, D.B., Amyotte, P.R., Dupuis, D.J., Pegg, M.J., Dahoe, A., de Heij, W.B.C., et al., 2001. The influence of injector design on the decay of pre-ignition turbulence in a spherical explosion chamber. J. Loss Prev. Process Ind. 14, 269–282.

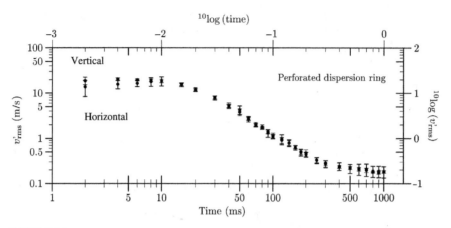

FIGURE 8.18

Space-averaged turbulent intensity in a 20-L sphere using the perforated dispersion ring device.

Reproduced with permission from Dahoe, A.E., Cant, R.S., Scarlett, B., 2001b. On the transient flow in the 20-liter sphere. J. Loss Prev. Process Ind. 14, 475–487.

tested three different dusts: Pittsburgh coal, Gilsonite (a naturally occurring organic mineral), and purple K (coated potassium carbonate). The dusts were tested at five concentrations. They also obtained light transmission data from the PRL 20-L vessel and the Fike 1-m^3 vessel. Performing 540 individual tests, they found that the uniformity of the dust concentrations was similar in all three vessels. The dust concentrations tested ranged from 25 to 350 g/m^3.

Du et al. performed flow visualization experiments in a transparent (plastic) Siwek 20-L sphere using wheat flour dust with the dust concentration ranging from 200 to 1000 g/m^3 (Du et al., 2015). They confirmed the uniformity of the dust dispersion process, but also noted that the turbulent vortex structure was somewhat attenuated at the highest dust concentration.

Dyduch and his colleagues performed turbulent intensity measurements in the 1-m^3 vessel (Dyduch et al., 2016). They used a bidirectional velocity probe to measure the turbulent intensity at different locations along an axis of the vessel and along a second axis perpendicular to the first. They found that the turbulent flow field in the 1-m^3 vessel is different from the 20-L vessel. Fig. 8.19 shows the turbulent intensity measurements for the 1-m^3 vessel and compares them with Dahoe et al. (2001a).

Their measurements indicated that the turbulence in the 1-m^3 vessel was homogeneous and isotropic starting around 600 ms, the time at which ignition is usually initiated (Fig. 8.19A). However, the magnitude of the turbulent intensity for the 20-L sphere is about four times greater than that for the 1-m^3 vessel at the time of ignition (60 ms for the 20-L sphere, 200 ms for the 1-m^3 vessel).

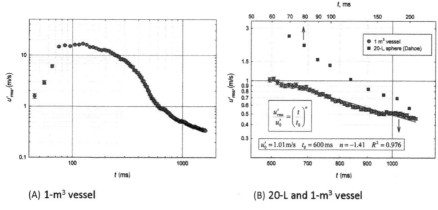

(A) 1-m³ vessel   (B) 20-L and 1-m³ vessel

FIGURE 8.19

Turbulent intensity measurements: (A) the 1-m³ vessel and (B) comparison of 20-L sphere measurements from Dahoe.

Reproduced with permission from Dyduch, Z., Toman, A., Adamus, W., 2016. Measurements of turbulence intensity in the standard 1-m³ vessel. J. Loss Prev. Process Ind. 40, 180–187.

8.6.3 CORRELATING DUST EXPLOSIBILITY TO TURBULENCE MEASUREMENTS

The turbulence measurements described in the previous section all involved cold flow tests (no combustion). Turbulence measurements have also been performed during combustion in deflagration vessels. These studies show that the turbulence due to dust dispersal influences the dust explosion parameters. Amyotte and Pegg conducted both cold flow turbulence measurements and dust deflagration tests with a 1.2-L cylindrical Hartmann bomb (Amyotte and Pegg, 1989). The deflagration tests were conducted with lycopodium powder with the dust concentration ranging from 100 to 1000 g/m³. They demonstrated that the K_{st} increased with increasing turbulent intensity. Their tests also highlighted the limitations of the Hartmann bomb due to its small volume and heat losses at the wall; the maximum K_{st} values obtained by from the Hartmann bomb were 4.4 times lower than values obtained in a 26-L sphere. In general, increasing levels of turbulence led to increasing values of the maximum rate of pressure rise.

Pu and his colleagues conducted a series of test programs to study the effects of turbulence on dust explosibility measurements (Pu et al., 1988). In one test program, they compared results from three vessels: a 6-L Hartmann bomb, a 26-L cylindrical bomb, and a 950-L spherical jet-stirred reactor. The turbulence measurements were done as cold flow tests using hot wire anemometry. They found that in all cases the turbulence could be characterized as high intensity and small scale (small integral scale values), a condition that was unlikely to be replicated in a typical accident setting. This implies that the test conditions were likely to overestimate the dust explosion parameters compared to what would be observed in an accidental deflagration.

Measuring turbulence in both methane−air and cornstarch−air deflagrations, they concluded that the integral scale of turbulence was on the order of the laminar flame thickness, and the turbulent intensity was larger than the laminar burning velocity. Within the range of conditions tested, the turbulent burning velocity was found to be a linear function of the turbulent intensity. They developed a correlation for the maximum turbulent burning velocity $S_{t,max}$ for the dust mixtures:

$$S_{t,max} = \frac{cK_{st}}{f(\Delta P_{max})} \tag{8.59}$$

$$f(\Delta P_{max}) = \Delta P_{max} \left[\frac{\Delta P}{P_0} + 1\right]^{1/\gamma} \left\{ 1 - \left(1 - \frac{\Delta P}{\Delta P_{max}}\right)\left(\frac{P_0}{\Delta P + P_0}\right)^{1/\gamma}\right\}^{2/3} \tag{8.60}$$

Here c is an empirical constant that is a function of the size of the test vessel ($c = 1$ for the 6-L and 950-L vessels, $c = 2$ for the 26-L vessel), ΔP_{max} is the maximum pressure rise, and ΔP is the overpressure at the inflection point of the pressure history corresponding to the maximum rate of pressure rise. Pu et al. comment that in their experiments $\Delta P \cong 0.5 \Delta P_{max}$.

In another test program, Pu et al. investigated the turbulence generated in a 20-L sphere (Pu et al., 1990, 1991). They characterized the turbulence generated by a perforated annular ring as having a decay period of 150−300 ms, an integral scale with a range of 1.2−1.5 cm, and a turbulent intensity with a range from 0.5 to 3 m/s. Deflagration tests with aluminum and cornstarch powders supported a linear correlation of the turbulent burning velocity with increasing levels of turbulent intensity. The turbulent burning velocity increased significantly with decreasing particle size, but was not influenced as much by the chemical composition of the dust.

8.7 SURVEY OF CONFINED DUST DEFLAGRATION BEHAVIOR

In Chapter 6 we reviewed single particle combustion and examined how chemical composition and particle size influenced the particle burning rate. In Chapter 7 we considered a wide variety of heterogeneous flame studies with an emphasis on identifying the differences between organic and metallic solid fuels. This was motivated partly by the observation that single particles and heterogeneous flames behave differently from flammable gas flames, and partly by the diversity of combustion diagnostic measurements published in the scientific literature.

In confined deflagration studies, the range of combustion diagnostics available to the investigator is more limited. There is a wealth of publications featuring the standard deflagration parameters, $(dP/dt)_{max}$, K_{st}, P_{max}, as a function of dust concentration. But the deeper insights obtained from high-speed photography or spectroscopy are often not feasible due to the high particle loading and large optical thickness of ignitable dust clouds. In some ways, the deflagration test vessel is treated like a black box with the dust samples as the input and the deflagration parameters as the only output. It is important to bear in mind, however, that the insights derived from single particle combustion and unconfined flame propagation are directly relevant to these confined deflagration studies. Where evident, these parallels will be indicated.

This review of confined dust deflagration behavior will be somewhat brief. Our attention will be primarily on studies that illustrate the relationship between explosion testing results ($K_{st}, P_{max}, t_{Pmax}$) and fuel volatility, equivalence ratio (dust concentration), mean particle size, and particle size distribution. This has been the emphasis of the scientific literature and we will take advantage of these studies. I will further restrict this review to deflagration studies conducted in either a 20-L or 1-m^3 test vessel. Older studies conducted with the Hartmann bomb or other test vessels are still relevant. Explosibility data obtained using the older (pre-1980) test procedures can be used in a comparative sense relative to an appropriate reference combustible dust (such as Pittsburgh bituminous coal). The older data are a useful source of insight, but they are *not* appropriate for design purposes.

The dust explosion intensity, $(dP/dt)_{max}$, is the most direct measurement that best indicates the speed of the deflagration wave, but its value is dependent on the vessel volume. Therefore, much of the data presented will be in terms of the K_{st} value (deflagration index) instead of the maximum rate of pressure rise. Even with the potential deficiencies of the cube root scaling law, the deflagration index is still one of the most popular measures of dust explosibility classification (Barton, 2002). This classification scheme is based on testing in a 1-m^3 vessel, so there should be no concern about misclassification of a marginally explosible dust ($K_{st} < 50$ bar-m/s) (Myer et al., 2013) (Table 8.2).

The symbol "St" is an abbreviation for the word *staub* which is the German word for dust. This classification scheme does not have a fundamental interpretation, but it is used in some deflagration venting calculation algorithms. A key point here is that, depending on its mean particle size, a specific combustible dust could satisfy any one of these four categories.

Following the plan of Chapter 6 and Chapter 7 we will then survey both organic and metallic solids. To establish a reference case for deflagration

Table 8.2 Dust Hazard Classification Based on the K_{st} Value

Dust Hazard Class	K_{st} (bar-m/s)
St 0	0
St 1	>0–200
St 2	>200–300
St 3	>300

parameters, we will first consider a flammable gas mixture of methane and air. There are definite differences in the deflagration behavior between a flammable gas mixture and a combustible solid. We will then proceed to coal dust and other, lower volatility, carbonaceous dusts. Next we will examine the behavior of biomass dust and grain dust. We will complete this section with an overview of aluminum, iron, and other metallic solid dusts. As we proceed with this discussion, we will progress through data sets that illustrate the effect of fuel volatility, dust concentration, mean particle size, and particle size distribution.

8.7.1 FLAMMABLE GASES VERSUS COMBUSTIBLE DUSTS

We will first consider the deflagration characteristics of methane in air. Dahoe and de Goey performed a comprehensive study of the deflagration parameters for methane in air at a range of equivalence ratios as determined in a 20-L test vessel (Dahoe and de Goey, 2003). Typical pressure histories are shown in Fig. 8.20.

As the concentration increases in the fuel lean region (the upper graph of Fig. 8.20), the maximum pressure increases, the rate of pressure rise increases, and the time to maximum pressure decreases. In the fuel rich region (the lower graph of Fig. 8.20), as the fuel concentration increases, the maximum pressure decreases, the maximum rate of pressure rise decreases, the time to maximum pressure increases.

The corresponding deflagration parameters are presented in Fig. 8.21. In this format, it is more evident that both the maximum pressure and the deflagration index pass through a maximum in the neighborhood of the stoichiometric concentration.

Inspection of Fig. 8.21 shows that the methane—air system generates a broad range of maximum pressure and deflagration index values. The deflagration parameters have well-defined boundaries at the low and high ends of the fuel concentration range; these boundaries correspond to the lower and upper flammability limits for the methane—air system.

Fig. 8.22 is a comparison of laminar burning velocities extracted by Dahoe and de Goey from their data sets with estimated burning velocities obtained by other investigators.

Some of the burning velocity measurements were obtained from stationary burners and some from combustion bombs, so there is a bit of variance in the data just due to the differences in experimental techniques. A further complication is the estimation of the laminar burning velocity in cases where the flame becomes turbulent. The burning velocity data in Fig. 8.22 exhibit a maximum near the stoichiometric concentration (or slightly richer) and well-defined boundaries corresponding to the lower and upper flammability limits. The sharpness of these boundaries only occurs with flammable gases and vapors. They do not occur with combustible dusts.

Fig. 8.23 presents deflagration data for the methane—air system and compares it with two combustible dusts: polyethylene dust (~100% volatiles) and high

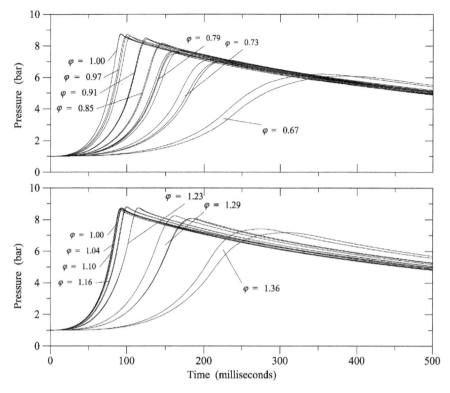

FIGURE 8.20

Pressure histories for methane–air deflagrations in a 20-L sphere; the upper graph is for fuel lean concentrations and the lower graph is for fuel rich concentrations.

Reproduced with permission from Dahoe, A.E., de Goey, L.P.H., 2003. On the determination of laminar burning velocity from closed vessel gas explosions. J. Loss Prev. 16, 457–478.

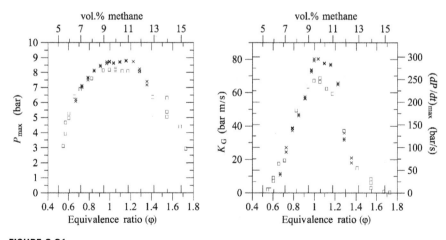

FIGURE 8.21

Deflagration parameters P_{max} and K_G for methane–air mixtures as a function of the equivalence ratio.

Reproduced with permission from Dahoe, A.E., de Goey, L.P.H., 2003. On the determination of laminar burning velocity from closed vessel gas explosions. J. Loss Prev. 16, 457–478.

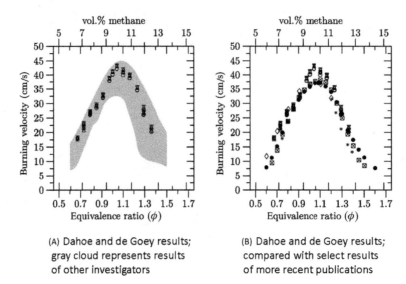

(A) Dahoe and de Goey results; gray cloud represents results of other investigators

(B) Dahoe and de Goey results; compared with select results of more recent publications

FIGURE 8.22

Laminar burning velocities for methane-air mixtures obtained by estimation from a confined deflagration model versus estimates obtained by other investigators.

Reproduced with permission from Dahoe, A.E., de Goey, L.P.H., 2003. On the determination of laminar burning velocity from closed vessel gas explosions. J. Loss Prev. 16, 457–478.

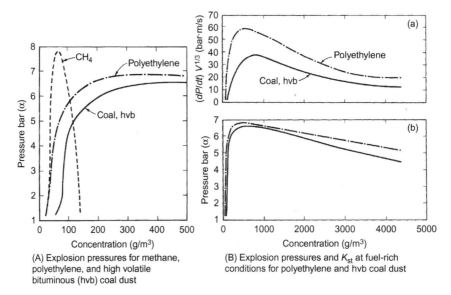

(A) Explosion pressures for methane, polyethylene, and high volatile bituminous (hvb) coal dust

(B) Explosion pressures and K_{st} at fuel-rich conditions for polyethylene and hvb coal dust

FIGURE 8.23

Comparison of methane deflagration data with two combustible dusts: polyethylene and high volatile bituminous coal.

Reproduced with permission from Cashdollar, K.L., 2000. Overview of dust explosibility characteristics. J. Loss Prev. Process Ind. 13, 189–199.

volatility bituminous (hvb) coal dust (37% volatiles) (Cashdollar, 2000). The left side of the figure includes the data for methane. The right side of the figure shows the extent of fuel rich region in which a deflagration was ignited.

The stoichiometric concentrations for methane, polyethylene, and coal are approximately 70, 81, and 102 g/m^3 (see Table 1 in Slatter et al., 2015). By inspection of Fig. 8.23, the upper limit of flammability for methane is roughly equal to an equivalence ratio equal to 2. For polyethylene and coal dust, the upper limit is not established even at equivalence ratios greater than 40. This absence of a well-defined upper flammability limit imposes a challenging constraint on dust explosion safety.

There are not many observable differences between dust deflagrations between noncharring organics and charring organics. Since there are few studies with noncharring organics, we will proceed directly our two paradigms of charring organic solids: coal dust and biomass dust.

8.7.2 COAL AND OTHER CARBONACEOUS DUSTS

The deflagration behavior of coal dust has been the subject of intense scientific investigation for over 100 years (Rice, 1910). During this time frame, extensive test programs have been undertaken in various countries. The basic trends of coal dust deflagration behavior were established in early studies using the cylindrical Hartmann bomb (Hertzberg et al., 1979; Nagy and Verakis, 1983):

- P_{max} at first increases, then reaches an asymptotic value as the mean particle size decreases or as the volatiles content increases
- $(dP/dt)_{max}$ increases as the mean particle size decreases or as the volatiles content increases
- $t_{P_{max}}$ decreases as the mean particle size decreases or as the volatiles content increases.

These basic trends have been largely verified in later studies using larger, spherical vessels with the now generally accepted test methods. We will focus on just three of these studies.

Amyotte and his colleagues tested three run of mine coals from Canadian coal mines (Amyotte et al., 1991a). Fig. 8.24 depicts the deflagration parameters for a coal sample (Prince 1) with a mass mean diameter of 23 μm and a volatiles content of 36%.

These data display the basic trend of most combustible dusts: there is a low dust concentration value, the minimum explosive concentration, which yields the lowest explosion pressure and the lowest value of the maximum rate of pressure rise. The curves describing the deflagration parameters as a function of dust concentration extend far into the fuel rich condition.

For insight into the role of volatiles content, we turn our attention to the work of Continillo and his colleagues (Continillo et al., 1991). They tested seven different types of coal (six bituminous, one anthracite) and sieved each sample to an approximate mean particle size of 53 μm. The coals exhibited a range of volatiles contents. Fig. 8.25 summarizes the test data for the seven coals.

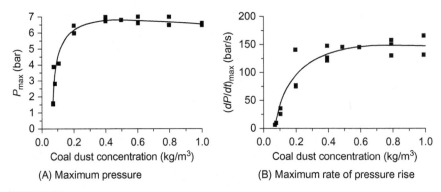

FIGURE 8.24

Deflagration parameters for Prince 1 bituminous coal (Nova Scotia, Canada).

Reproduced with permission from Amyotte, P., Mintz, K.J., Pegg, M.J., Sun, Y.-H., Wilkie, K.I., 1991a.
Laboratory investigation of the dust explosibility characteristics of three Nova Scotia coals. J. Loss Prev.
Process Ind. 4, 102–109.

Table 1 Proximate analysis

Coal sample	Volatiles	Ash	Fixed carbon
Russia	5.40	2.10	92.50
S. Africa	23.13	15.66	61.21
Sulcis no. 1	56.42	9.24	34.34
Snibston	41.20	4.70	54.10
Sulcis no. 2	44.57	16.06	39.37
N. Dakota	46.57	7.13	46.30
Montana	37.43	17.64	44.93

Table 2 Ultimate analysis

Coal sample	C	H	N	S	Ash	O
Russia	94.30	2.50	0.00	0.00	2.10	1.10
S. Africa	68.00	3.80	1.20	0.60	15.50	10.90
Sulcis no. 1	65.20	5.40	1.50	6.10	9.20	12.60
Snibston	78.40	5.00	1.50	1.00	4.80	9.30
Sulcis no. 2	57.44	5.31	1.69	7.04	16.06	12.46
N. Dakota	62.36	5.36	1.09	0.41	7.13	23.63
Montana	59.64	4.92	1.15	0.77	17.64	15.88

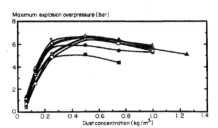

Figure 1 Maximum explosion overpressures versus nominal dust concentration for eight coal dusts at ambient initial conditions. ■, Russian anthracite; ▲, Sulcis lignite no. 1; ●, S. African coal; ×, Polish coal; +, Montana Rosebud; ○, Snibston; □, N. Dakota lignite; △, Sulcis lignite no. 2

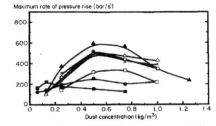

Figure 2 Maximum rates of pressure rise versus nominal dust concentration for eight coal dusts at ambient initial conditions. Symbols as in *Figure 1*

FIGURE 8.25

Proximate analysis, ultimate analysis, and the deflagration parameters for various coals measured in a 20-L sphere.

Reproduced with permission from Continillo, G., Crescitelli, S., Fumo, E., Napolitano, F., Russo, G., 1991.
Coal dust explosions in a spherical bomb. J. Loss Prev. Process Ind. 4, 223–229.

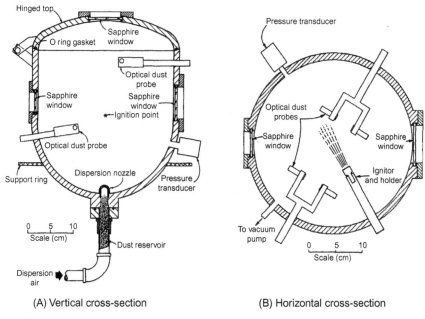

(A) Vertical cross-section   (B) Horizontal cross-section

FIGURE 8.26

Bureau of Mines 20-L deflagration test vessel.

Reproduced with permission from Cashdollar, K.L., 1996. Coal dust explosibility. J. Loss Prev. Process Ind.
9, 65– 76.

For a fixed dust concentration and particle size, both the explosion pressure and the maximum rate of pressure rise increase as the volatiles content increases. As a point of reference, at a dust concentration of approximately 500 g/m^3 the maximum rate of pressure rise varies by a factor of 3 solely due to the difference in volatiles content.

Cashdollar and Hertzberg were instrumental in developing a 20-L test vessel of their own design for use at the Bureau of Mines (Cashdollar and Hertzberg, 1985). This apparatus was specifically designed to facilitate their research in coal mine explosions and prevention. Fig. 8.26 shows a sketch of the device.

Cashdollar summarized the more recent work performed by the Bureau of Mines using this unique test vessel (Cashdollar, 1996). Fig. 8.27 is a summary of deflagration parameters as a function of dust concentration and mean particle size.

Cashdollar cautions that the turbulence level in the Bureau of Mines is much lower than that required by the standardized deflagration test procedures such as ASTM E1226. He estimated that the deflagration index values in Fig. 8.27 were too low by a factor of 3. The trends indicated by these data are sound. In particular,

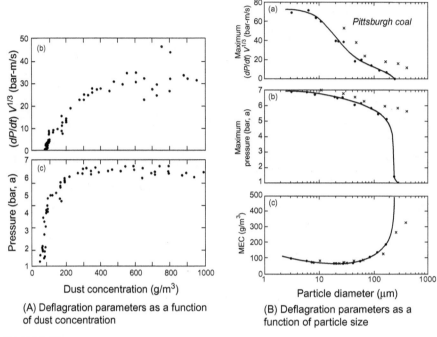

FIGURE 8.27

Coal dust deflagration parameters obtained from the Bureau of Mines 20-L test vessel.

Reproduced with permission from Cashdollar, K.L., 1996. Coal dust explosibility. J. Loss Prev. Process Ind. 9, 65–76.

the maximum pressure and the deflagration index increase with decreasing particle size. The data also indicate that the minimum explosive concentration decreases and approaches an asymptotic value as the particle size decreases.

We consider now two carbonaceous materials: activated carbon and graphite. Van der Wel et al. performed a series of 20-L sphere deflagration tests using activated carbon (van der Wel et al., 1991). They chose to work with activated carbon as an example of a material that depends on a heterogeneous chemical reaction for combustion (recall in Chapter 6 that carbon is converted to carbon monoxide at the particle surface and then the CO is oxidized as a homogenous reaction in the free stream). Fig. 8.28 summarizes the deflagration parameters for activated carbon at three different particle sizes.

The activated carbon deflagration data serve as an important reminder that differences in particle size alone can lead to changes in the dust hazard classification: the 20.6 μm dust is a Dust Hazard Class 1 and the 4.8 μm dust is (perhaps just barely) a Dust Hazard Class 2. Fig. 8.29 compares the deflagration index for the activated carbon with a particle size of 8.4 μm and a graphite powder with a particle size of 7.9 μm.

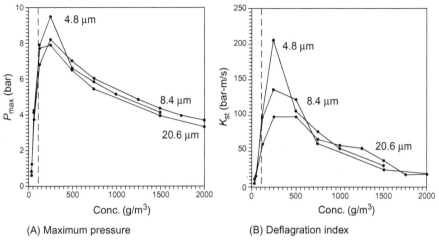

(A) Maximum pressure   (B) Deflagration index

FIGURE 8.28

Deflagration parameters for activated carbon measured in a 20-L sphere. The dashed line denotes the stoichiometric concentration of 100 g/m^3.

Reproduced with permission from van der Wel, P.G.J., Lemkowitz, S.M., Scarlett, B., van Wingerden, C.J.M., 1991. A study of particle factors affecting dust explosions. Part. Part. Sys. Character. 8, 90–94.

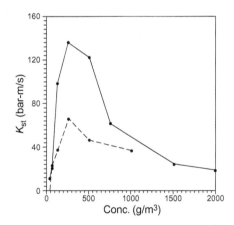

FIGURE 8.29

Comparison of the deflagration index for activated carbon and graphite.

Reproduced with permission from van der Wel, P.G.J., Lemkowitz, S.M., Scarlett, B., van Wingerden, C.J.M., 1991. A study of particle factors affecting dust explosions. Part. Part. Sys. Character. 8, 90–94.

The authors indicate that the primary physical difference between the activated carbon and the graphite samples is the large specific surface area of the activated carbon (estimated as 20–30 times larger than the graphite material). However, while the deflagration index for the activated carbon is larger than the values for

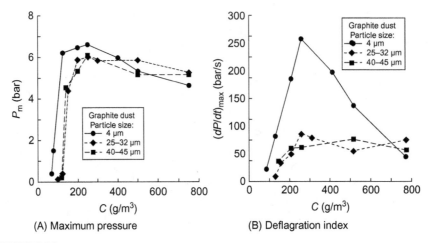

FIGURE 8.30

Graphite—air deflagration parameters for different particle sizes.

Reproduced with permission from Denkevits, A., Dorofeev, S., 2006. Explosibility of fine graphite and tungsten dusts and their mixtures. J. Loss Prev. Process Ind. 19, 174–180.

graphite, they are not 20–30 times greater. They suggest that there is an oxygen mass transfer limitation and only the largest pores of the activated carbon are accessible during the deflagration event.

Denkevits and Dorofeev also conducted an investigation on the deflagration behavior of graphite in a 20-L sphere (Denkevits and Dorofeev, 2006) (Fig. 8.30).

Compared to the bituminous coals, graphite yields similar explosion pressures but significantly lower maximum rates of pressure rise.

8.7.3 BIOMASS AND GRAIN DUST

The explosibility of biomass dust and grain dust has also received a great deal of attention over the last 100 years (Edwards and Leinbach, 1935; Jacobson et al., 1961; Aldis and Lai, 1979). Again, most of this work involved the use of the cylindrical Hartmann bomb. Furthermore, although very fine dust is relatively easy to work with, larger biomass particles can present a variety of fibrous, curvilinear shapes that make dust dispersion difficult. We will consider only two of these more modern studies and include a summary table of deflagration parameters for a variety of biomass dusts.

Cork, a form of biomass, is harvested from the inner bark of the cork oak. Pilão et al. conducted deflagration tests with cork dust to better understand the dust explosion hazards of the Portuguese cork industry (Pilão et al., 2006). Cork dust has a volatiles content of more than 90% and has a porous microstructure (Silva et al., 2005). The deflagration tests were conducted in a 22.7-L test vessel. Dust samples were sieved into six distinct fractions. The deflagration

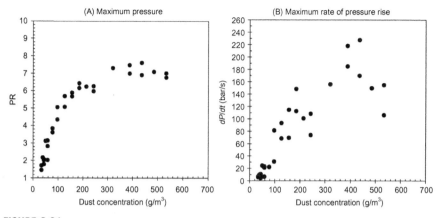

FIGURE 8.31

Deflagration parameters for cork dust (Pilão et al., 2006).

parameters for the fraction with a mass median diameter of 71.3 μm are shown in Fig. 8.31.

The dispersion of the data is an indication of the challenges one faces both with the biological variability of biomass and the unusual particle morphology of cork dust.

Researchers at the University of Leeds have been very active in the development of test hardware and procedures to accommodate the complex bulk material handling characteristics of biomass (Sattar et al., 2012; Sattar et al., 2013, 2014; Medina et al., 2015). Their deflagration studies have demonstrated the wide range of explosibility properties presented by biomass. Table 8.3 summarizes the deflagration parameters measured by these researchers as well as others (see the original paper for the various citations).

Most of these materials would be classified as Dust Hazard Class 1. The same is true for most bituminous coals.

Fig. 8.32 compares two softwood dusts (woody biomass) with two bituminous coals. The volatiles content of the Southern pine and Norway spruce were both approximately 79%. The volatiles content of the Kellingley and Columbian coals were 29% and 34%, respectively. The median diameter for both of the wood dust samples was on the order of 200 μm, and for both of the coal samples it was on the order of 30 μm.

The deflagration indices for the biomass materials are bounded by the coal samples. Based on the data presented in this chapter thus far, it is reasonable to infer that the deflagration parameters for coal and biomass can vary significantly depending on their volatiles content and mean particle size. With such a great degree of variability, if your application involves coal or biomass, it will most likely be necessary to determine the deflagration parameters for your specific material.

Table 8.3 Dust Deflagration Parameters for Select Biomass Materials

Fuel	K_{st} (bar-m/s)	P_{max} (bar)	Particle Size (μm)
Cork	179	7.2	$d_{90} = 280$
Walnut shells dust	105	9.4	$d_{90} = 311$
Pine nut shells dust	61	8.9	$d_{90} = 439$
Pistachio shells dust	82	9.3	$d_{90} = 240$
Wood (unspecified)	87	8.8	$d_{90} = 700$
Bark (unspecified)	98	9.7	$d_{90} = 700$
Forest residue	84	9.1	$d_{90} = 500$
Spanish pine	23	8.2	$d_{90} = 500$
Barley straw	58	9.3	$d_{90} = 500$
Miscanthus	31	8.1	$d_{90} = 350$
Sorghum	28	8.2	$d_{90} = 650$
Rape seed straw	32	8.2	$d_{90} = 500$
Wood dust (beech and oak mix)	136	7.7	$d_{90} = 125$
Forest residue (bark and wood)	92	9.1	$d_{50} = 275$
Wood dust	87	7.8	—
Wood dust, chipboard	102	8.7	$d_{90} = 43$
Wheat grain dust	112	9.3	<500
Olive pellets	74	10.4	—
Cellulose	66	9.3	125
British Columbia wood pellets	146	8.1	<63
Nova Scotia wood pellets	162	8.4	<63
Southern yellow pine wood pellets (USA)	98	7.7	<63
Wood dust	208	9.4	—
Fibrous wood	149	8.2	<75
Sawdust	115	9.0	—
Dry Douglas fir & Western red cedar	43	8.5	250
Dry mountain pine & Lodgepole pine	40	8.8	200
Dry spruce & pine & fir	51	8.2	$d_{90} = 200$
Southern pine	105	9.0	$d_{90} = 739$
Norway spruce	95	9.2	$d_{90} = 603$

8.7.4 ALUMINUM AND IRON DUSTS

The Bureau of Mines published several publications on the explosibility of metal dusts (Hertzberg et al., 1992; Cashdollar, 1994; Cashdollar and Zlochower, 2007). Our interest here is in aluminum, a volatile metallic solid, and iron, a nonvolatile metallic solid. We will consider three studies involving aluminum, but only one involving iron.

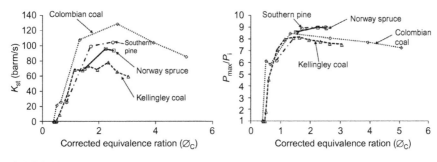

FIGURE 8.32

Deflagration parameters obtained from a 1-m³ test vessel.

Reproduced with permission from Medina, C.H., MacCoitir, B., Sattar, H., Slatter, D.J.F., Phylaktou, N.H., Andrews, G.E., et al., 2015. Comparison of the explosion characteristics and flame speeds of pulverized coals and biomass in the ISO standard 1-m³ dust explosion equipment. Fuel 151, 91–101.

Table 8.4 Aluminum Powder Characteristics

Powder	Particle Diameter (μm)	Specific Surface Area (m²/g)
Atomized		
RE 400	7.0	0.868
RE 1131	12.1	0.620
RE 1231	17.4	0.595
RE 1511	28.4	0.501
Flake		
RE 40XD	4.3	6.52
RE 30XD	5.7	5.52
RE 4301	12.4	4.70
RE 3XD	20.7	1.72

Reproduced with permission from Ogle, R.A., 1986. A new strategy for dust explosion research: a synthesis of combustion theory, experimental design and particle characterization. Ph.D. thesis, University of Iowa, Iowa City, IA.

    While there have been numerous deflagration studies on aluminum dusts, few have been conducted in either the 20-L sphere or the 1-m³ vessel. One of these early studies was performed by Ogle (1986). Tests were conducted with four atomized and four flake aluminum powders, all obtained from Reynolds Aluminum. The flake powders contained a stearic acid coating. The particle size distribution of each powder was measured by light scattering and the specific surface area was determined by nitrogen gas adsorption. The particle diameter corresponds to the mean volume diameter. Table 8.4 summarizes the particle characterization data.

    The effect of particle size on the deflagration pressure history is shown in Fig. 8.33. These tests were conducted at near-stoichiometric conditions.

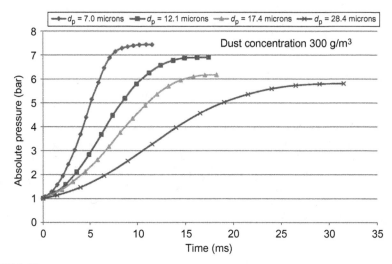

FIGURE 8.33

Deflagration pressure histories for four atomized aluminum powders. The tests were conducted in a 20-L sphere.

Reproduced with permission from Ogle, R.A., 1986. A new strategy for dust explosion research: a synthesis of combustion theory, experimental design and particle characterization. Ph.D. thesis, University of Iowa, Iowa City, IA.

As the particle size decreases, the maximum rate of pressure rise increases and the time to maximum pressure decreases. The effect of particle size on the deflagration parameter is depicted in Fig. 8.34.

As expected, the K_{st} values increase as the mean particle size decreases. An additional important observation from Fig. 8.34 is that the aluminum powder samples span all three (reactive) dust hazard classes. This emphasizes the point that particle size plays an important role in dust explosion severity.

Another deflagration parameter to consider is the time to maximum pressure $t_{P_{max}}$. This is a measure of the time for the flame to sweep through the test vessel. The time to maximum pressure decreases with particle diameter (Fig. 8.35). This is consistent with the interpretation that smaller particles have a shorter time scale for combustion.

Thus far the trends in the deflagration parameters have been analyzed only in terms of the mean particle diameter. The work of Dufaud and his colleagues provides additional insights into the role of the particle size distribution (Dufaud et al., 2010). They tested four aluminum powder samples median diameters of 7, 11, 27, and 42 μm. Fig. 8.36 shows the deflagration parameters from their 20-L test vessel.

They noted that the fine fraction, which they defined as the 10th percentile of the particle size distribution, had a significant impact on the maximum rate of pressure rise. Fig. 8.37 shows the particle size distributions tested and the time to maximum pressure (which they label as the combustion time).

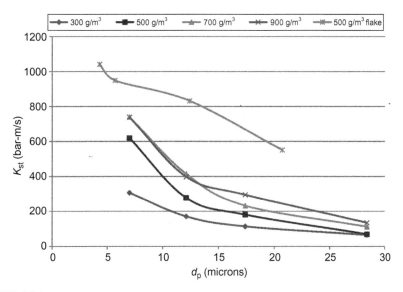

FIGURE 8.34

Deflagration index for four atomized aluminum powders and four aluminum flake
powders.

Reproduced with permission from Ogle, R.A., 1986. A new strategy for dust explosion research: a synthesis of
combustion theory, experimental design and particle characterization. Ph.D. thesis, University of Iowa, Iowa
City, IA.

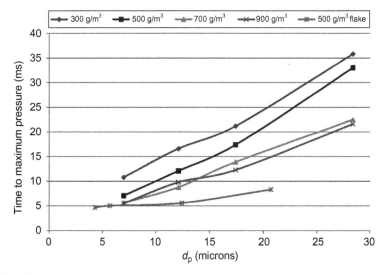

FIGURE 8.35

Time to maximum pressure versus particle diameter for atomized and flake aluminum
powders.

Reproduced with permission from Ogle, R.A., 1986. A new strategy for dust explosion research: a synthesis of
combustion theory, experimental design and particle characterization. Ph.D. thesis, University of Iowa, Iowa
City, IA.

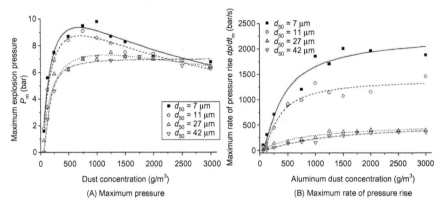

FIGURE 8.36

Deflagration parameters from Dufaud et al. 20-L test vessel.

Reproduced with permission from Dufaud, O., Traore, M., Perrin, L., Chazelet, S., and Thomas, D., 2010. Experimental investigation and modelling of aluminium dusts explosions in the 20 L sphere. J. Loss Prev. Process Ind. 23, 226–236.

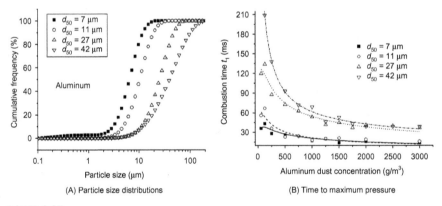

FIGURE 8.37

Particle size distributions tested and the time to maximum pressure (labeled as the combustion time).

Reproduced with permission from Dufaud, O., Traore, M., Perrin, L., Chazelet, S., and Thomas, D., 2010. Experimental investigation and modelling of aluminium dusts explosions in the 20 L sphere. J. Loss Prev. Process Ind. 23, 226–236.

The data support the interpretation that the finest particles have the greatest influence on the combustion rate.

Castellanos et al. studied the effect of the particle size distribution on the deflagration parameters by blending different aluminum powders to create modified particle size distributions (Castellanos et al., 2014). Specifically, they

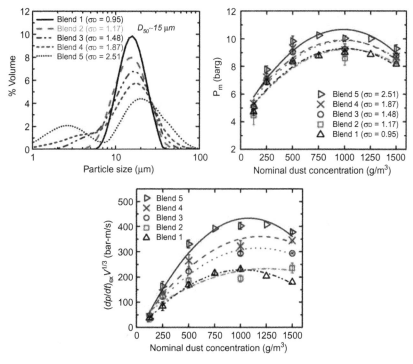

FIGURE 8.38

Results of tests in a 36-L vessel according to the ASTM E1226 procedure.

Reproduced with permission from Castellanos, D., Carreto-Vazquez, V.H., Mashuga, C.V., Trottier, R., Mejia, A.F., Mannan, M.S., 2014. The effect of particle size polydispersity on explosibility characteristics of aluminum dust. Powder Technol. 254, 331–337.

systematically modified the spread (variance) of the distribution. They defined the spread in the distribution according to the polydispersity index which they defined as $\sigma_d = (d_{90} - d_{10})/d_{50}$. They performed their tests in a 36-L vessel according to the ASTM E1226 procedure. Fig. 8.38 summarizes their results in terms of the blended particle size distributions tested and the deflagration parameters, P_{max} and K_{st}.

Even though each of the five powder blends had the same median diameter, their K_{st} values were sensitive to the polydispersity of the distribution.

Cashdollar and Zlochower reported on deflagration tests conducted with two types of iron dust (Cashdollar and Zlochower, 2007). The Fe-1 sample had a median diameter of 4 μm and the Fe-2 sample had a mean diameter of 45 μm. The deflagration parameters are summarized in Fig. 8.39.

The stoichiometric concentration for iron is about 650 g/m³. The deflagration parameters for iron dust are quite small compared to aluminum. Although the

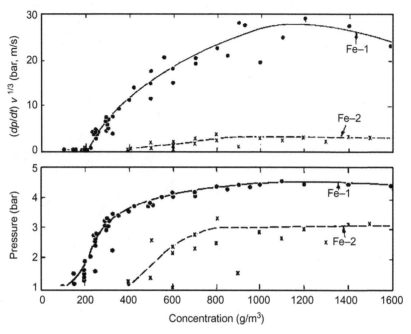

FIGURE 8.39

Deflagration data obtained for two different iron powders from the Bureau of Mines 20-L test vessel.

Reproduced with permission from Cashdollar, K.L., Zlochower, I.A., 2007. Explosion temperatures and pressures of metals and other elemental dust clouds. J. Loss Prev. Process Ind. 20, 337–348.

thermodynamic properties of aluminum combustion are different from iron combustion, these results are still an indication of the effect that fuel volatility (aluminum) can have on the severity of a dust explosion.

8.7.5 SUMMARY OF CONFINED DUST DEFLAGRATION STUDIES

For a given deflagration test vessel and ignition source, combustible dusts tend to exhibit a definite minimum explosive concentration, but not a maximum concentration limit. The maximum pressure and maximum rate of pressure rise typically occur at concentrations that are greater than stoichiometric, sometimes two to three times greater. For a fixed dust concentration, fuels with greater volatiles content tend to give greater explosion pressures and rates of pressure rise. Smaller particles burn faster than large particles, and therefore smaller particles in a monosized dust sample yield larger maximum rates of pressure rise. Smaller particles in a monosized dust sample will yield greater pressures in an asymptotic manner, ie, progressing from large particle sizes to smaller, the explosion pressure will at first increase and then approach asymptotic limit. Polydisperse dust samples yield more complex behavior, but the smaller particles will dominate the deflagration behavior as their relative amount increases.

8.8 EXPLOSION CONCENTRATION LIMITS

Explosion concentration limits, or simply explosion limits, are the concentration extremes that define the boundaries of combustion behavior. The explosion limits to be discussed are the minimum explosible concentration (MEC), the maximum explosible concentration, the limiting oxygen concentration (LOC), and the inert solids loading for explosion suppression.

The measurement of flammability concentration limits in gases and vapors has been an important part of combustion research for over 200 years (Britton, 2002; Britton et al., 2005, 2016; Crowl and Jo, 2009). These measurements play an important role in the safe handling and use of flammable gases and vapor (Zabetakis, 1965; Britton, 2002; Crowl, 2003, pp. 18−37 and Appendix A). The references cited in this paragraph have data on a large number of flammable gas−air mixtures. For the lower flammable limit (LFL), the Le Chatelier mixing rule is a useful first approximation for predicting the LFL of mixtures (Mashuga and Crowl, 2000).

The establishment of a flammability diagram for a combustible dust is a more problematic task. The difficulties include the norm of incomplete combustion, the somewhat arbitrary definition of a minimum explosible concentration, and the absence of a definitive maximum explosible concentration limit. Fig. 8.40 illustrates the basic explosibility diagram for a combustible dust.

The LOC and MEC points are indicated as distinct points. The locus of MEC as a function of oxygen concentrations is indicated by the solid line. The absence of a locus of maximum explosible concentrations is indicated by the dashed line.

From a practical perspective, test methods have been developed to provide useful data to assist the safety engineer, and these methods are founded on fundamental principles. But the task of establishing concentration limits is simply more difficult for combustible dusts that it is for flammable ·gases. One of the features that complicates this task is the common occurrence of incomplete combustion.

8.8.1 EVIDENCE OF INCOMPLETE COMBUSTION IN DUST DEFLAGRATIONS

For many organic combustible dusts, the dust concentration range typically spanned in deflagration testing (250, 500, 750, 1000 g/m^3) is mostly in the fuel rich regime. By definition, deflagration tests with dust concentrations in the fuel rich condition should result in incomplete combustion. But it has been found that

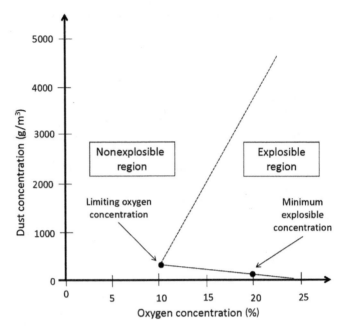

FIGURE 8.40

Generic explosibility diagram for a combustible dust.

for organic dusts even fuel lean concentrations can result in incomplete combustion. To a first approximation, the measured explosion pressure should correspond to the spatial-averaged burnt gas temperature. From this temperature, one can then calculate the percentage conversion of the fuel. The results of such calculations indicate that the combustion process is often incomplete even in the fuel lean range (Hertzberg et al., 1982; Dastidar and Amyotte, 2004). In Chapter 6 we discussed the physical processes, chemical reactions, and microstructural features that played a role in single particle combustion. These attributes play a direct role in determining the explosibility limits of a combustible dust. Flammable gas—air mixtures set a benchmark for combustion performance because the fuel and oxidizer commingle at the molecular scale. Combustible dust—air mixtures exhibit slower combustion performance because rate processes slow the rate of commingling. This is why even fuel lean mixtures of combustible dusts often exhibit incomplete combustion.

A maximum pressure measurement that is substantially lower than the constant volume explosion pressure is an indication of incomplete combustion (Hertzberg et al., 1986; Lee et al., 1992). This is especially evident in charring organic solids. Recall from Chapter 6 that investigators have found that in coal combustion volatiles combustion was 10 times faster than char oxidation. Consistent with some of the earliest scientific investigations of coal mine explosions, Bureau of Mines investigators determined that postdeflagration solid residues indicated incomplete combustion both in laboratory and experimental mine settings (Faraday and Lyell, 1845; Rice, 1910; Hertzberg et al., 1982; Cashdollar et al., 1992; Cashdollar et al., 2007; Man et al., 2011). In fuel lean coal dust deflagrations, the volatiles burn and leave behind the char (in the form of coke).

Investigators at the University Leeds have performed numerous studies on many different forms of biomass and found compelling evidence of incomplete combustion (Sattar et al., 2012; Medina et al., 2013; Saeed et al., 2015; Slatter et al., 2015). Other investigators have performed postdeflagration headspace analysis of gaseous combustion products with coal, carbonaceous materials, and cornstarch; these results also indicate incomplete combustion of the dust (Nagy and Verakis, 1983, pp. 39–41; van der Wel et al., 1991; Jarosinski et al., 1993).

One of the fundamental goals of dust explosion dynamics is to relate dust deflagration parameters to the conversion of fuel into combustion products. Numerous investigations with a wide variety of materials indicate that the incomplete combustion of the fuel caused by finite rate processes acts to confound the effort.

8.8.2 MINIMUM EXPLOSIBLE CONCENTRATION

The MEC is the lowest dust concentration in air that will support a deflagration. The operational definition for MEC is that it is the lowest dust concentration in air that yields a pressure ratio PR $\geq$ 2 (ASTM E1515, 2007). The pressure ratio is defined as

$$\mathrm{PR} = \frac{P_{\mathrm{ex}} - \Delta P_{\mathrm{igniter}}}{P_{\mathrm{ignition}}} \tag{8.61}$$

The ASTM standard defines the symbol P_{ex} as the maximum explosion pressure measured for the specific deflagration test, $\Delta P_{\mathrm{igniter}}$ is the pressure rise in the chamber due to the igniter by itself in air, and P_{ignition} is the absolute pressure at the time that the igniter is activated. Further discussion of this and other test methods for MEC can be found in the standard dust explosion references (Eckhoff, 2003, pp. 303–318, 518–525; Frank and Rodgers, 2012, Chapter 2).

A number of investigators have examined the role of test vessel size, igniter strength, test procedure, test criteria, and dust properties on the determination of MEC. Investigators compared the Bureau of Mines 20-L test vessel, the 20-L Siwek sphere, and the 1-m^3 vessel (Cashdollar et al., 1992; Chawla et al., 1996; Going et al., 2000). Testing coal, oil shale, and Gilsonite, they found that lower energy igniters (2500 J) in the 20-L test vessels gave comparable results in the determination of the MEC. Investigators at the University of Leeds developed a Hartmann tube procedure for testing biomass samples that were too bulky and fibrous for proper dispersal in a 20-L test vessel (Medina et al., 2013; Saeed et al., 2015). While the MEC values obtained were lower than those obtained from a 1-m³ vessel, they were able to use the Hartmann tube data to rank the reactivity of the biomass materials tested and proposed that this technique was easier and faster to use for screening test results.

The MEC can be understood from the perspective of a limit flame temperature. It has been observed by many investigators that the flame temperature of a flammable gas mixture at the LFL is much lower than the maximum adiabatic flame temperature (which usually occurs slightly richer than the stoichiometric concentration) (Mashuga and Crowl, 1999). Generally, the flame temperature at the LFL is around 1000−1500 K. Mashuga and Crowl explain this flame temperature as being influenced by the thermochemistry of the oxidation of carbon monoxide. For the combustion of hydrocarbons in air, the oxidation of carbon monoxide is the most exothermic portion of the reaction mechanism. This observation has been developed into a criterion for predicting the MEC (Hertzberg et al., 1986; Cashdollar et al., 1988; Wolanski, 1994; Dastidar and Amyotte, 2004). Clearly, one must be careful to develop temperature criteria appropriate for the type of combustible dust one is studying. Organics may well fall into the same temperature region, but metals may not.

The Bureau of Mines investigators explored the consequences of this model and found it to be reasonably successful for predicting the MEC of carbonaceous dusts (Hertzberg et al., 1986; Cashdollar et al., 1988). They also noted, however, that it was the combustion of volatiles that determined the MEC, not char oxidation. Wolanski suggested the following criterion based on the assumption of constant properties and a recommendation to use $\Delta T = 1400$ K (Wolanski, 1994):

$$\text{MEC} = \frac{\rho_{\text{mix}} C_{\text{p,mix}} \Delta T}{\text{HHV}} = \frac{1.8 \times 10^6}{\text{HHV}}; \ \text{MEC}\left(\frac{\text{g}}{\text{m}^3}\right), \ \text{HHV}\left(\frac{\text{kJ}}{\text{kg}}\right) \tag{8.62}$$

Since we are interested in the initiation of combustion (the constant pressure phase), and not the entire course through a deflagration, the flame temperature used here is the flame temperature at constant pressure. Dastidar and Amyotte calculated the limit flame temperatures for several materials based on their MEC measurements (Dastidar and Amyotte, 2004). The results are summarized in Table 8.5.

Table 8.5 Limit Flame Temperatures and Minimum Explosible Concentrations

Dust	Limit Flame Temperature (K)	MEC (g/m³)
Aluminum	2000	80
Anthraquinone	1200	45
Polyethylene	1250	30
Cornstarch	1200	85
Coal	1450	80

From Dastidar, A.G., Amyotte, P.R., 2004. Using calculated adiabatic flame temperatures to determine dust explosion inerting requirements. Process Saf. Environ. Prog. 82(B2), 142−155.

The limit flame temperature concept seems to work well as an organizing concept for the MEC. However, to calculate the flame temperature for a given combustible dust, one must have the thermochemical data required for the chemical equilibrium calculations. See the paper by Dastidar and Amyotte for a good discussion on the estimation of thermochemical data based on analog compounds.

8.8.3 WHY IS THERE NO MAXIMUM EXPLOSIBLE CONCENTRATION FOR DUSTS?

In the survey of dust deflagration behavior, it was evident that there was no maximum explosible concentration. Unlike flammable gases which typically show a maximum explosible concentration (or upper explosible limit) in the range of an equivalence ratio between 3 and 4, combustible dusts can support deflagrations at equivalence ratios of 10 and higher. There are exceptions to this observation: sometimes a maximum explosible concentration is observed, sometimes it is not. The existence of an upper explosible limit may depend on the width of the particle size distribution with monosized particles, or at least very narrow size distributions yielding a definite upper concentration limit (Mintz, 1993). Broad size distributions may not present a definite upper limit. At this point in time, the relationship between the upper explosion limit and the particle size distribution must be considered a hypothesis subject to further investigation.

There have been attempts to establish a theoretical basis for the upper explosible limit (Nomura et al., 1984; Mintz, 1993). Although differing in the details, both studies are based on the same conceptual model of a single burning particle and its oxygen demand compared to the total oxygen available in the test vessel. The model of Nomura et al. is more intricate in its analysis in that it assumes that the particles fill the volume of the test vessel in a periodic array. The model of Mintz is simpler and will satisfy our objectives.

Assume monosized combustible dust particles dispersed uniformly in air. Define the optimal dust concentration, $\rho_{d,opt}$, as the dust concentration that yields the maximum deflagration pressure. If the particles react to completion, then the optimal dust concentration is simply the stoichiometric dust concentration, $\rho_{d,st}$. The condition of complete combustion is equivalent to complete conversion (consumption) of the single fuel particle, $x_p = 1$. The stoichiometric concentration is based on the assumption of complete conversion of the fuel particles. At fuel rich conditions—the region where the upper explosible limit will be found—oxygen is the limiting reactant. The maximum amount of fuel that can react is equal to the stoichiometric fuel mass. The remaining fuel is unreacted.

If the particles react only partially, then the optimal dust concentration will be larger than the stoichiometric concentration. A mass balance on the dust gives

$$\rho_{d,st} = x_p \rho_{d,opt} \rightarrow \rho_{d,opt} = \frac{\rho_{d,st}}{x_p}; \; \rho_{d,opt} \geq \rho_{d,st} \tag{8.63}$$

Note that for incomplete combustion, the optimal dust concentration must be greater than the stoichiometric dust concentration. This simple model describes the observed behavior; it does not explain it. Further investigations are required.

A final observation about dust deflagration behavior in the fuel rich limit. For dust—air mixtures, regardless of the dust concentration, the oxygen concentration in the gas phase is always 20.9%. This is not the case with flammable gases where the oxygen concentration is diminished by the increase in fuel concentration. One reason for the sharp, well-defined upper explosible limits on flammable gas—air mixtures is this inherent reduction in the oxygen concentration. For many flammable gas—air mixtures, the upper explosible limit (or upper flammable limit) corresponds to a reduction in oxygen concentration from 20.9% to something on the order of 10%. The multiphase mixture character of a combustible dust—air suspension breaks this oxygen concentration constraint so that the dust always sees an oxygen concentration of 20.9%.

8.8.4 LIMITING OXYGEN CONCENTRATION

The limiting oxygen concentration for a combustible dust is an important safety parameter as it presents an opportunity to prevent a dust explosion by displacing oxygen with an inert gas. Test methods have been developed to measure the LOC with good reproducibility (Going et al., 2000; Dastidar, 2016). The basic protocol for determining the LOC is to progressively lower the oxygen content of a dust—air mixture at a fixed dust concentration until the ignition event fails to initiate a deflagration. The dust concentration chosen for the test is usually the dust that corresponds to the maximum pressure.

Earlier studies have demonstrated that the maximum pressure and maximum rate of pressure rise decrease almost linearly with decreasing oxygen concentration (Continillo et al., 1991; Wilén et al., 1998). For many organic combustible dusts, the LOC falls within the range of 5—10%, but for combustible metals the LOC can be even lower (NFPA 652, 2016).

8.8.5 INERT SOLIDS FOR EXPLOSION SUPPRESSION

The idea of using rock dust to inhibit coal dust explosions in underground coal mines took hold in the United States in the 1920s. (Cashdollar et al., 2010). Machines were developed that could spray rock dust—usually limestone—onto the walls, ceiling, and floor of the mine. It is a fair assessment that investigators were not sure how rock dusting inhibited dust explosions. But the Bureau of Mines had experimental mines at their disposal and were able to conduct hundreds of full-scale mine tests to research a number of mining safety issues including rock dusting. Although the theoretical basis was yet to be established, the empirical proof of the effectiveness was enough to persuade lawmakers to pass laws requiring the use of rock dust in underground coal mines.

The most pervasive use of inert solids to inhibit dust explosions comes from the coal mining industry. The Bureau of Mines investigators were heavily involved in evaluating rock dust as an inhibitor (Hertzberg et al., 1984, 1986; Man and Teacoach, 2009; Harris et al., 2010, 2015; Cashdollar et al., 2010). Much of this work built on the concept of the limit flame temperature enhanced by the recognition that the volatiles content of coal was a significant variable in deflagration behavior. This work included both laboratory work with the Hartmann bomb and the 20-L test vessel and also a significant number of full-scale tests in the experimental mines. One of the conclusions from this work was that limestone rock dust primarily worked to inhibit deflagrations by acting as a heat sink (ie, the rock dust absorbs heat from the deflagration). However, there was also evidence that a small fraction of the limestone rock dust was converted to calcium oxide (an endothermic chemical reaction) thus increasing slightly the effectiveness of the rock dust as a heat sink.

In Canada, Amyotte and his colleagues performed extensive work on the inerting of coal dust (Amyotte et al., 1991b; Amyotte et al., 1992, 2005; Dastidar et al., 1997; Dastidar et al., 2001). They found that the effectiveness of rock dust increased with decreasing limestone particle size. They tried different rock dusting materials and found that limestone, dolomite, and magnesite all exhibited similar effectiveness against coal dust deflagrations. They also investigated the scale-up of laboratory test data with full-scale mine data and concluded that rock dusting requirements did not scale quantitatively with laboratory data. Instead, they recommended that the laboratory data should be used as a qualitative guide for designing full-scale experiments.

Amyotte and his industrial colleagues have worked extensively on developing a thermal theory of deflagration inhibition (Amyotte and Pegg, 1992; Dastidar et al., 1999; Chatrathi and Going, 2000; Dastidar and Amyotte, 2002, 2004; Amyotte, 2006). The theory is an elaboration of the limit flame temperature. Readers interested in this topic are encouraged to read the original papers.

The inhibition of dust explosions by adding an inert material to a combustible dust is not limited to underground coal mines. In many manufacturing operations, particulate debris containing a combustible dust is created. As an example, a safety consultant evaluated the byproduct created by a manufacturer of aluminum products. The manufacturing operation included a buffing step to polish the aluminum components. The buffing operation included the action of a buffing wheel with a buffing compound polishing the aluminum surface. Tests of the byproduct material collected from the buffing operation determined that the fluff material consisted of 6% aluminum. Deflagration testing of the buffing byproduct indicated that it was a Dust Hazard Class 1. To mitigate the risk of generating a combustible dust, the manufacturer ordered buffing pads with flame retardant added to the active surface of the buffing pads. The treated pads generated a buffing byproduct that had a significantly lower deflagration index (Myers, 2008).

8.9 CONTROLLING DUST EXPLOSION HAZARDS

Dust explosions can be prevented or their severity can be controlled. Our attention here is restricted to primary dust explosions. Secondary dust explosions will be the topic of Chapter 9. The systematic development of dust explosion hazard controls can be developed with reference to the dust explosion pentagon: fuel, oxidizer, ignition source, dust dispersal, and confinement. There are several excellent books available on the subject of dust explosion protection, and the interested reader is encouraged to consult them (Amyotte, 2013; Bartknecht, 1989; Barton, 2002; CCPS/AIChE, 2005; Eckhoff, 2003; Nagy and Verakis, 1983; Frank and Rodgers, 2012). Because dust explosion protection is a very active area of research, the interested reader is also encouraged to consult the most current combustible dust safety standards on explosion protection and deflagration venting. There can be significant changes in these documents from one version to the next.

Fuel control can be based on a variety of tactics. In many applications, the combustible solid may be handled at a particle size so large that it cannot be readily dispersed of ignited. If the material can be managed in a way that no fines are produced, the hazards of combustible dust may be avoided. This tactic, however, is likely to have a low level of reliability. If combustible dusts are handled in an aerosol form, it may be permissible to simply dilute the dust to below its MEC. Consult applicable safety standards for the acceptable margin of safety required for this approach (such as 25% MEC). Alternatively, an additive may be added to a combustible dust to inhibit dust dispersion (eg, an organic oil or water) or combustion (inert solids).

Oxidant control is generally achieved by storing and handling a combustible dust in an inert atmosphere (or at least in an atmosphere below the dust's LOC). This strategy requires the use of gas-tight vessels and piping. The most common oxidizer is the oxygen present in air. In spaces with controlled atmospheres, consideration should be given to the measurement and control of oxygen levels to ensure that the oxygen level stays within its control limit. Measurements with an alarm to indicate loss of control may be a helpful added layer of protection.

Since the inert gas is a nonoxidizer, it is also likely a physical asphyxiant. Care must be taken to avoid the accidental release of an inert gas into an occupied space where personnel could become exposed to a potential suffocation hazard. Personnel should be trained and refreshed periodically on this hazard.

The control of ignition sources is a very challenging exercise because there are so many potential ignition sources in an industrial environment. Nevertheless, the disciplined and persistent control of ignition sources is a very important element in dust explosion protection. At a minimum, one should consider the hazard controls to adopt to manage the following ignition sources: electrical equipment and power distribution, static electricity, lightning, open flames, hot work (welding, torch cutting, grinding), smoking materials, exothermic chemical reactions (eg, smoldering piles), frictional heating, and metal sparks. Many combustible dust safety standards do not permit ignition control as the sole layer of protection.

It is difficult to prevent dust dispersal. In the storage and handling of bulk (particulate) solids, dust is often dispersed into a cloud. Discharging bulk solids into a silo, an intermediate bulk container, or even a bag will undoubtedly generate a dust cloud. Pneumatic conveying and pulsing baghouse filters create dust clouds. If your process equipment will generate a dust cloud, you will need to consider additional layers of protection to prevent a dust explosion. The dispersal of fugitive dust is a different matter. We will defer consideration of this issue to Chapter 9.

The final element of the dust explosion pentagon is confinement. An unconfined dust deflagration is a flash fire. A flash fire is undesirable, of course, but it has a small radius of influence compared to an explosion. In this context, I define an explosion as the sudden generation and release of high pressure gas that can perform mechanical work such as displacement of objects, the rupture of a confining structure, and the projection of missiles.

Confined deflagrations generate overpressure. The structure in which the deflagration occurs must either be strong enough to contain the pressure or the pressure must be controlled to a safe level. One strategy for controlling the deflagration pressure is to allow it to vent the flame and combustion products into a safe area away from personnel. Deflagration venting is a proven means of protecting solids processing equipment, storage vessels, and buildings by preventing the enclosure from experiencing the maximum deflagration pressure. The essential idea is to install a weak panel that will fail (open) at a pressure much lower than the failure pressure of the structure or enclosure. If a deflagration is ignited within the enclosure, the panel will fail providing a flow path that directs the combustion products out of the enclosure.

Deflagration venting is a well-established but complex technology. The design and implementation of deflagration venting is best left to specialists. Key design parameters needed for the design of a venting system are the maximum pressure and the deflagration index.

Designing equipment to tolerate the maximum explosion pressure is not a commonly employed safeguard. To do so successfully, the equipment must be designed as a pressure vessel. In many if not most circumstance, this is not an economical option. While deflagration venting is one means of preventing the pressure from exceeding the strength of the equipment enclosure, an alternative is to suppress (extinguish) the deflagration before the pressure becomes too great. Suppression technologies rely on a control loop concept. The suppression system consists of a sensor, a controller, and a suppression device. The suppression device is usually a canister of extinguishing agent (a particulate solid) which is explosively injected into the enclosure to extinguish the fireball. In the event that the sensor detects the ignition of a deflagration (via an infrared detector or pressure rise), the controller fires the suppression device.

A safeguard related to suppression is deflagration isolation. A process plant may consist of several pieces of equipment and storage vessels connected by ductwork. If a deflagration starts in one vessel, it may travel through the ductwork to other vessels, accelerating and increasing in overpressure to become increasingly hazardous. Deflagration isolation stops the propagation of the deflagration into adjacent pieces of equipment by denying passage through the ductwork. There are two basic technologies to prevent passage of the flame. One way is to fire a suppression canister in the ductwork to extinguish the flame. The other technology is to "close" the ductwork. In this case, the control loop explosively closes a valve (think of a guillotine blade) which prevents the deflagration from traveling through the ductwork. Deflagration isolation is a proven technology, but it does have its risks. For example, isolating a deflagration inside a piece of equipment means that this piece of equipment must be able to withstand the overpressure. Hence, deflagration isolation is usually a companion technology that supplements by venting or suppression.

8.10 SUMMARY

This chapter examined the challenges that arise in the modeling of unsteady dust flame propagation in a closed vessel. In many ways, this chapter was a synthesis of many of the concepts presented in the previous chapters on thermodynamics, single particle combustion, and flame propagation. An important difference in this chapter has been the pervasive influence of turbulence on flame behavior.

This chapter began with a description of some of the basic experimental observations caused by progressive combustion in an enclosure. We considered two important features in detail: the Flamm—Mache temperature gradient and the fractional pressure rise relation. On average, the explosion temperature of the burnt mixture is equal to the isochoric flame temperature. But the progressive combustion process creates a (Flamm-Mache) temperature gradient in the vessel with the highest temperature in the center of the vessel and the lowest temperature at the wall. Surprisingly, the fractional pressure rise relation derived for uniform combustion can also describe the pressure rise in progressive combustion.

We next considered the impact of the characteristic time and length scales of a confined dust deflagration on the underlying physical and chemical processes. Using the 20-L sphere as our reference system, we found that the particle diameter is a primary determinant of the relative importance of finite rate processes. For particle sizes smaller than 100 microns, velocity slip and gravitational settling can be neglected. We also confirmed that the pressure field in a 20-L sphere is uniform.

Following that we evaluated two representative types of these models: the adiabatic integral model and the isothermal integral model. At early times in a confined deflagration, the pressure rises with time-cubed. The significance of the K_{st} parameter and the cube root scaling law was examined next. The K_{st} parameter is a function of the initial pressure, the explosion pressure, and the burning velocity. These parameters, in turn, are functions of the fuel volatility, the dust concentration, and the particle size. The importance of experimental conditions on dust explosibility parameters was illustrated through the influence of dust dispersion and igniter strength. This led us to a consideration of some of the key features of the turbulent burning velocity.

Then some representative experimental results for organic and metallic solids were presented to illustrate the effects of fuel volatility, dust concentration, mean particle size, and the particle size distribution. Generally speaking, the dust explosion maximum pressure and maximum rate of pressure rise increase with increasing fuel volatility, increasing dust concentration, and decreasing particle size. The range of explosion concentration limits was analyzed from the perspective of a limiting energy balance in the dust flame. The significance of incomplete combustion in deflagration testing was indicated. The mechanism of flame suppression by inert solids was described. Finally, a systematic approach to controlling dust deflagration hazards was reviewed based on the dust explosion pentagon.

The focus of this chapter has been the primary dust deflagration. In the laboratory, the primary dust deflagration is a controlled event which presents the opportunity to measure and evaluate the destructive power of a combustible dust. In Chapter 9 we will consider flame acceleration, a fluid dynamic phenomenon that can magnify the destructive power of a dust deflagration.

REFERENCES

Aldis, D.F., Lai, F.S., 1979. Review of Literature Related to Engineering Aspects of Grain Dust Explosions, vol. 1375. Department of Agriculture, Science and Education Administration, Washington, DC.

Amyotte, P., 2006. Solid inertants and their use in dust explosion prevention and mitigation. J. Loss Prev. Process Ind. 19, 161–173.

Amyotte, P., 2013. An Introduction to Dust Explosions. Butterworth Heinemann, Elsevier, New York.

Amyotte, P., Pegg, M.J., 1989. Lycopodium dust explosions in a Hartmann bomb; effects of turbulence. J. Loss Prev. Process Ind. 2, 87–94.

Amyotte, P., Pegg, M.J., 1992. Dust explosion prevention by addition of thermal inhibitors. Plant/Operations Prog. 11, 166–173.

Amyotte, P., Chippett, S., Pegg, M.J., 1989. Effects of turbulence on dust explosions. Prog. Energy Combust. 14, 293–310.

Amyotte, P., Mintz, K.J., Pegg, M.J., Sun, Y.-H., Wilkie, K.I., 1991a. Laboratory investigation of the dust explosibility characteristics of three Nova Scotia coals. J. Loss Prev. Process Ind. 4, 102–109.

Amyotte, P., Mintz, K.J., Pegg, M.J., Sun, Y.-H., Wilkie, K.I., 1991b. Effects of methane admixture, particle size and volatile content on the dolomite inerting requirements of coal dust. J. Hazard. Mater. 27, 187–203.

Amyotte, P., Mintz, K.J., Pegg, M.J., 1992. Effectiveness of various rock dusts as agents of coal dust inerting. J. Loss Prev. Process Ind. 5, 196–199.

Amyotte, P., Mintz, K.J., Pegg, M.J., 2005. Effect of rock dust particle size on suppression of coal dust explosions. Trans. IChemE 73 (Part B), 89−100.

ASTM E1226, 2012. Standard Test Method for Explosibility of Dust Clouds. ASTM International, West Conshohocken, PA.

ASTM E1515, 2007. Standard Test Method for Minimum Explosible Concentration of Combustible Dusts. ASTM International, West Conshohocken, PA.

Bartknecht, W., 1981. Explosions: Course, Prevention, Protection. Springer-Verlag, Berlin.

Bartknecht, W., 1989. Dust Explosions: Course, Prevention, Protection. Springer-Verlag, Berlin.

Barton, K., 2002. Dust Explosion Prevention and Protection. Gulf Professional Publishing, Elsevier, New York.

Beyer, W.H. (Ed.), 1976. CRC Standard Mathematical Tables. 24th ed. CRC, Boca Raton, FL.

Bozier, O., Veyssière, B., 2010. Study of the mechanisms of dust suspension generation in a closed vessel under microgravity conditions. Microgravity. Sci. Technol. 22, 233−248.

Bradley, D., Mitcheson, A., 1976. Mathematical solutions for explosions in spherical vessels. Combust. Flame 26, 201−217.

Brinzea, V., Mitu, M., Movileanu, C., Musuc, A., Razus, D., 2010. Expansion coefficients and normal burning velocities of propane-air mixtures by the closed vessel technique. Anal. Univ. Bucuresti Chimie 19, 31−37.

Brinzea, V., Mitu, M., Movileanu, C., Razus, D., Oancea, D., 2011. Deflagration parameters of stoichiometric propane−air mixture during the initial stage of gaseous explosions in closed vessels. Rev. Chim. 62, 201−205.

Britton, L.G., 2002. Two hundred years of flammable limits. Process Saf. Prog. 21, 1−11.

Britton, L.G., Cashdollar, K.L., Fenlon, W., Frurip, D., Going, J., Harrison, B.K., et al., 2005. The role of ASTM E27 methods in hazard assessment. Part II: flammability and ignitibility. Process Saf. Prog. 24, 12−28.

Britton, L.G., Clouthier, M.P., Harrison, B.K., Rodgers, S.A., 2016. Limiting oxygen concentrations of gases. Process Saf. Prog. 35, 107−114.

Brown, H.H., 1917. Inflammability of carbonaceous dusts. J. Ind. Eng. Chem. 9, 269−275.

Cashdollar, K.L., 1994. Flammability of metals and other elemental dust clouds. Process Saf. Prog. 13, 139−145.

Cashdollar, K.L., 1996. Coal dust explosibility. J. Loss Prev. Process Ind. 9, 65−76.

Cashdollar, K.L., 2000. Overview of dust explosibility characteristics. J. Loss Prev. Process Ind. 13, 189−199.

Cashdollar, K.L., Hertzberg, M., 1985. 20-l explosibility test chamber for dusts and gases. Rev. Sci. Instruments 56, 596−602.

Cashdollar, K.L., Weiss, E.S., Greninger, N.B., Chatrathi, K., 1992. Laboratory and large-scale dust explosion research. Plant/Oper. Progr. 11, 247−255.

Cashdollar, K.L., Chatrathi, K., 1992. Minimum explosible dust concentrations measured in 20-L and 1-m^3 chambers. Combust. Sci. Technol. 87, 157−171.

Cashdollar, K.L., Zlochower, I.A., 2007. Explosion temperatures and pressures of metals and other elemental dust clouds. J. Loss Prev. Process Ind. 20, 337−348.

Cashdollar, K.L., Hertzberg, M., Zlochower, I.A., 1988. Effect of volatility on dust flammability limits for coals, Gilsonite, and polyethylene. Twenty-Second Symposium (International) on Combustion. The Combustion Institute, Elsevier, New York, pp. 1757−1765.

Cashdollar, K.L., Weiss, E.S., Montgomery, T.G., Going, J.E., 2007. Post-explosion observations of experimental mine and laboratory coal dust explosions. J. Loss Prev. Process Ind. 20, 607−615.

Cashdollar, K.L., Sapko, M.J., Weiss, E.S., Harris, M.L., Man, C.-K., Harteis, S.P., et al., 2010. Report of Investigations 19679. Recommendations for a New Rock Dusting Standard to Prevent Coal Dust Explosions in Intake Airways. U.S. Department of Health and Human Services, National Institute of Occupational Safety and Health, Atlanta, GA

Castellanos, D., Skjold, T., Carreto, V., Mannan, M.S., 2011. Correlating turbulence flow field in dust explosion vessels of different size. 14th Annual Symposium, Mary Kay O'Connor Process Safety Center. Texas A&M University, College Station, TX.

Castellanos, D., Carreto-Vazquez, V.H., Mashuga, C.V., Trottier, R., Mejia, A.F., Mannan, M.S., 2014. The effect of particle size polydispersity on explosibility characteristics of aluminum dust. Powder Technol. 254, 331–337.

CCPS/AIChE, 2005. Guidelines for Safe Handling of Powders and Bulk Solids. American Institute of Chemical Engineers. John Wiley & Sons, New York.

Chatrathi, K., 1994. Dust and hybrid explosibility in a 1 m^3 spherical chamber. Process Saf. Prog. 13, 183–189.

Chatrathi, K., Going, J., 2000. Dust deflagration extinction. Process Saf. Prog. 19, 146–153.

Chawla, N., Amyotte, P.R., Pegg, M.J., 1996. A comparison of experimental methods to determine the minimum explosible concentration of dusts. Fuel 75, 654–658.

Cloney, C.T., Amyotte, P.R., Khan, F.I., Ripley, R.C., 2014. Development of an organizational framework for studying dust explosion phenomena. J. Loss Prev. Process Ind. 30, 228–235.

Conti, R.S., Cashdollar, K.L., Liebman, I., 1982. Improved optical probe for monitoring dust explosions. Rev. Sci. Instrument. 53, 311–313.

Continillo, G., 1989. Two-zone model and a distributed parameter model of dust explosions in closed vessels. Arch. Combust. 9, 79–94.

Continillo, G., Crescitelli, S., Fumo, E., Napolitano, F., Russo, G., 1991. Coal dust explosions in a spherical bomb. J. Loss Prev. Process Ind. 4, 223–229.

Cross, J., Farrer, D., 1982. Dust Explosions. Plenum Press.

Crowl, D.A., 2003. Understanding Explosions. Center for Chemical Process Safety, American Institute of Chemical Engineers. John Wiley & Sons, New York.

Crowl, D.A., Jo, Y.-d, 2009. A method for determining the flammable limits of gases in a spherical vessel. Process Saf. Prog. 28, 227–236.

Dahoe, A.E., 2005. Laminar burning velocities of hydrogen-air mixtures from closed vessel gas explosions. J. Loss Prev. Process Ind. 18, 152–166.

Dahoe, A.E., de Goey, L.P.H., 2003. On the determination of laminar burning velocity from closed vessel gas explosions. J. Loss Prev. 16, 457–478.

Dahoe, A.E., Zevenbergen, J.F., Lemkowitz, S.M., Scarlett, B., 1996a. Dust explosions in spherical vessels: the role of flame thickness in the validity of the 'cube–root law'. J. Loss Prev. Process Ind. 9, 33–44.

Dahoe, A.E., Zevenbergen, J.F., Verheijen, P.J.T., Lemkowitz, S.M., Scarlett, B., 1996b. Dust explosions in spherical vessels: prediction of the pressure evolution and determination of the burning velocity and flame thickness. Seventh International Colloquium on Dust Explosions, Bergen, Norway pp. 5.67–5.86, 23–26 June.

Dahoe, A.E., Cant, R.S., Scarlett, B., 2001a. On the decay of turbulence in the 20-liter explosion sphere. Flow Turbul. Combust. 67, 159–184.

Dahoe, A.E., Cant, R.S., Scarlett, B., 2001b. On the transient flow in the 20-liter sphere. J. Loss Prev. Process Ind. 14, 475–487.

Dahoe, A.E., Skjold, T., Roekaerts, D.J.E.M., Pasman, H.J., Eckhoff, R.K., Hanjalic, K., et al., 2013. On the application of the Levenberg–Marquardt method in conjunction with an explicit Runge–Kutta and implicit Rosenbrock method to assess burning velocities from confined deflagrations. Flow Turbul. Combust. 91, 281–317.

Dastidar, A.G., Amyotte, P.R., Pegg, M.J., 1997. Factors influencing the suppression of coal dust explosions. Fuel 7, 663–670.

Dastidar, A.G., 2016. ASTM E2931: A new standard for the limiting oxygen concentration of combustible dusts. Process Saf. Prog. 35, 159–164.

Dastidar, A.G., Amyotte, P.R., 2002. Determination of minimum inerting concentrations for combustible dusts in a laboratory-scale chamber. Trans. IChemE 80 (Part B), 287–297.

Dastidar, A.G., Amyotte, P.R., 2004. Using calculated adiabatic flame temperatures to determine dust explosion inerting requirements. Process Saf. Environ. Prog. 82 (B2), 142–155.

Dastidar, A.G., Amyotte, P.R., Going, J., Chatrathi, K., 1999. Flammability limits of dusts: minimum inerting concentrations. Process Saf. Prog. 18, 56–63.

Dastidar, A.G., Amyotte, P.R., Going, J., Chatrathi, K., 2001. Inerting of coal dust explosions in laboratory- and intermediate-scale chambers. Fuel 80, 1593–1602.

Denkevits, A., Dorofeev, S., 2006. Explosibility of fine graphite and tungsten dusts and their mixtures. J. Loss Prev. Process Ind. 19, 174–180.

Di Benedetto, A., Russo, P., 2007. Thermo-kinetic modelling of dust explosions. J. Loss Prev. Process Ind. 20, 303–309.

Dorsett Jr., H.G., Jacobson, M., Nagy, J., Williams, R.P., 1960. Report of Investigations 5624. Laboratory Equipment and Test Procedures for Evaluating Explosibility of Dusts. U.S. Department of the Interior, Bureau of Mines.

Du, B., Huang, W., Liu, L., Zhang, T., Li, H., Ren, Y., et al., 2015. Visualization and analysis of dispersion process of combustible dust in a transparent Siwek 20-L chamber. J. Loss Prev. Process Ind. 33, 213–221.

Dufaud, O., Traore, M., Perrin, L., Chazelet, S., Thomas, D., 2010. Experimental investigation and modelling of aluminium dusts explosions in the 20 L sphere. J. Loss Prev. Process Ind. 23, 226–236.

Dyduch, Z., Toman, A., Adamus, W., 2016. Measurements of turbulence intensity in the standard 1 m^3 vessel. J. Loss Prev. Process Ind. 40, 180–187.

Eckhoff, R., 1987. Measurement of explosion violence of dust clouds. In: Carhart, H.W. (Ed.), Proceedings of the International Symposium on the Explosion Hazard Classification of Vapors, Gases, and Dusts. Publication NMAB-447, National Academy of Sciences, Washington, DC.

Eckhoff, R., 2003. Dust Explosions in the Process Industries, third ed. Gulf Professional Publishing, Elsevier, New York.

Edwards, P.W., Leinbach, L.R., 1935. Explosibility of Agricultural and Other Dusts as Indicated by Maximum Pressure and Rates of Pressure Rise. U.S. Department of Agriculture Tech. Bull. 490, Washington, DC.

Ellis, O.C.C., Wheeler, R.V., 1927. XXVI. The movement of flame in closed vessels: correlation with development of pressure. J. Chem. Soc. 153–158.

Ellis, O.C.C., Morgan, E., 1934. The temperature gradient in flames. Trans. Faraday Soc. 30, 287–298.

Eschenbach, R.C., Agnew, J.T., 1958. Use of the constant-volume bomb technique for measuring burning velocity. Combust. Flame 2, 273–285.

Evans, A.A., 1991. Deflagrations in spherical vessels: a theoretical analysis of the pressure−time relationship. Trans. IChemE 69, 90−96.

Evans, A.A., 1994. Deflagrations in spherical vessels: a comparison among four approximate burning velocity formulae. Combust. Flame 97, 429−434.

Fairweather, M., Vasey, M.W., 1982. A mathematical model for the prediction of overpressures generated in totally confined and vented explosion. Nineteenth Symposium (International) on Combustion. The Combustion Institute, Elsevier, New York, pp. 645−653.

Faraday, M., Lyell, C., 1845. On the subject of the explosion in the Haswell Collieries and on the means of preventing similar accidents. Philos. Mag. 26, 16−35.

Farrell, J.T., Johnston, R.J., and Androulakis, I.P. "Molecular structure effects on laminar burning velocities at elevated temperature and pressure," SAE Paper 2004-01-2936, 2004.

Field, P., 1989. Dust Explosions. Elsevier, New York.

Fiock, E.F., Marvin Jr., C.F., 1937. The measurement of flame speeds. Chem. Rev. 21, 367−387.

Fiock, E.F., Marvin Jr., C.F., Caldwell, F.R., Roeder, C.H., 1940. Flame speeds and energy considerations for explosions in a spherical bomb. NACA Report No. 682. National Advisory Committee on Aeronautics.

Frank, W.L., Rodgers, S.A., 2012. NFPA Guide to Combustible Dusts. National Fire Protection Association, Quincy, MA.

Garforth, A.M., 1976. Unburnt gas density measurement I a spherical combustion bomb by infinite-fringe laser interferometry. Combust. Flame 26, 343−352.

Garforth, A.M., Rallis, C.J., 1976. Gas movement during flame propagation in a constant volume bomb. Acta Astronaut. 3, 879−888.

Garforth, A.M., Rallis, C.J., 1978. Laminar burning velocity of stoichiometric methane−air: pressure and temperature dependence. Combust. Flame 31, 53−68.

Going, J.E., Chatrathi, K., Cashdollar, K.L., 2000. Flammability limit measurements for dusts in 20-L and 1-m^3 vessels. J. Loss Prev. Process Ind. 13, 209−219.

Guenoche, H., 1964. E. Flame propagation in tubes and in closed vessels. In: Markstein, G.H. (Ed.), Nonsteady Flame Propagation. Pergamon Press Oxford.

Harris, R.J., 1989. The Investigation and Control of Gas Explosions in Buildings and Heating Plant. E&FN Spon, London.

Harris, M.L., Weiss, E.S., Man, C.-K., Harteis, S.P., Goodman, G.V., Sapko, M.J., 2010. Rock dusting requirements in underground coal mines. Proceedings of the 13th U.S./North American Mine Ventilation Symposium. MIRARCO-Mining Innovation, Sudbury, ON, Canada, pp. 267−271.

Harris, M.L., Sapko, M.J., Zlochower, I.A., Perera, I.E., Weiss, E.S., 2015. Particle size and surface area effects on explosibility using a 20-L chamber. J. Loss Prev. Process Ind. 37, 33−38.

Hartmann, I., Nagy, J., 1957. Venting dust explosions. Ind. Eng. Chem. 49, 1734−1740.

Hertzberg, M., 1987. Report of Investigations 9095. A Critique of the Dust Explosibility Index: an Alternative for Estimating Explosion Probabilities. U.S. Department of the Interior, Bureau of Mines.

Hertzberg, M., Cashdollar, K.L., Opferman, J.J., 1979. Report of Investigations 8360. The Flammability of Coal Dust−Air Mixtures: Lean Limits, Flame Temperatures, Ignition Energies, and Particle Size Effects. U.S. Department of the Interior, Bureau of Mines.

Hertzberg, M., Cashdollar, K.L., Ng, D.L., Conti, R.S., 1982. Domains of flammability and thermal ignitability for pulverized coals and other dusts: particle size dependences and microscopic residue analyses. Nineteenth Symposium (International) on Combustion. The Combustion Institute, Elsevier, New York, pp. 1169–1180.

Hertzberg, M., Cashdollar, K.L., Zlochower, I.A., Ng, D.L., 1984. Inhibition and extinction of explosions in heterogeneous mixtures. Twentieth Symposium (International) on Combustion. The Combustion Institute, Elsevier, New York, pp. 1691–1700.

Hertzberg, M., Zlochower, I.A., Cashdollar, K.L., 1986. Volatility model for coal dust flame propagation and extinguishment. Twenty-first Symposium (International) on Combustion. The Combustion Institute, Elsevier, New York, pp. 325–333.

Hertzberg, M., Zlochower, I.A., Cashdollar, K.L., 1992. Metal dust combustion: explosion limits, pressures, and temperatures. Twenty-Fourth Symposium (International) on Combustion. The Combustion Institute, Elsevier, New York, pp. 1827–1835.

Hopkinson, B., 1906. Explosions of coal-gas and air. Proc. Royal Soc. Lond. A 77, 387–413.

Ishihama, W., Enomoto, H., 1973. New experimental method for studies of dust explosions. Combust. Flame 21, 177–186.

Jacobson, M., 1961. Explosibility of Agricultural Dusts. US Dept. of the Interior, Bureau of Mines, Washington, DC, Report of Investigations 5753.

Jarosinski, J., Pu, Y.K., Bulewicz, E.M., Kauffman, C.W., Johnson, V.G., 1993. Some fundamental characteristics of cornstarch dust–air flames. Prog. Astronaut. Aeronaut. Volume 152, 119–135.

Jo, Y.-D., Crowl, D.A., 2009. Flame growth model for confined gas explosion. Process Saf. Prog. 28, 141–146.

Jo, Y.-D., Crowl, D.A., 2010. Explosion characteristics of hydrogen–air mixtures in a spherical vessel. Process Saf. Prog. 29, 216–223.

Jost, W., 1946. Explosion and Combustion Processes in Gases. McGraw-Hill, New York.

Kalejaiye, O., Amyotte, P.R., Pegg, M.J., Cashdollar, K.L., 2010. Effectiveness of dust dispersion in the 20-L Siwek chamber. J. Loss Prev. Process Ind. 23, 46–59.

Kansa, E.J., Perlee, H.E., 1976. Report of Investigations 8163. Constant-Volume Flame Propagation: Finite-Sound-Speed Theory. U.S. Department of the Interior, Bureau of Mines.

Kauffman, C.W., Srinath, S.R., Tai, C.S., 1987. Needs in dust explosion testing, Publication NMAB-447. In: Carhart, H.W. (Ed.), Proceedings of the International Symposium on the Explosion Hazard Classification of Vapors, Gases, and Dusts. National Academy of Sciences, Washington, DC

Knapton, J.D., Stobie, I.C., Krier, H., 1973. Burning rate studies of fuel air mixtures at high pressures. Combust. Flame 21, 211–220.

Kumar, R.K., Bowles, E.M., Mintz, K.J., 1992. Large-scale dust explosion experiments to determine the effects of scaling on explosion parameters. Combust. Flame 89, 320–332.

Lautkaski, R., 2005. Pressure Rise in Confined Gas Explosions. VTT Nuclear Processes, Report No. PRO1/P1026/05, Finland.

Lee, J.H.S., 1987. On the classification of flammable gases, vapors and dusts, Publication NMAB-447. In: Carhart, H.W. (Ed.), Proceedings of the International Symposium on the Explosion Hazard Classification of Vapors, Gases, and Dusts. National Academy of Sciences, Washington, DC

Lee, J.H.S., Zhang, F., Knystautas, R., 1992. Propagation mechanisms of combustion waves in dust–air mixtures. Powder Technol. 71, 153–162.

Lee, J.H.S., Peraldi, O., Knystautas, R., 1993. Combustion of dust suspension in microgravity. In: 31st Aerospace Sciences Meeting & Exhibit, Reno, NV.

Lewis, B., von Elbe, G., 1934. Determination of the speed of flames and the temperature distribution in a spherical bomb from time-pressure explosion records. J. Chem. Phys. 2, 283−290.

Lewis, B., von Elbe, G., 1987. Combustion, Flames, and Explosions of Gases, third ed. Academic Press, New York.

Luijten, C.C.M., de Goey, L.P.H., 2007. New, accurate analytical relations for fractional pressure rise, laminar burning velocity, and the cubic root law in constant volume combustion. In: Proceedings of the Third European Combustion Meeting.

Luijten, C.C.M., Doosje, E., van Oijen, J.A., de Goey, L.P.H., 2009a. Impact of dissociation and end pressure on determination of laminar burning velocities in constant volume combustion. Int. J. Thermal Sci. 48, 1206−1212.

Luijten, C.C.M., Doosje, E., de Goey, L.P.H., 2009b. Accurate analytical models for fractional pressure rise in constant volume combustion. Int. J. Thermal Sci. 48, 1213−1222.

Man, C.K., Teacoach, K.A., 2009. How does limestone rock dust prevent coal dust explosions in coal mines? Mining Eng. 61, 69−73.

Man, C.K., Harris, M.L., Weiss, E.S., 2011. Analysis of post-explosion residues for estimating flame travel during coal dust deflagrations. Sci. Technol. Energetic Mater. 72, 136−140.

Mannan, S. (Ed.), 2005. Chapter 17: Explosion in Lee's Loss Prevention in the Process Industries, third ed. Elsevier, New York.

Manton, J., von Elbe, G., Lewis, B., 1953. Burning-velocity measurements in a spherical vessel with central ignition. Fourth Symposium (International) on Combustion. The Combustion Institute, Elsevier, New York, pp. 358−363.

Mashuga, C.V., Crowl, D.A., 1999. Flammability zone prediction using calculated adiabatic flame temperatures. Process Saf. Prog. 18, 127−134.

Mashuga, C.V., Crowl, D.A., 2000. Derivation of Le Chatelier's mixing rule for flammable limits. Process Saf. Prog. 19, 112−117.

Medina, C.H., Phylaktou, H.N., Sattar, H., Andrews, G.E., Gibbs, B.M., 2013. The development of an experimental method for the determination of the minimum explosible concentration of biomass powders. Biomass Bioenergy 53, 95−104.

Medina, C.H., MacCoitir, B., Sattar, H., Slatter, D.J.F., Phylaktou, N.H., Andrews, G.E., et al., 2015. Comparison of the explosion characteristics and flame speeds of pulverised coals and biomass in the ISO standard 1-m3 dust explosion equipment. Fuel 151, 91−101.

Mercer, D.B., Amyotte, P.R., Dupuis, D.J., Pegg, M.J., Dahoe, A., de Heij, W.B.C., et al., 2001. The influence of injector design on the decay of pre-ignition turbulence in a spherical explosion chamber. J. Loss Prev. Process Ind. 14, 269−282.

Metghalchi, M., Keck, J.C., 1980. Laminar burning velocity of propane−air mixtures at high temperature and pressure. Combust. Flame 38, 143−154.

Metghalchi, M., Keck, J.C., 1982. Burning velocities of mixtures of air with methanol, iso-octane and indolene at high pressure and temperature. Combust. Flame 48, 191−210.

Mintz, K.J., 1993. Upper explosive limit of dusts: experimental evidence for its existence under certain circumstances. Combust. Flame 94, 125−130.

Mintz, K.J., 1995. Problems in experimental measurements of dust explosions. J. Hazard. Mater. 42, 177−186.

Myers, T.J., 2008. Reducing aluminum dust explosion hazards: case study of dust inerting in an aluminum buffing operation. J. Hazard. Mater. 159, 72−80.

Myers, T.J., Ibarreta, A., Bucher, J., & Marr, K., 2013. Assessing the hazard of marginally explosible dust. In: American Institute of Chemical Engineers, 2013 Spring Meeting, 9th Global Congress on Process Safety, San Antonio, TX, April 28—May 1.

Myers, T.J., Ibarreta, A., 2013. Tutorial on combustible dust. Process Saf. Prog. 32, 298–306.

Nagy, J., Verakis, H.C., 1983. Development and Control of Dust Explosions. CRC Press, Boca Raton, FL.

Nagy, J., Conn, J.W., Verakis, H.C., 1969. Report of Investigations 7279. Explosion Development in a Spherical Vessel. U.S. Department of the Interior, Bureau of Mines.

Nagy, J., Seiler, E.C., Conn, J.W., Verakis, H.C., 1971. Report of Investigations 7507. Explosion Development in Closed Vessels. U.S. Department of the Interior, Bureau of Mines.

NFPA 652, 2016. Standard on the Fundamentals of Combustible Dust. National Fire Protection Association, Quincy, MA.

Nomura, S., Tanaka, T., 1992. Theoretical analysis of dust explosions. Powder Technol. 71, 189–196.

Nomura, S.-I., Tanaka, T., 1980. Prediction of maximum rate of pressure rise due to dust explosion in closed spherical and nonspherical vessels. Ind. Eng. Chem. Process Design Dev. 19, 451–459.

Nomura, S.-I., Torimoto, M., Tanaka, T., 1984. Theoretical upper limit of dust explosion in relation to oxygen concentration. Ind. Eng. Chem. Process Design Dev. 23, 420–423.

O'Donovan, K.H., Rallis, C.J., 1959. A modified analysis for the determination of the burning velocity of a gas mixture in a spherical constant volume combustion vessel. Combust. Flame 3, 201–214.

Ogle, R.A., 1986. A New Strategy for Dust Explosion Research: a Synthesis of Combustion Theory, Experimental Design and Particle Characterization. Ph.D. thesis. University of Iowa, Iowa City, Iowa.

Ogle, R.A., Beddow, J.K., Chen, L.-D., Butler, P.B., 1988. An investigation of aluminum dust explosions. Combust. Sci. Technol. 61, 75–99.

Oppenheim, A.K., Maxson, J.A., 1993. Thermodynamics of combustion in an enclosure. Prog. Astronaut. Aeronaut. Volume 152, 365–382.

Perlee, H.E., Fuller, F.N., Saul, C.H., 1974. Report of Investigations 7839. Constant-Volume Flame Propagation. U.S. Department of the Interior, Bureau of Mines.

Pilão, R., Ramalho, E., Pinho, C., 2006. Overall characterization of cork dust explosion. J. Hazard. Mater. B133, 183–195.

Proust, C., Accorsi, A., Dupont, L., 2007. Measuring the violence of dust explosions with the "20 L sphere" and with the standard "ISO 1-m^3 vessel": systematic comparison and analysis of the discrepancies. J. Loss Prev. 20, 599–606.

Pu, Y.K., Jarosinki, J., Tal, C.S., Kauffman, C.W., Sichel, M., 1988. The investigation of the feature of dispersion induced turbulence and its effects on dust explosions in closed vessels. Twenty-second Symposium (International) on Combustion. The Combustion Institute, Elsevier, New York, pp. 1177–1787.

Pu, Y.K., Jarosinski, J., Johnson, V.G., Kauffman, C.W., 1990. Turbulence effects on dust explosions in the 20-liter spherical vessel. Twenty-third Symposium (International) on Combustion. The Combustion Institute, Elsevier, New York, pp. 843–849.

Pu, Y.K., Li, Y.-C., Kauffman, C.W., Bernal, L.P., 1991. Determination of turbulence parameters in closed explosion vessels. Prog. Astronaut. Aeronaut. Dyn. Deflagr. Reactive Syst. Heterogeneous Combust. Volume 132, 107–123.

Pu, Y., Podfilipski, J., Jarosinski, J., 1998. Constant volume combustion of aluminum and cornstarch dust in microgravity. Combust. Sci. Technol. 135, 255–267.

Rallis, C.J., Garforth, A.M., 1980. The determination of laminar burning velocity. Prog. Energy Combust. Sci. 6, 303–329.

Rallis, C.J., Tremeer, G.E.B., 1963. Equations for the determination of burning velocity in a spherical constant volume vessel. Combust. Flame 7, 51–61.

Razus, D., Movileanu, C., Brinzea, V., Oancea, D., 2006a. Explosion pressures of hydrocarbon–air mixtures in closed vessels. J. Hazard. Mater. 135, 58–65.

Razus, D., Oancea, D., Movileanu, C., 2006b. Burning velocity evaluation from pressure evolution during the early stage of closed-vessel explosions. J. Loss Prev. Process Ind. 19, 334–342.

Razus, D., Brinzea, V., Mitu, M., Movileanu, C., Oancea, D., 2012. Burning velocity of propane–air mixtures from pressure–time records during explosion in a closed spherical vessel. Energy Fuels 26, 901–909.

Rice, G.S., 1910. The Explosibility of Coal Dust. U.S. Geological Survey, Bulletin 425, Washington, DC.

Saeed, K., 2009. Evaluation of the fractional pressure rise with mass fraction burned in a closed vessel combustion. In: Proceedings of the European Combustion Meeting.

Saeed, K., Stone, C.R., 2004a. The modelling of premixed laminar combustion in a closed vessel. Combust. Theory Model. 8, 721–743.

Saeed, K., Stone, C.R., 2004b. Measurements of the laminar burning velocity for mixtures of methanol and air form a constant-volume vessel using a multizone model. Combust. Flame 139, 152–166.

Saeed, M.A., Medina, C.H., Andrews, G.E., Phylaktou, H.N., Slatter, D., Gibbs, B.M., 2015. Agricultural waste pulverised biomass: MEC and flame speeds. J. Loss Prev. Process Ind. 36, 308–317.

Sami, M., Annamalai, K., Wooldridge, M., 2001. Co-firing of coal and biomass fuel blends. Prog. Energy Combust. Sci. 27, 171–214.

Sapko, M.J., Furno, A.L., Kuchta, J.M., 1976. Report of Investigations 8176. Flame and Pressure Development of Large-Scale CH_4–Air–N_2 Explosions. U.S. Department of the Interior, Bureau of Mines.

Sattar, H., Slatter, D., Andrews, G.E., Gibbs, B.M., Phylaktou, H.N., 2012. Pulverised biomass explosions: investigation of the ultra-rich mixtures that give peak reactivity. In: Proceedings of the IX International Seminar on Hazardous Process Materials and Industrial Explosions (IX ISHPMIE), Krakow.

Sattar, H., Huescar, C.M., Phylaktou, H.N., Andrews, G.E., Gibbs, B.M., 2013. Calibration of a 10-L volume dust holding pot for the 1 m^3 standard vessel, for use in low-bulk-density biomass explosibility testing. In: Proceedings of the Seventh International Seminar on Fire & Explosion Hazards, Providence, RI.

Sattar, H., Andrews, G.E., Phylaktou, H.N., Gibbs, B.M., 2014. Turbulent flame speeds and laminar burning velocities of dusts using the ISO 1-m^3 dust explosion method. Chem. Eng. Trans. 36, 157–162.

Silva, S.P., Sabino, M.A., Fernandes, E.M., Correlo, V.M., Boesel, L.F., Reis, R.L., 2005. Cork: properties, capabilities and applications. Int. Mater. Rev. 50, 345–365.

Singh, J., 1988. Explosion propagation in non-spherical vessels; simplified equations and applications. J. Loss Prev. Process Ind. 1, 39–45.

Slatter, D.J.F., Sattar, H., Medina, C.H., Andrews, G.E., Phylaktou, H.N., Gibbs, B.M., 2015. Biomass explosion testing: accounting for the post-test residue and implications on the results. J. Loss Prev. Process Ind. 36, 318–325.

Smith, J.M., Van Ness, H.C., Abbott, M.M., 1996. Introduction to Chemical Engineering Thermodynamics, fifth ed. McGraw-Hill, New York.

Tufano, V., Crescitelli, S., Russo, G., 1983. Overall kinetic parameters and laminar burning velocity from pressure measurements in closed vessels. Combust. Sci. Technol. 31, 119−130.

van den Bulck, E., 2005. Closed algebraic expressions for the adiabatic limit value for the explosion constant in closed volume combustion. J. Loss Prev. Process Ind. 18, 35−42.

van der Wel, P.G.J., Lemkowitz, S.M., Scarlett, B., van Wingerden, C.J.M., 1991. A study of particle factors affecting dust explosions. Part. Part. Sys. Character. 8, 90−94.

van der Wel, P.G.J., van Veen, J.P.W., Lemkowitz, S.M., Scarlett, B., van Wingerden, C.J.M., 1992. An interpretation of dust explosion phenomena on the basis of time scales. Powder Technol. 71, 207−215.

Wallis, G.B., 1969. One-dimensional Two-phase Flow. McGraw-Hill, New York.

Wilén, C., Rautalin, García-Torrent, J., Conde-Lázardo, E., 1998. Inerting biomass dust explosions under hyperbaric working conditions. Fuel 77, 1089−1092.

Wolanski, P., 1994. Minimum explosive concentration of dust−air mixtures. Proceedings of the Sixth International Colloquium on Dust Explosions. Northeastern University Press, Shenyang, PRC, pp. 206−219.

Zabetakis, M.G., 1965. Flammability Characteristics of Combustible Gases and Vapors. U.S. Department of the Interior, Bureau of Mines, Bulletin 627.

Zeldovich, Y.B., Barenblatt, G.I., Librovich, V.B., Makhviladze, G.M., 1985. The Mathematical Theory of Combustion and Explosions, translated from Russian by McNeill, D., Consultants Bureau, Plenum Publishing Corporation, New York.

Zhen, G., Leuckel, W., 1997. Effects of ignitors and turbulence on dust explosions. J. Loss Prev. Process Ind. 10, 317−324.

Zucrow, M.J., Hoffman, J.D., 1976. Gas Dynamics, vol. 1. John Wiley & Sons, New York.

Dust flame acceleration effects

Flame acceleration increases the hazard level or severity of a dust deflagration. Three consequences of flame acceleration were described in Chapter 1: secondary explosions, confined flame acceleration, and pressure piling. Each of these phenomena has the potential to cause a greater level of damage than a simple confined dust deflagration. Secondary explosions occur when fugitive dust deposits have accumulated on floors and other horizontal surfaces. These dust deposits can become suspended by a primary explosion and provide a means for propagating flame and overpressure over great distances. The entrainment of dust becomes more effective with faster flame speeds which, in turn, increases the secondary explosion pressure.

Flame acceleration effects are most pronounced in confined geometries, especially if the confinement is in the form of a one-dimensional channel or conduit. Examples of such flow channels include underground mine passageways, tunnels, grain elevator galleys, corridors, pipes, and ductwork. The greatest confinement hazard occurs with a combustible dust mixture is ignited at the closed end of a conduit and the deflagration is constrained to one-dimensional motion toward the open end. In this physical setting, the accelerating flame could potentially transition to a detonation, a particularly powerful type of explosion.

Pressure piling occurs when an explosion in one vessel is propagated via a conduit to a second vessel. The flame propagation pressurizes the unburnt gas ahead of the flame thus magnifying the final explosion pressure in the second vessel. The magnitude of the overpressure in the unburnt gas becomes increasingly larger as the flame accelerates.

In this chapter, we will first review the physical phenomena underlying dust flame acceleration. Then each of these three phenomena will be described with reference to simple models and a survey of experimental studies. Where possible, the performance of flame acceleration models will be compared with experimental data. Methods for preventing secondary dust explosions will conclude this chapter.

9.1 DUST FLAME ACCELERATION AND SHOCK WAVES

Fundamentally, flame acceleration is caused by turbulence (Bjerketvedt et al., 1997; Eckhoff, 2003, pp. 351−373). Turbulence leads to faster burning rates with increasing flame speeds. By itself, flame acceleration increases the hazard level of a deflagration because it reduces the time available for active safeguards such as deflagration isolation or suppression to function. But if flame acceleration is confined by a conduit, it will lead to the formation of a shock wave that can generate damaging overpressures. We will first consider how turbulence causes flame acceleration. Then we will present a conceptual model for shock wave formation in a one-dimensional channel and derive the property equations for both normal and reflected shocks. We conclude this section with a derivation of a simple acoustic model for estimating the overpressure caused by flame acceleration (the weak shock approximation) and an overview of the anatomy of blast waves.

The effects of flame acceleration are most easily studied from the perspective of compressible fluid dynamics, sometimes called gas dynamics. In compressible fluid dynamics, it is generally assumed that the flow velocity is comparable to the speed of sound in the fluid. At these speeds, the effects of transport phenomena can be neglected in many circumstances. The primary velocity gradient is in the direction of flow. Thus, the velocity profile in such flows often resembles plug flow, that is, there are no transverse velocity gradients. Consistent with general practice, we will assume that the gas obeys the ideal gas equation of state, that the thermodynamic properties are constant, and that any combustible dust−air mixture can be treated as a homogeneous mixture.

9.1.1 TURBULENT FEEDBACK IN COMBUSTION FLOWS

Turbulence causes a positive feedback loop that amplifies the mass burning rate of fuel and leads to the acceleration of the flame speed (Bjerketvedt et al., 1997). In concept, the mechanism is simple. As the flame sweeps through the unburnt mixture, it releases heat which causes a rise in temperature. The temperature increase causes a decrease in gas density thereby leading to an expansion of the burnt mixture. The expansion of the burnt mixture accelerates the flame speed. The acceleration of the flame speed causes the flame surface to become unstable and wrinkled. Eventually, the flame becomes turbulent and increasingly wrinkled. The wrinkling of the flame surface increases the flame volume and thus increases the mass burning rate of the flame. The increase in mass burning rate is accompanied by an increase in the rate of heat release which, in turn, increases the rate of gas expansion. The increased rate of gas expansion is manifested by an increase in the burnt gas velocity and the flame speed. This positive feedback loop is illustrated in Fig. 9.1.

If the traveling flame continues to find fresh fuel as it advances down the conduit, it is possible that the deflagration could transition into a detonation, an especially severe type of explosion discussed in Section 4.2.

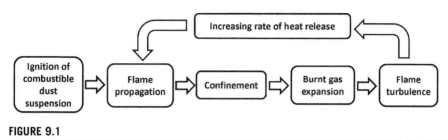

FIGURE 9.1

Turbulent combustion feedback loop.

9.1.2 EXPANSION WAVES, COMPRESSION WAVES, AND SHOCK WAVES

A fundamental property of a compressible fluid is that pressure changes are transmitted as a disturbance traveling at the speed of sound of the fluid. This disturbance can be thought of as sound wave. If the time scale of the flow is much larger than the acoustic time scale, the pressure change occurs in a spatially uniform manner (see Section 4.1.4), and the wave-like behavior of pressure disturbances can be neglected. If the flow time scale is comparable to the acoustic time scale, then pressure changes are not spatially uniform, and the behavior of the pressure waves must be considered. There are two types of pressure waves to consider, *expansion waves* and *compression waves*.

One way to visualize an expansion wave is portrayed in Fig. 9.2. Consider a piston in a tube open on both ends to the ambient environment. Ambient air at ambient temperature and pressure fills the tube on either side of the piston.

At time zero, the piston is withdrawn from the tube at a velocity equal to the speed of sound in the ambient fluid. On the withdrawal side of the tube, the displacement of the piston has created an infinitesimal void (vacuum) at a less-than-ambient pressure. The air adjacent to that infinitesimal void is at a higher pressure than the void and so this air flows toward the piston to fill the void and reestablish pressure equilibrium. The motion of that air occurs at the same speed as the piston, the speed of sound. The infinitesimal slice of air that has moved to fill the void is now at a lower pressure than the air column next to it. And so the next infinitesimal slice of air is set in motion at the speed of sound to fill the next void. Thus a disturbance has been set into motion with each adjacent infinitesimal slice of air moving to fill the void created by its neighbor. This disturbance of falling pressure is an *expansion wave*. Fig. 9.2B shows the pressure profile at three increasing times, the initial condition t_0 and two subsequent times, t_1 and t_2. Fig. 9.2C depicts the trajectory of the expansion wave and the piston in the $z - t$ plane.

Consider now a different situation. A combustible dust mixture dispersed uniformly in a horizontal tube with one end closed and the other end open to the atmosphere. This time there is no piston. Assume that a dust deflagration

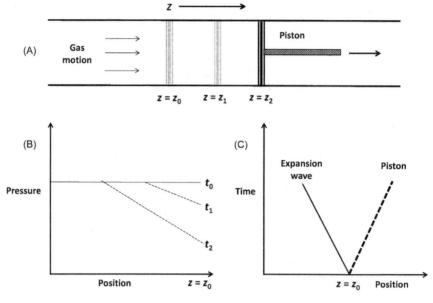

FIGURE 9.2

Expansion wave caused by withdrawal of piston from a tube: (A) physical situation;
(B) axial profile of pressure at different times; and (C) wave diagram showing expansion
wave and piston trajectories.

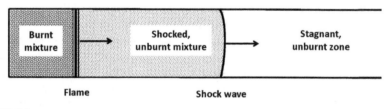

FIGURE 9.3

Flame chasing the shock wave.

is ignited at the closed end of the tube. Neglecting heat losses or incomplete
combustion, the ideal expansion ratio ρ_u/ρ_b for a typical combustible dust is in
the range of 8−10 (Frank and Rodgers, 2012, pp. 54−55). The high pressure, low
density gas expands outwards providing the driving force for accelerating the
flame down the tube toward the open end. The accelerating flame can be idealized as a piston that pushes the unburnt mixture causing it to accelerate as well.
This flame acceleration process causes the formation of a shock wave. Fig. 9.3
illustrates this physical situation.

The region ahead of the shock wave has zero velocity and is therefore called the stagnant zone. As the shock wave passes through a portion of the stagnant zone, the stagnant gas is shocked (compressed) and accelerated toward the moving shock wave. In this one-dimensional geometry, the only velocity gradient is along the axis of the conduit. There are no transverse gradients as those are caused by viscosity, and for simplicity, we are neglecting viscous effects.

It is also interesting to note that in the absence of friction, the particle velocity of the gas behind the shock is equal to the flame speed. Do not let the term "particle velocity" confuse you. The particle velocity does not refer to the velocity of combustible dust particles. The term refers to the gas velocity behind a shock wave. The particle in "particle velocity" is actually a material point within the fluid. This terminology is commonly used when discussing the physics of wave motion.

But why does a shock wave form at all? The answer is that this is one of the defining characteristics of a compressible fluid. In a compressible fluid such as air, a transient mechanical disturbance is propagated as a pressure pulse (the pressure pulse is accompanied by a density and velocity pulse of similar shape). A pressure pulse with a magnitude slightly greater than the ambient pressure in a compressible fluid propagates like a wave traveling at the speed of sound. This type of pressure pulse is called a *compression wave*. Broadly speaking, there are three kinds of compression waves that propagate in a compressible medium: acoustic waves, nonlinear waves, and shock waves. The distinction between the three types of waves is the magnitude of the disturbance $\hat{P}_{wave}$ defined in nondimensional terms as $\hat{P}_{wave} = (P - P_\infty)/P_\infty$. P is the magnitude of the disturbance measured and P_∞ is the magnitude of the ambient pressure. For acoustic waves $\hat{P}_{wave} \to 0$ and for shock waves the disturbance is finite and discontinuous. Nonlinear waves fall in between acoustic waves and shock waves.

Shock waves are caused by the coalescence of acoustic waves. One of the simplest flow configurations that can give rise to shock wave is the accelerating motion of a piston in a tube. Consider the situation illustrated in Fig. 9.4. The tube is open on both ends. The gas is uniform in its properties. The circumference of the piston forms a frictionless seal with the wall of the tube.

The piston is initially at rest (Fig. 9.4A). As it accelerates into motion, it pushes the gas ahead of it thus imparting motion to the gas. We will assume that the speed of the piston is less than the speed of sound in the stagnant gas. With each infinitesimal displacement of the piston, an infinitesimal disturbance (acoustic wave) is generated (Fig. 9.4B). Each disturbance compresses the gas as it travels toward the stagnant region. For an ideal gas, each infinitesimal compression slightly raises the temperature of the gas. The speed of sound for an ideal gas a_g is proportional to the square root of its temperature:

$$a_g = \sqrt{\frac{\gamma R_g T}{M}} \tag{9.1}$$

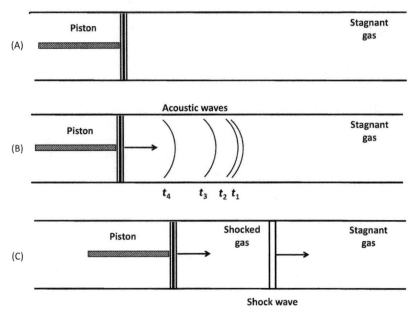

FIGURE 9.4

Formation of a shock wave by the coalescence of acoustic waves: (A) initial condition with piston at rest; (B) piston in motion generating successive acoustic waves ($t_4 > t_3 > t_2 > t_1$); and (C) shock wave formation.

This means that each passing sound wave is traveling faster than the previous wave. The faster waves catch up with the slower waves and form a larger wave with an ever-increasing amplitude. In this manner, the infinitesimal overpressure of individual sound waves grows into the large, finite, discontinuous pressure rise of a shock wave (Fig. 9.4C). Very good discussions of shock wave formation can be found in the compressible fluid dynamics books cited in the references (Thompson, 1972, Chapters 7 and 8; Courant and Friedrichs, 1976, Chapter III; Landau and Lifshitz, 1959, Chapters IX and X).

9.1.2.1 Normal shock wave relations

In Chapter 4, we introduced the notion of shock waves as a means of discussing the properties of the Chapman–Jouget detonation. It will be helpful to revisit shock waves and discuss the computation of their properties. We will consider only *normal shocks*, that is, planar shock waves that propagate with the flow normal to the shock surface. Shocks that propagate at an inclined angle to the flow are called *oblique shocks*. Oblique shocks are especially important in the consideration of shock wave interactions with solid surfaces; these important considerations are beyond our scope. The interested reader is encouraged to refer to any of the gas

(A) Laboratory coordinates

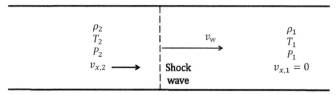

(B) Shock-fixed coordinates

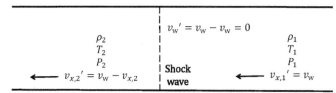

FIGURE 9.5

Shock wave properties relative to laboratory and shock-fixed coordinate frames of reference.

dynamics texts cited in the references (but especially see Zucrow and Hoffman, 1976, Chapter 7; Thompson, 1972, Chapter 7; Kinney and Graham, 1985).

The properties of a shock wave are related to a coordinate frame of reference fixed on the shock wave. The definition sketch is shown in Fig. 9.5 (the same sketch appeared as Fig. 4.4). Throughout this discussion of shock waves, we will take advantage of the simplification offered by assuming that the fluid obeys the ideal gas equation of state and has constant physical properties. The reader is encouraged to perform these derivations as the discussion proceeds. Although obtaining these results are a straightforward exercise in algebra, one must be careful in the bookkeeping of the various terms to avoid errors.

We begin by writing the jump conditions for mass, momentum, and energy across the shock wave relative to the shock-fixed coordinate system.

$$\text{Mass} \quad \dot{m}''_x = \rho_1 v'_{x,1} = \rho_2 v'_{x,2} \tag{9.2}$$

$$\text{Momentum:} \quad P_1 + \rho_1 v'_{x,1} = P_2 + \rho_2 v'_{x,2} \tag{9.3}$$

$$\text{Energy:} \quad h_1 + \frac{1}{2}\left(v'_{x,1}\right)^2 = h_2 + \frac{1}{2}\left(v'_{x,2}\right)^2 \tag{9.4}$$

This derivation follows the presentation of Churchill (1980, pp. 72–77). The enthalpy term in the energy equation can be restated in terms of pressure and density:

$$h_2 - h_1 = C_p(T_2 - T_1) = \frac{C_p \mathcal{M}}{R_g}\left(\frac{P_2}{\rho_2} - \frac{P_1}{\rho_1}\right) = \frac{\gamma}{\gamma - 1}\left(\frac{P_2}{\rho_2} - \frac{P_1}{\rho_1}\right) \tag{9.5}$$

As was done in Chapter 4, the mass and momentum equations can be combined to give the *Rayleigh* equation:

$$\dot{m}_x'' = \rho_1 v_{x,1}' = \rho_2 v_{x,2}' = \left(\frac{P_2 - P_1}{\frac{1}{\rho_1} - \frac{1}{\rho_2}} \right)^{1/2} \tag{9.6}$$

Combining Eqs (9.5) and (9.6) and solving for the pressure ratio gives the *Rankine–Hugoniot* equation for a shock wave:

$$\frac{P_2}{P_1} = \frac{\beta \left(\frac{\rho_2}{\rho_1} \right) - 1}{\beta - \left(\frac{\rho_2}{\rho_1} \right)} \tag{9.7}$$

The constant β is defined as $\beta \equiv (\gamma + 1)/(\gamma - 1)$. Additional shock relations can be formulated by solving Eq. (9.7) for the density ratio and invoking the mass jump condition:

$$\frac{v_{x,1}'}{v_{x,2}'} = \frac{\rho_2}{\rho_1} = \frac{1 + \beta \left(\frac{P_2}{P_1} \right)}{\beta + \left(\frac{P_2}{P_1} \right)} \tag{9.8}$$

From the ideal gas equation of state, the following relations can be derived:

$$\frac{T_2}{T_1} = \frac{\beta + \left(\frac{P_2}{P_1} \right)}{\beta + \left(\frac{P_1}{P_2} \right)} = \frac{\beta - \left(\frac{\rho_1}{\rho_2} \right)}{\beta - \left(\frac{\rho_2}{\rho_1} \right)} = \frac{\beta - \left(\frac{v_{x,2}'}{v_{x,1}'} \right)}{\beta - \left(\frac{v_{x,1}'}{v_{x,2}'} \right)} \tag{9.9}$$

One of the major benefits of the ideal gas and constant property assumptions is that if one shock property ratio is known, then the others can be calculated.

These property ratio equations can be recast into forms depending only on the incoming Mach number M_1. Our starting point is to write the Mach numbers for the flow before and after the shock in terms of pressure and density.

$$M_1^2 = \frac{\rho_1 \left(v_{x,1}' \right)^2}{\gamma P_1} = \frac{\rho_1 v_w^2}{\gamma P_1}; \quad M_2^2 = \frac{\rho_2 \left(v_{x,2}' \right)^2}{\gamma P_2} \tag{9.10}$$

Note that the Mach numbers are evaluated in terms of the shock-fixed coordinate frame. Consider the expressions for M_1^2. Substituting the Rankine–Hugoniot expression for $\rho_1 v_{x,1}'$ and Eq. (9.8) for the density ratio and then simplifying yields

$$M_1^2 = \frac{\gamma - 1}{2\gamma} \left[1 + \beta \left(\frac{P_2}{P_1} \right) \right] \tag{9.11}$$

Similarly,

$$M_2^2 = \frac{\gamma - 1}{2\gamma} \left[1 + \beta \left(\frac{P_1}{P_2} \right) \right] \tag{9.12}$$

These relations can now be used to write the property ratios in terms of the incoming Mach number M_1.

$$\frac{P_2}{P_1} = 1 + \frac{2\gamma}{\gamma + 1} \left(M_1^2 - 1 \right) \tag{9.13}$$

$$\frac{v'_{x,2}}{v'_{x,1}} = \frac{\rho_1}{\rho_2} = 1 - \frac{2}{\gamma + 1} \left(1 - \frac{1}{M_1^2} \right) \tag{9.14}$$

$$\frac{T_2}{T_1} = \left[1 + \frac{2\gamma}{\gamma + 1} \left(M_1^2 - 1 \right) \right] \left[1 - \frac{2}{\gamma + 1} \left(1 - \frac{1}{M_1^2} \right) \right] \tag{9.15}$$

$$M_2^2 = \frac{1 + \dfrac{\gamma - 1}{\gamma + 1} \left(M_1^2 - 1 \right)}{1 + \dfrac{2\gamma}{\gamma + 1} \left(M_1^2 - 1 \right)} \tag{9.16}$$

Thus, if the incoming Mach number is specified (or if the velocity is measurable) and the initial fluid state (P_1, T_1) is known, all of the downstream properties can be determined.

There are two limiting cases for shock behavior: weak shocks and strong shocks. Different authors use different criteria, but one criterion is based on the limiting values of the incoming Mach number, which are $M_1 \to 1$ for weak shocks and $M_1 \to \infty$ for strong shocks. Table 9.1 summarizes the property ratios for weak and strong shocks.

These limiting values are for a gas that behaves like an ideal gas with constant properties. In the limit of weak shocks, the fluid properties are unchanged. In the limit of strong shocks, the postshock temperatures and pressures increase without bound. For a real gas, this limiting value would not be attained due to the temperature dependence of the physical properties and dissociation effects. Note, however, that for strong shocks the postshock Mach number and the density ratio attain finite limiting values.

The following example indicates how to use the property relations in a shock wave problem.

Table 9.1 Comparison of Limiting Property Ratios for Weak and Strong Shocks

Shock Strength	M_1	M_2	$\dfrac{\rho_2}{\rho_1} = \dfrac{v_{x,1}}{v_{x,2}}$	$\dfrac{T_2}{T_1}$	$\dfrac{P_2}{P_1}$
Weak	1	1	1	1	1
Strong	∞	$\sqrt{\dfrac{\gamma - 1}{2\gamma}}$	$\dfrac{\gamma + 1}{\gamma - 1}$	∞	∞

EXAMPLE 9.1

A dust deflagration generates a normal shock wave that is propagating with a velocity of 450 m/s into ambient air. Assume that the behavior of air is adequately characterized as an ideal gas. The ambient temperature is 300 K, the pressure is 1.01 bar, the molar mass of air is 29, and the heat capacity ratio is 1.4. Compute the shocked air property ratios P_2/P_1, T_2/T_1, and $v'_{x,2}/v'_{x,1}$, and compute the air velocity behind the shock wave.

Solution:

The property ratios can be calculated directly from the normal shock relations if we have the incoming Mach number. The incoming Mach number is the shock velocity divided by the speed of sound of the ambient air.

$$a_1 = \sqrt{\frac{\gamma R_g T}{M}} = \sqrt{\frac{1.4(8314 \text{ J/kmol} \cdot \text{K})300 \text{ K}}{29 \text{ g/mol}}} = 347 \text{ m/s}$$

$$M_1 = \frac{v'_{x,1}}{a_1} = \frac{450 \text{ m/s}}{347 \text{ m/s}} = 1.30$$

The pressure ratio is calculated from the following expression:

$$\frac{P_2}{P_1} = 1 + \frac{2\gamma}{\gamma + 1}\left(M_1^2 - 1\right) = 1 + \frac{2(1.4)}{1.4 + 1}\left[(1.30)^2 - 1\right] = 1.81$$

This gives a shocked air pressure of 1.81 bar or an overpressure of 0.80 barg. The temperature ratio is calculated with the expression:

$$\frac{T_2}{T_1} = \left[1 + \frac{2\gamma}{\gamma + 1}\left(M_1^2 - 1\right)\right]\left[1 - \frac{2}{\gamma + 1}\left(1 - \frac{1}{M_1^2}\right)\right]$$

$$= \left[1 + \frac{2(1.4)}{1.4 + 1}\left((1.30)^2 - 1\right)\right]\left[1 - \frac{2}{1.4 + 1}\left(1 - \frac{1}{(1.30)^2}\right)\right] = 1.19$$

The shocked gas temperature is thus 357 K (84°C). The calculate the shocked air velocity, we first calculate the velocity ratio from the relation

$$\frac{v'_{x,2}}{v'_{x,1}} = 1 - \frac{2}{\gamma + 1}\left(1 - \frac{1}{M_1^2}\right) = 1 - \frac{2}{1.4 + 1}\left(1 - \frac{1}{(1.30)^2}\right) = 0.660$$

$$v'_{x,2} = 0.660 v'_{x,1} = 297 \text{ m/s}$$

$$v_{x,2} = v_w - v'_{x,2} = 450 - 297 = 153 \text{ m/s}$$

This completes the analysis.

There are many additional relations that can be derived for the description of shock behavior. One such relation, called the *Prandtl shock relation*, is helpful for analyzing reflected shock behavior. The Prandtl relation links the particle velocities v_1 and v_2 to a constant called the critical sound speed. It is based on the consideration of the energy jump condition across the stationary shock. Following the derivation of Anderson, we begin by writing the energy equation in terms of temperature (Anderson, 1982, pp. 50−51):

$$C_p T_1 + \frac{1}{2}\left(v'_{x,1}\right)^2 = C_p T_2 + \frac{1}{2}\left(v'_{x,2}\right)^2 \tag{9.17}$$

We invoke the following definitions and substitute into the energy equation:

$$C_p = \frac{R_g}{\mathcal{M}}\left(\frac{\gamma}{\gamma - 1}\right); \quad a = \sqrt{\frac{\gamma R_g T}{\mathcal{M}}} \tag{9.18}$$

$$\frac{a_1^2}{\gamma - 1} + \frac{v_1^2}{2} = \frac{a_2^2}{\gamma - 1} + \frac{v_2^2}{2} \tag{9.19}$$

The critical state, designated by an asterisk, is a reference state for the fluid in which the fluid is accelerated adiabatically to sonic velocity. Thus, the energy equation can be expressed in terms of the critical sound speed as

$$\frac{a_1^2}{\gamma - 1} + \frac{v_1^2}{2} = \frac{a_*^2}{2(\gamma - 1)} + \frac{a_*^2}{2} = \frac{\gamma + 1}{2(\gamma - 1)} a_*^2 \tag{9.20}$$

The critical sound speed a_* is a constant across the shock. Returning to the original form of the jump conditions for a normal shock, the mass equation can be divided into the momentum equation to give the intermediate result

$$\frac{P_1}{\rho_1 v_1} - \frac{P_2}{\rho_2 v_2} = v_2 - v_1 \tag{9.21}$$

The definition of the sound speed can be used to write Eq. (9.21) in terms of velocities

$$\frac{a_1^2}{\gamma v_1} - \frac{a_2^2}{\gamma v_2} = v_2 - v_1 \tag{9.22}$$

Using the critical sound speed form of the energy equation, we can write an expression for a_1^2 and a_2^2:

$$a_1^2 = \left(\frac{\gamma + 1}{2}\right)a_*^2 - \left(\frac{\gamma - 1}{2}\right)v_1^2 \quad and \quad a_2^2 = \left(\frac{\gamma + 1}{2}\right)a_*^2 - \left(\frac{\gamma - 1}{2}\right)v_2^2 \tag{9.23}$$

Substituting these expressions into Eq. (9.22) and simplifying yields the *Prandtl shock relation*:

$$a_*^2 = v_1 v_2 \tag{9.24}$$

This relation is often helpful in deriving shock properties. We will employ this relation as we consider the change in fluid properties caused by the reflection of a normal shock wave from a solid surface.

9.1.2.2 Reflected shock waves

When a normal shock wave impinges on a solid surface, a new shock wave, the *reflected shock*, is transmitted in the reverse direction of travel. We consider only the special case of normal impact. Reflected shocks are important because they are typically much stronger (greater overpressure) than the original shock. Thus, the greater overpressures of reflected shocks can create greater structural damage. Examples of reflected shocks are a shock wave in an unconfined state striking an enclosure wall or, in a confined state, a shock traveling through a duct striking an elbow joint.

Fig. 9.6 illustrates the physical situation. State 0 is the ambient condition, state 1 is the compressed state due to the incident shock, and state 2 is the compressed state due to the reflected shock.

It is interesting to observe that the gas behind the reflected shock is stagnant (gas velocity is zero) relative to the wall. The reflected shock propagates through the previously shocked gas. Hence, with a reflected shock, the ambient gas is shocked twice.

It is desired to calculate the properties of the reflected shock wave from a measured value of the incident shock velocity v_w or from the pressure ratio P_1/P_0. In this section, we will follow closely the derivation presented by O'Neill and Chorlton (1989, pp. 287–290; a similar derivation can be found in Courant and Friedrichs, 1948, pp. 152–154). We begin with the energy equation,

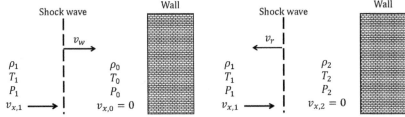

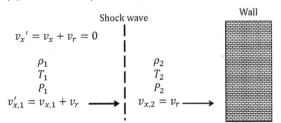

FIGURE 9.6

Shock wave properties for a shock reflected from a solid wall in both laboratory coordinates and shock-fixed coordinates.

Eq. (9.20), written in terms of the shock-fixed coordinates. We need two energy equations: one for the incident shock and one for the reflected shock:

$$\frac{v_w^2}{2} + \frac{a_0^2}{\gamma - 1} = \frac{1}{2}(v_w - v_1)^2 + \frac{a_1^2}{\gamma - 1} = \frac{\gamma + 1}{2(\gamma - 1)}a_*^2 \tag{9.25}$$

$$\frac{v_r^2}{2} + \frac{a_2^2}{\gamma - 1} = \frac{1}{2}(v_r + v_1)^2 + \frac{a_1^2}{\gamma - 1} = \frac{\gamma + 1}{2(\gamma - 1)}a_*^2 \tag{9.26}$$

The Prandtl shock relation provides an additional link between the two shock waves:

$$v_w(v_w - v_1) = a_*^2 = v_r(v_r + v_1) \tag{9.27}$$

The Prandtl relation is substituted into the two energy equations

$$\frac{1}{2}(v_w - v_1)^2 + \frac{a_1^2}{\gamma - 1} = \frac{1}{2}\left(\frac{\gamma + 1}{\gamma - 1}\right)v_w(v_w - v_1) \tag{9.28}$$

$$\frac{1}{2}(v_r + v_1)^2 + \frac{a_1^2}{\gamma - 1} = \frac{1}{2}\left(\frac{\gamma + 1}{\gamma - 1}\right)v_r(v_r + v_1) \tag{9.29}$$

The Mach numbers for each shock wave can be defined as follows:

$$M_1 = \frac{v_w - v_1}{a_1} \quad \text{and} \quad M_2 = \frac{v_r + v_1}{a_1} \tag{9.30}$$

Substituting the Mach number definitions into Eqs. 9.28 and 9.29 gives a quadratic equation in the incident and reflection Mach numbers:

$$M_1^2 + \frac{1}{2}(\gamma + 1)\frac{v_1}{a_1}M_1 - 1 = 0 \quad \text{and} \quad M_2^2 - \frac{1}{2}(\gamma + 1)\frac{v_1}{a_1}M_2 - 1 = 0 \tag{9.31}$$

We have two equations and two unknowns, M_1 and M_2. The coefficient in front of the linear term is a constant once the incident shock condition is specified. An easy way to solve this system of equations is to subtract the M_2 equation from the M_1 equation and simplify to get the intermediate result:

$$M_2 - M_1 = \frac{1}{2}(\gamma + 1)\frac{v_1}{a_1} \tag{9.32}$$

Substituting the intermediate result into either Mach number equation yields the solution $M_1 M_2 = 1$.

Applying the pressure ratio shock relation, one gets the following two equations:

$$\frac{P_1}{P_0} = \frac{2\gamma M_1^2 - (\gamma - 1)}{\gamma + 1} \quad \text{and} \quad \frac{P_2}{P_1} = \frac{2\gamma M_2^2 - (\gamma - 1)}{\gamma + 1} \tag{9.33}$$

Solving for the pressure ratio P_2/P_1 first, invoking the Mach number relation $M_1 M_2 = 1$, and then converting the pressure ratio equations to an expression for the excess pressure ratio $(P_2 - P_0)/(P_1 - P_0)$, one obtains the final desired result:

$$\frac{P_2 - P_0}{P_1 - P_0} = \frac{(3\gamma - 1)\dfrac{P_1}{P_0} + (\gamma + 1)}{(\gamma - 1)\dfrac{P_1}{P_0} + (\gamma + 1)} \tag{9.34}$$

We consider now two limiting cases of reflected shocks, weak shocks, and strong shocks.

$$\text{Weak shock limit} \qquad \frac{P_1}{P_0} \cong 1 \rightarrow \frac{P_2 - P_0}{P_1 - P_0} \cong 2 \qquad\qquad (9.35)$$

$$\text{Strong shock limit} \qquad \frac{P_1}{P_0} \gg 1 \rightarrow \frac{P_2 - P_0}{P_1 - P_0} \cong \frac{3\gamma - 1}{\gamma - 1} \cong 8 \; (\gamma = 1.4) \qquad (9.36)$$

Thus, even a weak reflected shock can double the overpressure applied to the target surface. The stronger the incident shock, the greater the reflected pressure ratio. Reflected shocks are one mechanism for the phenomenon of pressure piling.

EXAMPLE 9.2

Consider the following reflected shock wave problem. A dust deflagration propagating inside a dust collection duct generated a normal shock wave which strikes an elbow. The impact of the normal shock on the elbow creates a reflected shock. The undisturbed air inside the duct can be treated as an ideal gas and is at a temperature of 290 K, a pressure of 1.00 bar, a molar mass of 29, and the heat capacity ratio of air is 1.4. The incident shock wave has a velocity of 566 m/s when it strikes the elbow. Find the reflected shock pressure if the incident shock pressure is 2.5 bar.

Solution:

The reflected shock pressure is easily calculated from the following equation:

$$\frac{P_2 - P_0}{P_1 - P_0} = \frac{(3\gamma - 1)\frac{P_1}{P_0} + (\gamma + 1)}{(\gamma - 1)\frac{P_1}{P_0} + (\gamma + 1)} = \frac{(3(1.4) - 1)\frac{2.5}{1} + (1.4 + 1)}{(1.4 - 1)\frac{2.5}{1} + (1.4 + 1)} = 3.06$$

This is the ratio of shock overpressures or the reflected shock strength. The reflected shock pressure follows directly:

$$P_2 = 3.06 \, (P_1 - P_0) + P_0$$

We need the incident shock pressure and for that we need the incident Mach number. The sound speed is 341 m/s. With an incident shock velocity of 566 m/s, the Mach number is 1.66. The incident shock pressure is obtained from the equation

$$\frac{P_1}{P_0} = \frac{2\gamma M_i^2 - (\gamma - 1)}{\gamma + 1} = \frac{2(1.4)(1.66)^2 - (1.4 - 1)}{1.4 + 1} = 3.05$$

Substituting for P_1 and solving for P_2, we obtain $P_2 = 7.27$ bar. This is the desired result.

We will next examine a simpler linear model based on an acoustic approximation.

9.1.3 ACOUSTIC MODEL FOR THE OVERPRESSURE CAUSED BY ACCELERATING FLAMES

Despite the highly nonlinear behavior of shock waves, there are occasions when a weak shock wave can be adequately represented as an acoustic wave. The motivation for doing this is that it can simplify the analysis. We consider three aspects of the acoustic approximation: the acoustic time scale, the relationship between the shock overpressure and the particle velocity, and the relative importance of the static and dynamic pressure quantities.

The acoustic time scale t_{ac}, previously introduced in Chapter 8, is a natural measure of the time scale for wave interactions. The weak shock formula derived next is most valid for the first acoustic wave passage. The weak shock approximation breaks down with reflected waves or wave interactions. Since flame acceleration effects are greatest in one-dimensional geometries, the acoustic time scale is simply related to the flow length $\mathcal{L}$ by the equation $t_{ac} = \mathcal{L}/a$, where $\mathcal{L}$ is a length of the flow path. At an ambient temperature of 300 K (27°C), the sound speed in air is 347 m/s. As a numerical example, the radius of a 20-L test vessel is 0.168 m. The acoustic time scale is $t_{ac} = 0.168$ m/347m/s or 0.48 ms. The time to maximum pressure for the aluminum powder data sets discussed in Chapter 8 were on the order of 10–30 ms. This suggests that the weak shock wave generated by the deflagration would reflect and interact 20–60 times in the course of the deflagration. The multiple wave interactions tend to "smear" the spatial gradient in pressure; this results in a more or less uniform pressure field. The weak shock approximation is not appropriate for analyzing such data.

In early studies in experimental mines, researchers observed an approximate linear relationship between flame speed and the overpressure generated by the flame (Richmond and Liebman, 1975; Cybulski, 1975, pp. 86–88). A linear relationship between gas velocity and overpressure is an attribute of an acoustic wave. The acoustic approximation for a pressure disturbance can be derived by considering the following system (Thompson, 1972, pp. 184–187). Consider a fluid discontinuity propagating as a wave at the speed of sound, that is, a weak shock wave. The velocity of the gas ahead of the wave is v_1 and the velocity behind the wave is v_2. The physical situation is shown in Fig. 9.7. Comparing with Fig. 9.3, we are interested in the change that occurs across the weak shock. Zone 1 corresponds to the stagnant, unburnt region and zone 2 corresponds to the shocked, unburnt region. The flame that is chasing the shock is not shown in Fig. 9.7.

Since the wave is moving at the acoustic speed a, the velocity of the fluid entering the wave is $v'_{wave} = v_1 + a_1$. Writing the jump mass balance across the wave we get

$$\rho_2(v_2 - v'_{wave}) = \rho_1(v_1 - v'_{wave}) \tag{9.37}$$

$$\rho_2(v_2 - v_1 - a_1) = \rho_1(v_1 - v_1 - a_1) \tag{9.38}$$

$$\rho_1 a_1 = \rho_2(v_1 + a_1 - v_2) \tag{9.39}$$

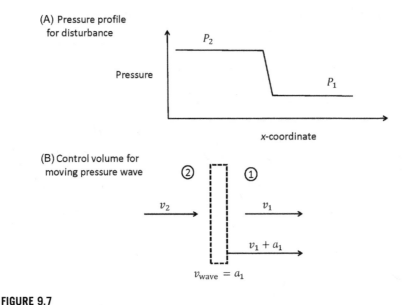

(A) Pressure profile for disturbance

(B) Control volume for moving pressure wave

FIGURE 9.7

Spatial profile and control volume for moving pressure disturbance.

After Thompson, P.A., 1972. Compressible-Fluid Dynamics. McGraw-Hill, New York.

Table 9.2 Tabulation of Weak Shock Overpressure as a Function of Flame Speed Calculated from the Acoustic Approximation for Air at 1.013 bars and 300 K

Δv(m/s)	0	50	100	200	300	400	500
ΔP(barg)	0	0.20	0.41	0.82	1.23	1.64	2.05

The jump momentum balance across the wave is

$$\rho_2 v_2(v_2 - v_1 - a_1) - \rho_1 v_1(v_2 - v_1 - a_1) = -(P_2 - P_1) \tag{9.40}$$

$$\rho_2 v_2(v_1 + a_1 - v_2) - \rho_1 v_1 a_1 = P_2 - P_1 \tag{9.41}$$

Substituting the mass balance relation into the momentum balance and simplifying yields the desired relation:

$$P_2 - P_1 = \rho_1 a_1(v_2 - v_1) \quad \text{or} \quad \Delta P = \rho_1 a_1 \Delta v \tag{9.42}$$

This expression can be used to estimate the overpressure generated across the weak shock wave if the particle velocity Δv is known. The particle velocity Δv is equal to the flame speed. In an experimental device, the flame speed can be measured by marking the time of flame passage with photodiodes or ionization sensors distributed at discrete points along the flow path. Alternatively, if pressure is measured, the flame speed can be calculated. Table 9.2 illustrates the order of magnitude of the overpressure developed as a function of flame speed.

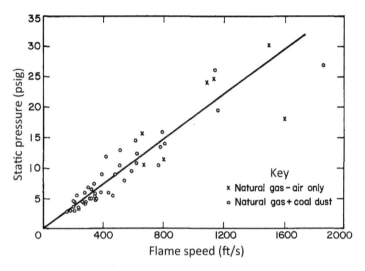

FIGURE 9.8

Experimental validation of the acoustic approximation for a weak shock wave. Data are from deflagration tests conducted in the Bureau of Mines experimental mine.

Public domain (Nagy, J., 1981 The Explosion Hazard in Mining. U.S. Department of Labor, Mine Safety and Health Administration, Informational Report 1119).

The calculation in Table 9.2 was based on the properties of air estimated at 1.013 bars and 300 K: $\rho_1 = 1.18 \text{ kg/m}^3$ and $a_1 = 347 \text{ m/s}$. Notice that the overpressure value is a gauge pressure, not the absolute pressure.

The acoustic approximation has been tested against experimental data from full-scale mine tests conducted by the U.S. Bureau of Mines (from Richmond and Liebman, 1975, as reported in Nagy, 1981). The length of the mine passageway was approximately 400 m in length and had a cross-sectional area of 5.58 m^2. Investigators tested both natural gas–air mixtures and hybrid mixtures of natural gas, coal dust, and air. The initiating event was a natural gas–air mixture deflagration at the closed end of the tunnel. The deflagration then propagated down the mine passageway, with the shock wave sweeping up coal dust off of the floor and the slower dust flame chasing the shock wave. Fig. 9.8 summarizes some of this flame speed–overpressure data and is plotted against the acoustic model.

The reader should note the magnitude of the overpressure and the flame speed. Remember that the pressures in Fig. 9.8 are not the isochoric explosion pressures. These pressures correspond to the overpressure generated by the weak shock wave. As a benchmark for comparison, Nagy indicated that the explosion pressures for these mixtures in laboratory tests varied from 100 to 115 psig with a starting condition of atmospheric pressure (Nagy, 1981, p. 57).

Table 9.3 Injury Potential and Structural Damage Criteria from Explosion Overpressures

Overpressure (psi)	Damage Effect
Human Injury	**Description**
0.6	Threshold injury from flying glass
2.4	Threshold eardrum rupture
10.0	Threshold lung hemorrhage
20.5	50% probability of fatality from direct blast effects (lung hemorrhage)
29.0	99% probability of fatality from direct blast effects (lung hemorrhage)
Property Damage	**Description**
0.5–1.0	Shattering of glass windows
1.0–2.0	Failure of wood siding panels (typical housing construction)
2.0–3.0	Shattering of nonreinforced concrete or cinder block wall panels
3.0–4.1	Collapse of self-framing steel panel buildings
5.0–7.0	Nearly complete destruction of houses
10.0	Probable total destruction of buildings

Data from NFPA 921, 2014. Guide for Fire and Explosion Investigations. National Fire Protection Association.

So the weak shock wave overpressures are only a fraction of the isochoric explosion pressure.

The agreement of the model with the experimental data is good for flame speeds up to roughly 400 m/s, which is about 10% greater than the speed of sound in the unburnt mixture (in terms of Mach number, $Ma \cong 1.1$). The acoustic approximation for the weak shock wave was further validated by Chi and Perlee in their computational study of flame propagation in a coal mine (Chi and Perlee, 1974). Chi and Perlee solved the unsteady, one-dimensional equations of change for a compressible fluid using the method of characteristics to simulate flame propagation and acceleration in a mine passageway. They found that the acoustic approximation predicted the pressures developed ahead of the flame within approximately 10%.

To place these numbers in perspective, consider the explosion overpressure damage criteria presented in Table 9.3 (NFPA 921, 2014, pp. 233–234).

Explosion damage mechanisms are actually very complex and are not easily reducible to simple overpressure criteria. However, overpressure damage criteria can be a useful order of magnitude guide for predicting damage effects. Comparing the magnitudes of Fig. 9.8 against the damage criteria in Table 9.2, it is evident that the deflagration tests in the experimental mine generated blasts with a potential to cause significant human injury and property damage.

EXAMPLE 9.3

Consider again the dust deflagration of Example 9.1. Calculate the overpressure of the normal shock wave generated by the flame using the acoustic approximation. Assume that the flame is propagating with a velocity of 150 m/s into ambient air (this was the velocity of the shocked air following the shock wave in Example 9.1). Assume that the behavior of air is adequately characterized as an ideal gas. The ambient temperature is 300 K, the pressure is 1.01 bar, the molar mass of air is 29, and the heat capacity ratio is 1.4.

Solution

In Example 9.1, we calculated the sound speed to be 347 m/s. The density of air at these conditions is

$$\rho = \frac{P\mathcal{M}}{R_g T} = \frac{1.01 \times 10^5 \text{ Pa}(29 \text{ kg/kmol})}{8314 \frac{\text{J}}{\text{kmol·K}}(300 \text{ K})} = 1.17 \text{ kg/m}^3$$

Using the acoustic approximation, the overpressure is given by the equation

$$\Delta P = \rho_1 \, a_1 \Delta v = (1.17 \text{ kg/m}^3)(347 \text{ m/s})(150 \text{ m/s}) = 0.61 \text{ barg}$$

In comparison, the normal shock relations gave a shocked air pressure of 1.62 bar or an overpressure of 0.61 barg. Thus, the overpressure calculated by the acoustic approximation is equal to the value calculated by the normal shock relation. The Mach number for this example is 1.23, indicating that even though we are at the outer bounds of the range of applicability for the acoustic approximation, it still works very well for this example.

When investigating the consequences of flame acceleration, it is important to distinguish between the static pressure P, which is the hydrostatic pressure of the fluid, and the dynamic pressure P_{dynamic}, which is the pressure contribution associated with fluid motion. Both types of pressure can cause injury or explosion damage, but the mechanisms by which they operate is different. The static pressure causes damage by imposing a static force on the surface of a structure such that the imposed pressure exceeds the strength of that component. The dynamic pressure is the "blast wind." The dynamic pressure can cause damage by momentum transfer from the blast wind to the object in question. For example, the dynamic pressure from an accelerating flame can accelerate objects from rest to some terminal velocity until the object collides with an obstacle. Typical magnitudes of these quantities will be discussed shortly.

Consider the propagation of an accelerating flame preceded by a weak shock in a one-dimensional channel like a coal mine passageway. With the accelerating flame, the initial pressure before combustion will be greater than atmospheric.

Thus, the final explosion pressure behind the flame will be greater than the explosion pressure measured in the laboratory. Recalling Eq. (3.50),

$$P_{ex} = P_0 \left(\frac{T_{ex}}{T_0} \right)$$

(9.43)

In this case, the initial pressure P_0 is no longer the atmospheric pressure; it is the pressure of the precompressed unburnt mixture region. So in the laboratory test vessel, if the flame temperature ratio is 10, the explosion pressure will be $P_{ex} = 10P_0$ or 10 atmospheres absolute pressure ($\Delta P = 9$ atm). If, however, the initial pressure due to flame acceleration is 1.2 atmospheres absolute, the explosion pressure will be 12 atmospheres absolute ($\Delta P = 11$ atm). This magnification effect is called pressure piling and will be discussed further in this chapter.

The dynamic pressures generated in the Bureau of Mines experimental mine were also capable of great destruction. Nagy observed that hurricane winds on the order of 150–200 miles per hour can do significant property damage (in the units of Fig. 9.7, that range is 220–290 ft/s or in SI units, 67–89 m/s) (Nagy, 1981, pp. 58–62). He further indicates that most coal dust explosions in mines generate at least 200 miles per hour (290 ft/s or 89 m/s) blast winds and will typically fail to propagate at 100 miles per hour (150 ft/s or 45 m/s). These particle velocities are enormous on a human scale and are capable of inflicting injury and structural damage. The blast wind exerts a drag force on objects and, if of sufficient magnitude, can accelerate and propel these objects as missiles. The dynamic pressure $P_{dynamic}$ and the drag force $\mathcal{F}_{drag}$ are related to the blast wind through the following equations:

$$P_{dynamic} = \frac{1}{2}\rho_g(\Delta v)^2; \quad \mathcal{F}_{drag} = \frac{1}{2}C_{drag}A_{object}\,\rho_g(\Delta v)^2$$

(9.44)

The drag coefficient C_{drag} depends on the geometry of the body and A_{object} is the area of the body normal to the blast wind. The other symbols have been previously defined.

EXAMPLE 9.4

Calculate the dynamic pressure and drag force on a mantrip (rail guided vehicle used for transportation of miners in underground coal mines) caused by a blast wind of 150 m/s. The drag coefficient is 0.3 and the projected area of the mantrip is 0.8 m². Assume air behaves as an ideal gas and use the usual properties for T = 300 K and P = 1.01 bar.

Solution

The dynamic pressure is given by the equation

$$P_{dynamic} = \frac{1}{2}\rho_g(\Delta v)^2 = \frac{1}{2}(1.17\ \text{kg/m}^3)(150\ \text{m/s})^2 = 0.132\ \text{barg}$$

The drag force is calculated as follows:

$$\mathcal{F}_{drag} = \frac{1}{2} C_{drag} A_{object} \rho_g (\Delta v)^2 = \frac{1}{2}(0.3)(0.8 \text{ m}^2)(1.17 \text{ kg/m}^3)(150 \text{ m/s})^2 = 3,160 \text{ N}$$

For a 1000-kg vehicle, this drag force could accelerate the vehicle at approximately 1/3 g, a value that is strong enough to be felt by the mantrip occupants, but less likely to cause injury. For a 100-kg man, this drag force would result in an acceleration of 3 g, a potentially hazardous acceleration.

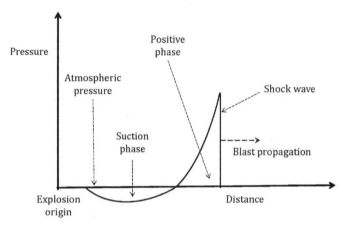

FIGURE 9.9

Blast wave features.

9.1.4 BLAST WAVES VERSUS SHOCK WAVES

In several places in this section, I have mentioned the term blast wind, and when talking about flame acceleration you will often see the term *blast wave*. The distinction between a blast wave and a shock wave is very important, especially when trying to predict or evaluate structural damage caused by an explosion. A blast wave is the moving shock wave plus the flow field that "chases" the shock. The shock portion of a blast wave is important because that is the source of the maximum overpressure that will strike a target. The flow field behind the shock is important for two reasons: first, the flow field is the transient blast wind that can impose a dynamic pressure load on a structure; second, the hydrostatic pressure of the flow field actually goes negative (less than atmospheric) (Courant and Friedrichs, 1948, pp. 416–417; Kinney and Graham, 1985, pp. 88–106; Crowl, 2003, pp. 54–58, 99–103). Fig. 9.9 depicts the pressure profile of a typical blast wave at a fixed point in time.

The blast wave is traveling from left to right. If the blast wave impacts a structure, it imparts a complex loading on it. The structure will first experience the shock overpressure (the positive phase of the blast wave) and the dynamic pressure of the blast wind. This is followed by the suction phase which draws a reverse flow of blast wind and a partial static pressure relative to atmospheric. The degree of structural damage depends not only on the strength of the structure but also on its dynamic response. A good introduction to the basics of blast wave dynamics and structural response can be found in the book by Kinney and Graham (1985).

If the wave is propagating in free space, the amplitude of the shock wave is greatest at the explosion origin and will continuously decrease with distance until it dissipates its energy as sound and heat.

9.1.5 SUMMARY

Flame acceleration effects magnify the hazard severity of a dust deflagration. The consequences of flame acceleration can lead to secondary dust explosions, confined dust flame acceleration leading to a possible detonation, and pressure piling. Turbulent combustion is the mechanism that promotes flame acceleration, and flame acceleration is the driving force behind shock wave formation. Shock waves are a powerful mechanism for transforming the chemical energy of combustion into destructive mechanical energy. The properties of normal shock waves and reflected shock waves were derived for an ideal gas with constant properties. We examined a simple acoustic wave model for weak shock waves which is a useful model for relating shock overpressure and blast wind (dynamic overpressure). We also briefly reviewed damage criteria based on the shock overpressure and examined the anatomy of a blast wave. We now turn to secondary explosions.

9.2 DUST CLOUD SUSPENSION AND SECONDARY DUST EXPLOSIONS

The term secondary dust explosion refers to a complex sequence of events beginning with an initial dust explosion (the primary dust explosion) rupturing its confining structure. The rupture of the confinement leads to the propagation of a blast wave (shock wave plus the following flow field) which entrains dust deposits and creates a low-lying dust cloud. The slower propagating flame then sweeps through the dust cloud accelerating in velocity and increasing in pressure. Many of the most devastating dust explosions have been secondary dust explosions (Eckhoff, 2003, pp. 7−10, 221−229; Frank and Rodgers, 2012, pp. 234−244; Taveau, 2012; Amyotte, 2013, pp. 128−137).

The physics behind a secondary explosion is extremely complex. To simplify the analysis of the phenomena involved, I have divided the problem into three constituent parts: the formation of the fugitive dust layer, the process of dust lofting and dust cloud formation, and flame propagation and acceleration.

9.2.1 **DUST LAYER PROPERTIES AND THICKNESS**

Fugitive dust emitted into the ambient atmosphere will eventually settle out and form a layered deposit on horizontal surfaces. The dust layer is a static state. To form an ignitable dust cloud, the dust layer must be dislodged from the substrate surface, dispersed into the surrounding air, and suspended long enough to achieve ignition. Eckhoff draws an interesting contrast between characterizing the state of a flammable gas mixture and the state of a combustible dust cloud (Eckhoff, 2003, p. 200). Once the flammable gas achieves a state of uniform mixing in the surrounding air, the thermodynamic state of the gas is specified by its chemical composition, temperature, and pressure. The state of the flammable gas mixture is static with time and remains well-mixed. The combustible dust cloud, on the other hand, is a dynamic state. It cannot form an ignitable mixture unless the dust particles are maintained in a dispersed state. The state of static equilibrium for a combustible dust cloud is the settling out of the particles into a stagnant dust layer deposit.

There are three different kinds of adhesive forces that can apply between particles or between a particle and a surface: the van der Waals force, electrostatic force, and capillary force (Hinds, 1999, Chapter 6; Eckhoff, 2003, Chapter 3). The van der Waals force arises from the random motion of electrons in the atoms of the particle. This random motion creates transient areas of concentrated charge on the particle surface called dipoles that are responsible for the attraction to the substrate surface or to neighboring particles. The electrostatic force arises due to the accumulation of charge on the particle. Particles of electrically insulating material at low levels of ambient humidity tend to accumulate and retain electrical charge. The capillary force arises from the attachment of two adjacent particles by a thin liquid film between them. Each of these three forces are proportional to the diameter of the particle. Their influence usually ends within a few particle diameters of separation. Generally, the van der Waals and capillary forces are greater than the electrostatic force.

There are essentially two types of dust layers: monolayers and multilayers. A *monolayer* in this context is a deposit of dust particles sufficiently dilute such that each particle is in contact with the surface and none of the particles is in contact, or within the range of influence, of its nearest neighbors. The distinction is based on the extent to which the substrate surface affects the particle resistance to removal. Imagine a clean, smooth horizontal surface. If particles deposit onto this surface, there will be a force of attraction between the particle and the surface. The smaller the particle, the stronger this adhesive force becomes in relation to other forces acting on the particle. Smaller particles are more difficult to detach from a substrate surface than larger particles.

For the purposes of this text, I define a *multilayer* deposit as a deposit with a sufficient surface concentration and number of layers such that only interparticle forces are at work. The substrate surface does not exert any influence on the nature of the force required to dislodge a particle from the dust deposit. If the substrate surface is rough, then the depth of dust layer must exceed the height of the surface asperities (microscopic protrusions extending above the mean surface level). The magnitude of interparticle forces is generally less than the adhesive force between a particle and the substrate surface.

The dust layer properties are a function of the chemical composition of the particulate material (bulk and surface composition), the particle size distribution, the ambient environment (especially temperature and humidity), and age. Depending on this combination of factors, dust layers can exhibit a range of responses to an imposed stress. The particles within the dust layer may not experience any force from adjacent particles other than their weight. The bulk material in this situation is said to be *cohesionless* or *noncohesive*. If an air stream of sufficient velocity blows across the surface of a cohesionless dust deposit, the particles tend to be dispersed from the dust deposit surface as a cloud of distinct particles entering the air stream.

On the other hand, interparticle forces can cause the particles to become attracted to each other and to resist separation. The bulk particulate material will resist a change in shape until an imposed stress exceeds the strength of the material causing the material to fail. If interparticle forces become significant, the bulk material is said to be *cohesive*. If an air stream of sufficient velocity blows across the surface of a cohesive dust deposit, the particles tend to be dislodged from the surface of the dust deposit in clumps which may then disintegrate in the air stream.

Two physical properties in particular govern the behavior of dust layers: the bulk density and the unconfined yield stress. A discussion on the measurement of these properties is available in several good books (Seville et al., 1997, Chapters 4 and 8; Schulze, 2008, Chapters 3, 4, and 6; Rhodes, 2008, Chapter 10) The significance of powder technology to combustible dust behavior has been discussed by Eckhoff (2009b). The bulk density of a dust deposit may change over time. If the deposit was formed by freely settling fugitive dust, then the bulk density of the dust deposit will resemble its "as-poured" or aerated bulk density. With time, random vibration of the horizontal surface or other mechanical disturbances can cause the deposit to become compacted leading to a slightly larger density. The compaction of the dust deposit may be caused by elastic or plastic deformation of the particles leading to an increase in the points of contact between particles thus increasing the particle surface area available for exerting adhesive forces on neighboring particles.

The mechanical property of interest for dust deposits is the unconfined yield stress. In a very informal sense, the unconfined yield stress is related to the force that must be applied to a dust deposit to cause the deposit to rupture or break apart into two or more deposit segments. The greater the cohesiveness of the dust, the greater its unconfined yield stress. Compaction of the dust deposit may lead to greater yield stress, or adsorption of water or other airborne contaminant may lead to greater cohesion. Cohesive dust deposits tend to resist the detachment of individual particles and instead tend to fail in clumps. The clump detaches from the main body of the powder mass by slipping along a failure plane (the mathematical analysis of the stress—strain behavior of powders is discussed extensively in books on soil mechanics).

9.2.2 **DUST LOFTING AND DUST CLOUD FORMATION**

Bagnold published one of the first comprehensive studies of dust lofting, dust cloud formation, and settling (Bagnold, 1954). His book was based on his studies the fundamental fluid dynamic processes involved in the formation of sand dunes in the desert. The book has maintained its popularity because much of it is still relevant today. In addition to his use of field observations and measurements, Bagnold developed a portable wind tunnel that could be placed on the desert surface to measure erosion processes. An engineer by training, he interpreted his observations in terms of the fluid dynamics of particle motion and turbulent boundary layer theory. One of the early observations made by Bagnold was the notion that natural sands exhibited certain types of motion depending on the particle size. These modes of particle motion were surface creep, saltation, and suspension. Fig. 9.10 illustrates these modes of motion.

As a rough indication of the mode of motion, particles with a maximum diameter of 500−1000 μm are limited to surface creep, particles with a diameter in the range from 100 to 500 μm undergo saltation, and suspension occurs for particles smaller than 50 μm (Lyles, 1988). Shao gives a similar, but different, range of particle sizes: surface creep for particle diameters greater than 500 μm, saltation between 70 and 500 μm, and suspension for less than 70 μm (Shao, 2008, pp. 131−132). In summary, for combustible dusts in practical environments, all three types of particle motion are potentially significant.

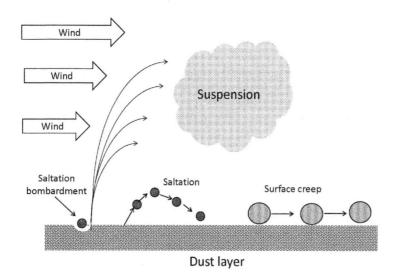

FIGURE 9.10

Modes of particle motion with a dust layer.

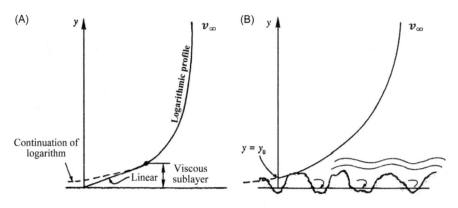

FIGURE 9.11

Turbulent boundary layer velocity profile for a smooth surface (A) and a rough surface (B).

Reproduced with permission from Kundu, P.K., Cohen, I.M., Dowling, D.R., 2016. Fluid Mechanics,
Sixth Edition. Academic Press, New York.

9.2.2.1 Fluid dynamics of dust entrainment

Turbulent boundary layer theory will be a common theme in the next two sections, so we will take a quick detour to introduce some of the basic ideas through simple mathematical models. Most studies of dust entrainment are conducted as confined flows such as in pipes or wind tunnels. So we will also consider the velocity distribution in pipe flow. As a flow enters a tube, the frictional resistance of viscosity is imposed close to the wall. At some distance downstream from the tube's entrance, the flow in the boundary layer will become turbulent. The thickness of the velocity profile varies with the free stream velocity, the properties of the fluid, and the distance from the tube entrance. The exact nature of the velocity profile will depend on whether the surface is smooth or rough. If the wall surface is actually a dust layer and the fluid is air, the thickness of the boundary layer can become small enough to exert different magnitudes of friction on different particles depending on their size. The turbulent velocity profile can be divided into three distinct regions: the viscous sublayer (sometimes called the laminar sublayer), the buffer layer, and the logarithmic layer (Fig. 9.11).

It is customary to define a dimensionless velocity v^+ and a dimensionless coordinate y^+

$$v^+ \equiv \frac{v_\infty}{\sqrt{\tau_0/\rho}}; \quad y^+ \equiv \frac{\sqrt{\tau_0/\rho}}{(\mu/\rho)} \cdot y \qquad (9.45)$$

In these definitions, v_∞ is the free stream velocity, τ_0 is the shear stress at the plate surface, and ρ is the density of the fluid. The quantity $\sqrt{\tau_0/\rho}$ has units of velocity and is called the friction velocity. The friction velocity can be calculated from the Blasius relations:

$$\tau_0 = 0.0225 \, \rho v_\infty^2 \left(\frac{(\mu/\rho)}{v_\infty r_w} \right)^{1/4} \qquad (9.46)$$

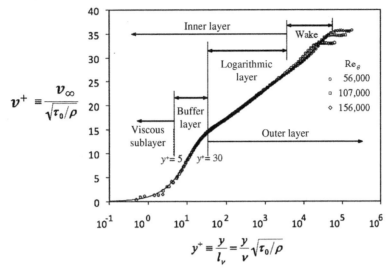

FIGURE 9.12

The logarithmic velocity correlation for turbulent boundary layer flow in a smooth tube.

Reproduced with permission from Kundu, P.K., Cohen, I.M., Dowling, D.R., 2016. Fluid Mechanics,
Sixth Edition. Academic Press, New York.

For turbulent flow in smooth tubes, the different regions of the velocity profile are described by the following equations (Welty et al., 1984, pp. 193–196; Kundu et al., 2016, pp. 648–659):

$$\text{Turbulent core, } y^+ \geq 30; \quad v^+ = 5.5 + 2.5 \ln y^+ \tag{9.47}$$

$$\text{Buffer layer, } 30 \geq y^+ \geq 5; \quad v^+ = -3.05 + 5 \ln y^+ \tag{9.48}$$

$$\text{Laminar sublayer, } 5 \geq y^+ \geq 0; \quad v^+ = y^+ \tag{9.49}$$

The velocity correlations are plotted in Fig. 9.12.

Bagnold used these concepts to correlate his field observations and experiments on sand movement (Bagnold, 1954). As an example of their utility, consider the following illustration, Fig. 9.13, of a turbulent boundary layer with a free stream velocity of 30 m/s (Corn and Stein, 1965).

The profiles of two particles are shown within the boundary layer. The particle with a 10-μm diameter fits within the laminar sublayer and the larger particle, 100-μm diameter, is mostly outside the laminar sublayer and well within the buffer layer. At this location, one would predict that the 100-μm particle can be suspended into the air stream and the 10-μm particle cannot.

A useful rule of thumb for estimating the maximum particle diameter that cannot be removed by an airflow velocity is given in the review paper by Ziskind et al. (1995). The concept is that if the particle lies completely within the laminar sublayer, then it will not most likely be removed by the flow. They

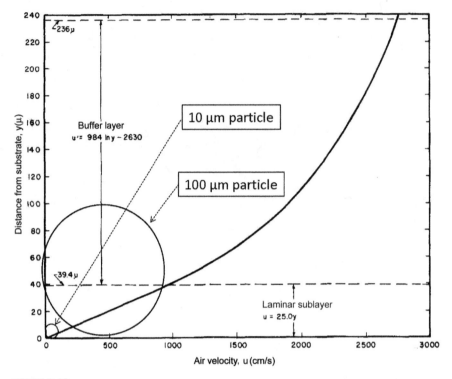

FIGURE 9.13

Turbulent boundary layer structure with a free stream velocity of 30 m/s showing two
particle profiles with diameters of 10 and 100 μm.

Reproduced with permission from Corn, M., Stein, F., 1965. Re-entrainment of particles from a plane
surface. Amer. Ind. Hyg. Assoc. J. 26, 325–336.

define a dimensionless particle diameter, d_p^+, in terms of the Reynolds number
for the flow:

$$d_p^+ = 0.19887 \, \text{Re}^{7/8} \left(\frac{d_p}{d_{\text{pipe}}} \right) \tag{9.50}$$

If $d_p^+ < 5$, the particle is below the laminar sublayer and will not likely be
removed by the airflow.

9.2.2.2 Laboratory studies of monolayer and multilayer detachment

The creation of a dust cloud from a dust layer requires first detachment of the
particles from surface. The force required for detachment of particles has been
the subject of several empirical studies (Corn, 1961a, 1961b; Corn and Stein,
1965; Ziskind et al., 1995; Boor et al., 2013a, 2013b). Early on investigators saw
the value of distinguishing between dust deposits consisting of monolayers versus
multilayers. The distinction has to do with the nature of the adhesive forces

responsible for the adhesion of the particles to a surface versus the forces that are responsible for interparticle adhesion.

Beginning with investigations from the 1930s, Corn reviewed the literature on single particle adhesion to solid surfaces and noted that the force of adhesion was a function of many factors such as particle size and shape, the chemical composition of the particle, the chemical composition of the surface, the ambient humidity, the topography of the surface, and the presence or absence of electric charge (Corn, 1961a). He described the use of a microbalance technique for measuring the adhesion force of a single quartz particle on a glass surface under controlled humidity conditions (Corn, 1961b). He observed that the adhesion force increased with increasing particle size, increasing humidity, and decreasing surface roughness. Although these test conditions seem far removed from a dusty industrial environment, these carefully controlled experiments produced important insights into single particle adhesion onto solid surfaces.

In another study, Corn and Stein advanced from single particle experiments to monolayers of particles (Corn and Stein, 1965). They used two techniques, centrifugal force and a high-velocity air stream, to remove monolayers of adhered particles from surfaces. They compared the adhesion of spherical glass particles with nonspherical fly ash particles on both glass and stainless steel surfaces. They used an ultracentrifuge to obtain the desired centrifugal forces and placed the test module with the particles adhered to a test substrate surface inside a container to preclude exposing the particles to aerodynamic forces. The high-velocity air stream tests were conducted in a test cell with a rectangular cross-section. They found that the factors that governed single particle adhesion were predictive of monolayer adhesion.

Corn and Stein observed that increased exposure time did not increase the fraction of particles removed from a surface by centrifugal force, but did increase when using high-velocity air. They used the measured removal force obtained by the ultracentrifuge tests to predict, using turbulent boundary layer theory, the air stream velocity needed to attain a drag force of the same magnitude. They reasoned that higher free stream velocities were required to create a thinner laminar sublayer which exposed smaller particles to the turbulent eddies so that they would be entrained into the free stream flow.

It has been found that particle detachment with multilayer deposits requires less force than monolayers of the same particles (Boor et al., 2013a). In a review of 29 different studies of dust deposition in ventilation ducts and flooring surfaces, Boor et al. found that multilayer deposits required less force to detach particles and resuspend them compared to monolayer deposits. They also indicated that for particles with a diameter less than 100 μm, the multilayers formed a cohesive mass that was prone to detachment as agglomerates (clumps) rather than as individual particles.

Working with a reference standard called Arizona Test Dust, they created both monolayer and multilayer dust deposits on two test surfaces, linoleum flooring and galvanized sheet metal. Using a wind tunnel, they then determined the detachment rates over a range of velocities. For monolayers, they used air velocities at 25, 50, and 75 m/s. For multilayer deposits, the air velocities tested were 2.5, 5.0, 7.5, 10.0, 12.5,

15, 20, and 25 m/s. Boor et al. found that the air velocity required to detach mono-layer particles was an order of magnitude greater than that required to detach multilayer deposits (50 m/s to detach the monolayer particles vs 5 m/s to detach the multilayer particles, Boor et al., 2013b). They couched their explanation of their observations in terms of turbulent boundary layer theory. They concluded from both their experiments and those of other investigators that the slow detach-ment of monolayer particles was due to the occasional penetration of turbulent bursts from the fully turbulent sublayer. As the multilayer height increased to a level greater than the laminar sublayer of the turbulent boundary layer, the det-achment of agglomerates became easier as the uppermost portion of the multi-layer deposit was continuously exposed to the eddies of the turbulent sublayer.

EXAMPLE 9.5

Assume a primary dust explosion in a grain elevator tunnel has propagated a shock wave with a particle velocity of 30 m/s. There is a multilayer deposit of grain dust on the tunnel floor with a broad size distribution having a range of 1–1,000 μm. Use the rule of thumb derived by Ziskind et al. to determine the maximum particle diameter of combustible dust that cannot be lofted if the height of the tunnel is 2 m. The ambient conditions are 25°C and 1.01 bar.

Solution

The rule of thumb for particle lofting is that if the dimensionless particle diameter $d_p^+ < 5$, then particles of this size or smaller cannot be lofted because they are below the height of the laminar sublayer. The dimension-less particle diameter is defined by the following equation:

$$d_p^+ = 0.19887 \, \text{Re}^{7/8} \left(\frac{d_p}{d_{\text{pipe}}} \right)$$

If the particle velocity is 30 m/s, the value of the Reynolds number is

$$\text{Re} = \frac{\rho_g v_x d_{\text{pipe}}}{\mu_g} = \frac{\left(1.1766 \, \text{kg/m}^3\right)\left(30 \, \text{m/s}\right)\left(2 \, \text{m}\right)}{1.853 \times 10^{-5} \text{Pa} \cdot \text{s}} = 3.81 \times 10^6$$

The density and viscosity data for air were obtained from Kays and Crawford (1980, p. 388). Solving for the particle diameter and substituting the values for the remaining parameters into the dimensionless particle diameter criterion give the following result:

$$5 < 0.19887 \, \text{Re}^{7/8} \left(\frac{d_p}{d_{\text{pipe}}} \right) = 0.19887 \left(3.81 \times 10^6\right)^{7/8} \left(\frac{d_p}{2 \, \text{m}} \right); \quad d_p < 8.77 \times 10^{-5} \text{m}$$

Particles smaller than 88 μm are below the laminar sublayer and will not be lofted into the free stream flow field.

9.2.2.3 Wind tunnel studies of dust suspension

Investigators interested in combustible dust behavior tended to work with larger quantities of dust subjected to air flows in wind tunnels, "explosion tunnels," shock tubes, and full-scale mines. These experimental settings are less controlled but more realistic than the laboratory settings described in the above section. Some of the earliest work on the lofting of combustible dust was performed by Dawes in the UK Safety in Mines Research and Testing Branch (Dawes, 1950, 1952a, 1952b). In his review of the literature, Dawes indicated that work on this problem extended back to the 1930s. Dawes first worked out the mechanics of dust layer behavior using the principles of solid mechanics and derived criteria for dust layer rupture. He related the rupture criteria to the hydrodynamic stresses imposed on a dust layer by a flowing air stream. He then tested these ideas by measuring the dust lofting response of 23 noncombustible powders (mostly varieties of limestone) in a wind tunnel. His wind tunnel was 1.8 m in length and had a rectangular cross-section 15 cm in width and 7.5 cm in height. Air velocities ranged from 20 to 80 m/s.

His test specimens were dust layers of rectangular shape measuring 18 cm in length, 5.0 cm in width, and 0.6 cm in height. He observed two distinct types of dust lofting behavior: erosion and denudation. Fig. 9.14 illustrated both types of dust lofting behavior using limestone or limestone-like powders.

Erosion is the removal of particles from the top surface of a powder mass in a continuous stream. In Fig. 9.14, the dust entrainment profile is suggestive of boundary layer flow. Cohesionless dusts tended to loft by erosion. Denudation is the removal of clumps of material from the top surface of powder mass. The clumps typically disintegrated almost immediately in the air stream. Cohesive dusts tended to loft by denudation. Dawes worked primarily with flat rectangular piles of powders with a height of approximately 6 mm. Fig. 9.15 shows the erosion of a deposit with an airflow of approximately 30 m/s.

Dawes obtained some general agreement of his models with the test data, but the research program was cut short and he was unable to complete it. His primary contribution was his finding that cohesive dust layers lofted more easily (at lower air stream velocities) than noncohesive layers.

Singer and his colleagues at the U.S. Bureau of Mines embarked on a similar series of investigations using a wind tunnel, an "explosion tunnel" and a full-scale experimental mine (Singer et al., 1969, 1972, 1976). They tested anthracite dust, bituminous (Pittsburgh seam) coal dust, and rock dust in both monolayers and in piles. The wind tunnel was 3 m in length with a rectangular cross-section 7.6 cm in width and an adjustable height of 2.5 or 5.0 cm. The airflow was set at either 156 or 270 m/s, an order of magnitude greater than the velocities used in Dawes's experiments.

In their monolayer experiments, Singer et al. used coal dust ranging 6−70 μm in particle diameter placed onto glass slides (Singer et al., 1969). They found that larger particles were detached more easily than small particles.

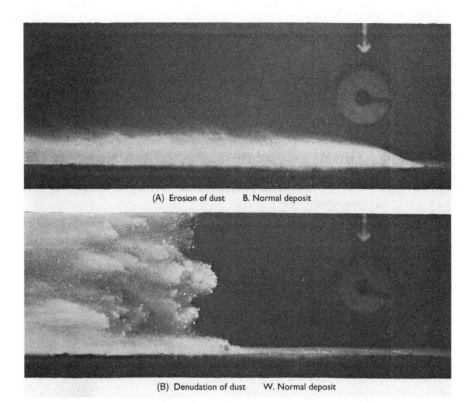

(A) Erosion of dust   B. Normal deposit

(B) Denudation of dust   W. Normal deposit

FIGURE 9.14

Photographs of wind tunnel experiment showing two different modes of dust lofting:
(A) erosion and (B) denudation. Airflow was approximately 30 m/s from right to left.
Public domain (Dawes, J.G., 1952a. Dispersion of Dust Deposits by Blasts of Air: Part I, Ministry of Fuel and
Power, Safety in Mines Research Establishment. Research Report No. 36).

For the multilayer tests, they formed conical piles of material. They noted that with conical piles of limestone dust with a height of 6.4 mm, dust lofting was by first by erosion followed by denudation. Fig. 9.16 illustrates a typical result from their tests.

They noted the occurrence of bluff body vortices on the leeward side of the piles. These vortices enhanced dust lofting rates. They also experimented with deposits of two distinct layers: one coal dust and one rock dust. One interesting finding was that when they tested rock dust on top of coal dust (the desired configuration in a coal mine), the rock dust had a tendency to slide on the surface of the coal dust leading to higher than expected lofting rates. Based on their wind tunnel studies, they offered the rule of thumb that dusts with an average particle diameter greater than 100 μm behaved as cohesionless bulk materials, while dusts with a particle diameter less than 100 μm were cohesive.

In 1976, Fletcher of the UK Safety in Mines Research Establishment published two papers on his investigations of the role played by boundary layer

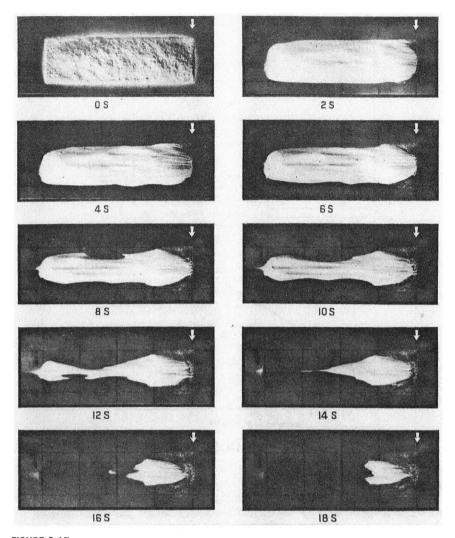

FIGURE 9.15

Photographs of dust deposit looking downwards. Wind tunnel test for erosion of limestone dust deposit by airflow of approximately 30 m/s. Airflow was from right to left. The arrow indicates the leading edge of the dust deposit.

Public domain (Dawes, J.G., 1952a. Dispersion of Dust Deposits by Blasts of Air: Part I, Ministry of Fuel and Power, Safety in Mines Research Establishment. Research Report No. 36).

flows on dust suspension (Fletcher, 1976b, 1976c). He indicated that previous investigations had demonstrated that with increasing air velocities the dust layers were entrained by denudation, that is, segments of the dust layer would lift off of the surface and disintegrate in the free stream. The entrainment depended on the length of the deposit: short deposits could be completely removed with one

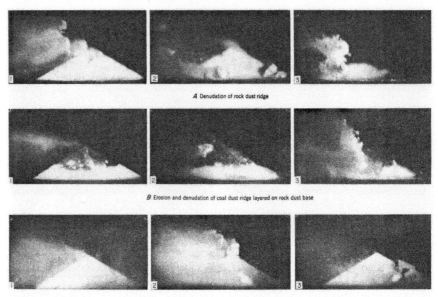

A Denudation of rock dust ridge

B Erosion and denudation of coal dust ridge layered on rock dust base

C Incipient erosion and denudation of rock dust ridge

FIGURE 9.16

Wind tunnel test with limestone powder exhibiting both erosion and denudation processes. Side view with airflow going from right to left.

Public domain (Singer, J.M., Greninger, N.B., Grumer, J., 1969. Some aspects of the aerodynamics of formation of float coal dust deposits. U.S. Department of Commerce, Bureau of Mines, Report of Investigations 7252).

denudation event, but with longer deposits (0.3 m or longer) the denudation occurred at the leading edge of the dust layer with only a fraction of the dust layer length removed. He also noted that thicker deposits had a greater tendency to denude than thin deposits. He predicted, based on turbulent boundary layer theory, that for a fixed air velocity, the erosion rate would be initially high but would diminish with time falling to zero. In his experiments, he estimated the friction velocity to be $\sqrt{\tau_0/\rho} = 0.062\, v_\infty$.

Fletcher observed the effects of saltation on dust entrainment and remarked that after finishing a test the surface of the dust layer was often pitted. When the dust in the vicinity of a pit was removed, a large particle, typically greater than 500 μm in diameter, was found buried in the pit. This observation was consistent with the notion of saltation bombardment, where a larger particle undergoing saltation would, upon impact with the dust layer surface, disperse smaller particles into the air stream. These observations were consistent with the saltation dynamics first observed and modeled by Bagnold (1954).

Owens published a more sophisticated fluid dynamic model of saltation (Owen, 1964). Other researchers have followed suit incorporating more physics into their models (Fairchild and Tillery, 1982). Shao and his colleagues published two papers extending the theory of saltation to include the effect of saltation bombardment (Shao et al., 1993; Lu and Shao, 1999).

The threshold friction velocity (or pickup velocity) has been the subject of both experimental and theoretical investigations. Shao and Lu investigated its importance in natural systems (Shao and Lu, 2000). Kalman and his colleagues studied it for application to pneumatic conveying (Kalman et al., 2005; Rabinovich and Kalman, 2009).

9.2.2.4 Explosion tunnel and shock tube studies of dust suspension

We are especially interested in the detachment and suspension of dust particles by the flow field caused by a shock wave passing over a dust deposit followed by the transient boundary layer flow. There have been a number of such studies using a variety of shock tube geometries, dusts, and test conditions. On the one hand, the one constant in the majority of these studies is the recognition of the role that the turbulent boundary layer plays in dust entrainment. On the other hand, the exact nature of the shock wave−dust layer interaction is controversial. Specifically, the question debated in many of the studies is the degree to which shock wave−dust layer interactions contribute to the vertical dust flux. We consider only a few of these studies to highlight some specific observations. A more complete survey of these studies can be found in the report by Ural (2011a).

Singer and his colleagues then explored the validity of their observations at larger scales using both an "explosion tunnel" and a full-scale experimental mine (Singer et al., 1976). The explosion tunnel was a steel pipe measuring 0.61 m in diameter and 49.7 m in length. They dispersed coal dust and rock dust along test sections of the explosion tunnel. They used methane deflagrations to initiate a blast wave to travel down the length of the pipe. They used a variety of instrumentation to monitor the behavior of the blast wave and its effect on dust entrainment including pressure transduces, hot wire anemometers, flame detectors, and high-speed photography. They observed as many as 10 acoustic oscillations in their tests, but observed that the dispersal of dust always occurred with the passage of the first wave. Fig. 9.17 is a wave diagram of a methane−air flame propagating down the mine passageway.

They also observed that the dimensionless rate of regression of the dust layer $v_{regress}/v_\infty$ scaled with the dimensionless friction velocity v_*/v_∞. The dust dispersion results were similar to results obtained from the previous laboratory wind tunnel tests.

The full-scale mine tests were conducted in a test mine measuring 400 m in length and having a rectangular cross-section of 2.0 m high and 2.9 m in width. Differences were noted in the dispersion behavior of the dusts. One of the reasons for this difference was the formation of large-scale vortices superimposed on the turbulent flow. The investigators opined that these vortices played a significant role in enhancing the entrainment of dust. Such large-scale vortices have been subsequently studied by Ben-Dor and their role in dust entrainment has been demonstrated (Ben-Dor, 1995).

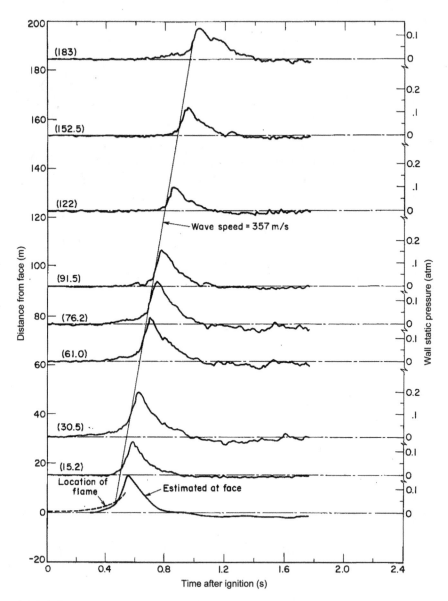

FIGURE 9.17

Wave diagram showing the wave trajectory in the experimental mine and the pressure history at each pressure transducer location. Deflagration was a 7.5% methane–air flame in 4.8 m compartment.

Public domain (Singer, J.M., Harris, M.E., Grumer, J., 1976. Dust dispersal by explosion-induced airflow. U. S. Department of Commerce, Bureau of Mines, Report of Investigations 8130).

Gerrard investigated the interaction of a shock wave with a dust layer using a shock tube with a circular cross-section measuring 3.8 cm in diameter and 4.1 m in length (Gerrard, 1963). He used two dust samples obtained from the UK Safety in Mines Research Establishment, the same source used by Dawes (1952a,b). The maximum particle diameter for these two powders was estimated to be less than 60 μm. He also tested a third dust: powdered blackboard chalk. The dust samples were a noncohesive limestone dust and a cohesive-precipitated carbonate; the dust layer depth was varied from 1.0 to 12.7 mm. The dust samples were placed on a flat plate with no compression of the dust. Using air as his working fluid, he subjected dust samples to shocks with Mach numbers of 1.2 and 1.6. He used Schlieren photography to record the interaction of the shock wave with the dust layer in the earliest stages of the interaction.

Gerrard hypothesized that the total dust dispersion event was a result of two interrelated processes: the interaction of the shock wave with the dust layer and the boundary layer flow that formed behind the shock. The shock wave interacted with the dust layer in two different ways. First, Gerrard reasoned that because the pressure behind the shock wave was greater than the preshock condition, then there must be a transient flow of air into the porous dust layer (Dawes made a similar inference in his 1950 report). A second feature of the interaction was based on his review of the Schlieren photographs which indicated that the shock wave was curved down toward the interface between the shock and the dust layer. He opined that the shock interaction resulted in waves reflecting from the base of the deposit that would produce an upwards vertical impulse to the particles thereby initiating their vertical motion. He also observed that there was a time lag from the passage of the shock to the initial formation of the suspended dust cloud. The dust cloud grew in height with increasing time after the passage of the shock wave. He also observed that consolidating the bulk density of the dust by compression to twice its original aerated bulk density resulted in a dust cloud height equal to 10% of the height produced with the unconsolidated dust.

Fletcher performed an expanded study of shock wave interactions with dust layers and identified sources of experimental error in Gerrard's work (Fletcher, 1976a). Fletcher indicated Gerrard missed certain physical phenomena due to a poor resolution in some of Gerrard's photographs. Fletcher challenged Gerrard's interpretation of the influence of the shock on the motion of the dust layer. Fletcher performed a number of vertical shock tube experiments to determine the velocity of propagation of disturbances through various granular materials. In a horizontal shock tube, he then performed experiments with dust layers ranging in thickness from 3.2 to 22.4 mm. Based on his experiments of the propagation velocity, Fletcher argued that the initial vertical velocity imparted to the dust layer particles was caused not by shock wave interactions but by the fluidization caused by the inflow of air into the dust layer and boundary layer flow (both physical mechanisms recognized but not quantified by Gerrard).

Mirels developed a very detailed model of the turbulent boundary layer behind a shock wave to account for the interaction of the boundary layer flow with the dust layer (Mirels, 1984). His model was based on a blowing parameter incorporated into the integral form of the turbulent boundary layer equations. This approach enabled him to predict the vertical flux of the dust and the concentration profile of the dust cloud. He was able to compare his model with experimental data from two studies performed by others, and noted that the predicted vertical flux of dust (the erosion rate) was within a factor of 2 of the measured rates. This may seem to be a rather large error, but its importance is that it represents an attempt at a priori prediction of erosion rates.

Federov and his colleagues have performed a number of theoretical and experimental investigations of dust suspension by shock waves (Gosteev and Federov, 2002; Federov, 2004). Their work has focused on understanding the various forces acting on a single particle and how these contribute to the dust suspension process. Federov has also noted the potential enhancement of the dust suspension process by large-scale features like vortices imposed on the fine-scale turbulent structures (as noted by Singer et al., 1976 and Ben-Dor, 1995).

Finally, Ural's recent investigation of dust suspension is notable for its practical contribution (Ural, 2011a, 2011b, 2011c). Based on his review of the literature, he has recommended of an algorithm for calculating the erosion rate from a dust layer based on the free stream air velocity and particle characteristics. His erosion rate (the entrained mass flux) is given by the following empirical correlation equations:

$$\dot{m}_d'' = 0.002 \cdot \rho v_\infty \left[v_\infty^{1/2} - \left(\frac{v_{th}^2}{v_\infty^{3/2}} \right) \right] \tag{9.51}$$

$$v_{th} = 0.46 \cdot \rho_s^{1/3} \tag{9.52}$$

Here $\dot{m}_d''$ is the mass flux of dust entrained into suspension (kg/m^2 · s), v_∞ is the free stream velocity (m/s), v_{th} is the threshold velocity, ρ is the gas density (kg/m^3), and ρ_s is the solid density of the dust particles (kg/m^3). Fig. 9.18 is a plot of the threshold velocity for four types of particles.

The entrained mass flux is analogous to a mass transfer flux of particles from the dust deposit surface into the free stream of the flow field. Since dust explosion parameters are reported in terms of dust concentration, it would be convenient to have an algorithm for converting the entrained dust flux into a dust concentration. In keeping with our emphasis on simple conceptual models, this can be accomplished most easily for the case where the free stream velocity is constant for a specific duration in time and spatially uniform over the dust deposit surface. The final piece of this puzzle is an estimated height of entrainment. If this sounds like a lot of guesswork, your intuition is correct. However, empirical studies do support certain broad generalizations about the dust entrainment process. The objective of this very simple model is to simply provide insight. Deeper, more accurate analysis will likely require more sophisticated mathematical modeling tools. The next example illustrates the simple approach. Ural provides additional examples of how to incorporate dust entrainment flux calculations using certain classic fluid dynamic flow situations [Ural, 2011c].

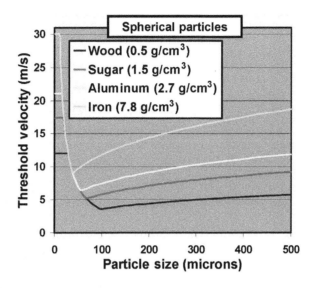

FIGURE 9.18

Threshold velocity for wood, sugar, aluminum, and iron particles.

Reproduced with permission from Ural, E., 2011c. Toward Estimating Entrainment Fraction for
Dust Layers. Springer.

EXAMPLE 9.6

Consider the following combustible dust hazard scenario. A machine shop processes aluminum alloy components including grinding and polishing operations. The facility owner has not instituted a formal housekeeping program for managing combustible dust and, as a result, aluminum dust has accumulated on the floor of the workspace. The area of the dust deposit is 100 m^2 in a room with a rectangular floor plan measuring $5 \text{ m} \times 20 \text{ m}$. The ceiling height is 6 m. Doors and windows in the workspace are normally closed.

Assume that a primary explosion is initiated at one end of the 5 m wide wall. This explosion has propagated a shock wave with a particle velocity of 30 m/s. The solid density of the aluminum alloy particles is 2700 kg/m^3 and the ambient conditions are 25°C and 1.01 bar. The MEC for this dust is 45 g/m^3.

Use Ural's algorithm to calculate the threshold velocity, the optimal particle size, and entrainment flux for this situation (Ural 2011a). Then estimate the duration of the entrainment flux using an acoustic time scale argument and the entrainment height based on the empirical observations of Tamanini and Ural, (1992). The ultimate objective of this analysis is to determine if the dust cloud created by the primary explosion is at an ignitable concentration ($\rho_{\text{dust}} \geq \text{MEC}$).

Solution

This analysis is essentially a quasi-steady, one-dimensional flow analysis. The quasi-steady component arises from the assumption that highly transient phenomena—the motion of a blast wave and its ability to entrain dust off of the floor—can be modeled as steady phenomena over a specified duration of time. The one-dimensional flow approximation is necessary to keep the problem simple. Given the "boxy" geometry of the workplace, a three-dimensional flow analysis would be more accurate.

To calculate the entrainment flux we begin by calculating the threshold velocity using Eq. 9.52.

$$v_{th} = 0.46 \cdot \rho_s^{1/3} = 0.46 \, (2700 \text{ kg/m}^3)^{1/3} = 6.41 \text{ m/s}$$

The minimum dust particle diameter that can be entrained by the free stream velocity is given by the optimal diameter relation:

$$d_{p,optimal} = 7.9 \times 10^{-4} \cdot \rho_s^{-1/3} = 7.9 \times 10^{-4} (2700 \text{ kg/m}^3)^{-1/3} = 5.67 \times 10^{-5} \text{ m}$$

The minimum entrained particle diameter is 57 μm. The corresponding mass flux of entrained aluminum dust is

$$\dot{m}_d'' = 0.002 \cdot \rho v_\infty \left[v_\infty^{1/2} - \left(\frac{v_{th}^2}{v_\infty^{3/2}} \right) \right]$$

$$= 0.002 \, (1.1766 \text{ kg/m}^3) \, (30 \text{ m/s}) \left[(30 \text{ m/s})^{1/2} - \left(\frac{(6.41 \text{ m/s})^2}{(30 \text{ m/s})^{3/2}} \right) \right]$$

$$= 0.369 \text{ kg/m}^2 \cdot \text{s}$$

We will estimate the duration of the entrainment flux by assuming that it is equal to the travel time for the incident shock wave plus the travel time for the first reflected shock wave. After those two events, the particle velocity is assumed to be zero. In reality the free stream particle velocity behind the shock wave will decay exponentially over a time period of several acoustic time scales. So this assumption is oversimplifying the true fluid dynamic phenomena. The duration of the steady flow field is calculated as

$$t_{duration} = \frac{L_{incident} + L_{reflected}}{a} = \frac{20 \text{ m} + 20 \text{ m}}{350 \text{ m/s}} = 0.114 \text{ s}$$

The total mass of dust entrained m_d is

$$m_d = \dot{m}_d'' \, A_{floor} \, t_{duration} = 0.369 \text{ kg/m}^2 \cdot \text{s} \, (100 \text{ m}^2)(0.114 \text{ s}) = 4.21 \text{ kg}$$

Tamanini and Ural performed a number of secondary dust explosion tests in a long, narrow gallery. They observed that the entrainment height was typically 1/3 of the gallery height [Tamanini and Ural, 1992]. Adopting this empirical observation, the average dust concentration in the workplace area is calculated to be

$$\rho_d = \frac{m_d}{A_{\text{floor}}H_{\text{cloud}}} = \frac{4.21 \text{ kg}}{100 \text{ m}^2(2 \text{ m})} = 211 \text{ g/m}^3$$

The estimated dust concentration dispersed by the blast wave of the primary explosion is greater than the MEC. This means that the slower moving deflagration wave from the primary explosion could ignite this dust cloud and initiate a secondary dust explosion.

9.2.3 SECONDARY DUST EXPLOSIONS: UNCONFINED FLAME PROPAGATION AND OVERPRESSURE

In Section 3.9 we considered a simple thermodynamic model for the final pressure attained in a secondary dust explosion. In this section we consider the fluid dynamic processes involved and begin with the assumption that a dust deposit has been dispersed into the environment. The dust cloud forms a continuous, combustible dust cloud with a dust concentration above its minimum explosible concentration. The dust cloud has been ignited and now we want to calculate the flame speed and overpressure as a function of time and position.

Our goal here is to understand the flame propagation and overpressure characteristics of the secondary explosion. Synthesizing all of the elements of a secondary dust explosion is an intimidating assignment. With so many diverse spatial and temporal scales, and with such nonlinear physics, the complete treatment must rely on computational fluid dynamics. It is difficult to develop a simple analytical model for this purpose in part because of the lack of confinement. Unconfined dust flame propagation with central ignition will result in spherical propagation of a leading shock wave followed by a flame. In addition, turbulence plays an essential role in the combustion process, and the presence of obstacles can dominate the development of the turbulent flow field. We will highlight here an approach based on the acoustic approximation. Most of the studies mentioned in this section have been developed for unconfined gas explosion analysis. So the application of these techniques to dust deflagrations needs to be approached with caution. We will close this discussion with a simple method for evaluating the potential for a secondary dust explosion.

Some investigators have examined the use of the acoustic approximation to simulate the propagation of a weak shock wave driven by a flame (Taylor, 1946; Strehlow, 1981). The advantage is that this approach does yield closed form solutions. They are limited to small overpressures in the acoustic limit. Several nonlinear models relating flame speed and overpressure have been proposed. The work of Strehlow and his colleagues and the work of Williams are especially noteworthy (Strehlow, 1975; Strehlow et al., 1979; Williams, 1976, 1983). These models are important contributions to the study of blast waves, but they do not yield analytical solutions; they require numerical solution. More will be said about nonlinear flame acceleration models in Chapter 10.

We will consider one example of an analytical solution for unconfined flame propagation. Strehlow adopted an acoustic approximation for pressure wave created by a propagating spherical flame (Strehlow, 1981). The physical problem is to consider the sound wave (think "weak shock wave") emitted by the injection of gas mass into the ambient environment as a point source. The propagation of the flame and the pressure disturbance are assumed to be one dimensional; we therefore select spherical coordinates for our governing equations. We begin by writing the acoustic wave equation in terms of the velocity potential ϕ (Thompson, 1972, pp. 159−163):

$$r^2 \frac{\partial^2 \phi}{\partial t^2} = a_0^2 \left[\frac{\partial}{\partial r} \left(r^2 \frac{\partial \phi}{\partial r} \right) \right] \tag{9.53}$$

The velocity potential must satisfy the following two constraints:

$$v_r = \frac{\partial \phi}{\partial r}; \quad P - P_0 = -\rho_0 \frac{\partial \phi}{\partial t} \tag{9.54}$$

Strehlow adopted the Stokes simple source model to describe the pressure disturbance (Lighthill, 1987, pp. 17−23):

$$P - P_0 = \frac{\ddot{m}}{4\pi} \left(t - \frac{r}{a_0} \right) \tag{9.55}$$

where $\ddot{m}$ is the mass addition source strength defined as the time derivative of the rate of mass addition.

The mass source term can be rewritten in terms of the volume rate of addition. For a deflagration, the source term becomes

$$\dot{m}(t) = \rho_0 \dot{V}(t) = \rho_0 \left(\frac{V_b - V_u}{V_u} \right) \frac{d}{dt} [S_u(t) \cdot A_f(t)] \tag{9.56}$$

where $S_u(t)$ is the burning velocity, $A_f(t)$ is the flame area, and $(V_b - V_u)/V_u$ is the volume increase due to combustion per unit volume of unburnt gas. The volume ratio is related to the energy of combustion by the relation:

$$\frac{P - P_0}{P_0} = \frac{q}{4\pi a_0^2 r} [S_u(t) \cdot A_f(t)]; \, q = \frac{(\gamma - 1)Q}{P_0 V_0} \tag{9.57}$$

For a constant velocity flame with spherical flame growth,

$$\frac{P - P_0}{P_0} = \frac{qS_u}{4\pi a_0^2 r} \left(\frac{dA_f}{dt} \right); \quad A_f = 4\pi r_f^2; \quad \frac{dr_f}{dt} = S_b = S_u \left(\frac{V_b}{V_u} \right) \tag{9.58}$$

Substituting the additional relations into the overpressure equation

$$\frac{P - P_0}{P_0} = \frac{2qS_u^2 r_f}{a_0^2 r}\left(\frac{V_b}{V_u}\right); \quad r \geq r_f \qquad (9.59)$$

The acoustic pressure on the unburnt side of the flame can be found from Eq. (9.58) by setting $r \geq r_f$ and substituting the definition of q:

$$\left(\frac{P-P_0}{P_0}\right)_u = 2\gamma\left(1 - \frac{V_u}{V_b}\right)\left(\frac{V_b}{V_u}\right)^2 M_{S_u}^2 \qquad (9.60)$$

Recognizing that $r_f = S_b\, t$ and that beyond the flame $t \to [t - (r/c_0)]$, Eq. (9.59) becomes

$$\left(\frac{P-P_0}{P_0}\right)_u = 2\gamma\left(1 - \frac{V_u}{V_b}\right)\left(\frac{V_b}{V_u}\right)^3 M_{S_u}^3\left(t - \frac{r}{a_0}\right) \qquad (9.61)$$

This is the desired result. This equation gives the far field acoustic overpressure generated by the flame from the secondary explosion. The ratio of burnt gas volume to the unburnt gas volume is simply the expansion ratio. Thus, the model requires for input values the burning velocity and the expansion ratio. This model is restricted to unconfined flame propagation. A similar closed-form solution overpressure to flame speed for *confined* flame propagation is a much more difficult task due to the interaction of compressibility and turbulent effects.

9.2.4 SUMMARY

In this section, we have reviewed some of the physical properties of dust deposits and how these influence dust entrainment. The next topic covered was the relationship between turbulent boundary layer theory and the physics of particle detachment and entrainment. Dusts with a large particle size (larger than roughly 100 microns) tend to behave like cohesionless powders and are entrained as a dispersion of individual particles. This entrainment process is called erosion. Fine dusts (particle size smaller than roughly 100 microns) are entrained in clumps which subsequently disintegrate in the turbulent air stream. This entrainment process is called denudation. A sampling of empirical studies was reviewed to get a sense of some of the important empirical observations about dust entrainment and suspension. The broad diversity of experimental techniques—from microbalances to full-scale experimental mines—leads us to a simple conclusion: the entrainment of dust deposits is a complex phenomenon.

A simple acoustic model was derived to describe the overpressure field generated by an unconfined deflagration. The model depends on the burning velocity and the expansion ratio as inputs.

9.3 CONFINED DUST FLAME ACCELERATION

It is well established that igniting a flammable gas mixture at the closed end of a tube can result flame acceleration and damaging overpressures. Under the right conditions, the accelerating flame can transition to a detonation (deflagration to detonation transition or DDT). Since the pressure ratio across a detonation wave can be in the range of 15–20 (see Section 4.2), detonation waves represent an extreme hazard to be avoided.

The flame acceleration process, and especially it transition from a deflagration to a detonation depend on a host of physical and chemical factors including turbulence, gas dynamics, and chemical kinetics. The nature of the turbulent flow field can be influenced by the conduit geometry and roughness and is especially sensitive to the type, number, and configuration of obstacles to the flow. In short, obstacles can have a profound influence on the turbulent flow field. The gas dynamic behavior is influence by the geometry of the conduit, by the presence of turns in the flow (eg, pipe elbows), and by the physical properties (density and sound speed) of any obstacles in the flow. Chemical kinetics is important because the combustion reactions must be fast enough to sustain combustion as the time scale of the flow decreases with acceleration. The book by Lee and a number of recent papers review the key features to flame acceleration and DDT (Lee, 2008; Kerampran et al., 2000; Ciccarelli and Dorofeev, 2008; Dorofeev, 2011; Proust, 2015).

A key requirement to bear in mind with confined dust flame acceleration is that this phenomenon is possible only if there is combustible dust deposits inside the conduit. If there is no combustible dust, flame acceleration stops. The accelerated dust flame will propagate a shock wave into the "clean" section of the conduit and the shock will diminish in strength until it either dies out due to viscous dissipation or finds a point of release into the ambient environment. Having already covered the properties of dust deposits and dust entrainment, we will next survey what has been learned from confined dust flame acceleration investigations. Most of these studies were motivated by coal mine safety as coal mine passageways present an (unfortunately) excellent opportunity for dust flame acceleration.

9.3.1 SURVEY OF EXPERIMENTAL METHODS AND SELECT RESULTS

In 1961, Artingstall (UK Safety in Mines Research Establishment) published a study on coal dust flame acceleration in mine passageways (Artingstall, 1961). He performed an analysis of one-dimensional coal dust deflagrations and detonations using the Chapman–Jouguet analysis method, insights from wave diagrams, and the method of characteristics. Artingstall concluded that for stoichiometric coal dust–air mixtures, the maximum deflagration speed was 1100 m/s and the Chapman–Jouguet detonation velocity was 2300 m/s. He asserted that the quoted deflagration speed was a maximum because it was the speed of sound in the burnt gas. He compared his model with measurements of coal dust explosion experiments conducted in experimental mines and obtained good agreement with flame speed–overpressure comparisons up to flame speeds of 600 m/s. At higher flame speeds, on the order of 1000 m/s, the measured overpressures were too low.

Rae reported on coal dust explosion tests conducted in an experimental mine (Rae, 1973). Tests were conducted with coal dust having 36% volatiles content and 85% passing a 65-μm sieve. The stoichiometric concentration of the coal dust was calculated to be 300 g/m^3 (based on volatiles) and the constant volume explosion pressure was 8 bars. Coal dust was deposited on the floor and various ignition sources were used. One interesting observation made was that the initiating explosion had to achieve an overpressure of at least 12 kPa in order to propagate beyond the ignition volume of the mine passage. Based on the acoustic approximation, this corresponds roughly to a flame speed of 30 m/s. Pressure wave oscillations associated with weak explosions (low overpressures) also caused oscillating flame venting at the open end of the tunnel. With strong explosions (high overpressures), he reported flame speeds ranging from 1000 to 1300 m/s.

Cybulski and his colleagues published an authoritative book on coal dust explosion and suppression (Cybulski, 1975). The book summarized international research activities on coal dust explosion safety spanning a 50-year period. Numerous tests conducted in the Experimental Mine Barbara (400 m gallery, typical) demonstrated flame acceleration effects with large flame speeds and overpressures. Pressure wave oscillations were a common observation. The book is a treasure trove of empirical data and test methods related to coal mine safety.

Richmond and his colleagues at the U.S. Bureau of Mines conducted a number of tests in the experimental mine (Richmond and Liebman, 1975; Richmond et al., 1979). Although various Bureau of Mines investigators had performed dust explosion tests at this facility since the 1910s, the work of Richmond et al. was notable for the diversity of instrumentation used in their work. Their instrumentation included pressure transducers, flame sensors (infrared and visible light), Pitot tubes to measure total pressure, drag probes, high-speed photography, thermocouples, and optical dust probes. The experimental mine measured approximately 400 m in length with a rectangular cross-section of 5.6 m^2. As discussed earlier in this chapter, they used their flame speed−overpressure measurements to test and validate the acoustic approximation formula. Fig. 9.19 is a wave diagram of the propagation of a coal dust flame in the experimental mine.

Their experiments indicated successful propagation of an explosion required an initial flame speed on the order of 50−100 m/s. They were also able to validate some of the wave interaction phenomena predicted by Artingstall such as the reflection of the first compression wave from the open end of mine as an expansion wave.

In the first paper, Richmond and Liebman investigated coal dust explosion behavior in a single entry (passageway) mine configuration. In the second paper, Richmond and colleagues investigated coal dust explosion behavior in a more complicated mine configuration, a mine with two entries (passageways) with three cross-cuts (connections between entries) (Richmond et al., 1979). One entry was used for ignition and was the primary pathway for flame propagation. The cross-cuts provided venting into the other entry but did not diminish the overpressure by very much. In addition to the instrumentation used before, they added gas sampling devices. Examination of gas sampling data suggested significant combustion of volatiles but not of char. In testing explosion suppressants, they found that some performed better in the experimental mine than laboratory results would have predicted.

Kauffman and his colleagues conducted accelerated dust flame tests using laboratory-scale setups (Kauffman et al., 1984; Srinath et al., 1987). Their flame

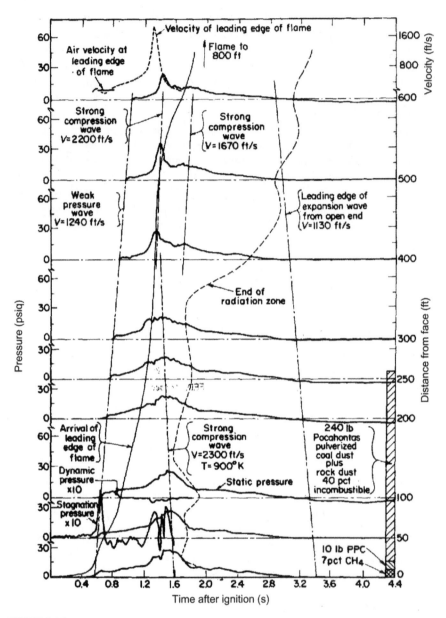

FIGURE 9.19

Wave diagram for the propagation of a coal dust flame in the U.S. Bureau of Mines experimental mine.

Reproduced with permission from Richmond, J.K., Liebman, I., 1975. A physical description of coal mine explosions. In: Fifteenth Symposium (International) on Combustion. The Combustion Institute. Elsevier, pp. 115–126.

acceleration tube was 36.6 m in length with a circular cross-section measuring 29.9 cm in diameter. They performed tests with cornstarch, two types of mixed grain dust, and oil shale dust. Tests were conducted by depositing dust uniformly along the length of the tube. They employed a variety of dust loadings and ultimately measured flame speeds ranging from 100 to 750 m/s. Their test results confirmed the hypothesis that smaller particle size dusts propagated stronger secondary explosions (faster flame speeds, higher overpressures) than coarse dusts. They also explored the use of turbulence generating grids and demonstrated some enhancement in the flame speed as a function of dust concentration, but the trends with enhanced turbulence were not consistent.

Pu et al. investigated the role of turbulence generators in dust flame acceleration (Pu et al., 1988). They compared three different types of turbulence generation: the turbulence of the dust dispersion device, turbulence grids, and flow obstacles. Their test setup involved two flame acceleration tubes. The smaller tube was a 1-m length tube with a square cross-section of 5 cm × 5 cm. This tube had two windows for Schlieren photography and the obstacle area could be adjusted to give 20%, 40%, and 60% blockage ratios. Pressure was measured by a pressure transducer and flame position was measured by ionization probes. Comparison tests were run with methane−air mixtures and cornstarch−air mixtures. Fig. 9.20 shows Schlieren photographs of gas and dust flame propagation with flow obstacles set to give a blockage ratio of 40%.

The presence of the obstacles reduced the explosion pressure of the methane mixture compared to the test without obstacles. It was suggested in the paper that the reason was the additional heat sink effect of the flow obstacle plates (composed of metal). For the dust flame, the effect was the opposite. The dust flame was accelerated and the explosion pressure was greater. The authors suggested that mass burning rate of the cornstarch was significantly slow than methane in the absence of obstacles. With the obstacles, the mass burning rate of cornstarch was enhanced by the recirculation zones between each obstacle plate to such a degree that it overcame any heat loss effect due to the obstacle plates. Furthermore, the enhancement of the rate of combustion due to the obstacles, inferred from the maximum rate of pressure rise, was at least 10 times greater than the turbulence generated by dispersion or by the presence of the grids. The authors suggested that the obstacle blockage caused flame acceleration mainly by the effects of nonuniform mean velocity and not by increasing the turbulent intensity.

Tamanini and Ural reported on a series of dust flame acceleration experiments conducted in two test galleries: one was an intermediate-scale setup and the other was full-scale setup (Tamanini and Ural, 1992). The intermediate-scale setup was 6.1 m in length and had a square cross-section of 0.3 m in width. The full-scale setup measured 24.4 m in length and square cross-section of 2.4 m in width. They used cornstarch as their combustible dust and deployed it uniformly on the gallery floor. In the full-scale gallery, they measured flame speeds of 50−90 m/s and overpressures as high as 0.37 barg. In the absence of obstacles, the dust entrained from the floor filled the bottom third of the gallery height.

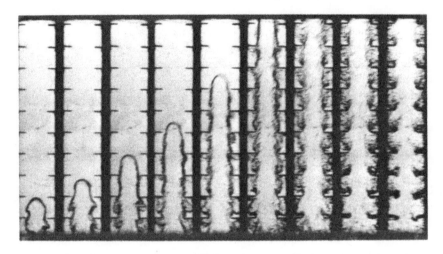

(A) Methane

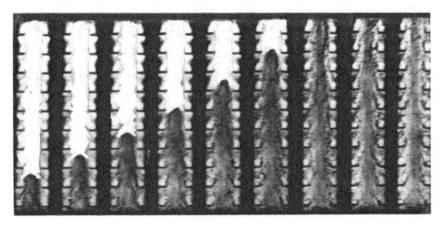

(B) Cornstarch

FIGURE 9.20

Comparison of gas and dust flame propagation in tubes with flow obstacles (blockage ratio 40%). Gas mixture is 6.5% methane in air and dust is cornstarch at 500 g/m³. Spacing between obstacles is 25 mm. Frame rate for Schlieren photography is (A) 640 frames/s and (B) 500 frames/s.

Reproduced with permission from Pu, Y.K., Mazurkiewicz, J., Jarosinski, J., Kauffman, C.W., 1988.
Comparative study of the influence of obstacles on the propagation of dust and gas flames.
In: Twenty-Second Symposium (International) on Combustion. The Combustion Institute, Elsevier,
pp. 1789–1797.

EXAMPLE 9.7

This example is an alternative simple model for a secondary dust explosion by Epstein (2007). The value of Epstein's method is that it provides a simple conceptual model for a secondary explosion. His approach yields an estimate of the minimum flame speed required to create a dust cloud for the propagation of a secondary dust explosion. It does not, however, produce a relation between the overpressure and flame speed of the secondary explosion.

The physical setting is a long rectangular gallery closed at both ends. A primary dust explosion is initiated at one end of the gallery. The objective of the analysis is to determine the minimum strength of the shock wave—stated in terms of flame speed—that will entrain sufficient dust into a dust cloud with a dust concentration equal to or in excess of its MEC.

Solution

The ultimate objective of Epstein's analysis was to determine the strength of the shock wave needed to create a dust cloud with an ignitable dust concentration. The problem analysis is based on a quasi-steady, one-dimensional flow analysis of the flame propagation and dust entrainment process.

Similar to the quasi-steady approach used in Example 9.6, Epstein began with calculating the time duration of dust entrainment. He related the duration to the acoustic time scale,

$$t_{acoustic} = \frac{L_{travel}}{a}$$

The distance of travel for the acoustic wave was the length of the gallery. The dust entrainment flux was modeled with an expression that was inspired by the vertical erosion flux model of Bagnold, [1954]. It featured a velocity cubed functional dependency:

$$\dot{m}_d'' = k_{ent} v_x^3 = k_{ent} S_b^3; \quad k_{ent} = 2.0 \times 10^{-4} \frac{\text{kg} \cdot \text{s}^2}{\text{m}^5}$$

The flame speed was substituted for the gas particle velocity and the coefficient k_{ent} was evaluated in a previous study. It is worthwhile to suggest caution at this point as Bagnold's flux model is thought to be suited for lower free stream velocities and larger particle sizes more relevant to soil erosion than combustible dust entrainment.

The entrained dust mass was calculated as the entrainment flux times the dust deposit area (assumed to be equal to the floor area) times the duration of the entrainment event:

$$m_d = \dot{m}_d'' A_{floor} t_{duration} = \frac{3}{2} \dot{m}_d'' A_{floor} t_{acoustic}$$

He posited that the time duration was represented by three incident-reflection cycles of the first acoustic wave emitted by the flame motion. This hypothesis was the source of the factor of 3 in the equation. The factor of ½ arose from the average exposure time of an area element of the dust deposit to the gas motion. The maximum value of this exposure would be $t_{acoustic}$, corresponding to the area element in contact with the flame front at the beginning of flame motion. The minimum exposure time would be zero, corresponding to the area element at the opposite end of the length of travel.

The dust cloud concentration caused by entrainment was then written in terms of the product of the entrainment flux and the duration of entrainment divided by the volume of the dust cloud:

$$\rho_d = \frac{m_d}{A_{floor} H_{cloud}} = \frac{3k_{ent}\, S_b^3 L_{travel}}{aH_{cloud}}$$

Epstein then asserted that $H_{cloud} = H_{ceiling}/3$, relying on the empirical observation by Tamanini and Ural (1994). This last equation can be converted into a criterion for the propagation of a secondary dust explosion by requiring $\rho_d \geq$ MEC. Thus, this final equation can be solved to give a minimum flame speed requirement for a propagating secondary dust explosion:

$$S_b \geq \left[\frac{(MEC)\, aH_{cloud}}{3k_{ent}\, L_{travel}}\right]^{1/3}$$

Citing experimental data on secondary dust explosion tests conducted with cornstarch [Kumar et al., 1992], Epstein gives a numerical example which corroborates the reasonable, if approximate, nature of this analysis. Given a sound speed of 340 m/s and a MEC for cornstarch of 40 g/m³, he calculated a minimum flame speed of 7.2 m/s to propagate a secondary dust explosion, a result in rough agreement with the test results of Kumar et al.

9.3.2 DEFLAGRATION TO DETONATION TRANSITION

There are essentially two ways to create a dust detonation wave: (1) cause a deflagration to detonation transition by flame acceleration or (2) directly initiate a detonation. In this section, we consider the dust DDT phenomenon. Throughout this book, we have seen that combustion of purely gaseous systems is generally easier to investigate that combustion of dust mixtures. Even in flammable gas systems, it is difficult to predict if and when the DDT phenomenon will occur (Lee, 2008, pp. 15−17). In this section, we will review some studies that demonstrate that DDT can occur in combustible dust systems. Proust has reviewed the literature on dust explosions in pipes including DDT and detonations (Proust, 1996). Proust indicates that the length to diameter ratio, *L/D*, is an important factor in determining the combustion wave behavior in flame acceleration and DDT.

Gardner et al. demonstrated DDT in coal−air mixtures (Gardner et al., 1986). They performed flame propagation experiments in a 0.6-m-diameter circular duct with a length of 42 m (making the *L/D* = 70). The initiating event was a coal dust deflagration in a 20-m³ chamber attached to the horizontal duct. They achieved DDT with two samples of a UK bituminous coal with a volatiles content of 33.5%, one with 65% fines ($d_p < 71 \mu$m) and the other with 87% fines. The coal was blown into the test section and ignited. The average concentration of coal dust was estimated at 330 g/m³. The two successful test results were maximum flame speeds of 1400 and 2000 m/s with maximum explosion pressures of 24.8 and 33.3 bars, respectively.

Li and his colleagues used a flame acceleration tube measuring 70 m in length and a circular cross-section with a diameter of 30 cm (*L/D* = 233) (Li et al., 1993, 1995; Sichel et al., 1995). They deposited combustible dust on the floor of the tube and used a hydrogen−oxygen deflagration as the ignition source. They used a nominal dust concentration of 500 g/m³ and tested three combustible dusts: corn, cornstarch, and Mira Gel (a processed starch product). They achieved DDT with dust concentrations in the range of 250−500 g/m³, with detonation pressures and velocities that compared favorably with Chapman−Jouguet detonation calculations.

Zhang et al. have reviewed DDT and detonation waves in combustible dust systems (Zhang et al., 2001). A key finding of their review is that most combustible dusts are relatively difficult to detonate compared to flammable gases. Therefore, combustible dusts require more initial conditions more conducive to DDT. They offer a few rules of thumb on conditions that seem to favor DDT in pipes or conduits. They suggest that the pipe diameter must be between 0.1 and 1.0 m, and the length to diameter ratio needs to be on the order of 100 or more. For example, they demonstrated a DDT with cornstarch having a mean particle diameter of 10 μm and a dust concentration of 1000 g/m³ ($\Phi = 1.72$). Fig. 9.21 shows the wave diagram for the DDT.

The detonation tube used for this test had a 0.3 m diameter and a 37 m length giving it an *L/D* = 123. The initial pressure for the cornstarch test was atmospheric. When they attempted to achieve DDT with cornstarch in a 0.1-m tube, they found that they had to increase the initial pressure to 2.5 bars. Also in this paper, the authors provided data on two other dusts that underwent successful DDT: anthraquinone and aluminum. The anthraquinone achieved DDT in a 0.1-m-diameter detonation tube, but the aluminum dust required the 0.3 m tube.

9.3.3 THE CHAPMAN−JOUGET DETONATION

The Chapman−Jouguet (CJ) detonation state is important because it is, in certain circumstances, the maximum wave velocity and overpressure that can be achieved in a fuel−air mixture. The constraints on this statement are steady flow of a

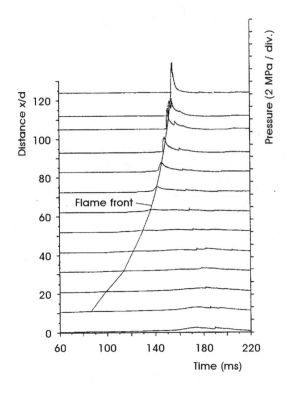

FIGURE 9.21

Wave diagram for DDT of cornstarch in a 0.3-m-diameter detonation tube and length of 37 m.

Reproduced with permission from Zhang, F., Grönig, H., van de Ven, A., 2001. DDT and detonation waves in dust–air mixtures. Shock Waves 11, 53–71.

reacting mixture in chemical equilibrium. However, just like a deflagration, a detonation can be overdriven, it can become unstable with a change in boundary conditions, or it can decelerate to a high-speed deflagration due to nonequilibrium processes. These examples of real world behavior can be exceedingly difficult to predict, whereas the CJ detonation state is a fairly straightforward (though sometimes highly computational) calculation. The true value of the CJ detonation is that it is a reference case to be used for comparison to real flows. As such it represents an *estimate* of the worst possible case for human injury or structural damage.

In flammable gas mixtures, the direct initiation of a detonation requires an order of magnitude more ignition energy than the initiation of a deflagration. The situation is similarly true for dust detonations, although it may be even more difficult. In this section, we will discuss the direct initiation of dust detonations and compare some typical results with CJ detonation calculations as previously presented in Chapter 4. The structure of detonation waves is considerably more complex than the situation as portrayed by the CJ detonation.

Strehlow, Nettleton, and Lee have written comprehensive discussions about the three-dimensional nature of the detonation wave structure, and the interested reader is encouraged to consult these references as well as others listed in the References section (Strehlow, 1984, Chapters 9 and 13; Nettleton, 1987; Lee, 2008; Zhang, 2009).

We will simply observe here without proof that there is a characteristic length scale called the "cell size" which is associated with the detonation structure. The cell size is a function of the chemical reactions driving the detonation. Detonable mixtures of combustible dust have larger cell sizes than detonable mixtures of flammable gases (Zhang et al., 1992, 2001; Zhang, 2009). The larger cell size implies that combustible dusts need more coaxing to enter into a detonation state. This is generally thought to be because of slower rates of kinetics and transport processes. Suitable inducements to detonation are larger diameter detonation tubes with longer lengths, higher concentrations of oxygen or another oxidizer, the addition of a flammable gas to create a hybrid mixture, or larger initiation energies.

While there is a large assortment of combustible dusts, dust detonations have been demonstrated with only a select few: cornstarch, anthraquinone, coal, grain, and aluminum (Organic solids: Cybulski, 1975, pp. 100−101; Wolanski, 1991; Zhang and Grönig, 1993; Li et al., 1995; Aluminum: Strauss, 1968; Tulis and Selman, 1982; Zhang et al., 2009). These successful detonation experiments were usually conduced with fuel-rich dust concentrations. There is no question that combustible dusts can undergo detonation under carefully controlled laboratory conditions. There is also evidence of the rare occurrence of a coal dust detonation in an experimental coal mine passageway. But to reiterate the dust detonation rule of thumb, the conduit diameter must be between 0.1 and 1.0 m, and the length to diameter ratio needs to be on the order of 100 or more.

9.3.4 SUMMARY

This section has examined dust flame acceleration, the deflagration to detonation transition, and the dust detonation. The common theme through these different phenomena is the appearance and strengthening of a shock wave precursor followed by a flame. Two measures of the hazard of the accelerating flame are the flame speed and the overpressure. The accelerating flame is a blast wave. The shock wave, the flame, and the following flow are all part of the destructive potential of the blast wave. Compared to hydrostatic loads, blast waves lead to very different and very hazardous loads on people and structures.

9.4 PRESSURE PILING

Pressure piling occurs when a deflagration travels from one process vessel to another via a connecting pipe or duct. The scenario begins with a deflagration in the first process vessel. Even if the first vessel is vented, flame can enter the connecting

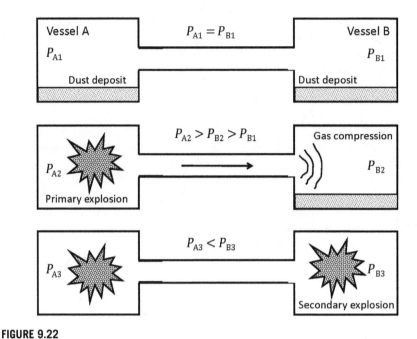

FIGURE 9.22

Pressure piling scenario.

conduit and travel toward the second process vessel. We have already seen that the propagation of a flame through a conduit leads to the formation of a shock wave ahead of the flame. The acceleration of the flame causes a precompression of the gas in the connected process vessel. Assume that there is an ignitable dust concentration within the second vessel and a dust deposit along the length of the conduit. When the flame enters the second vessel and ignites the dust within, the deflagration pressure within the second vessel will not be the explosion pressure associated with that dust concentration. The final explosion pressure will be magnified by the precompression effect. The physical situation is depicted in Fig. 9.22.

Earlier in this chapter, we considered a simple thermodynamic model of the pressure piling scenario assuming flame propagation in a one-dimensional channel. The situation is different when the one-dimensional channel is connected to two vessels as shown in Fig. 9.19. It seems reasonable to expect that the ratio of the two volumes will be important in determining the magnitude of the pressure piling effect. We must also consider the extent to which venting from Vessel A will influence the pressure developed in Vessel B.

The initial condition consists of two vessels connected by a single pipe. The total volume of Vessel A is assumed to be its volume plus the volume of the connecting pipe. The vessels and pipe are at atmospheric pressure (P_{A1}, P_{B2}) and ambient temperature (T_{A1}, T_{B2}), and contain a quantity of combustible dust sufficient to yield an ignitable concentration if uniformly dispersed in the enclosed system. We assume that a constant volume deflagration occurs in Vessel A characterized by $P_{A2} = P_{ex}$ and $T_{A2} = T_{ex}$. A portion of the combustion products then transfer into the Vessel B raising both the pressure and temperature in Vessel B. These elevated values are P_{B2} and T_{B2}. Next, a deflagration occurs in Vessel B resulting in an explosion pressure P_{B3} and temperature T_{B3} greater than the standard constant volume explosion parameters due to the elevated pressure and temperature conditions that existed in Vessel B prior to its ignition.

To make this model tractable, we will assume that the thermodynamic properties for a constant volume explosion are known, that the quantity of dust is specified, and that the reduction in pressure in Vessel A due to transferring material into Vessel B is specified. The unburnt and burnt dusty gas mixtures are assumed to obey the ideal gas equation of state. The vessel walls are assumed to be rigid, impermeable, and adiabatic. To further simplify the analysis, we will neglect the changes in the initial temperature and consider only the effect of pressure. Invoking Eq. (3.50),

$$P_{B_3} = P_{B_2}\left(\frac{T_{ex}}{T_0}\right) \tag{9.62}$$

In this case the initial pressure P_{B_2} is the pressure of the precompressed unburnt mixture region. So in the laboratory test vessel, if the flame temperature ratio is 10, the explosion pressure will be $P_{ex} = 10P_{B_2}$. Thus, if the precompressed atmosphere in Vessel B is at a pressure of 20% above atmospheric, then the explosion pressure is $P_{B3} = 10\,P_{B2} = 10(1.2)\,P_0 = 12$ atm. In this greatly simplified example, the final explosion pressure is a linear function of the initial pressure. A more general thermodynamic model for pressure piling will be nonlinear and presumably, less tractable.

Pressure piling is a more complicated phenomenon than a secondary explosion or flame acceleration in a one-dimensional conduit. In its most general form, the physical situation may include a host of geometric complications: a process vessel with a complex geometry and filled with flow obstacles; the connecting conduit may have a variable cross-section, may incorporate connections that change direction, or split the flow; the receiving vessel may also be of complex shape and internal features. Flame propagation, turbulence, and gas dynamic may all play an important role in determining the final pressures attained in the vessels and conduit. The flame that enters the secondary vessel behaves like a jet flame ignition source, which is a very large ignition stimulus compared to the minimum ignition energy for typical combustible dusts. We will consider next a short survey of representative investigations on this complex phenomenon.

Singh surveyed some of the earlier studies on pressure piling as well as describing his own experimental studies (Singh, 1994). He noted the key geometric parameters that influence pressure pilings: the volume ratio of the vessels, the diameter of the connecting conduit, and the volume of the primary vessel. For example, he observed that in a test with a fixed volume ratio of 14, doubling the volume of the primary vessel more than doubled the peak pressure in the secondary vessel. He also indicated the importance of the ignition location; moving from

Table 9.4 Deflagration Parameters for Dusts Tested (van Wingerden et al., 1995)

Dust	P_{max}(bar)	K_{st}(bar · m/s)
Maize starch	7.4	145
Peat	7.6	118
Silicon	7.1	140
Polypropylene	6.5	130
Wheat	5.2	55

central ignition to edge ignition increased the final pressure in the primary vessel by a factor of 1.8. Most of Singh's work was conducted with methane–air mixtures in laboratory-scale vessels.

Singh offered an insightful description of the pressure piling process. Assume that both the primary and secondary vessels are filled with a flammable gas mixture at an initial temperature and pressure. With central ignition in the primary vessel, flame spreads radially outwards causing the pressure to rise. The pressure difference between the primary and secondary vessels causes a flow of unburnt mixture through the conduit and into the secondary vessel. This flow continues until the flame travels through the conduit and into the secondary vessel. But at the moment of ignition in the secondary vessel, the secondary vessel is no longer at the initial pressure. It is slightly elevated due to the inflow and compression of unburnt mixture from the primary vessel. This elevated initial pressure results in a magnified final explosion pressure in the secondary vessel. The actual final pressure in the secondary vessel will be somewhat less than predicted by thermodynamics because of reverse from the secondary vessel through the conduit and toward the primary vessel. Singh closes his discussion with several empirical calculation methods.

A pressure piling study with full-scale equipment was performed on a dust filter–conduit–process vessel system using five different combustible dusts: maize starch, peat, silicon, polypropylene, and wheat (van Wingerden et al., 1995). The deflagration properties of the powders tested are summarized in Table 9.4.

The powders were introduced into the flow system by adjusting a metered feed rate using a blower as the prime mover. They tested a range of nominal dust concentrations from 30 to 2000 g/m³. The airflow was stopped and the feeder valve blocked off just before ignition. The volume of the first vessel (the dust filter) was 5.8 m³, the volume of the second vessel was 2 m³, and the connecting conduit was 0.15 m in diameter and 22 m in length (volume equal to 1.6 m³). Pressure was monitored in both vessels and the conduit. They tested the response of the interconnected system to explosion venting in the first vessel. Vent sizes were changed in some tests. The propensity for pressure piling varied with the dust reactivity (the K_{st} value) and dust concentration. Fig. 9.23 shows two typical results, one test with insufficient and one with sufficient venting.

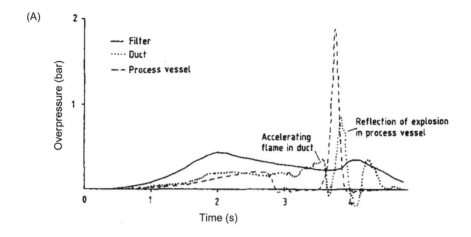

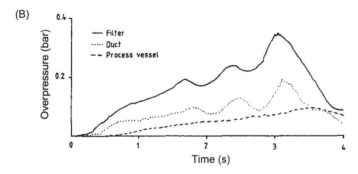

FIGURE 9.23

Pressure records at three locations in the integrated system: (A) maize starch at 1000 g/m^3, 0.04 m^2 vent on dust filter, 0.25 m^2 vent on process vessel and (b) wheat dust at 650 g/m^3, 0.08 m^2 vent on dust filter, 0.25 m^2 process vessel. Note the difference in scale for the pressure coordinate.

Reproduced with permission from van Wingerden, K., Pedersen, G.H., Eckhoff, R.K., 1995. Violence of dust explosions in integrated systems. Process Saf. Prog. 14, 131–138.

The propagation of pressure waves is apparent in both tests. The magnitude of the pressure spike due to pressure piling in the process vessel in test (A) is quite striking and is indicative of inadequate venting. The result in test (B) indicates good venting.

In general, the investigators indicated that at low to intermediate dust concentrations they observed the deflagration accelerating through the conduit leading to high peak pressures in the conduit and the process vessel. At high dust concentrations, they saw slower accelerations suggesting flame extinction within the

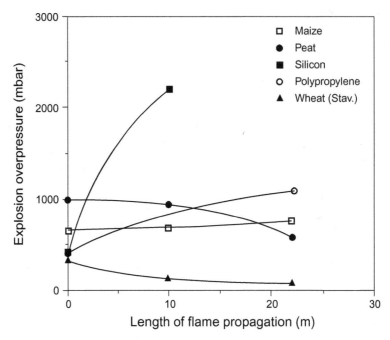

FIGURE 9.24

Overpressure versus flame length in the conduit between dust filter and process vessel. Results for all five dusts are shown, but the test with silicon was conducted without the process vessel.

Reproduced with permission from van Wingerden, K., Pedersen, G.H., Eckhoff, R.K., 1995. Violence of dust explosions in integrated systems. Process Saf. Prog. 14, 131–138.

conduit. A surprising outcome of the work by van Wingerden and his colleagues was the diversity of deflagration behavior given the similarity of the deflagration parameters for the dusts tested. Fig. 9.24 summarizes the pressure development in this system when the dust filter had a 0.4-m^2 vent.

The overpressure and flame length in the conduit are shown for the optimum concentration tested. The wheat dust did not propagate flame in any of its tests. Wheat, maize, and peat all behaved in a similar fashion except there was a higher than expected level of variance in the peak pressures measured in the dust filter. Silicon was especially surprising since its deflagration parameters were within the same order of magnitude as the other dusts. The authors concluded that the K_{st} parameter does not capture all aspects of the reactivity of a combustible dust.

In addition to testing the effect of vent size on pressure reduction in the dust filter and the process vessel, they also investigated the effect of small vents on the conduit. They found that the conduit vents did reduce the pressure inside the conduit, however, the conduit vents did not prevent flame propagation through the conduit. Similar observations have been made by Tamanini and Ural (1992): pressure relief in a corridor/conduit does not prevent flame propagation, it just slows it down.

Lunn and his colleagues at the UK Health and Safety Executive published two studies on pressure piling in nonvented and vented systems (Lunn et al., 1996; Holbrow et al., 1996). In a third paper, they presented a series of guidelines and a decision flow chart to assist in the design of venting for connected process vessels (Holbrow et al., 1999). In the first paper, the investigators tested two combustible dusts, coal and toner (Lunn et al., 1996). The basic test setup was a primary vessel linked to a secondary vessel by a conduit. The vessel volumes ranged from 2 to 20 m^3 connected by a conduit with a length of 5 m. The conduit diameter was varied using 0.15, 0.25, and 0.5 m diameters.

They established that the vessel volumes, the vessel volume ratio, and the diameter of the conduit were all important variables affecting the magnitude of pressure piling. In their tests, the dust was injected into both vessels and additional dust was deposited along the length of the pipe. They concluded that pressure piling did not always occur; narrower pipe diameters tended to prevent pressure piling. Specifically, in their tests, pressure piling did not occur with 0.15-m-diameter pipe. Recall that the work by van Wingerden et al. successfully propagated pressure piling with 0.15-m pipe. There are a number of potential reasons for the conflicting results, but the most important point to draw from this conflict is that pressure piling is a complex phenomenon.

The investigators found that the magnitude of overpressure generated by pressure piling depended on the volume ratio of vessels, the absolute volume of the vessels, and the diameter of the connecting duct. The vessel volume ratio, V_2/V_1, was defined as the ratio of the volume of the second vessel, V_2, divided by the sum of the first vessel volume plus the volume of the connecting duct, V_1. They derived a thermodynamic model to correlate their test results, but the model equation contains errors and is not dimensionally correct.

Lunn et al. demonstrated that the pressure piling effect was more pronounced when the volume of the primary vessel was greater than the secondary vessel. If the pressure rise is greater in the second vessel compared to the first, then the greater pressure in the second vessel will drive combustion gases back toward the primary vessel. Fig. 9.25 and Fig. 9.26 illustrate this point in tests conducted with coal dust ($P_{max} = 7.7$ barg). Fig. 9.25 is the test result when the volume ratio is larger than unity: $V_2/V_1 = 4.01$.

Fig. 9.26 is the test result when the volume ratio is smaller than unity: $V_2/V_1 = 0.191$.

There is an analogy between this behavior and the reflection of acoustic waves in a closed tube, but the difference is that in pressure piling this back-and-forth action is bulk flow of combustion gases. Just to add more complication, in some

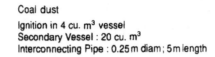

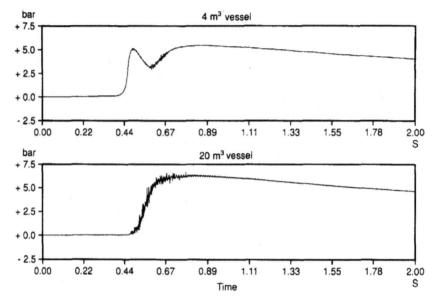

FIGURE 9.25

Coal dust explosion for large volume ratio.

Reproduced with permission from Lunn, G.A., Holbrow, P., Andrews, S., Gummer, J., 1996. Dust explosions in totally enclosed interconnected vessel systems. J. Loss Prev. Process Ind. 9, 45–58.

cases of pressure piling, acoustic waves can be superimposed on the bulk flow, further increasing the overpressure. They also found that the larger the diameter of the connecting duct, the greater the effect of back-venting and the greater the final explosion pressure. On the other hand, they found that at certain smaller diameters, pressure piling did not occur.

Based on the results from vented vessels, the investigators found that venting each vessel using conventional design methods may not be sufficient unless the maximum pressure upon venting was kept sufficiently low; their recommendation was that if should be less than 0.2 barg (Holbrow et al., 1996).

As a final note, the investigations by van Wingerden et al. and by Lunn et al. all involved having combustible dust suspended throughout the length of the connecting conduit. Valiulis et al. performed investigations on propagation of deflagrations from process equipment through empty conduits (Valiulis et al., 1999). They found that the propagation of flame in the conduit was limited due to the absence of dust deposits.

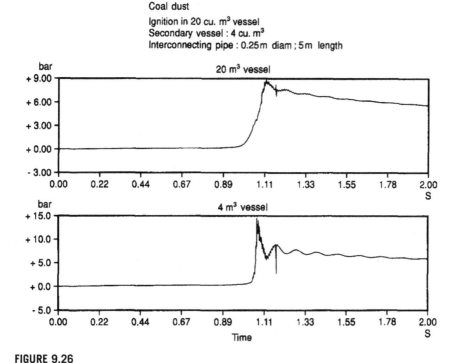

Coal dust
Ignition in 20 cu. m³ vessel
Secondary vessel : 4 cu. m³
Interconnecting pipe : 0.25m diam ; 5m length

FIGURE 9.26

Coal dust explosion for small volume ratio.

Reproduced with permission from Lunn, G.A., Holbrow, P., Andrews, S., Gummer, J., 1996. Dust explosions in totally enclosed interconnected vessel systems. J. Loss Prev. Process Ind. 9, 45–58.

9.5 PREVENTING OR CONTROLLING SECONDARY DUST EXPLOSIONS

Combustible dusts command respect; they can present significant fire and explosion hazards. To prevent serious injury or property damage, these hazards must be identified, evaluated, and controlled. The potential for flame acceleration magnifies these hazards. Flame acceleration can occur in open space, in confined one-dimensional channels, and between connected process equipment. Several of the more general books on combustible dust hazards provide excellent discussions on the prevention and control of flame acceleration effects and I encourage the interested reader to consult them (Amyotte, 2013; Bartknecht, 1989; Barton, 2002; CCPS, 2005; Eckhoff, 2003; Nagy and Verakis, 1983; Frank and Rodgers, 2012). Dust explosion protection is a very active area of research. Individuals charged with managing combustible dust hazards at their workplace should study the most current combustible dust safety standards for guidance on good practices for the prevention of secondary explosions and other flame acceleration effects.

Probably the most common flame acceleration hazard is the secondary dust explosion. In this discussion, we will focus on the primary safeguard for preventing secondary dust explosions: the control of fugitive dust. For all practical purposes, the fugitive dust control strategy protects people from injury as well as the building and its contents from fire or blast damage. The safeguards for preventing pressure piling—venting, suppression, and isolation—were discussed in Section 8.9.

The processing and handling of combustible particulate materials can generate combustible dust. In addition to the safety hazards presented by combustible dusts, they can also present health hazards due to respiratory exposure. For example, many nontoxic dusts are respiratory irritants. For these reasons, there is a strong incentive to prevent the occurrence of fugitive dust in the workplace.[1] Furthermore, there are well-established technologies for controlling dust in process plants and the workplace. So it may come as a surprise that there are still many instances of secondary dust explosions wreaking havoc in the workplace. The solution to this problem seems simple: prevent accumulations of fugitive dust or clean them up when they occur. But in practice, the problem is much more complex.

If the presence of combustible dust cannot be eliminated, then one can implement engineering controls to eliminate or reduce dust leaks, and for the dust recirculated within the process, one can install dust control systems (Cooper and Alley, 2002, Chapters 3–8). But no matter how much capital and effort is invested into a process plant, no technology is perfect. Dust will be generated during processing and handling of particulate materials, and some of it will escape. This is fugitive dust, and it can be managed with administrative controls such as a housekeeping program.

A housekeeping program for fugitive combustible dust is a policy with a set of procedures intended to keep the quantity of fugitive dust within certain specified guidelines. Conceptually, the workplace can be divided into discrete areas based on function or building construction. The housekeeping requirements, equipment, and procedures can then be established for each area and responsibilities for implementing the program can be assigned. Some level of documentation verifying the implementation of the program may be appropriate, and, as with any administrative control, routine monitoring and auditing of the program should occur.

Generally speaking, there are two kinds of housekeeping activities: routine cleaning and spill response. The routine cleaning should be based on some type of measurable (observable) criterion. For example, the NFPA combustible dust standards offer specific accumulated dust layer thicknesses or a quantity of dust per designated area (NFPA 654, 2013; Rodgers and Ural, 2011; Rodgers, 2012). Likewise, threshold quantities should be specified for different levels of spill

---

[1] I define fugitive dust to be dust which has escaped the process equipment and entered into the ambient environment.

response activity. The NFPA standards tend to be hazard based. Risk-based approaches are also an option (Ramirez et al., 2015).

The development of action levels for housekeeping, whether prescriptive or risk-based, should be pragmatic. Anecdotal evidence suggests that thin dust layers are not the cause of catastrophic secondary dust explosions; the problem is thick dust layers that measure centimeters in depth (Cholin, 2013). Such accumulations tend to indicate the absence of any housekeeping program.

9.6 SUMMARY

This chapter examined flame acceleration effects and their role in magnifying the hazards of dust deflagrations. Flame acceleration effects were divided into three categories: secondary dust explosions, confined dust flame acceleration, and pressure piling. Dust cloud suspension processes were considered from both theoretical and experimental perspectives. Combustible dusts can be expected to exhibit a range of entrainment behavior: fine dusts become lofted as clumps of material and coarse dusts as a dispersion of individual particles. The next section considered mechanisms for confined dust flame acceleration and the transition from deflagration to detonation. The relevance of the Chapman–Jouguet detonation model as a limiting condition was also explained and compared with experimental data. The phenomenon of pressure piling in interconnected vessels was presented. Finally, the last section illustrated the use of fuel control to prevent secondary dust explosions.

REFERENCES

Amyotte, P., 2013. An Introduction to Dust Explosions. Butterworth Heinemann, Elsevier, New York.

Anderson, J.D., 1982. Modern Compressible Flow with Historical Perspective. McGraw Hill, New York.

Artingstall, G., 1961. On the relation between flame and blast in coal-dust explosions. Ministry of Fuel and Power Safety in Mines Research Establishment, Research Report No. 204.

Bagnold, R.A., 1954 reprint. The Physics of Blown Sand and Desert Dunes, 2005 edition. Dover Publications, Mineola, NY.

Bartknecht, W., 1989. Dust Explosions: Course, Prevention, Protection. Springer-Verlag, Berlin.

Barton, K., 2002. Dust Explosion Prevention and Protection. Gulf Professional Publishing, Elsevier, New York.

Ben-Dor, G., 1995. Dust entrainment by means of a planar shock induced vortex over loose dust layers. Shock Waves 4, 285–288.

Bjerketvedt, D., Bakke, J.R., van Wingerden, K., 1997. Gas explosion handbook. J. Hazard. Mater. 52, 1–150.

Boor, B.E., Siegel, J.A., Novoselac, A., 2013a. Monolayer and multilayer particle deposits on hard surfaces: literature review and implications for particle resuspension in the indoor environment. Aeros. Sci. Technol. 47, 831−847.

Boor, B.E., Siegel, J.A., Novoselac, A., 2013b. Wind tunnel study on aerodynamic particle resuspension from monolayer and multilayer deposits on linoleum flooring and galvanized sheet metal. Aeros. Sci. Technol. 47, 848−857.

CCPS/AIChE, 2005. Guidelines for Safe Handling of Powders and Bulk Solids. American Institute of Chemical Engineers, New York, 2005.

Chi, D.N.H., Perlee, H.E., 1974. Mathematical study of a propagating flame and its induced aerodynamics in a coal mine passageway. U.S. Department of Commerce, Bureau of Mines, Report of Investigations 7908.

Cholin, J.M., 2013. How much dust is too much dust? Fire Protect. Eng. Fourth Quarter 26−38.

Churchill, S.W., 1980. The Practical Use of Theory in Fluid Flow. Book 1: Inertial Flows. Etaner Press, Thornton, PA.

Ciccarelli, G., Dorofeev, S., 2008. Flame acceleration and transition to detonation in ducts. Prog. Energy Combust. Sci. 34, 499−550.

Cooper, C.D., Alley, F.C., 2002. Air Pollution Control: a Design Approach, third ed. Waveland Press, Long Grove, IL.

Corn, M., 1961a. The adhesion of solid particles to solid surfaces, I. A review. J. Air Pollution Control Assoc. 11, 523−528.

Corn, M., 1961b. The adhesion of solid particles to solid surfaces, II. J. Air Pollution Control Assoc. 11, 566−584.

Corn, M., Stein, F., 1965. Re-entrainment of particles from a plane surface. Am. Ind. Hyg. Assoc. J. 26, 325−336.

Courant, R., Friedrichs, K.O., 1948 reprint. Supersonic Flow and Shock Waves, 1976 edition. Springer Verlag, Berlin.

Crowl, D.A., 2003. Understanding Explosions. Center for Chemical Process Safety. American Institute of Chemical Engineers, New York.

Cybulski, W., 1975. No. 54001 .Coal Dust Explosions and Their Suppression, Vol. 73. Bureau of Mines, US Department of the Interior and the National Science Foundation

Dawes, J.G., 1950. Notes on the physics of dust dispersion. Ministry of Fuel and Power, Safety in Mines Research Establishment, Research Report No. 3.

Dawes, J.G., 1952a. Dispersion of Dust Deposits by Blasts of Air: Part I. Ministry of Fuel and Power, Safety in Mines Research Establishment, Research Report No. 36.

Dawes, J.G., 1952b. Dispersion of Dust Deposits by Blasts of Air: Part II. Ministry of Fuel and Power, Safety in Mines Research Establishment, Research Report No. 49.

Dorofeev, S., 2011. Flame acceleration and explosion safety applications. Proc. Combust. Inst. 33, 2161−2175.

Eckhoff, R., 2009b. Understanding dust explosions—the role of powder science and technology. J. Loss Prev. Process Ind. 22, 105−116.

Eckhoff, R., 2003. Dust Explosions in the Process Industries, third ed. Gulf Professional Publishing, Elsevier, New York.

Epstein, M., 2007. Primary dust explosion strength required to initiate secondary dust explosion. FAI Process Safety News. Fauske and Associates, Inc., Winter.

Fairchild, C.I., Tillery, M.I., 1982. Wind tunnel measurements of the resuspension of ideal particles. Atmos. Environ. 16, 229−238.

Federov, A.V., 2004. Mixing in wave processes propagating in gas mixtures (review). Combust. Explos. Shock Waves 40, 17−31.

Fletcher, B., 1976a. The interaction of a shock with a dust deposit. J. Phys. D Appl. Phys. 9, 197−202.

Fletcher, B., 1976b. The erosion of dust by an airflow. J. Phys. D Appl. Phys. 9, 913−924.

Fletcher, B., 1976c. The incipient motion of granular materials. J. Phys. D Appl. Phys. 9, 2471−2478.

Frank, W.L., Rodgers, S.A., 2012. NFPA Guide to Combustible Dusts. National Fire Protection Association, Quincy, MA.

Gardner, B.R., Winter, R.J., Moore, M.J., 1986. Explosion development and deflagration-to-detonation transition in coal dust/air suspensions. Twenty-first Symposium (International) on Combustion, The Combustion Institute. Elsevier, New York, pp. 335−343.

Gerrard, J.H., 1963. An experimental investigation of the initial stages of the dispersion of dust by shock waves. Br. J. Appl. Phys. 14, 186−192.

Gosteev, Y.A., Fedorov, A.V., 2002. Calculation of dust lifting by a transient shock wave. Combust. Explos. Shock Waves 38, 322−326.

Hinds, W.C., 1999. Aerosol Technology: Properties, Behavior, and Measurement of Airborne Particles, second ed. Wiley-Interscience, New York.

Holbrow, P., Andrews, S., Lunn, G.A., 1996. Dust explosions in interconnected vented vessels. J. Loss Prev. Process Ind. 9, 91−103.

Holbrow, P., Lunn, G.A., Tyldelsey, A., 1999. Dust explosion protection in linked vessels: guidance for containment and venting. J. Loss Prev. Process Ind. 12, 227−234.

Kalman, H., Satran, A., Meir, D., Rabinovich, E., 2005. Pickup (critical) velocity of particles. Powder Technol. 160, 103−113.

Kauffman, C.W., Srinath, S.R., Tezok, F.I., Nicholls, J.A., Sichel, M., 1984. Turbulent and accelerating dust flames. Twentieth Symposium (International) on Combustion. The Combustion Institute, Elsevier, New York.

Kerampran, S., Desbordes, D., Veyssiere, B., 2000. Study of the mechanisms of flame acceleration in a tube of constant cross section. Combust. Sci. Technol. 158, 71−91.

Kinney, G.F., Graham, K.J., 1985. Explosive Shocks in Air, second ed. Springer, Berlin.

Kundu, P.K., Cohen, I.M., Dowling, D.R., 2016. Fluid Mechanics, sixth edition. Academic Press, New York.

Landau, L.D., Lifshitz, E.M., 1959. Fluid Mechanics. Pergamon Press, Oxford.

Lee, J.H.S., 2008. The Detonation Phenomenon. Cambridge University Press, Cambridge.

Li, Y.-C., Alexander, C.G., Wolanski, P., Kaufman, C.W., Sichel, M., 1993. Experimental investigations of accelerating flames and transition to detonation in layered grain dust. Prog. Astronaut. Aeronaut. 154, 170−184.

Li, Y.-C., Harbaugh, A.S., Alexander, C.G., Kauffman, C.W., Sichel, M., 1995. Deflagration to detonation transition fuelled by dust layers. Shock Waves 5, 249−258.

Lighthill, J., 1987. Waves in Fluids. Cambridge University Press, Cambridge.

Lu, H., Shao, Y., 1999. A new model for dust emission by saltation bombardment. J. Geophys. Res. 104 (16), 827−16,842.

Lunn, G.A., Holbrow, P., Andrews, S., Gummer, J., 1996. Dust explosions in totally enclosed interconnected vessel systems. J. Loss Prev. Process Ind. 9, 45−58.

Lyles, L., 1988. Basic wind erosion processes. Agric. Ecosyst. Environ. 22/23, 91−101.

Mirels, H., 1984. Blowing model for turbulent boundary-layer dust ingestion. AIAA J. 22, 1582−1589.

Nagy, J., 1981. The Explosion Hazard in Mining. U.S. Department of Labor, Mine Safety and Health Administration 1119, Informational Report.

Nagy, J., Verakis, H.C., 1983. Development and Control of Dust Explosions. CRC Press, Boca Raton, FL.

Nettleton, M.A., 1987. Gaseous Detonations: Their Nature, Effects and Control. Chapman and Hall, London.

NFPA 654, 2013. Standard for the Prevention of Fire and Dust Explosions from the Manufacturing, Processing, and Handling of Combustible Particulate Solids. National Fire Protection Association, Quincy, MA.

NFPA 921, 2014. Guide for Fire and Explosion Investigations. National Fire Protection Association, Quincy, MA.

O'Neill, M.E., Chorlton, F., 1989. Viscous and Compressible Fluid Dynamics. Ellis Horwood, Chichester.

Owen, P.R., 1964. Saltation of uniform grains in air. J. Fluid Mech. 20, 225−242.

Proust, C., 1996. Dust explosions in pipes: a review. J. Loss Prev. Process Ind. 9, 267−277.

Proust, C., 2015. Gas flame acceleration in long ducts. J. Loss Prev. Process Ind. 36, 387−393.

Pu, Y.K., Mazurkiewicz, J., Jarosinski, J., Kauffman, C.W., 1988. Comparative study of the influence of obstacles on the propagation of dust and gas flames. Twenty-Second Symposium (International) on Combustion. The Combustion Institute, Elsevier.

Rabinovich, E., Kalman, H., 2009. Pickup velocity from particle deposits. Powder Technol. 194, 51−57.

Rae, D., 1973. Initiation of weak coal-dust explosions in long galleries and the importance of the time dependence of the explosion pressure. Fourteenth Symposium (International) on Combustion. The Combustion Institute, Elsevier, New York.

Ramirez, J.R., Ogle, R.A., Carpenter, A.R., 2015. Risk evaluation and the NFPA 654 layer depth criteria for dust explosion and flash fire hazards. In: Proceedings of the ASME 2015 International Mechanical Engineering Congress & Exposition, IMECE2015, Houston, TX, November 13−19.

Rhodes, M., 2008. Introduction to Particle Technology, second ed. John Wiley & Sons, New York.

Richmond, J.K., Liebman, I., 1975. A physical description of coal mine explosions. Fifteenth Symposium (International) on Combustion. The Combustion Institute, Elsevier, New York.

Richmond, J.K., Liebman, I., Bruszak, A.E., Miller, L.F., 1979. A physical description of coal mine explosions. Part 2. Seventeenth Symposium (International) on Combustion. The Combustion Institute, Elsevier, New York.

Rodgers, S.A., 2012. Application of the NFPA 654 dust layer thickness criteria— recognizing the hazard. Process Saf. Prog. 31, 24−35.

Rodgers, S.A., Ural, E.A., 2011. Practical issues with marginally explosible dusts— evaluating the real hazard. Process Saf. Prog. 30, 266−279.

Schulze, D., 2008. Powders and Bulk Solids: Behavior, Characterization, Storage and Flow. Springer, Berlin.

Seville, J.P.K., Tüzün, U., Clift, R., 1997. Processing of Particulate Solids. Blackie Academic and Professional, London.

Shao, Y., 2008. Physics and Modelling of Wind Erosion, 2nd revised and expanded edition. Springer, Berlin.

Shao, Y., Lu, H., 2000. A simple expression for wind erosion threshold friction velocity. J. Geophys. Res. 105 (22), 437−22,443.

Shao, Y., Raupach, M.R., Findlater, P.A., 1993. Effect of saltation bombardment on the entrainment of dust by wind. J. Geophys. Res. 98 (12), 719−12,726.

Sichel, M., Kauffman, C.W., Li, Y.-C., 1995. Transition from deflagration to detonation in layered dust explosions. Process Saf. Prog. 14, 257−265.

Singer, J.M., Greninger, N.B., Grumer, J., 1969. Some Aspects of the Aerodynamics of Formation of Float Coal Dust Deposits. U.S. Department of Commerce, Bureau of Mines, Report of Investigations 7252.

Singer, J.M., Cook, E.B., Grumer, J., 1972. Dispersal of Coal and Rock Dust Deposits. U. S. Department of Commerce, Bureau of Mines, Report of Investigations 7642.

Singer, J.M., Harris, M.E., Grumer, J., 1976. Dust Dispersal by Explosion-Induced Airflow. U.S. Department of Commerce, Bureau of Mines, Report of Investigations 8130.

Singh, J., 1994. Gas explosions in inter-connected vessels: pressure piling. Process Saf. Environ. Protect. 72, 220−228.

Srinath, S.R., Kauffman, C.W., Nicholls, J.A., Sichel, M., 1987. In: Cashdollar, K.L., Hertzberg, M. (Eds.), Secondary dust explosions, Industrial Dust Explosions: ASTM STP 958. American Society of Testing and Materials, West Conshohocken, PA, pp. 90−106.

Strauss, W.A., 1968. Investigation of the detonation of aluminum powder−oxygen mixtures. AIAA J. 6, 1753−1756.

Strehlow, R.A., 1975. Blast waves generated by constant velocity flames: a simplified approach. Combust. Flame 24, 257−261.

Strehlow, R.A., 1981. Blast wave from deflagrative explosions: an acoustic approach. AIChE Loss Prev. 14, 145−152.

Strehlow, R.A., 1984. Combustion Fundamentals. McGraw-Hill, New York.

Strehlow, R.A., Luckritz, R.T., Adamczyk, A.A., Shimp, S.A., 1979. The blast wave generated by spherical flames. Combust. Flame 35, 297−310.

Tamanini, F., Ural, E.A., 1992. FMRC studies of parameters affecting the propagation of dust explosions. Powder Technol. 71, 135−151.

Taveau, J., 2012. Secondary dust explosions: how to prevent them or mitigate their effects. Process Saf. Prog. 31, 36−50.

Taylor, G.I., 1946. The air wave surrounding an expanding sphere. Proc. Royal Soc. Lond. A Math. Phys. Sci. 186, 273−292.

Thompson, P.A., 1972. Compressible-Fluid Dynamics. McGraw-Hill, New York.

Tulis, A.J., Selman, J.R., 1982. Detonation tube studies of aluminum particles dispersed in air. Nineteenth Symposium (International) on Combustion. The Combustion Institute, Elsevier, New York.

Ural, E., 2011a. Towards Estimating Entrainment Fraction for Dust Layers. The Fire Protection Research Foundation, Quincy MA.

Ural, E., 2011b. A new strawman methodology to predict combustible dust entrainment from layers. 7th Global Congress on Process Safety. American Institute of Chemical Engineers, Chicago, IL.

Ural, E., 2011c. Towards Estimating Entrainment Fraction for Dust Layers. Springer, Berlin.

Valiulis, J.V., Zalosh, R.G., Tamanini, F., 1999. Experiments on the propagation of vented dust explosions to connected equipment. Process Saf. Prog. 18, 99−106.

Welty, J.R., Wicks, C.E., Wilson, R.E., 1984. Fundamentals of Momentum, Heat, and Mass Transfer, third ed. John Wiley & Sons, New York.

Williams, F.A., 1976. Qualitative theory of nonideal explosions. Combust. Sci. Technol. 12, 199–206.

Williams, F.A., 1983. Estimation of pressure fields in combustion of vapour clouds. J. Fluid Mech. 127, 429–442.

van Wingerden, K., Pedersen, G.H., Eckhoff, R.K., 1995. Violence of dust explosions in integrated systems. Process Saf. Prog. 14, 131–138.

Wolanski, P., 1991. Deflagration and detonation combustion of dust mixtures. Dynamics of Deflagrations and Reactive Systems: Heterogeneous Combustion, Progress in Astronautics and Aeronautics 132, 3–31.

Zhang, F., 2009. Detonation of gas-particle flow. In: Zhang, F. (Ed.), Shock Wave Science and Technology Library: Heterogeneous Detonation. Springer-Verlag, Berlin, pp. 87–168.

Zhang, F., Grönig, H., 1993. Detonability of organic dust–air mixtures. Prog. Astronaut. Aeronaut. 154, 195–215.

Zhang, F., Greilich, P., Grönig, H., 1992. Propagation mechanism of dust detonations. Shock Waves 2, 81–88.

Zhang, F., Grönig, H., van de Ven, A., 2001. DDT and detonation waves in dust–air mixtures. Shock Waves. 11, 53–71.

Zhang, F., Gerrard, K., Riley, R.C., 2009. Reaction mechanism of aluminum-particle–air detonation. J. Propul. Power. 25, 845–858.

Ziskind, G., Fichman, M., Gutfinger, C., 1995. Resuspension of particulates form surfaces to turbulent flows—review and analysis. J. Aerosol Sci. 26, 613–644.

Zucrow, M.J., Hoffman, J.D., 1976. Gas Dynamics, Volume 1. John Wiley & Sons, New York.

Comprehensive dust explosion modeling

10

This chapter summarizes some of the more recent efforts to model dust explosion phenomena as multiphase, turbulent, chemically reacting flows. Our emphasis in this chapter will be on computational fluid dynamics (CFD). Dust flames are governed by physical and chemical phenomena that present difficult challenges to overcome. In Chapter 3 through Chapter 9, I have attempted to present simple, useful models that approximate some of the key features of dust combustion. These simple models can be used to develop insight into and predict the consequences of smoldering, flash fires, primary and secondary dust explosions, the deflagration to detonation transition, and pressure piling.

But models, by definition, are approximations of reality. The art of mathematical modeling is to develop models of physical system behavior with a sufficient degree of realism that the models will yield useful predictions without including unnecessary details. The simple models presented earlier in this text are useful, but their utility was obtained by decoupling the interplay between chemical kinetics, multiphase flow, turbulence, and thermal radiation. Decoupling these finite rate processes is what made the models tractable to analytical solutions. But the coupling of these finite rate transport processes with chemical kinetics is what makes dust explosions so fascinating. If we wish to explore some of this more complex terrain to better understand the coupling of different phenomena, we must resort to the methods of CFD.

There is a large body of work devoted to the application of CFD to multiphase, turbulent, chemically reacting flows. Much of this work is focused on the study of problems in energy conversion, propulsion, chemical reactor design, and environmental emissions control. The application of CFD to dust deflagrations for safety applications represents a much smaller body of work. There may be many reasons for this, but at least one significant reason for this is the difficulty of specifying the physical and chemical properties of a polydisperse dust deposit aging in an industrial setting subject to the random fluctuations of its ambient environment. The ambient environment can influence the behavior of the dust deposit through vibration (leading to consolidation), temperature cycling, humidity, and airborne contaminants. Another challenge is that accidental dust deflagrations are not engineered systems. In technology development, systems can be specifically engineered to operate in a reliable and reproducible manner. Accidents are not laboratory experiments; they are uncontrolled and (usually) poorly instrumented events. Therefore, to recreate a combustible dust accident scenario, one must be able to specify the values of the dominant variables and parameters so that the CFD model can generate reasonably accurate simulations.

Finally, as has been emphasized throughout this book, the relative contribution of a finite rate process to the overall rate of combustion of a combustible dust depends significantly on the particle size. The size dependence of the dust explosion parameters is straightforward, in concept, with monodisperse powders. The particle size dependence is fixed. When analyzing the combustion behavior of a polydisperse combustible dust, it can seem, when considering the full range of the particle size distribution, that everything is important and nothing can be neglected.

Despite these challenges, there is a small international community of combustion specialists who have made important contributions to the application of CFD to investigate combustible dust hazards. We will review a small sampling of these more comprehensive studies. To keep this review reasonable in size and scope, many excellent combustion studies must be omitted. As always, the interested reader is encouraged to consult Eckhoff's publications for a broader introduction to the technical literature on this topic (Eckhoff, 2003, Chapters 4 and 9; Eckhoff, 2009).

Partly due to the sparseness of this literature, we will take a case study approach and consider the application of CFD to four specific combustible dust problems: unconfined dust flames, confined propagating dust flames, accelerating dust flames, and flame propagation in industrial accident scenarios. For each of these problems, we will consider a first case study as a reference case. The reference case will be selected based on the simplicity of the model and by virtue of the fact that it represents an early CFD endeavor. The comparison case will be a more recent CFD study that has benefited from the advances in both computer hardware and software. We will then briefly review subsequent studies in which the investigators have introduced more realism into their models.

The first section outlines many challenges of modeling combustible dust hazards with CFD. Section 10.2 considers unconfined dust flame propagation. The problem of confined unsteady dust flame propagation follows in Section 10.3. The investigation of accelerating dust flames is presented in Section 10.4, and Section 10.5 examines the use of CFD in accident investigation and safety analysis for industrial settings. A summary section completes the chapter.

10.1 CFD FOR COMBUSTIBLE DUST PROBLEMS

In the last 50 years, there have been great advances in the theoretical and empirical underpinnings of combustion science (Buckmaster et al., 2005; Westbrook et al., 2005; Law, 2007; Peters, 2009). Powerful spectroscopic techniques have enabled the experimentalist to measure the concentrations of short-lived chemical intermediates; laser diagnostics have enabled the measurement of the turbulent fluctuations of velocity, temperature, and species concentration; nonlinear mathematical analysis has revealed deep insights into combustion dynamics; and CFD has enabled detailed studies of the coupled interactions of chemical kinetics, transport phenomena, and fluid dynamics.

CFD is the application of numerical methods to the solution of fluid dynamics problems. As a discipline, CFD is roughly 60 years old (ENIAC and UNIVAC appeared in 1952−1956). John von Neumann, a brilliant mathematician who was keenly interested in fluid dynamics problems, was one of the scientists who was instrumental in the development of the modern computer (Westbrook et al., 2005). From the 1950s to the 1980s, a popular dissertation topic for engineering graduate students was to select a fluid dynamics problem, develop a numerical method for its solution, write and debug a computer program (usually in FORTRAN) for the implementation of the numerical method, perform a variety of simulations to "solve" the fluid dynamics problem, and write up the results as a contribution towards their dissertation project.

The combined advancement of numerical methods and computer hardware/software in the last five decades has given rise to an enormous growth in the practical application of CFD. As early as the 1970s, entrepreneurs identified a potentially lucrative market for offering comprehensive CFD software that would allow investigators to bypass the need to develop specialized software for each new problem of interest. The commercial CFD market developed at an almost exponential rate. With continued improvements and cost reductions in computer software and hardware, CFD became a practical tool in industry, government, and academia. One industry pundit recently observed that in 2015 the annual revenue for CFD vendors surpassed the billion dollar (US$) mark (Hanna, 2015). Fig. 10.1 shows the growth in CFD software sales starting from its infancy in the early 1980s to today.

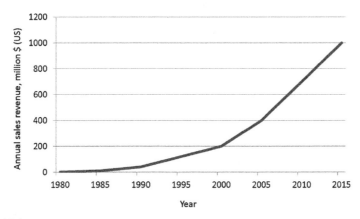

FIGURE 10.1

Annual sales revenue for CFD software.

Data from Hanna, K., 2015. CFD breaks the $billion barrier! Mentor Graphics blog. Available at: https://blogs.mentor.com/khanna/blog/2015/03/26/cfd-breaks-the-billion-barrier/ (posted 26.03.15.).

The explosion of CFD applications has created a strong demand for powerful yet easy to use CFD software, and the market has responded with the development of very powerful software products. CFD simulation software has become easy to use, affordable, and spans an extraordinary range of physical and chemical phenomena.

The great advantage of CFD is that it liberates the investigator from the limitations of overly simplistic assumptions, applied to a flow field with either planar, cylindrical, or spherical symmetry. CFD makes it possible to tackle arbitrary geometries, variable physical properties, and diverse physical and chemical phenomena. An impressive array of numerical simulation tools has been developed for modeling multiphase, turbulent, chemically reacting flows (Anderson et al., 1984; Oran and Boris, 2005; Versteeg and Malalasekera, 2007; Sirignano, 2010, Chapters 8 and 10; Crowe et al., 2012, Chapter 9; Kuo and Acharya, 2012a, Chapter 6; Kuo and Acharya, 2012b, Chapter 7). With the right CFD software, computer hardware, and sufficient time and money, it is tempting to think that any problem can be tackled.

Given the great advances in combustion science and CFD, what makes dust explosion modeling so difficult? The difficulty lies in accurate problem specification, complex geometries, and diverse length and time scales. Additionally, CFD can generate solutions to problems, but the validity and the accuracy of the solution may still require careful consideration and evaluation. There is much science—and art—in the practical application of CFD to realistic combustible dust problems.

10.1.1 PROBLEM SPECIFICATION

The CFD analysis of a combustible dust scenario in an industrial setting requires the accurate specification of several types of information:

- Problem geometry, initial conditions, and boundary conditions
- Combustible dust particle size distribution
- Combustible dust physical and chemical properties
- Model parameters for all relevant physical and chemical rate processes
- Description of the ambient environment.

The accuracy of the problem specification is challenged by numerous factors: the complex geometries of industrial equipment and workplaces, the large variance often encountered in dust particle size distributions, the variability of combustible dust physical and chemical properties, the uncertainty in determining the model parameters for finite rate process, and the often uncontrolled ambient environment. If one is able to provide accurate input data for a CFD simulation, the resulting output can be useful. If one must make approximations or assumptions to fill the input data gaps, the simulation output must be checked carefully for reasonableness. Recall the caution stated in Section 1.8: as one increases the degree of model realism, one also increases the opportunity for uncertainty in the simulation output to grow as well.

10.1.2 MODELING STRATEGY

In any calculation, there is a potential for uncertainty to enter into the result. Uncertainty is not a problem, it is simply an attribute of computation that must be managed. The best approach for managing the uncertainty of the modeling activity is to employ a systematic strategy. The strategy suggested here is not a recipe to be administered in a rote, unthinking fashion. Instead, one must adapt the strategy for the particular problem at hand. Some problems may require an additional emphasis on one aspect and other problems may require emphasis on another. The key is to follow a systematic, structured approach.

Another argument in favor of taking a structured, slightly formal approach to CFD analysis as the assignment often involves multiple participants and stakeholders. A more formalized approach facilitates communication and understanding between the various parties. Table 10.1 summarizes the CFD modeling strategy.

Defining the system of interest includes specifying the equipment or building volume of interest, the combustible dust and its properties, and the physical environment. The study objectives should be explicitly stated so that a consensus is formed between the modeling participants and stakeholders. The CFD analysis must then select the model scale and resolution needed to achieve the study objectives. This is where negotiation is sometimes needed to balance staffing, resources, schedule, and budget. The next step is to build the model and run test cases to determine if the model is performing as expected. Model verification refers to the activity of evaluating the model for mathematical accuracy, ie, to test whether the model solving the mathematical problem satisfactorily. Model validation, on the other hand, is the task of testing the model predictions against empirical data. If satisfied by the verification and validation efforts, then run the

Table 10.1 CFD Modeling Strategy

1. Define the system of interest
2. Identify study objectives
3. Select model scale and resolution based on study objectives
4. Build the model and run test cases
5. Verify the model against mathematical benchmarks
6. Validate the model against empirical measurements
7. Perform desired simulations
8. Examine the results for reasonableness and sensitivity
9. Interpret the results and know when to stop
10. Document the results

Adapted in part from Oran, E.S., Boris, J.P., 1987. Numerical Simulation of Reactive Flow. Elsevier, New York (Chapters 1–3); Oberkampf, W.L., Trucano, T.G., 2002. Verification and validation in computational fluid dynamics. Prog. Aerospace Sci. 38, 209–272; Skjold, T., Pedersen, H.H., Bernard, L., Middha, P., Narasimhamurthy, V.D., Landvik, T., et al., 2013b. A matter of life and death: validating, qualifying and documenting models for simulating flow-related accident scenarios in the process industry. Chem. Eng. Trans. 31, 187–192.

desired simulations. Evaluate the results for reasonableness (do the results match physical intuition?) and sensitivity (how is the solution affected by small changes in the parameters?). Then interpret the results in light of the study objectives to ensure that the original intent of the assignment has been satisfied. This is a good time to ask if additional work is warranted or if it is time to close out the assignment. This decision may be more influenced by business considerations instead of scientific concerns. Finally, document the results so that, if necessary, someone unfamiliar with the work could execute it and obtain similar results.

This suggested modeling strategy should not be viewed as a recipe that must be followed each and every time without fail. Some situations will call for modifications to the strategy and one should be flexible in that regard. The important point is to view the CFD results with a healthy dose of skepticism and evaluate their reasonableness. The simple models developed in Chapters 3−9 should be helpful in this regard.

10.1.3 EQUATIONS OF CHANGE FOR TURBULENT MULTIPHASE CHEMICALLY REACTING FLOWS

There are numerous books and literally thousands of scientific papers devoted to the proper formulation of the equations of change for turbulent multiphase chemically reacting flows. The field of CFD is too dynamic, too broad, and too deep to do it justice with a short review. Instead, I will refer the reader to a small sample of references available on the topic (Crowe et al., 2012; Kuo and Acharya, 2012a, b; Oran and Boris, 2005; Wilcox, 2006; Versteeg and Malalasekera, 2007).

There is a classification scheme for multiphase flow models that is in common usage. The scheme divides multiphase flow models into two classes: Eulerian−Eulerian models and Eulerian−Lagrangian models. The terms Eulerian and Lagrangian are used here in the fluid dynamics sense, ie, the Eulerian form of the equations of change are written for a fixed coordinate system. The Lagrangian form of the equations of change are written for a coordinate frame that is fixed to a fluid particle and moves with the local fluid velocity. When one speaks of an Eulerian−Eulerian multiphase model, the first "Eulerian" refers to the fluid and the second refers to the particle phase. This is sometimes referred to as the two-fluid model. With an Eulerian−Eulerian multiphase model, two sets of governing equations are required: one for the fluid phase and one for the particle "phase." In a sense, the particle phase is being treated as a cloud of particles.

In the Eulerian−Lagrangian formulation, the fluid phase equations are written in the Eulerian form, but the particle phase is written in the Lagrangian form. There are different ways to formulate the Lagrangian approach. One way is to identify a computational particle at each location within the flow field and to track the motion and behavior of the computational particle as it moves through the flow field (Crowe et al., 2012, Chapter 8).

Finally, the selection of a turbulence model is an important and sometimes difficult decision. The $k-\varepsilon$ model, a so-called two-equation model, is perhaps the most popular choice for combustion flows (Westbrook et al., 2005). The literature on alternative formulations is extensive and the reader is referred to the literature for more information (Pope, 2005; Wilcox, 2006).

10.1.4 BASIC CFD CONCEPTS FOR TURBULENT MULTIPHASE CHEMICALLY REACTING FLOWS

A CFD code solves the equations of change by numerical methods. This is accomplished by transforming a continuous medium into a collection of discrete points that looks like a mesh or a grid. Said another way, the partial differential equations of change are transformed into a system of coupled algebraic equations. Once that transformation has been accomplished, the algebraic equations are solved by reference to the initial conditions and boundary conditions. Several methods have been developed to generate the discrete equations that represent the equations of change. The more important of these methods are the finite difference method, the finite volume method, the finite element method, the boundary element method, and the spectral element method. I will briefly discuss two of these techniques, the finite difference method, because of its simplicity, and the finite volume method, because this method dominates the CFD market (Versteeg and Malalasekera, 2007).

The finite difference method is based on the approximation of a continuous derivative with a discrete finite difference. Consider steady, one-dimensional heat conduction with a steady source term. The system is a solid bar of length L. Fig. 10.2 illustrates how a continuous medium is transformed into a discretized body and then into a finite difference domain.

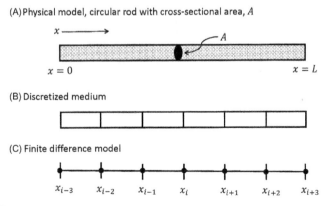

(A) Physical model, circular rod with cross-sectional area, A

$x \longrightarrow$

A

$x = 0$ $x = L$

(B) Discretized medium

(C) Finite difference model

x_{i-3} x_{i-2} x_{i-1} x_i x_{i+1} x_{i+2} x_{i+3}

FIGURE 10.2

Finite difference representation of a material body: (A) the original continuous medium which is a rod with a circular cross-section, (B) the same body now divided into discrete segments, and (C) the discrete pieces collapsed into points on a number line representing the x-coordinate.

Once the domain of interest has been discretized, it is possible to write the governing equations as finite differences. This can be done in one, two, or three spatial dimensions and it can be done in time as well. To illustrate the process, I will show how a derivative becomes a finite difference. Consider the energy equation for the rod. The transport terms are heat conduction and the heat source. Assume constant thermal properties and a constant heat source term. This illustration will be discretized in one spatial dimension, x. The conversion from a spatial derivative to a finite difference looks like this:

$$0 = -k\frac{d^2T}{dx^2} + \dot{Q} \rightarrow \frac{T_{i+1} + T_{i-1} - 2T_i}{(\Delta x)^2} = \frac{\dot{Q}}{k} \tag{10.1}$$

$$T_{i+1} = 2T_i - T_{i-1} + \left(\frac{\dot{Q}}{k}\right)(\Delta x)^2 \tag{10.2}$$

In Eq. (10.1), I have transformed the second derivative into a second-order finite difference (called a centered finite difference). I then solved Eq. (10.1) for the unknown variable, T_{i+1}. Eq. (10.2) says that the value of the temperature at the unknown node, T_{i+1}, is equal to the sum of three terms: the twice the value of the temperature at the i-th node, $2T_i$, minus the value at the $i-1$ node, plus an incremental contribution of the heat source. The algorithm for solving a complete heat conduction problem is to start at the first node and keep marching along until you reach the final grid point. I will stop at this point and simply say that some decisions must be made about how to deal with the boundary conditions. Once that is sorted out, the solution of the problem is a straightforward exercise in linear algebra. Implementing a finite difference algorithm to solve a heat conduction problem is relatively easy. Most modern heat transfer textbooks teach finite differences for solving heat conduction problems. Moving from heat conduction in a solid to turbulent multiphase chemically reacting flow in a compressible fluid becomes a very complicated affair. Consider this example nothing more than a quick peek behind the curtain.

The finite volume technique builds on to the foundation established by finite differences. There is an important difference. In the finite volume method, the governing differential equation is first converted into an integral form of the conservation law. The integrated form is then discretized.

Taking a closer look, we first begin with the volume integration over the volume of the cell and apply Gauss' divergence theorem (Versteeg and Malalasekera, 2007, pp. 24−26):

$$\int_{\Delta V} \frac{d}{dx}\left(k\frac{dT}{dx}\right)dV + \int_{\Delta V} \dot{Q}dV = \int_A \left(k\frac{dT}{dx}\right)dA + \int_{\Delta V} \dot{Q}dV \tag{10.3}$$

This volume ΔV is called the control volume for the cell. The integral is then discretized in the following manner. Consider the physical system of a rod with a circular cross-section and a finite length. As with the finite difference technique, we discretize the rod into a series of individual segments and visualize a point at

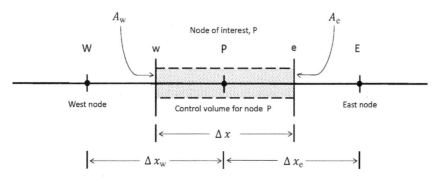

FIGURE 10.3

Definition sketch for the finite volume technique. The node of interest, designated P, is centered at the midpoint of the control volume. The nodes to the left and right of P are designated W (west) and E (east). The volume of the cell is $\Delta V = A \cdot \Delta x$.

the center of each discrete segment. Contrary to the finite difference method, we keep the individual segments of the rod in our sketch along with the nodes. Fig. 10.3 illustrates the discretization scheme.

We now take the integral form of the energy equation and discretize it as follows:

$$\int_A \left(k\frac{dT}{dx} \right) dA + \int_{\Delta V} \dot{Q} dV = \left(kA\frac{dT}{dx} \right)_e - \left(kA\frac{dT}{dx} \right)_w + \dot{Q}_{avg}\Delta V \qquad (10.4)$$

We now evaluate the derivatives as centered finite differences. The following definitions apply for the thermal conductivity:

$$k_w = \frac{1}{2}(k_W + k_P); \quad k_e = \frac{1}{2}(k_P + k_E) \qquad (10.5)$$

The conductive heat fluxes are calculated as centered finite differences:

$$\left(kA\frac{dT}{dx} \right)_e = k_e A_e \left(\frac{T_E - T_P}{\Delta x_e} \right); \quad \left(kA\frac{dT}{dx} \right)_w = k_w A_w \left(\frac{T_P - T_E}{\Delta x_w} \right) \qquad (10.6)$$

To simplify this derivation, we have assumed that the nodes are equally spaced. This assumption can be relaxed. For the sake of convenience, we will assume that the heat source is a constant. Assembling the definitions for the centered finite difference quantities and substituting into the energy equation, we get the desired result:

$$\left[\left(\frac{k_e}{\Delta x} \right) A_e + \left(\frac{k_w}{\Delta x} \right) A_w - \dot{Q} \right] T_P = \left[\left(\frac{k_w}{\Delta x} \right) A_w \right] T_w + \left[\left(\frac{k_e}{\Delta x} \right) A_e \right] T_e + \dot{Q}_{avg}\Delta V \qquad (10.7)$$

Finally, the boundary conditions are applied, and then the equations are solved by linear algebra. Again I have brushed aside several important details (like how to solve these equations) to simply illustrate the fact that differential equations which cannot be solved directly can be approximated by finite differences and solved approximately.

A final observation can be made about these numerical techniques. The accuracy of the numerical solution depends on the grid size. The smaller the grid size, the better the resolution. There is a point, however, at which the improvement in the numerical solution is not worth the additional computational effort. This assumes that the numerical method is capable of convergence. Another issue that can be a source of frustration is instability. Instability is the failure of the numerical solution to converge to a constant value; instead, it oscillates with ever-increasing amplitude until it finally diverges to infinity. These and many other issues are the discussed at length in CFD books; the interested reader is urged to consult these references for further information (Versteeg and Malalasekera, 2007; Oran and Boris, 2005; Anderson et al., 1984).

Modern CFD codes require an enormous amount of input data, especially for complex geometries. Most CFD codes rely on preprocessing software to formulate the input parameters and problem geometry. Similarly, CFD codes generate enormous amounts of output. Often the best way to decipher the output data is through graphics and animation. Specialized software called postprocessing software is available to assist the analyst in this regard.

10.2 CFD STUDIES OF UNCONFINED DUST FLAME PROPAGATION

A very basic introduction to gas phase flame propagation was presented in Chapter 4. Great advances have been made in the combustion modeling of gas phase systems and are discussed at length in the references cited in Section 4.3. The key attributes of interest in premixed flame propagation are the burning velocity and the flame thickness. We saw how the balance between chemical kinetics and transport phenomena influenced these attributes. In Chapter 7 we considered at length some of the theoretical and experimental results for both stationary and propagating unconfined dust flames. The discrete nature of particles introduced interphase transport and heterogeneous kinetics effects that are not present in gas phase combustion.

The CFD studies selected for the comparison between the reference case and the comparison cases afford an opportunity to better understand the role that chemical kinetics can play in the propagation of a laminar dust flame. Both cases formulate the governing equations as a dispersed multiphase flow. Turbulence is not explicitly modeled in these studies.

10.2.1 REFERENCE CASE: LAMINAR COAL DUST FLAME PROPAGATION

The reference case is the early experimental and CFD study performed by Smoot and his colleagues (Smoot and Horton, 1977). This was an investigation of the flame propagation characteristics of coal dust—air mixtures with the objective of comparing the CFD predictions with experimental measurements obtained from a stationary burner. They used a dispersed phase multiphase flow formulation and neglected thermal radiation exchange through the dust cloud. They used the unsteady one-dimensional planar flow form of the governing equations for the CFD model. They cast their model in an Eulerian—Lagrangian formulation. The governing equations for the gas phase were the continuity equation, the species equation, and the thermal energy equation. The momentum equation was not necessary as they assumed that the Mach number of the flow was small, $Ma \ll 1$. Thermodynamic data and transport properties were based on published values and updated for changes in temperature. They neglected the velocity slip between the gas phase and particles. To reduce the number of equations to be solved, they invoked a stream function to eliminate the mixture continuity equation.

The combustible dust was a Pittsburgh bituminous coal with a volatiles content of 50%. The model equations were solved using a finite difference technique developed by the authors of the study. Fig. 10.4 is an example of the flame structure predicted by their model.

The CFD model of Smoot et al. was solved by finite differences using custom software developed at their laboratory. Gas diffusion terms and the particle equations were solved explicitly while the kinetic terms and energy equation

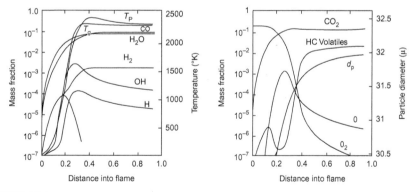

FIGURE 10.4

Species concentration, temperature, and particle diameter profiles in a coal dust flame, dust concentration 300 g/m³, mean particle diameter 33 μm.

Reproduced with permission from Smoot, L.D., Horton, M.D., and Williams, G.A., 1977. Propagation of laminar pulverized coal-air flames. Sixteenth (International) Symposium on Combustion, The Combustion Institute, Elsevier, New York, pp. 375–387.

were solved implicitly. The authors presented a total of nine simulations with one base case and eight parametric variations for comparison with the base case. They used 50 grid points to resolve the flame structure. The flame thickness was on the order of 10 mm and the burning velocity was on the order of 15 cm/s. At the time of the publication of this paper, there was not much empirical data on the detailed structure of a dust flame. The work of Smoot et al. indicated that laminar propagation of coal dust—air flames was a complex function of particle heating transients, devolatilization, volatiles combustion, thermal radiation, and gas phase kinetics.

10.2.2 COMPARISON CASE: IMPROVED SPATIAL AND TEMPORAL RESOLUTION

Qiao compared the flame propagation behavior of carbon powder with coal (bituminous coal with approximately 43% volatiles) (Qiao, 2012; Qiao and Xu, 2012). In the first study, she considered flame propagation in a spherical cloud of carbon dust. The particle size of the monodisperse carbon dust cloud was 30 μm. Similar to the work of Smoot et al., Qiao restricted her investigation to laminar flow for a dispersed multiphase system. Qiao, however, also included the effects of thermal radiation by solving the radiative transfer equation using the discrete ordinate method (see Modest, 2013, Chapter 17). She used CHEMKIN to compute thermodynamic properties, transport properties, and chemical reaction rates. Her reaction kinetics scheme was based on the GRI-Mech 3.0 database. Qiao cast the governing equations in an Eulerian—Eulerian model formulation.

Qiao built her own CFD model instead of using a commercial package. She solved her governing equations using finite differences with the diffusion terms represented with second-order central differences and the convective terms represented with a third-order QUICK scheme. She integrated the discretized governing equations using a routine called DASPK. She employed an adaptive mesh algorithm to allow for a higher level of resolution in the reaction zone. The smallest time step in her simulations was 10^{-7} seconds and the maximum number of grid points was 400; the smallest grid size was 45 μm. Her computational space was a sphere with a radius of 25 cm (compare this with a 20-L sphere radius of 16.8 cm). More details are available in the original paper.

The investigation of the carbon dust system involved varying the concentration of oxygen with the balance of the gas being nitrogen; the base case used 40% oxygen with a carbon dust concentration of 116 g/m^3 ($\phi = 1.12$). Fig. 10.5 shows the flame structure for the carbon dust—40% oxygen—60% nitrogen system.

The flame thickness for the base case simulation was approximately 10 mm. The flame speed was on the order of 35 cm/s, but in the early stages of flame propagation she observed oscillations in the flame speed. Fig. 10.6 shows the flame speed predicted oscillation.

She surmised that these flame speed oscillations were caused by the velocity slip ahead of the flame. According to her simulations, the velocity slip led to a periodic change in the local fuel concentration which changes, in turn, the local

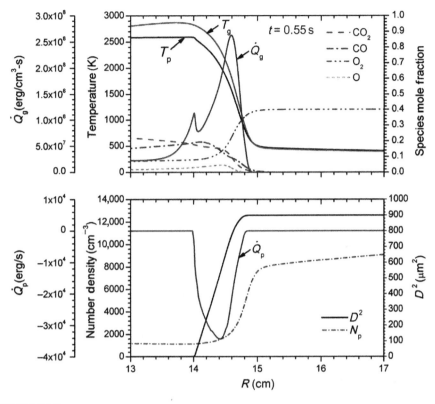

FIGURE 10.5

Flame structure for carbon dust—40% O_2—60% N_2 system.

Reproduced with permission from Qiao, L., 2012. Transient flame propagation process and flame-speed oscillation phenomenon in a carbon dust cloud. Combust. Flame 159, 673–685.

reaction rate driving the flame propagation. The influence of velocity slip on flame propagation was described in Section 7.7. Velocity slip in unconfined flame propagation has been observed and measured (Sun et al., 1998, 2001, 2006a).

Qiao found that thermal radiation played an important role early in the flame propagation process, but then its role in flame propagation diminished in later times as the flame grew larger. It must be emphasized that while the oscillation in flame speed is predicted in her simulations, it has not been observed experimentally in the carbon dust—oxygen—nitrogen system. However, flame speed oscillations have been observed in other combustible dust systems such as freely propagating aluminum dust—air flames (Julien et al., 2015).

Qiao and Xu conducted a companion study with coal (bituminous coal with approximately 43% volatiles) in oxygen enriched air mixtures (30% O_2 and 70% N_2) (Qiao and Xu, 2012). Fig. 10.7 shows the flame speed as a function of radial position.

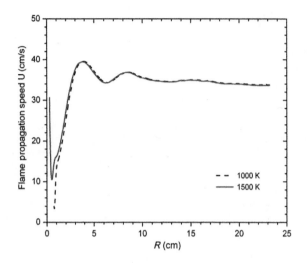

FIGURE 10.6

Flame speed oscillation for carbon dust–oxygen–nitrogen system.

Reproduced with permission from Qiao, L., 2012. Transient flame propagation process and flame-speed oscillation phenomenon in a carbon dust cloud. Combust. Flame 159, 673–685.

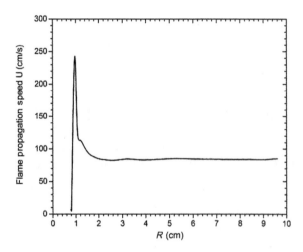

FIGURE 10.7

Flame speed of coal dust flame as a function of radial position.

Reproduced with permission from Qiao, L., Xu, J., 2012. Detailed numerical simulations of flame propagation in coal-dust clouds. Combust. Theory Model. 16, 747–773.

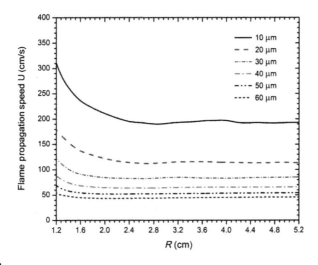

FIGURE 10.8

Flame speed as a function of radial position for coal dust flames with different particle diameters.

Reproduced with permission from Qiao, L., Xu, J., 2012. Detailed numerical simulations of flame propagation in coal-dust clouds. Combust. Theory Model. 16, 747–773.

Qiao and Xu used a CFD code very similar to that used by Qiao for the carbon dust simulations. As in the carbon dust simulations, velocity slip during flame propagation was predicted for the coal−30% O_2−70% N_2 system. However, flame speed oscillations were not observed in this system. Their interpretation of the simulation results indicated to them that the rate of flame propagation was dominated by the rates of devolatilization and volatiles combustion.

Qiao and Xu explored the sensitivity of the flame speed to changes in particle size. Fig. 10.8 shows the flame speed as a function of radial position for several particle diameters. As expected, the flame speed increased with decreasing particle diameter.

There are both similarities and differences between the work of Smoot et al. and Qiao and Xu. The formulation of the governing equations is similar, but Qiao and Xu had the benefit of almost 35 years of additional research on the reaction kinetics for coal combustion. However, the flame structure predicted by Qiao and Xu is very similar to that proposed by Smoot et al. We must be cautious in these comparisons because Qiao and Xu considered an oxygen enriched gas stream whereas Smoot et al. considered air. Also, Qiao and Xu cast the flame propagation rate in terms of flame speed while Smoot et al. reported the burning velocity. Finally, perhaps the most telling difference in these two studies is the scale of resolution. Qiao and Xu were able to use a finer grid size resulting in greater refinement in the flame structure. This finer scale of resolution and the incorporation of velocity slip enabled these investigators to show the effect of volatiles combustion on the flame speed development, with coal dust clouds creating a more stable flame speed while the carbon dust cloud exhibited the oscillating flame speed.

EXAMPLE 10.1

Consider an unconfined laminar coal dust flame with an initial dust concentration that is approximately stoichiometric. Compare the laminar burning velocity predictions using the models of Ogle et al. (see Section 7.3.2) and Ballal and Lefebvre (see Section 7.4.1) with the CFD predictions of Smoot et al. (1977) and Qiao and Xu (2012). Assume an initial temperature of 300 K, a dust flame temperature of 2500 K, $\rho_d = 1.30$ kg/m^3 and $C_{pm} = 1100$ j/kg·K. The dusty gas mixture is assumed to obey the ideal gas equation of state.

Solution

Using the simple thermal theory model of Ogle et al., the burning velocity is estimated with the following equations:

$$S_u = \frac{\sigma T_f^4}{\rho_{m,0} C_{p,m}(T_f - T_u)}$$

$$S_u = \frac{(5.67 \times 10^{-8} \text{ W/m}^2 \cdot \text{k}^4)(2500)^4}{(1.30 \text{ kg/m}^3)(1100 \text{ J/kg·K})(2500 - 300)} = 70 \text{ cm/s}$$

From Fig. 7.6, the Ballal model predicts, depending on the coal volatility, a laminar burning velocity in the range of 10–20 cm/s. Smoot et al. predicted a range from 6 to 17 cm/s with their CFD model (with particle diameters ranging from 10 to 33 μm). Qiao and Xu predicted a *flame speed*, S_b, of 80–190 cm/s (for particle diameters of 30 and 10 μm, respectively). The laminar burning velocity can be estimated from the relation

$$S_u = S_b\left(\frac{\rho_b}{\rho_u}\right) = S_b\left(\frac{T_u}{T_b}\right) = (80 - 190 \text{ cm/s})\left(\frac{300 \text{ K}}{2500 \text{ K}}\right) = 9.6 - 23 \text{ cm/s}$$

Compared to the Ballal model and the two CFD models, the thermal theory of Ogle et al. overestimates the burning velocity by an order of magnitude. This suggests that this thermal theory oversimplifies the physical and chemical rate processes involved in coal combustion. Ballal's model is in reasonable agreement with the two CFD models. It would appear that the main difference between the two CFD models is the improvement in flame structure resolution with Qiao and Xu's model.

10.2.3 SUMMARY OF SELECT INVESTIGATIONS OF UNCONFINED DUST FLAME PROPAGATION

The body of work for the CFD analysis of unconfined dust flames is largely represented by studies of one-dimensional, laminar planar flames of coal dust–air mixtures. This flow configuration is of great industrial importance as it establishes a benchmark for comparison with more complex burner configurations (think of utility boiler fireboxes). The 1979–1985 time frame witnessed several important advances in the theory of multiphase laminar flows (Smoot and Pratt, 1979, Chapter 13; Krazinski et al., 1979; Kansa and Perlee, 1980; Slezak et al., 1985). In addition to the multiphase flow formulation, the efforts of these investigators included the incorporation of more sophisticated models for chemical kinetics and thermal radiation. This period was also marked by the exploration of various numerical methods for the solution of the model equations. Although most of the studies formulated their equations in a finite difference scheme they differed in the method of numerical integration employed. Each of these studies highlighted limitations on the numerical solutions imposed by the stiffness of the differential equations.

The next important advance in the modeling of unconfined dust flame propagation was the incorporation of the $k-\varepsilon$ turbulence model. The advantage of the $k-\varepsilon$ model is that it is a predictive model that does not rely as heavily on calibration with experimental data as the simpler turbulence models like the mixing length model. Smoot and Smith devoted an entire book to the simulation of turbulent multiphase chemically reacting flows for the study of pulverized coal combustion (Smoot and Smith, 1985).

Wingerden and his colleagues introduced a CFD model specifically designed to simulate combustible dust hazard scenarios, called DESC (Dust Explosion Simulation Code), (van Wingerden, 1996). This model was a modified version of a CFD code called FLACS (Flame Acceleration Simulator) which was developed for the simulation of gas explosions. The FLACS code was based on a finite volume technique with a weighted upwind/central differencing scheme for the convective terms. DESC was capable of handling complex geometries and had been tested against standardized hazard scenarios with corn starch, wheat dust, and lycopodium.

Kosinski and his colleagues developed a CFD code for simulating dust explosions (Kosinski et al., 2002). Their model was an Eulerian–Eulerian formulation. More will be said of this model in the flame acceleration section. Spijker et al. have explored the use of the open-source CFD code OpenFOAM and have presented some preliminary results on dust flame propagation (Spijker et al., 2013).

10.3 CFD STUDIES OF CONFINED DUST DEFLAGRATIONS

The unconfined premixed gaseous flame is a staple of the combustion literature. The propagation of gaseous flames in a closed volume was a popular topic of study in the early 20th century, but by the 1950s combustion diagnostics for unconfined flames were becoming powerful tools for probing flame structure. These diagnostic tools do not work as well in closed vessels and as a result, combustion studies in closed vessels have become increasingly rare. The trend for combustible dusts has been in the opposite direction. The 20-L sphere and 1-m^3 vessel have become the international standard for studying dust flames. This has led to an awakening of interest in CFD studies of confined flame propagation.

First, a few observations on the mathematical analysis of premixed flame propagation in closed vessels. Several investigators have examined the problem of a flame propagating through a spherical vessel with central ignition. In these studies, it is usually assumed that the flame is a surface of discontinuity (flame sheet) whose behavior is subject to the jump conditions of fluid dynamics. This model is often referred to as a hydrodynamic model. One advantage of the hydrodynamic model is that it collapses into the integral model for a confined deflagration (Sivashinsky, 1974, 1979; Takeno and Iijima, 1981; Continillo, 1989). Another practical application of the hydrodynamic model is its utility as a model for simulating the piston—cylinder arrangement of an internal combustion engine (Carrier et al., 1980; Oppenheim and Maxson, 1993). The limitations of the hydrodynamic model and its mathematical behavior have been investigated using scaling arguments and activation energy asymptotics (Majda and Sethian, 1985; Matalon, 1995; Kelley et al., 2012).

Ramos studied the simulation of flame propagation in a spherical vessel as a benchmark case to compare the performance of nine different numerical algorithms (Ramos, 1983). Ramos studied the combustion of a propane—air mixture and found that good accuracy could be achieved with 82 grid points with a dimensionless step size of 0.0125 and a time step of 10^{-5} seconds. Kono et al. also studied spherical flame propagation with the propane—air system (Kono et al., 1985). They used a CFD code available from the Los Alamos National Laboratory. They subdivided their spatial domain into 30 grid points (dimensionless step size of 0.033) and a time step of 5×10^{-6} seconds.

10.3.1 REFERENCE CASE: MULTIPHASE MIXTURE WITH MIXING LENGTH TURBULENCE MODEL

Ogle et al. developed a model for unsteady, one-dimensional flame propagation of aluminum dust in the 20-L sphere (Ogle et al., 1988). The model was used to compare with dust deflagration experiments conducted with a 20-L sphere for the purpose of estimating kinetic parameters for an aluminum combustion mechanism postulated by the authors. The governing equations were based on the multiphase mixture formulation with velocity slip and temperature lag neglected. Turbulence was modeled with the mixing length hypothesis; the eddy diffusivity was estimated using a turbulent premixed flame correlation (Abdel-Gayed et al., 1979). Thermal radiation transport through the cloud of particles was modeled with the diffusion approximation. The ideal gas was selected for the equation of state and gas viscosity was allowed to vary according to the Chapman gas approximation,

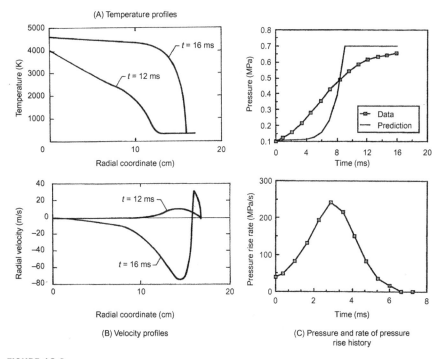

(A) Temperature profiles

(B) Velocity profiles

(C) Pressure and rate of pressure rise history

FIGURE 10.9

Simulation results for an aluminum dust deflagration in a 20-L sphere: (A) temperature profiles at two times; (B) mixture velocity at two times; and (C) pressure history and rate of pressure rise history.

Reproduced with permission from Ogle, R.A., Beddow, J.K., Chen, L.-D., Butler, P.B., 1988. An investigation of aluminum dust explosions. Combust. Sci. Technol. 61, 75–99.

$\rho\mu$ = constant, and the Lewis number was assumed to be unity. Thermodynamic properties were assumed to be constant. The pressure throughout the vessel was assumed to be spatially uniform as the Mach number of the flame was very small.

To simplify the governing equations, two transformations were used: an adiabatic transformation which removed the compressible work term ($\partial P/\partial t$) from the energy equation and a Lagrangian transformation was used to remove the convective terms from the equations of change and to satisfy the continuity equation. The model equations were formulated as explicit finite difference equations. The finite difference equations were integrated using a dimensionless grid size of 0.025 with 81 grid points, and the time step was 10^{-7} seconds. The computations were performed on an IBM 3033 mainframe computer and typically required 140 minutes of CPU time.

Fig. 10.9 shows some typical results for a simulation of an aluminum dust deflagration. This simulation was for an aluminum dust with a volume mean diameter of 12.1 μm (determined by light scattering) and a dust concentration of 300 g/m^3

($\phi = 0.98$). The temperature profile exhibits the Flamm—Mache effect, but the center temperature is too high due to the assumption of constant properties and neglecting dissociation effects. The gas velocity profile exhibited a strong inward velocity of burnt gas. Several criteria were used to try to match the simulation pressure history to the experimental data. The illustration in Fig. 10.9 shows a matching attempt based on the maximum rate of pressure rise. The combination of stiff differential equations, the inherent limitations of the numerical methods, and the constraints on computer CPU time hindered the effort to optimize the simulation of the combustion dynamics in a 20-L sphere.

10.3.2 COMPARISON CASE: MULTIPHASE MIXTURE WITH $K-\varepsilon$ TURBULENCE MODEL

The investigations by Fan et al. and Bind and his colleagues afford an opportunity to witness how progressive advances in CFD have improved our ability to simulate dust deflagrations (Fan et al., 1993; Bind et al., 2011, 2012). Fan et al. used the same governing equations as Ogle et al. with the same parameter values for reaction kinetics. They solved the equations with finite differences and integrating over time with the method of fractional lines. They used an adaptive gridding algorithm to follow the flame so as to get a higher level of resolution of the flame structure. While the figures in their paper indicate a much better resolution of the flame structure, they provided very little detail about their numerical simulation results.

In addition to their CFD analysis, Fan et al. also conducted deflagration tests in 20-L and 50-L spheres. They reported that they tested two aluminum powders: one with a mean diameter of 10 μm and the other with a diameter of 15 μm, at a dust concentration of 300 g/m^3. They indicated that the maximum explosion pressure attained was 10 bars for the 10 μm powder and 9 bars for the 15 μm powder. They also tested the solid residue from the aluminum deflagration tests and found 81% conversion for the 10 μm powder and 75% conversion for the 15 μm powder. These results are generally consistent with expectations.

Bind and his colleagues performed CFD analyses with a commercial CFD package called Fluent 6.3.26 (Bind et al., 2011, 2012). They employed the governing equations for multiphase flow and incorporated turbulence with the $k-\varepsilon$ model, and simulated aluminum dust deflagrations in a 20-L sphere using the chemical kinetics model of Ogle et al. The dust concentration was fixed at 500 g/m^3. They too used an adaptive gridding algorithm to resolve the fine structure of the dust flame. Fig. 10.10 shows some typical results.

In a subsequent study, they achieved better agreement using kinetic parameters based on an analysis of Beckstead's burnout time correlation for aluminum particles (Bind et al., 2012; Beckstead, 2005). In this later study, they also demonstrated successful simulation of starch dust explosion data from a 10.3-m^3 cylindrical vessel.

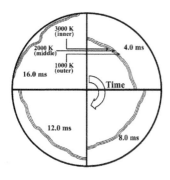

(A) Flame position versus time

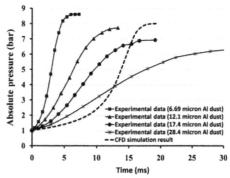

(B) Experimental and model pressure histories

FIGURE 10.10

Aluminum dust deflagration simulations with a dust concentration of 500 g/m^3: (A) flame position at four distinct times and (B) pressure histories for four powders (data from Ogle et al., 1988) and the CFD simulation.

Reproduced with permission from Bind, V.K., Roy, S., Rajagopal, C., 2011. CFD modeling of dust explosions: rapid combustion in a 20-L apparatus. Can. J. Chem. Eng. 89, 663–670.

EXAMPLE 10.2

In Chapter 7, deflagration wave efficiency was defined as the condition where the deflagration time scale, $t_{wave} = \delta/S_u$, was equal to the burnout time of a single particle. Consider the pressure history data for aluminum deflagrations shown in Fig. 10.10. Assume that the deflagration time scale is equal to the time to maximum pressure. For each of the mean particle diameters, calculate the burnout time for a single particle assuming diffusion-controlled combustion. Use the burn rate coefficient for aluminum $K = 250 \ (\mu m)^2/ms$.

Solution

The burnout time for the diffusion-controlled burning of a single particle is given by $\tau = d_{p,0}^2/K$. The table below summarizes the calculations.

$d_{p,0}(\mu m)$	$t_{P_{max}}(ms)$	$\tau \ (ms)$	$t_{P_{max}}/\tau$
6.69	7.1	0.179	397
12.1	12.1	0.586	20.7
17.4	17.4	1.21	14.4
28.4	33.0	3.23	10.2

For all four particle sizes, the ratio of the time to maximum pressure is at least 1 order of magnitude greater than the burnout time for a single particle. At least one interpretation of these results is that the macro-scale transport and kinetic processes dominate the confined deflagration behavior compared to single particle processes.

10.3.3 SUMMARY OF SELECT INVESTIGATIONS OF CONFINED DUST DEFLAGRATIONS

It may seem surprising the degree to which the 20-L sphere has been a focal point for confined dust deflagration modeling. There are perhaps two reasons for this. First, since the 20-L sphere has become a standardized deflagration test device, there is a pressing need to understand its performance characteristics. A sound CFD analysis offers an opportunity to test assumptions about the performance of the apparatus. For example, with CFD it is possible to understand the dynamics of the flame propagation process or to test the uniformity of the dust dispersal process prior to ignition. A second reason for its popularity is that investigators have discovered that even a powerful CFD code can be challenged by the physics and chemistry occurring during a dust deflagration in a 20-L volume. We will review here several CFD analyses for confined deflagration analysis. In all cases, the flame Mach number is small and so flame acceleration effects can be neglected.

Continillo developed a model for coal dust deflagrations in a spherical chamber with a 5-cm radius (volume of 0.52 L). He assumed one-dimensional laminar flow with central ignition (Continillo, 1988, 1989). He used an Eulerian–Lagrangian formulation for multiphase flow to allow for velocity slip and temperature lag. He assumed the gas satisfied the ideal gas equation of state. His model neglected species diffusion, thermal radiation transport through the dust cloud, heat losses through the vessel wall, and dissociation effects. His model predictions of the maximum pressure were high compared to typical experimental data (data obtained from typical literature values, not from direct testing); he concluded that this was caused by a combination of the assumptions of no heat loss at the wall and complete combustion.

He solved this model with a finite difference technique and used an adaptive gridding algorithm to allow for better spatial resolution of the flame structure. The advantage of his Lagrangian formulation for the particle phase is that it allowed him to explicitly account for particle diameter effects. For example, Continillo noted that his model indicated that velocity slip caused a local enrichment in the fuel particle concentration at the beginning of the deflagration. This is because at the beginning of the deflagration the flame moves outward from the center toward the wall. Due to their inertia, the particles accelerate more slowly than the gas and so the fuel concentration is larger than the initial condition. At the end of the deflagration, as the flame approaches the wall the gas velocity is directed toward the center. The particles, on the other hand, are still traveling toward the wall causing an enrichment of the local dust concentration at the wall. Thus, Continillo predicted that velocity slip would cause a nonuniform particle concentration profile to form during the deflagration. This prediction is consistent with the empirical studies of dust flame propagation discussed in Sections 7.2 and 7.7.

Kjäldman reported on a CFD study in which he compared model predictions with dust deflagration tests performed in a 20-L vessel (Kjäldman, 1992). He used an Eulerian–Lagrangian formulation for multiphase flow and the $k-\varepsilon$ turbulence model. Deflagration tests were performed with peat having a volatiles content of 70% on a dry basis. He used a commercially available CFD code called PHOENICS81. He modeled the test vessel as a cylinder with a coarse grid of 10×10 cells. He found that the presence of moisture had a significant impact on computing times. Dry dust simulations took 76 minutes on a Cyber 180/840 computer and moist dust simulations took 129 minutes. His simulations included both closed vessels and vented vessels. The vented vessel simulations took 222 minutes of computing time.

In Sections 1.5 and 8.6, the concept of "overdriven" deflagrations was discussed. This can occur if the igniter energy is so great that it distorts the measured deflagration parameters. The 1-m^3 vessel is relatively immune to this problem; it is instead an issue primarily with the 20-L sphere. Cloney et al. studied the nature of overdriving (Cloney et al., 2014). Following other investigators, they indicated that overdriving can be divided into two contributions: igniter-induced pressure rise and preconditioning. Fig. 10.11 illustrates the effect of the igniter contribution to the pressure rise.

Preconditioning was defined as a change in the initial conditions of the test vessel caused by the release of the igniter energy. In the standard deflagration test, the dust is dispersed into the vessel and then, after a time delay, the igniter is initiated. The igniter releases its energy as a miniature detonation blast that drives a shock wave toward the vessel wall. The shock wave and the blast wind following it can accelerate the dispersed particles and disturb the uniformity of the mixture. Fig. 10.12 depicts the various phenomena caused by the detonation of the igniter.

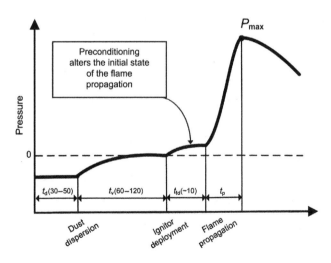

FIGURE 10.11

Example of igniter-induced overpressure.

Reproduced with permission from Cloney, C.T., Amyotte, P.R., Khan, F.I., Ripley, R.C., 2014. Development of an organizational framework for studying dust explosion phenomena. J. Loss Prev. Process Ind. 30, 228–235.

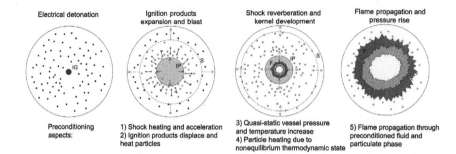

FIGURE 10.12

Sequence of events triggered by an igniter in a deflagration test vessel.

Reproduced with permission from Cloney, C.T., Amyotte, P.R., Khan, F.I., Ripley, R.C., 2014. Development of an organizational framework for studying dust explosion phenomena. J. Loss Prev. Process Ind. 30, 228–235.

The authors used a CFD analysis to explore the effects of igniter energy released into a 20-L sphere and a 1-m³ sphere. They modeled the igniter disturbance as an unsteady one-dimensional spherical flow. The gas was treated as an ideal gas. The Eulerian—Eulerian formulation of the multiphase equations of change was adopted for this problem. The equations were solved using a finite volume technique.

The base case calculation considered a 5-kJ igniter initiated in polyethylene dust cloud with a dust concentration of 500 g/m³. Their simulations confirmed the empirical observation that preconditioning can significantly affect the test environment of a 20-L sphere but will not affect the test environment of a 1-m³ vessel.

Several investigators have applied CFD to the specific problem of determining the quality of the dust dispersion process in a 20-L sphere or other laboratory device (Di Benedetto et al., 2013; Di Sarli et al., 2014; Salamonowicz et al., 2015; Murillo et al., 2013). Of these studies, the one by Murillo et al. was the only one to not use FLUENT; Murillo and his colleagues used a specialty CFD code modified for their specific problem. The physical domain of these investigations was relatively small compared to the size of an industrial grain elevator. Yet because of the scale of resolution that was required to discern the detailed structure of the turbulent mixing, these investigations employed hundreds of thousands of grid elements and required hours of CPU time to achieve a solution. These investigations serve to remind us that as we approach industrial-sized hazard scenarios, sacrifices in the scale of resolution will have to be made to keep the computational problem manageable.

10.4 CFD STUDIES OF DUST FLAME ACCELERATION

Flame acceleration is possible only if the fluid medium exhibits some degree of compressibility. The measure of fluid compressibility is the sound speed; compressible fluids have a finite sound speed and incompressible fluids have an infinite sound speed. The hallmark feature of compressible fluid dynamics is the formation of compression waves in the form of sound waves, nonlinear pulses, or shock waves. Depending on the geometry, the initial conditions, and the boundary equations, the formation of expansion waves, contact surfaces, and other wave inter-actions can be expected. The governing equations for compressible fluid dynamics are partial differential equations with an accumulation term, a convection (inertial) term, and source terms. If diffusional processes can be neglected, the compressible fluid flow equations can be reduced to ordinary differential equations by transform-ing them into the characteristic equations. The characteristic equations enable one to track the motion of waves and their interaction with surfaces. This mathematical technique is called the method of characteristics (Courant and Friedrichs, 1948, Chapter II; Thompson, 1972, Chapter 8). Perhaps the greatest limitation of the method is that its use is restricted to problems with two independent variables.

Before the advent of CFD codes, the method of characteristics was a popular technique for solving compressible fluid dynamic problems. Although it remains a powerful conceptual framework for thinking about wave motion in compressible fluids, it has become less popular due to the ability of many CFD codes to handle these issues well. Before moving into dust flame acceleration, we consider briefly two examples of gas flame acceleration effects analyzed through the method of characteristics by Bureau of Mines investigators.

Kansa and Perlee studied gas flame acceleration using the method of characteristics to study flame propagation in a spherical vessel (Kansa and Perlee, 1976). At first this geometry may seem an odd choice for studying flame acceleration, but the vessel under consideration was 3.6-m-diameter sphere. They simulated flame propagation in a spherical chamber with a radius of 1.8 m (volume $= 24.4$ m^3) for a burning velocity of 17.5 m/s (consistent with a turbulent burning velocity) and an expansion ratio $\rho_u/\rho_b = 10$. They assumed ideal gas behavior and assigned values for the heat capacity ratios as 1.4 and 1.2 for the unburnt and burnt gases, respectively. Note that the burning velocity and the expansion ratio are assigned values; the method of characteristics cannot independently calculate these values.

Kansa and Perlee tracked the flame trajectory as well as the formation and propagation of compression and expansion waves. Fig. 10.13 shows the trajec-tories of the flame and the various characteristic waves (compression and expansion waves) as they progress from the vessel center outward to the wall and back.

The subscripted symbols C and R stand for compression and rarefaction (expansion) waves. The parent compression wave is C, C_r is the parent compression wave reflected from the wall, C_{rt} is the parent compression wave reflected from the wall and transmitted through the flame, and C_{rtr} is the wave C_{rt} reflected from the center of the vessel. The method of characteristics makes it possible to track these wave trajectories and interactions directly. Wave interaction events are designated in Fig. 10.13 by the letters A, B, C, D, E, F, G, and H.

The flame acceleration effect is evident in Fig. 10.14. Event A, a compression wave C_r reflected off of the vessel wall, caused a deceleration in the flame speed.

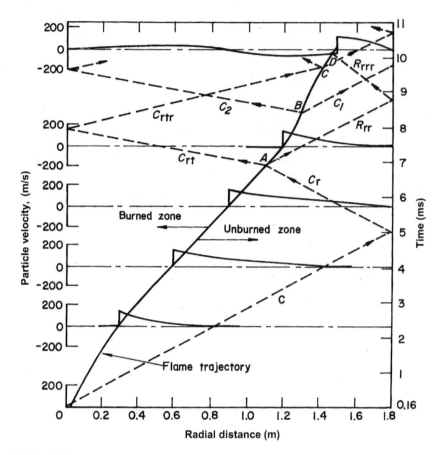

FIGURE 10.13

Axial velocity profiles at different times as a flame propagates through the 24.4 m³ spherical vessel. The flame trajectory and the first characteristic wave trajectory are shown.

Public domain from Kansa, E.J., Perlee, H.E., 1976. Constant-Volume Flame Propagation: Finite-Sound-Speed Theory. U.S. Department of the Interior, Bureau of Mines, Report of Investigations 8163.

The compression wave was traveling in the opposite direction as the flame, and so the compression wave reduced the flame speed and the particle velocity of the burnt gas. Event B is when the second compression wave C_2 generated at the center of the vessel caught up with the flame and accelerated the flame as C_2 passed through it.

The conclusion to draw from this analysis is that even a simple laboratory system can exhibit a diverse range of wave interactions which can accelerate or decelerate the flame motion.

Kansa and Perlee ultimately concluded from this study that the acoustic approximation for weak shock waves was appropriate for flame Mach numbers less than 0.2. They also concluded that the pressure oscillations frequently

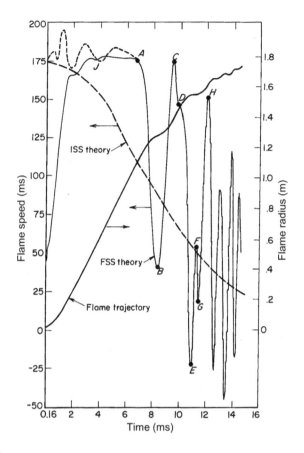

FIGURE 10.14

Calculated flame speed and flame position histories in the 24.4 m^3 spherical vessel. The acceleration and deceleration from wave interactions are evident. ISS (infinite sound speed) denotes the integral flame propagation model and FSS (finite sound speed).

Public domain from Kansa, E.J., Perlee, H.E., 1976. Constant-Volume Flame Propagation: Finite-Sound-Speed Theory. U.S. Department of the Interior, Bureau of Mines, Report of Investigations 8163.

observed in constant volume deflagration tests were likely the result of wave interactions with the vessel wall and also due to the finite speed of the gas motion generated by flame propagation. Finally, they observed that the averaged pressure history obtained by the method of characteristics matched the pressure history calculated from an integral flame propagation model.

Chi and Perlee applied the method of characteristics to flame propagation in a long channel similar to the Bureau of Mines explosion tunnel with ignition occurring at the closed end and the opposite end open (Chi and Perlee, 1974). The physical setting was a conduit 50 m in length with a circular cross-section having a diameter of 0.69 m. The burning velocity was specified as 7.4 m/s and both the heat capacity ratios and the expansion factor were assigned values consistent with a stoichiometric methane–air mixture. In this study, they found the acoustic approximation for the flame speed–overpressure relationship was accurate until the third and fourth compression waves interacted with the flame. After that, the flame and wave dynamics were too complex for the simple acoustic approximation.

10.4.1 REFERENCE CASE: MULTIPHASE FLOW MODEL

Clark and Smoot performed an investigation of accelerating coal dust flame propagation using an Eulerian–Eulerian multiphase flow model (Clark and Smoot, 1985). Their model was based on the unsteady one-dimensional equations of change with the Rankine–Hugoniot jump conditions to specify the properties of the coal dust flame. A schematic of their physical problem is presented in Fig. 10.15.

The objective of the modeling exercise was to calculate the pressure and velocity fields as a coal dust flame propagated down a conduit. The flow field was divided into four zones: the undisturbed reactants, the compressed unburnt reactants, the flame, and the combustion products. The effect of turbulence was modeled with a turbulent burning velocity. They assumed that the pressure wave generated by the propagating flame could entrain coal dust off of the floor and walls with sufficient turbulent mixing to form an ignitable dust cloud. For the description of the flow field created by the pressure wave, Clark and Smoot relied on the analytical solution presented in Cybulski's monograph (Cybulski, 1975, pp. 89–100). The overpressure and particle velocity of the pressure wave influence the turbulent flame speed which in turn influences the overpressure and particle velocity of the pressure wave. Thus, this model was capable of demonstrating how flame acceleration arose from the turbulent combustion feedback loop discussed in Chapter 9.

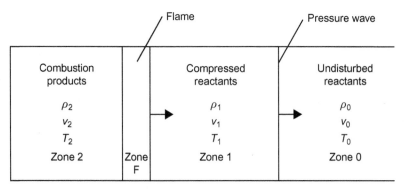

FIGURE 10.15

Schematic diagram of coal dust flame with three zones traveling in a one-dimensional passageway. Zone F denotes the flame region.

Reproduced with permission from Clark, D.P., Smoot, L.D., 1985. Model of accelerating coal dust flames, Combust. Flame 62, 255–269.

In the flame zone, the model assumed that the gas and particles were in dynamic equilibrium (no velocity slip). The laminar burning velocity of the coal dust–air mixture was obtained by an empirical correlation based on data from the University of Utah's coal research program. The correlation estimated the laminar burning velocity as a function of dust concentration, coal type, and particle size. A turbulent flame scaling relation was used to calculate the turbulent burning velocity.

The governing equations of the model were cast in terms of finite differences and solved by a fourth-order Runge–Kutta technique. They integrated the unsteady equations with a time step of 10^{-3} seconds until they achieved steady state. They performed several simulations for Pittsburgh bituminous coal with an assumed particle diameter of 33 μm.

They demonstrated the effect of turbulence on flame propagation by varying the conduit diameter. With a smaller diameter conduit the bulk gas velocity increased thus increasing the level of turbulence. This effect is shown in Fig. 10.16.

As expected, the higher level of turbulence associated with smaller conduit diameters also leads to higher pressures and burnt gas temperatures. They also demonstrated the effect of particle size by comparing the baseline simulation (33 μm) with smaller and larger sizes: 10 and 41 μm, respectively. Fig. 10.17 shows the effect of particle size on the predicted flame velocity (turbulent burning velocity) and pressure.

Their simulations reflect the expected behavior with smaller particles burning faster causing an increase in the turbulent burning velocity and combustion pressure.

The relationship between the pressure of the weak shock wave versus the particle velocity is presented in Fig. 10.18 for a conduit diameter of 2.67 m, the approximate diameter of the Bureau of Mines experimental mine. This figure includes measurements from Richmond and Liebman's work as well as others (Richmond and Liebman, 1975).

It is interesting to note that this model predicts the nonlinear pressure–velocity relation reasonably well up to velocities of 500 m/s. This nonlinear relation is *indirect* evidence that the model was accurately accounting for the effect of wave interactions on flame propagation. However, unlike the method of characteristics, the direct numerical integration of the equations of change does not present a convenient technique for observing and tracking wave interactions.

10.4.2 COMPARISON CASE: MULTIPHASE FLOW MODEL WITH THE $K-\varepsilon$ TURBULENCE MODEL

The Dust Explosion Simulation Code (DESC) is a commercially available CFD code specifically designed for the simulation of dust deflagration behavior (Skjold et al., 2007). It was developed during the time period 2002–2005 for a consortium of several European companies, academic institutions, and safety organizations. The code uses dust deflagration data obtained from the 20-L vessel for the dust combustion model. The multiphase mixture model is used to describe

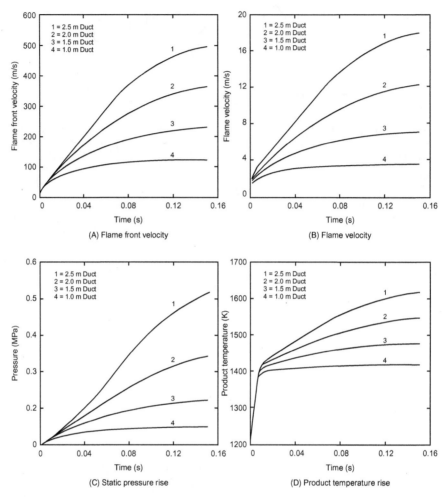

FIGURE 10.16

Simulation results for the effect of turbulence on the flame propagation by varying the conduit diameter.

Reproduced with permission from Clark, D.P., Smoot, L.D., 1985. Model of accelerating coal dust flames, Combust. Flame 62, 255–269.

the gas and particle phases and assumes velocity equilibrium (no velocity slip), so certain phenomena like dust settling cannot be directly modeled. Turbulence is simulated with the $k-\varepsilon$ model. The three-dimensional governing equations are solved using the finite volume technique. The DESC code has many features in common with another commercially available CFD code, FLACS. DESC uses commercially available preprocessing software for establishing the problem geometry and initial/boundary conditions, and it uses postprocessing software for data analysis and graphical presentations of results.

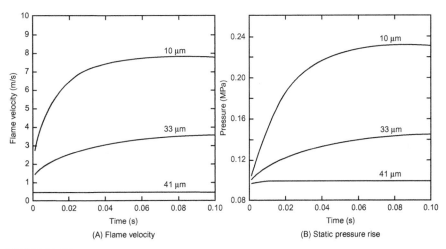

FIGURE 10.17

Accelerating coal dust flame behavior as a function of particle size.

Reproduced with permission from Clark, D.P., Smoot, L.D., 1985. Model of accelerating coal dust flames,
Combust. Flame 62, 255–269.

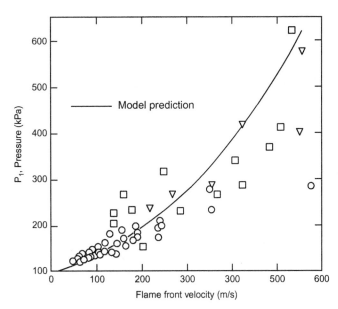

FIGURE 10.18

Comparison of model predictions versus experimental data for the relationship between flame speed and pressure. Experimental data denoted by circles, squares, and triangles. See original reference for data sources.

Reproduced with permission from Clark, D.P., Smoot, L.D., 1985. Model of accelerating coal dust flames,
Combust. Flame 62, 255–269.

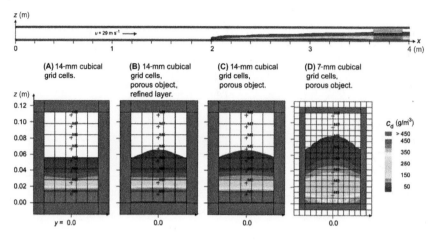

FIGURE 10.19

Cross-section of the laboratory wind tunnel used for dust lifting experiments. Case A is the base case, Case B includes a gridding refinement at the dust–air interface, Case C is a second gridding refinement, and Case D is a finer resolution of the grid spacing.

Reproduced with permission from Skjold, T., Eckhoff, R.K., Arntzen, B.J., Lebecki, K., Dyduch, Z., Klemens, R. et al., 2007. Simplified modelling of explosion propagation by dust lifting in coal mines. In: Proceedings of the Fifth International Seminar on Fire & Explosion Hazards, Edinburgh, UK.

Skjold and his collaborators demonstrated the performance of DESC with a coal dust accelerating flame problem (Skjold et al., 2007). One of the ways in which DESC was specialized for combustible dust hazard analysis was to incorporate a submodel for dust entrainment induced by the passage of a shock wave over a dust layer (sometimes called dust lifting in the literature). The dust lifting submodel was developed from tests in a laboratory wind tunnel and from multiphase flow simulations (Zydak and Klemens, 2007). Since the multiphase flow description in the model is based on the assumption of equilibrium flow, dust lifting was approximated as a diffusive flux with an assumed dust concentration ρ_d of 1 kg/m^3 and a vertical velocity v_z given by the empirical equation:

$$v_z = 0.004 \delta_d^{0.216} v_\infty^{1.742} d_p^{-0.053} \rho_s^{-0.160} A_p^{0.957} \qquad (10.8)$$

Here δ_d is the dust layer thickness in millimeters, v_∞ is the free stream gas velocity in m/s, d_p is the characteristic particle diameter in μm, ρ_s is the solid density of the dust particles in kg/m^3, and A_p is a dimensionless empirical constant.

Fig. 10.19 shows some characteristic results of dust lifting simulations intended to replicate experimental conditions. While accurately depicting the general trend, the results were still dependent on the grid size of the unit cells (comparing Case D, grid size 7 mm, with Case A, grid size 14 mm).

Fig. 10.20 illustrates some typical pressure histories from simulations of a series of six explosion tests in the 100-m-long experimental mine "Barbara" which is operated by the Warsaw University of Technology. This series of tests

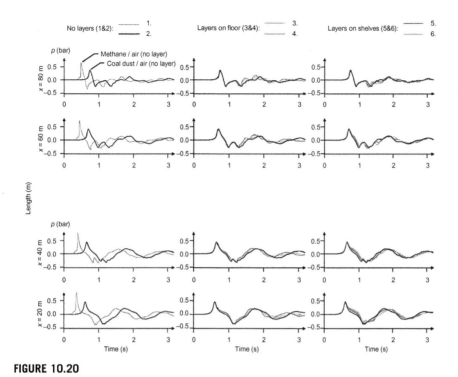

FIGURE 10.20

Pressure histories for the six primary simulations of the explosion tests conducted in the experimental mine "Barbara." The measurements correspond to four locations in the tunnel: 20, 40, 60, and 80 m from the closed end of the tunnel with a total length of 100 m.

Reproduced with permission from Skjold, T., Eckhoff, R.K., Arntzen, B.J., Lebecki, K., Dyduch, Z., Klemens, R. et al., 2007. Simplified modelling of explosion propagation by dust lifting in coal mines. In: Proceedings of the Fifth International Seminar on Fire & Explosion Hazards, Edinburgh, UK.

demonstrated the consequences of flame acceleration under different initial and boundary conditions. The combustible dust used in these tests was maize starch with an estimated K_{st} value of 150 bar-m/s and a mean particle diameter of 15 μm. In tests with fugitive dust, the dust was deployed on either the floor of the tunnel or shelves. Six primary test cases were simulated, with additional sensitivity cases based on higher or lower burning velocities (essentially a measure of the reactivity variation of the dust) resulting in a total of 14 simulations performed.

The six primary test cases simulated are described below.

- Case 1: Primary explosion of stoichiometric methane−air mixture with a volume of 10 m³, no maize starch dust
- Case 2: Primary explosion of maize starch at a dust concentration of 350 g/m³ and a volume of 10 m³, no methane, no fugitive maize starch dust
- Case 3: Primary explosion of maize starch at a dust concentration of 350 g/m³ and a volume of 10 m³, dust layer on the floor, base case value of the dust dispersability constant $A_p = 0.6$

- Case 4: Primary explosion of maize starch at a dust concentration of 350 g/m³ and a volume of 10 m³, dust layer on the floor, alternative value of the dust dispersability constant $A_p = 1.2$
- Case 5: Primary explosion of maize starch at a dust concentration of 350 g/m³ and a volume of 10 m³, dust layers on shelves, base case value of the dust dispersability constant $A_p = 0.6$
- Case 6: Primary explosion of maize starch at a dust concentration of 350 g/m³ and a volume of 10 m³, dust layers on shelves, alternative value of the dust dispersability constant $A_p = 1.2$.

The computational grid size was 0.2 m with cubical volume elements (cells) extending throughout the tunnel and projected 10 m beyond the tunnel opening. Note that the mean particle size of the coal used in these simulations was 18 µm. Thus, the length scale ratio of the grid cell size to the particle diameter was roughly 10^4 meaning that the resolution of the model was not at the individual particle level but more at the scale of a cloud of coal dust particles.

Comparing Cases 1 and 2, the methane primary explosion yields a higher explosion pressure and is faster than the maize starch dust primary explosion. The pressure histories did not appear to exhibit characteristics significantly different from Case 2. However, the flame extension was quite different as seen in Fig. 10.21.

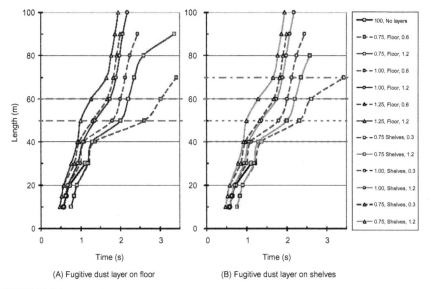

FIGURE 10.21

Flame length predicted by DESC for different cases of fugitive dust deployment and different levels of dust reactivity. The key indicates the reactivity level and the dispersability constant.

Reproduced with permission from Skjold, T., Eckhoff, R.K., Arntzen, B.J., Lebecki, K., Dyduch, Z., Klemens, R. et al., 2007. Simplified modelling of explosion propagation by dust lifting in coal mines. In: Proceedings of the Fifth International Seminar on Fire & Explosion Hazards, Edinburgh, UK.

As expected, higher dust dispersability and higher reactivity result in longer flame extensions.

The authors acknowledged that despite the success in simulating many of the important features of a dust explosion in an industrial-scale system, there are practical limitations to the use of DESC for real applications in explosion investigation or hazard analysis. The limitations fall into two categories: parameter uncertainty and computational effort. In practical settings, the input parameters for the model are often not known or are specified only approximately. It can be especially difficult to specify the physical and combustion characteristics of the dust. The challenge with computational effort is the conflicting desires to have a small enough grid size that can resolve the fine structure of the flow field without overwhelming the investigator's computer resources with millions of computational cells.

EXAMPLE 10.3

Consider Example 9.6 again. This time, use the dust entrainment correlation used in the DESC CFD code. Assume that a primary explosion has propagated a shock wave with a particle velocity of 30 m/s. The solid density of the aluminum alloy particles is 2700 kg/m^3 and the ambient conditions are 25°C and 1.01 bar. The depth of the accumulated dust layer is 5 mm and the mean particle diameter is 60 μm. The value of the dimensionless constant is $A_p = 1.2$. Calculate the mass flux of the entrained dust and compare this with the result obtained in Example 9.6 using Ural's method.

Solution

We begin by calculating the vertical velocity of the entrained dust:

$$v_z = 0.004\, \delta_d^{0.216}\, v_\infty^{1.742}\, d_p^{-0.053}\, \rho_s^{-0.160}\, A_p^{0.957}$$
$$= 0.004\,(5\text{ mm})^{0.216}\,(30\text{ m/s})^{1.742}\,(60\ \mu m)^{-0.053}\,(2700\text{ kg/m}^3)^{-0.160}\,(1.2)^{0.957}$$
$$= 0.574\text{ m/s}$$

The entrained mass flux of dust $\dot{m}_d''$ is based on an assumed entrained dust concentration of 1 kg/m^3:

$$\dot{m}_d'' = \rho_d\, v_z = (1\text{ kg/m}^3)\,(0.574\text{ m/s}) = 0.574\text{ kg/m}^2\cdot\text{s}$$

This compares reasonably well with Ural's method which resulted in a mass flux of 0.369 kg/m^2-s

---

10.4.3 SUMMARY OF OTHER ACCELERATING DUST FLAME STUDIES

Bielert and Sichel formulated a multiphase mixture model based on the Euler equations (the equations of change with no diffusion or heat conduction) and solved them using Godunov's method (Bielert and Sichel, 1999). They assumed both dynamic and thermal equilibriums between the gas and particle phases. They specified an empirical expression for the laminar burning velocity for corn starch and applied a turbulence correlation to convert it to a turbulent burning velocity. They compared the performance of the model against experimental data obtained by others to study flame propagation in a dust conveying system. The model was able to simulate certain flame propagation phenomena observed in the test program including flame acceleration to supersonic speeds and oscillating flame speeds.

Krause and Kasch simulated the dust filling process in a full-scale silo using the commercially available code FLUENT 4 (Krause and Kasch, 2000). The experiments, performed by others, were performed with corn starch dust in a 45-m^3 silo with a rectangular cross-section. Krause and Kasch computed the dust three-dimensional concentration profiles at two different filling rates. They described their work as "cold flow" simulations, ie, simulations of dust−air mixture without ignition. They concluded that for the base case filling rate the dust concentration in the location of the ignition source (as specified in the experimental study) had achieved the optimum concentration of 400−600 g/m^3. In the experimental study, this was a particularly violent deflagration. The second, more rapid filling rate was a less violent deflagration in the experimental study, and the CFD simulations predicted that the dust concentration in the location of the ignition source was over 1000 g/m^3, a fuel-rich condition that could explain the lower explosion pressure.

Radandt et al. performed a very clever flame acceleration study of corn starch dust deflagrations in a series of cylindrical enclosures with varying height to diameter (H/D) ratios using both numerical and experimental methods (Radandt et al., 2001). They developed an Eulerian−Lagrangian multiphase flow model and incorporated the $k-\varepsilon$ turbulence model. A more detailed discussion of their model formulation can be found in the paper by Zhong and Deng (2000). The range of H/D ratios tested are 1, 2, 4, 6, 8, 10, and 15. The intent of the study was to determine if flame acceleration effects were observable in the larger H/D ratios. It was determined that this indeed was the case. Their CFD results were in reasonable agreement with their experimental results which indicated that for $H/D > 8$ the rate of pressure rise was much greater than the rate of pressure rise for vessels with $H/D < 8$. For vessels with $H/D < 8$, the rate of pressure rise was fairly consistent with $H/D = 1$. Thus, flame acceleration effects were determined to be important for cylindrical vessels with $H/D > 8$. This study by Radandt et al. is an excellent example of how CFD analysis can supplement an experimental investigation and provide a deeper level of understanding of the combustion phenomena involved.

Jian et al. studied flame propagation through a pipe connecting two vessels (Jian et al., 2009). The primary vessel had a volume of 9.63 m^3, the secondary vessel had a volume of 4.4 m^3, and they were connected by a pipe 29.3 m in length and a diameter of 0.3 m. They performed flame propagation experiments with potato starch dust and measured the pressure history in both vessels and tracked the flame motion through the pipe. They developed and used an Eulerian−Lagrangian model to simulate the accelerating dust flame in the connecting pipe. The simulation agreed well with the measured final flame front speed of 338 m/s.

Skjold and his colleagues have demonstrated the ability of the DESC code to simulate flame acceleration in cylindrical tubes with obstacles (Skjold et al., 2005a,b, 2014). In the first study, they used the flame acceleration experimental results obtained by Pu et al. with a flammability tube 1.86 m in length and 0.19 m in width (Pu et al., 1988). Skjold et al. tried three different grid sizes—28.5, 19.0, and 9.5 mm—and found that the smallest grid size gave the best results in comparison with the experimental data. In the later study, they used a custom-made flame acceleration tube with internal obstacles to generate both gas and dust propagating flames (Skjold et al., 2014). The tube measured 3.6 m in length and had a square cross-section with a width of 0.27 m. Their objective of the test program was to generate flame propagation data specifically for validating the FLACS and DESC models. FLACS did a better job of predicting flame propagation with initially turbulent propane−air mixtures compared to initially quiescent mixtures. FLACS reportedly underpredicted the rate of combustion in fuel-lean mixtures and overpredicted the rate of combustion in fuel-rich mixtures. For the maize starch flame propagation tests, the results tended to be a bit scattered. *In toto*, the errors between experimental and CFD-predicted peak pressures ranged from 20% to over 100%. The DESC code predicted the highest explosion pressures reasonably well. To improve the agreement of the model simulations with the experimental data, a radiative heat transfer submodel was added to the model. These studies are yet another example of how CFD analysis and experimental programs can provide a deeper understanding of combustible dust phenomena.

Vingerhoets et al. performed an experimental program and analyzed the results with DESC for a dust conveying system (Vingerhoets et al., 2013). The physical system consisted of a pipe connected to a vessel which was in turn connected to another pipe. A fan connected at the entrance to the inlet pipe provided the motive force to drive combustible dust through the inlet pipe, the vessel, and the outlet pipe. The investigators tried different pipe lengths and pipe diameters. Ignition always occurred in the vessel. Three dusts were tested: wood flour, phenolic dust, and corn starch. Dust concentrations were varied from 250, 500, and 750 g/m^3. In all, the investigators performed 266 DESC simulations: 94 baseline simulations and 172 sensitivity variations. In their simulations, dust flames propagated outward from the vessel to the open end of both pipes. Even with inlet air velocities of 30 m/s, the dust flame propagated upstream to the end of the pipe. The authors cautioned that the DESC simulations gave physically correct results but in some cases the quantitative agreement could be off by as much as 50% in predicting peak pressures and 20% in flame speeds.

10.5 CFD STUDIES FOR ACCIDENT INVESTIGATION AND SAFETY ANALYSIS

In this chapter, we have been examining the role of CFD analysis in progressively more complex problems involving dust explosion dynamics. Two trends should be apparent by now: first, technical advances over the last 50 years have made CFD analysis a very powerful tool for exploring dust deflagration behavior. Second, CFD analysis continues to suffer some of the same limitations imposed by time and length scale constraints imposed by combustion phenomena. While both software tools and computer hardware have become more powerful, the difficulties imposed by velocity slip, turbulent eddies, radiant heat losses, and chemical reaction rates continue to conspire to make the mathematical modeling of dust explosion dynamics a difficult task. The bigger the size of the problem setting, the bigger these challenges become.

The need for the effective management of combustible dust hazards can arise in many different types of physical settings, but in general, these physical settings can be broadly classified as industrial, agricultural, or mining. There are many characteristics of these settings that set them apart from the laboratory, but perhaps the most important features are their physical size and complexity. The characteristic length scale for these physical settings can range from 10 to 100 m for industrial (factories) and agricultural facilities (grain elevators). For underground coal mines, the length scale can reach 1000 m. If the mean diameter of a combustible dust particle is 100 μm, then this ratio of length scales becomes 10^4-10^8, and that is only in one dimension. For a three-dimensional problem, even with adaptive gridding algorithms the computational mesh size can number in the millions of cells. This geometric scale and complexity directly influence the evolution of the turbulent flow field and, in turn, the rate of combustion (Cant et al., 2004). The ability to apply a CFD analysis to a practical setting must face this difficult challenge.

For most of us with an interest in dust explosion dynamics, we will have neither the time nor the resources to build our own CFD code. A number of investigators have tackled complex dust explosion dynamics problems using commercially available CFD codes such as Star CCM, FLUENT, or DESC, or open-source CFD codes like OpenFOAM. The DESC code was developed specifically to analyze combustible dust hazards, and its developers have been very active in demonstrating and testing its capabilities. Hence, there will be many references to this particular CFD code. Contributions from investigators using other CFD codes will be noted as well. We shall now examine some of the contributions in applying CFD to three of these more challenging environments: explosion development in process vessels, explosions in interconnected vessels, and coal mines.

10.5.1 EXPLOSION DEVELOPMENT IN PROCESS VESSELS

Zhong and Deng reported on an Eulerian–Lagrangian multiphase model for simulating dust deflagration behavior (Zhong and Deng, 2000). The $k-\varepsilon$ turbulence model was incorporated into the CFD code. They applied this CFD code to the problem of a maize starch dust deflagration from a 12-m^3 silo. Their paper does not give a description of the numerical solution of the model, so it is not possible to determine the performance characteristics of the CFD code.

There have been several papers published on the application of either FLACS or DESC code to the analysis of explosion development and venting from process vessels (Siwek et al., 2004; Skjold et al., 2006a,b; Davis et al., 2011; Skjold, 2014). Siwek et al. characterized the turbulent velocity and solids concentration fields in four industrial spray dryers ranging in size from 6 to 310 m^3. They also performed deflagration experiments with two combustible dusts—maize starch and cellulose—in a test vessel with a volume of 43.3 m^3. They then performed explosion venting and suppression tests in the vessel. They used the FLACS code to test the ability of the CFD code to simulate the test conditions as well as evaluate its use for extrapolating the results to spray dryers of other sizes. They found the performance of FLACS to be satisfactory. Skjold and his colleagues followed suit with tests of the DESC code (Skjold et al., 2006a,b). Davis and his collaborators demonstrated the use of DESC for assistance in evaluating the risk of combustible dust deflagrations (Davis et al., 2011). Finally, Skjold demonstrated the ability of DESC to simulate a series of deflagration venting tests from a 236-m^3 silo (Skjold, 2014). The silo had an *H/D* ratio of approximately 6, so venting performance was very sensitive to the location of the ignition source in relation to the explosion vent. The closer the vent was to the ignition source, the lower the peak pressure during venting.

EXAMPLE 10.4

In one of their corn starch deflagration tests, Radandt et al. tested a cylindrical vessel with an elongation ratio of $H/D = 15$ and a volume of 1 m^3. The cloud was ignited at one end of the vessel. Overpressures were measured at three locations: the vessel center, the midpoint between the center to the end of the vessel, and the end of the vessel. The peak overpressure measurements at each location were approximately 7.0, 9.0, and 17.5 barg, respectively. The measured maximum pressure in a standard 1-m^3 deflagration test vessel was 9.4 barg. The measured pressure histories are shown in the figure below.

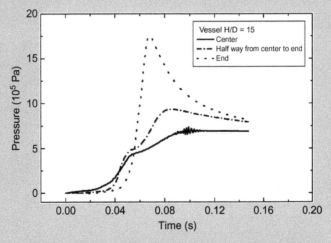

Pressure histories of corn starch deflagration measured at three locations.

Reproduced with permission from Radandt, S., Shi, J., Vogl, A., Deng, X.F., Zhong, S.J., 2001. Cornstarch explosion experiments and modeling in vessels ranged by height/diameter ratios. J. Loss Prev. Process Ind., 14, 495–502.

Are these data consistent with the occurrence of a detonation inside this test vessel?

Solution

The characteristics of a detonation are a large overpressure (approximately twice the constant volume explosion pressure) and an extremely rapid pressure rise. If the initial combustion event is a deflagration, then flame acceleration is required to achieve a detonation. This suggests three criteria for gauging whether a detonation occurred: the magnitude of the maximum pressure, the magnitude of the maximum rate of pressure rise, and evidence of flame acceleration.

The measured maximum pressure from the standardized test vessel provides us with our reference value of 9.4 barg (measured constant volume explosion pressure). The overpressure measured in the vessel center, 7.0 barg, was less than the maximum value suggesting incomplete combustion. The burn fraction implied by the center pressure measurement is

$$f = \frac{P_{measured} - P_0}{P_{max} - P_0} = \frac{7.0 \text{ barg}}{9.4 \text{ barg}} = 0.745$$

Thus, if the deflagration was adiabatic, then roughly 75% of the fuel participated in the deflagration. However, as noted by the authors of this study, the surface area for heat transfer increases with increasing H/D ratios. On the one hand, heat losses will tend to decrease deflagration overpressures, but on the other hand, flame acceleration will tend to increase the overpressure. Since incomplete combustion is common in dust deflagrations, the overpressure measured at the center seems reasonable.

The pressure measured at the three-quarters point (the midpoint between the center of the vessel and the end of the vessel), 9.0 barg, approached the maximum explosion pressure. If incomplete combustion is the rule in dust deflagration, the magnitude of this measurement suggests flame acceleration. The three pressure measurements are increasing in magnitude as they continue their advancement towards the end of the vessel. This trend is also consistent with flame acceleration.

The final pressure is more than twice the value of the center pressure measurement, and approximately twice the value of the pressure measured at three-quarters point. In Chapter 4, we learned that the C–J detonation pressure was approximately twice the value of the explosion pressure. The pressure measured at both the center and the three-quarters point was consistent with a detonation.

The rate of pressure rise is depicted in the figure above. Inspection of the pressure histories reveals that the maximum pressure is attained at the end of the vessel furthest from the ignition source before the center or three-quarters point. This suggests that the pressure rise inside the vessel is not hydrostatic (uniform in all directions). Instead, this observation suggests that there was a spatial discontinuity in the vessel pressure, further evidence of flame acceleration.

To further test these observations, consider the acoustic time scale for this deflagration test. Assume that the sound speed in the unburnt gas of the vessel is 350 m/s. The authors of this paper tell us that the volume of the vessel is 1 m^3. For a cylindrical vessel with an $H/D = 15$, the diameter is 8.49 cm and the length is 1.27 m. The acoustic time scale, $t_{acoustic} = 1.27$ m/350 m/s $= 3.6$ ms. The peak pressure at the end of the vessel occurred at roughly 30 ms after ignition. A compression wave could travel to the end of the vessel, reflect off of the vessel wall and travel back to the ignition location several times before the pressure at the end of the vessel achieved its maximum value. This indicates a relatively slow pressure rise compared to what one would expect with a detonation.

Even a slow detonation wave should have a speed on the order of 1000 m/s, giving a detonation time scale on the order of 1 ms. If a deflagration-to-detonation transition had occurred, the peak pressure should have been attained much earlier than 30 ms. Thus, the rate of pressure rise at the end of the vessel appears to be too slow to be a true detonation. Instead, this was more likely than not an accelerated deflagration.

10.5.2 INTERCONNECTED VESSELS

In Chapter 9, Flame Acceleration Effects, the hazard of pressure piling was discussed. The basic problem is to simulate flame propagation from one vessel to a second when they are connected by a pipe. Kosinski and Hoffmann developed and tested an Eulerian–Lagrangian 2D model to simulate the propagation of a dust flame from a primary vessel to a secondary vessel via a short pipe (Kosinski and Hoffmann, 2005, 2006). They implemented Godunov's method for the numerical solution of the governing equations.

This problem is more challenging than just flame propagation through a pipe because of the complications introduced by flame growth and propagation in the two vessels. For their initial conditions, they assigned the first vessel to be filled with hot pressurized combustion products and the second vessel to be filled with a combustible dust cloud at ambient conditions. Fig. 10.22 shows the problem sketch for the CFD analysis with the dimensions as shown for the two-dimensional rectangular geometry.

The hot gases then propagated through the connecting pipe and entered the second vessel. As the hot gas entered the second vessel, it expanded and pushed the unburnt dust cloud toward the walls. Fig. 10.23 shows the solids' concentration profile in the second vessel at three successive moments in time.

It was evident from the simulation results that the larger channel height resulted in a faster displacement rate of the unburnt solids in the second vessel.

This problem has also been the focus of attention of Skjold and his colleagues at the University of Bergen in Norway (Skjold et al., 2005a,b; Skjold, 2007a,

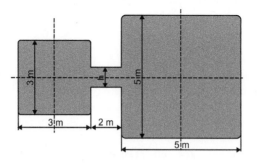

FIGURE 10.22

Computational domain for the pressure piling problem.

Reproduced with permission from Kosinski, P., Hoffmann, A.C., 2005. Dust explosions in connected vessels: mathematical modeling. Powder Technol. 155, 108–116.

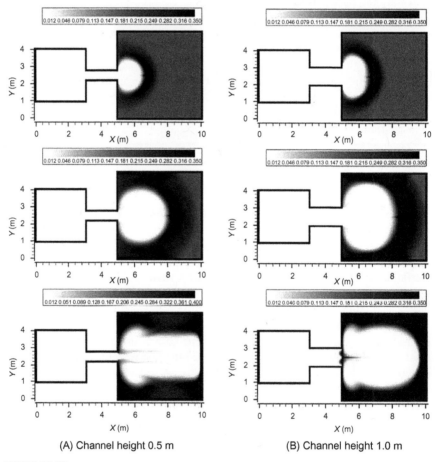

(A) Channel height 0.5 m (B) Channel height 1.0 m

FIGURE 10.23

The effect of channel height on the speed of displacement of the unburnt solids in the second vessel. The solids' concentration fields in the second vessel are shown at three distinct times (from top to bottom): 7.5, 12.5, and 27.5 milliseconds.

Reproduced with permission from Kosinski, P., Hoffmann, A.C., 2005. Dust explosions in connected vessels: mathematical modeling. Powder Technol. 155, 108–116.

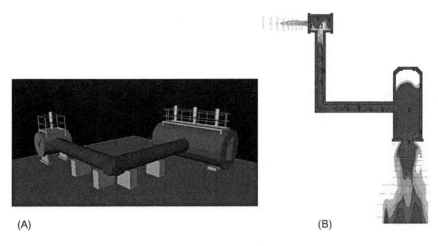

(A)   (B)

FIGURE 10.24

(A) The three-dimensional rendering of the interconnected vessels from DESC.

(B) A graphical image of deflagration venting from both vessels.

Reproduced with permission from Skjold, T., Eckhoff, R.K., Arntzen, B.J., Lebecki, K., Dyduch, Z., Klemens, R. et al., 2007. Simplified modelling of explosion propagation by dust lifting in coal mines. In: Proceedings of the Fifth International Seminar on Fire & Explosion Hazards, Edinburgh, UK.

2010). In their studies, they applied the DESC code to a problem for which there was experimental data. The system involved a primary vessel with a volume of 20 m³, a secondary vessel with a volume of 2 m³, and a connecting pipe with an elbow. The length of each pipe segment was 6.4 m with the elbow at the midpoint giving a total pipe length of 12.8 m. Two pipe diameters were used: 0.25 and 0.50 m. The layout of the system is shown in Fig. 10.24.

The experimental program consisted of a total of 34 deflagration tests. Four types of combustible dust were tested: two types of potato starch, coal, and silicon. Explosions were transmitted more frequently with the larger diameter pipe (0.5 m). When flame propagation failed, it tended to fail at the elbow of the piping. Only 13 of the 34 tests produced a pressure piling effect in the second vessel. Three of these tests using coal dust were selected for simulation with DESC.

A total of 14 simulations were conducted to calibrate the value of the laminar burning velocity and to evaluate the effect of changing the ignition location and delay time. All of these simulations used a 0.01 m cubical grid spacing. This resulted in a total of 730,000 grid cells. A final simulation was performed with a grid size of 0.05 m to test the grid size dependence of the solution; this simulation required 3.5 million grid cells.

The best values selected predicted the peak pressures in the primary vessel to within 10% in two of the three selected cases. But in the secondary vessel, the error in the predicted peak pressures ranged from 25% to over 100%. The change in grid size resulted in changes in the predicted peak pressures of roughly 5% indicating that the original grid size was adequate for the simulations.

Although a portion of the error in the predicted peak pressures could be attributed to deficiencies in the parameter estimation for the combustion submodel (the laminar burning velocity), a greater source of error was likely the operational and instrumentation issues with the deflagration tests themselves. For example, the deflagration from the primary vessel could propagate to the secondary vessel only if the primary dust deflagration could disperse the dust layer in the connecting pipe. The failure to disperse the dust layer in the pipe resulted in the failure of the primary deflagration to propagate into the second vessel. The difficulty in performing full-scale dust deflagration tests with industrial-scale equipment cannot be overstated. As Skjold indicates in his paper, if full-scale tests are to be useful for CFD model validation, they must be conducted with redundant instrumentation, replicate tests should be performed on the baseline test conditions (at least three times), and the determination of combustion rate parameters must be based on representative samples of the combustible dust (Skjold, 2010).

10.5.3 UNDERGROUND COAL MINES

Underground coal mines present a particularly hazardous environment. Although bituminous coal dust itself is usually a hazard classification of St 1, there are two other factors which magnify this hazard: the presence of fugitive methane gas and the extreme level of confinement imposed by the walls of the gallery (especially in room-and-pillar mining). By the 1920s, it was realized that the application of inert rock dust to the floor, walls, and roof of the mine could limit flame propagation and suppress coal dust explosions. Rock dusting requirements at that time were largely determined empirically (Cybulski, 1975; Nagy, 1981). Even the most recent changes in the United States were largely based on empirical tests in the laboratory and in experimental mines (Cashdollar et al., 2010). The reason for the change in rock dusting requirements was technological: the degree of mechanization of coal mining had changed considerably since the 1920s and with this change came the increased generation of "float coal dust," coal dust with a particle size smaller than 75 μm.

CFD studies offer the possibility of developing a deeper understanding of how inerting works and to help us predict under what conditions it might fail. An early CFD study was undertaken at the U.S. Bureau of Mines. Edwards and Ford developed a CFD model for rock dust entrainment during a coal dust explosion (Edwards and Ford, 1988). They built a multiphase mixture model which modeled turbulence with an eddy diffusivity. Dust entrainment was modeled with a friction velocity correlation. The model was solved by the finite differences method. The model was used to evaluate experimental mine test data with the goal of using the model to predict the conditions under which a coal dust explosion test would be propagating (insufficient rock dust) or nonpropagating (sufficient rock dust). The model did demonstrate some success in that regard, but the population of test results was rather limited. It is not apparent that this model played a significant role in the Bureau's future work.

As discussed earlier in this chapter, DESC has been successfully used to demonstrate its applicability to dust flame acceleration problems in mine galleries (Skjold et al., 2007; Skjold, 2007b). Although future work remains to be done, it would appear that DESC, with some modification, could be applied successfully to rock dusting questions.

Additionally, FLACS, the progenitor of DECS, has been used to demonstrate oscillating behavior of forward and reverse flows of combustion products caused by the complex network of mine entries and cross-cuts (Davis et al., 2015). There are certain key features that have been traditionally used in the investigation of underground coal mine accidents and, as shown by Davis et al., some of these features are based erroneously on one-dimensional flow intuition in a complex, two-dimensional flow network. This is just one example of how CFD tools can provide insight for accident investigation in the complex environment of the underground coal mine.

Kosinski has also contributed to our understanding of shock wave dynamics and dust entrainment (Kosinski, 2008). He used an Eulerian–Lagrangian CFD model to investigate explosion suppression by a cloud of inert particles. The gridding of the computational domain required a little over 120,000 cells to simulate a gallery 8 m in length. His simulations demonstrated the effectiveness of suppression increased with increasing inert dust cloud concentration and decreasing inert dust particle size.

Houim and Oran have developed an Eulerian–Eulerian CFD code for simulating dust layer entrainment (Houim and Oran, 2015a,b,c, 2016). They have been able to apply this model to study the effectiveness of inert dust deployment. Their highly detailed CFD model like Kosinski's demonstrated the effect of inert dust on quenching the flame. However, in some of their simulations the grid size was as small as 174 μm. This resulted in some extremely fine-scale detail in the development of vortices in the simulated flow field (recall from Chapter 9 the experimental observations of vortices in the dust entrainment experiments performed by the U.S. Bureau of Mines). Note that even at this remarkably fine level of resolution, the cell size was still almost six times larger than the 30 μm particle size used in the simulations. The computational demands of this model preclude (for now) the routine use of it as an analytical tool for investigating the optimal design of inert dust deployment. Typical simulations over a length of 20 m required 2 million grid cells and over 70,000 CPU hours to perform. Thus, this very fine scale of resolution of the flow field comes with a large computational price that is likely not yet feasible in an industrial project.

Finally, the work of Collecutt and his colleagues is noteworthy in that they used an open-source CFD code, OpenFOAM, to design an active (electronically initiated) coal dust explosion suppression system (Collecutt et al., 2009; Proud et al., 2015). The design problem was to specify the characteristics of a water spray barrier that would trigger if a dust flame was detected. Their calculations required 300,000 grid cells. They used their model to design the nozzle array, spray flow rate, and spray droplet size. The prototype design was tested and successfully demonstrated in a half-scale mine.

10.6 SUMMARY

Many investigators have begun to apply the tools of modern combustion science to the study of dust explosion dynamics. There have been significant advances in both experimental and theoretical methods. In the laboratory, combustion measurement and diagnostic tools like flow visualization, laser Doppler anemometry, calorimetry, and various forms of spectroscopy have revealed the interplay of fundamental physical and chemical processes in dust combustion phenomena. This wealth of empirical data is just beginning to be understood in the context of modern combustion science. Without question, one of the most powerful advancements in dust explosion dynamics has been the application of CFD to dust combustion. Depending on the specific hazard scenarios under consideration, there is an intricate relationship among multiphase flow, turbulence, thermal radiation, and chemical kinetics.

While these combustion science-based investigations have resulted in deeper insights into the fundamental aspects of dust deflagrations, the application of these insights to real industrial settings remains very challenging. These challenges arise not so much from the lack of theoretical or experimental tools but rather from the challenge of describing explosion consequences measured on a length scale on the order of meters when the combustion phenomena occur at a length scale of microns. The difficulties are magnified by the broad variance one typically encounters in measuring the physical and chemical properties of real dusts and powders exposed to the ambient industrial environment.

There have been some very important technological advances in the measurement and control of combustible dust hazards. The current industrial state of the art focuses heavily on representative sampling of dusts, the measurement of the combustion hazard properties using standardized laboratory equipment, and the application of this information to empirical design methods for explosion safeguards. This current state of the art represents a quantum leap in improvement from the state of the art 30 years ago. But modern combustion science offers an opportunity to learn so much more.

Analyzing dust combustion phenomena from the perspective of modern combustion science is a challenging assignment. However, there are many powerful tools at your disposal to help in this effort. Simple combustion models, sound laboratory measurements, and the judicious application of CFD can reveal important insights that can enhance one's ability to identify, evaluate, and control combustible dust hazards.

REFERENCES

Abdel-Gayed, R.G., Bradley, D., McMahon, M., 1979. Turbulent flame propagation in premixed gases: theory and experiment. Seventeenth Symposium (International) on Combustion. The Combustion Institute, Elsevier, New York.

Anderson, D.A., Tannehill, J.C., Pletcher, R.H., 1984. Computational Fluid Mechanics and Heat Transfer. Hemisphere Publishing Corporation, New York.

Beckstead, B.W., 2005. Correlating aluminium burning times. Combust. Explos. Shock Waves 41, 533−546.

Bielert, U., Sichel, M., 1999. Numerical simulation of dust explosions in pneumatic conveyors. Shock Waves 9, 125–139.

Bind, V.K., Roy, S., Rajagopal, C., 2011. CFD modeling of dust explosions: rapid combustion in a 20-L apparatus. Can. J. Chem. Eng. 89, 663–670.

Bind, V.K., Roy, S., Rajagopal, C., 2012. A reaction engineering approach to modeling dust explosions. Chem. Eng. J. 201, 625–634.

Buckmaster, J.D., Clavin, P., Liñán, A., Matalon, M., Peters, N., Sivashinsky, G., et al., 2005. Combustion theory and modeling. Proc. Combust. Inst. 30, 1–19.

Cant, R.S., Dawes, W.N., Savill, A.M., 2004. Advanced CFD and modeling of accidental explosions. Annu. Rev. Fluid Mech. 36, 97–119.

Carrier, G.F., Fendell, F.E., Feldman, P.S., 1980. Nonisobaric flame propagation. Dynamics and Modelling of Reactive Systems. Academic Press, New York, pp. 333–351.

Cashdollar, K.L., Sapko, M.J., Weiss, E.S., Harris, M.L., Man, C.-K., Harteis, S.P., et al., 2010. Recommendations for a New Rock Dusting Standard to Prevent Coal Dust Explosions in Intake Airways U.S. Department of Health and Human Services, National Institute of Occupational Safety and Health, Report of Investigations, 19679.

Chi, D.N.H., Perlee, H.E., 1974. Mathematical Study of a Propagating Flame and Its Induced Aerodynamics in a Coal Mine Passageway U.S. Department of Commerce, Bureau of Mines, Report of Investigations, 7908.

Clark, D.P., Smoot, L.D., 1985. Model of accelerating coal dust flames. Combust. Flame 62, 255–269.

Cloney, C.T., Amyotte, P.R., Khan, F.I., Ripley, R.C., 2014. Development of an organizational framework for studying dust explosion phenomena. J. Loss Prev. Process Ind. 30, 228–235.

Collecutt, G., Humphreys, D., Proud, D., 2009. CFD simulation of underground coal dust explosions and active explosion barriers. In: Seventh International Conference on CFD in the Minerals and Process Industries, Melbourne, Australia.

Continillo, G., 1988. Numerical study of coal dust explosions in spherical vessels. Preprints of papers. American Chemical Society, Division of Fuel Chemistry, 33 CONF-8806317.

Continillo, G., 1989. Two-zone model and a distributed parameter model of dust explosions in closed vessels. Arch. Combust. 9, 79–94.

Courant, R., Friedrichs, K.O., 1948. Supersonic Flow and Shock Waves. Springer Verlag, Berlin (reprint 1976 edition).

Crowe, C.T., Schwarzkopf, J.D., Sommerfield, M., Tsuji, Y., 2012. Multiphase Flows With Droplets and Particles, second ed. CRC Press, New York.

Cybulski, W., 1975. Coal Dust Explosions and Their Suppression. Vol. 73. No. 54001. Bureau of Mines, US Department of the Interior and the National Science Foundation.

Davis, S.G., Hinze, P.C., Hansen, O.R., Wingerden, K. van, 2011. Does your facility have a dust problem: methods for evaluating dust explosion hazards. J. Loss Prev. Process Ind. 24, 837–846.

Davis, S.G., Engel, D., Wingerden, K. van, 2015. Complex explosion development in mines: case study-2010 Upper Big Branch Mine explosion. Process Saf. Prog. 34, 286–303.

Di Benedetto, A., Russo, P., Sanchirico, R., Di Sarli, V., 2013. CFD simulations of turbulent fluid flow and dust dispersion in the 20 liter explosion vessel. AIChE J. 59, 2485–2495.

Di Sarli, V., Russo, P., Sanchirico, R., Di Benedetto, A., 2014. CFD simulations of dust dispersion in the 20-L vessel: effect of nominal dust concentration. J. Loss Prev. Process Ind. 27, 8–12.

Eckhoff, R., 2003. Dust Explosions in the Process Industries, third ed. Gulf Professional Publishing, Elsevier, New York.

Eckhoff, R., 2009. Review article: dust explosion prevention and mitigation, status and developments in basic knowledge and in practical application. Int. J. Chem. Eng. Volume 2009, 1–12, article ID 569825.

Edwards, J.C., Ford, K.M., 1988. Model of Coal Dust Explosion Suppression by Rock Dust Entrainment U.S. Department of Commerce, Bureau of Mines, Report of Investigations, 9206.

Fan, B.C., Ding, D.Y., Tank, M.J., Pu, Y.K., Hu, S., 1993. An aluminium dust explosion in a spherical closed vessel. In: Proceedings of the Fifth International Colloquium on Dust Explosions, Pultusk, Poland, pp. 21–31, April.

Hanna, K., 2015. CFD breaks the $billion barrier! Mentor Graphics blog. Available at: https://blogs.mentor.com/khanna/blog/2015/03/26/cfd-breaks-the-billion-barrier/ (posted 26.03.15).

Houim, R.W., Oran, E.S., 2015a. Numerical simulation of dilute and dense layered coal-dust explosions. Proc. Comb. Inst. 35, 2083–2090.

Houim, R.W., Oran, E.S., 2015b. Structure and flame speed of dilute and dense layered coal-dust explosions. J. Loss Prev. Process Ind. 36, 214–222.

Houim, R.W., Oran, E.S., 2015c. Effect of radiation on the propagation of planar coal dust flames in air. In: 25th International Colloquium on the Dynamics of Explosions and Reactive Systems, Leeds, UK.

Houim, R.W., Oran, E.S., 2016. A multiphase model for compressible granular-gaseous flows: formulation and initial tests. J. Fluid Mech. 789, 166–220.

Jian, W., Xinguang, L., Shengjun, Z., Radandt, S., Fuli, W., Chunli, R., 2009. Flame propagation through potato starch/air mixture in pipe of interconnected vessels. In: Proceedings of the 8th International Conference on Measurement and Control of Granular Materials, pp. 376–380.

Julien, P., Vickery, J., Whiteley, S., Wright, A., Goroshin, S., Frost, D.L., et al., 2015. Freely-propagating flames in aluminum dust clouds. Combust. Flame 162, 4241–4253.

Kansa, E.J., Perlee, H.E., 1976. Constant-Volume Flame Propagation: Finite-Sound-Speed Theory. U.S. Department of the Interior, Bureau of Mines, Report of Investigations, 8163.

Kansa, E.J., Perlee, H.E., 1980. A transient dust-flame model: application to coal-dust flames. Comb. Flame 38, 17–36.

Kelley, A.P., Bechtold, J.K., Law, C.K., 2012. Premixed flame propagation in a confining vessel with weak pressure rise. J. Fluid Mech. 691, 26–51.

Kjäldman, L., 1992. Numerical flow simulation of dust deflagrations. Powder Technol. 71, 163–169.

Kono, M., Tsukamoto, T., Iinuma, K., 1985. Fundamental study on the modeling of flame propagation in constant volume vessels. In: International Symposium on Diagnostics and Modeling of Combustion in Reciprocating Engines, Tokyo, Japan, September 4–6.

Kosinski, P., 2008. Numerical investigation of explosion suppression by inert particles in straight ducts. J. Hazard. Mater. 154, 981–991.

Kosinski, P., Hoffmann, A.C., 2005. Dust explosions in connected vessels: mathematical modeling. Powder Technol. 155, 108–116.

Kosinski, P., Hoffmann, A.C., 2006. An investigation of the consequences of primary dust explosions in interconnected vessels. J. Hazard. Mater. A137, 752−761.

Kosinski, P., Klemens, R., Wolanski, R., 2002. Potential of mathematical modelling in large-scale dust explosions. J. Phys. IV 12, 125−132.

Krause, U., Kasch, T., 2000. The influence of flow and turbulence on flame propagation through dust−air mixtures. J. Loss Prev. Process Ind. 13, 291−298.

Krazinski, J.L., Buckius, R.O., Krier, H., 1979. Coal dust flames: a review and development of a model for flame propagation. Prog. Energy Combust. Sci. 5, 31−71.

Kuo, K., Acharya, R., 2012a. Applications of Turbulent and Multiphase Combustion. John Wiley & Sons, New York.

Kuo, K., Acharya, R., 2012b. Fundamentals of Turbulent and Multiphase Combustion. John Wiley & Sons, New York.

Law, C.K., 2007. Combustion at a crossroads: status and prospects. Proc. Comb. Inst. 31, 1−29.

Majda, A., Sethian, J., 1985. The derivation and numerical solution of the equations for zero Mach number combustion. Combust. Sci. Technol. 42, 185−205.

Matalon, M., 1995. Flame propagation in closed vessels. In: Buckmaster, J., Takeno, T. (Eds.), Modelling in Combustion Science. Springer, Berlin, pp. 161−175.

Modest, M.F., 2013. Radiative Heat Transfer, third ed. Academic Press, New York.

Murillo, C., Dufaud, O., Bardin-Monnier, N., Lopez, O., Munoz, F., Perrin, L., 2013. Dust explosions: CFD modeling as a tool to characterize the relevant parameters of the dust dispersion. Chem. Eng. Sci. 104, 103−116.

Nagy, J., 1981. The Explosion Hazard in Mining. U.S. Department of Labor, Mine Safety and Health Administration. Informational Report 1119.

Oberkampf, W.L., Trucano, T.G., 2002. Verification and validation in computational fluid dynamics. Prog. Aerospace Sci. 38, 209−272.

Ogle, R.A., Beddow, J.K., Chen, L.-D., Butler, P.B., 1988. An investigation of aluminum dust explosions. Combust. Sci. Technol. 61, 75−99.

Oppenheim, A.K., Maxson, J.A., 1993. Thermodynamics of combustion in an enclosure. Prog. Astronaut. Aeronaut. 152, 365−382.

Oran, E.S., Boris, J.P., 1987. Numerical Simulation of Reactive Flow. Elsevier, New York.

Oran, E.S., Boris, J.P., 2005. Numerical Simulation of Reactive Flow, second ed. Cambridge University Press, Cambridge.

Peters, N., 2009. Multiscale combustion and turbulence. Proc. Combust. Inst. 32, 1−25.

Pope, S.B., 2005. Turbulent Flows. Cambridge University Press.

Proud, D.J., Collecutt, G.R., Humphreys, D.R., 2015. CFD modelling of coal dust explosions and suppression systems. 27th International Conference on Parallel Computational Fluid Dynamics. Montreal, Canada.

Pu, Y.K., Mazurkiewicz, J., Jarosinski, J., Kauffman, C.W., 1988. Comparative study of the influence of obstacles on the propagation of dust and gas flames. Twenty-Second Symposium (International) on Combustion. The Combustion Institute, Elsevier, New York.

Qiao, L., 2012. Transient flame propagation process and flame-speed oscillation phenomenon in a carbon dust cloud. Combust. Flame 159, 673−685.

Qiao, L., Xu, J., 2012. Detailed numerical simulations of flame propagation in coal-dust clouds. Combust. Theory Model. 16, 747−773.

Radandt, S., Shi, J., Vogl, A., Deng, X.F., Zhong, S.J., 2001. Cornstarch explosion experiments and modeling in vessels ranged by height/diameter ratios. J. Loss Prev. Process Ind. 14, 495−502.

Ramos, J.I., 1983. Numerical studies of laminar flame propagation in spherical bombs. AIAA J. 21, 415–422.

Richmond, J.K., Liebman, I., 1975. A physical description of coal mine explosions. Fifteenth Symposium (International) on Combustion, The Combustion Institute. Elsevier, New York, pp. 115–126.

Salamonowicz, Z., Kotowski, M., Polka, M., Barnat, W., 2015. Numerical simulation of dust explosion in the spherical 20-L vessel. Bull. Polish Acad. Sci. Tech. Sci. 63, 289–293.

Sirignano, W.A., 2010. Fluid Dynamics and Transport of Droplets and Sprays, second ed. Cambridge University Press, Cambridge.

Sivashinsky, G.I., 1974. On propagation of a flame in a closed vessel. Israel J. Technol. 12, 317–321.

Sivashinsky, G.I., 1979. Hydrodynamic theory of flame propagation in an enclosed volume. Acta Astronaut. 6, 631–645.

Siwek, R., Wingerden, K. van, Hansen, O.R., Sutter, G. Kubainsky, C., Schwartzbach, O.R., et al., 2004. Dust explosion venting and suppression of conventional spray dryers. In: Eleventh International Symposium on Loss Prevention and Safety Promotion in the Process Industries, Praha, Czech Republic.

Skjold, T., 2007a. Review of the DESC project. J. Loss Prev. Process Ind. 20, 291–302.

Skjold, T., 2007b. Simulating the effect of release of pressure and dust lifting on coal dust explosions. In: 21st International Colloquium on the Dynamics of Explosions and Reactive Systems, Poitiers, France.

Skjold, T., 2010. Flame propagation in dust clouds: challenges for model validation. In: Proceedings of the Eighth International Seminar on Fire & Explosion Hazards, Yokohama, Japan.

Skjold, T., 2014. Simulating vented maize starch explosions in a 236 m^3 silo. Fire Saf. Sci. 11, 1469–1480.

Skjold, T., Arntzen, B.J., Hansen, O.R., Taraldset, O.J., Storvik, I.E., Eckhoff, R.K., 2005a. Simulating dust explosions with the first version of DESC. Process Saf. Environ. Protect. 83 (B2), 151–160.

Skjold, T., Pu, Y.K., Arntzen, B.J., Hansen, O.R., Storvik, I.E., Taraldset, O.J., et al., 2005b. Simulating the influence of obstacles on accelerating dust and gas flames. In: Twentieth International Colloquium on the Dynamics of Explosions and Reactive Systems (ICDERS), Montreal, Canada.

Skjold, T., Arntzen, B.J., Hansen, O.R., Storvik, I.E., Eckhoff, R.K., 2006a. Simulation of dust explosions in complex geometries with experimental input from standardized tests. J. Loss Prev. Process Ind. 19, 210–217.

Skjold, T., Larson, O., Hansen, O.R., 2006b. Possibilities, limitations, and the way ahead for dust explosion modelling. In: HAZARDS XIX, Manchester, UK, March 28–30, 2006, pp. 282–297.

Skjold, T., Eckhoff, R.K., Arntzen, B.J., Lebecki, K., Dyduch, Z., Klemens, R., et al., 2007. Simplified modelling of explosion propagation by dust lifting in coal mines. In: Proceedings of the Fifth International Seminar on Fire & Explosion Hazards, Edinburgh, UK.

Skjold, T., Pedersen, H.H., Bernard, L., Middha, P., Narasimhamurthy, V.D., Landvik, T., et al., 2013b. A matter of life and death: validating, qualifying and documenting models for simulating flow-related accident scenarios in the process industry. Chem. Eng. Trans. 31, 187–192.

Skjold, T., Castellanos, D., Olsen, K.L., Eckhoff, R.K., 2014. Experimental and numerical investigation of constant volume dust and gas explosions in a 3.6-m flame acceleration tube. J. Loss Prev. Process Ind. 30, 164–176.

Slezak, S.E., Buckius, R.O., Krier, H., 1985. A model of flame propagation in rich mixtures of coal dust in air. Combust. Flame 59, 251–265.

Smoot, L.D., Horton, M.D., 1977. Propagation of laminar pulverized coal-air flames. Prog. Energy Combust. Sci. 3, 235–258.

Smoot, L.D., Horton, M.D., Williams, G.A., 1977. Propagation of laminar pulverized coal–air flames. Sixteenth (International) Symposium on Combustion. The Combustion Institute. Elsevier, New York, pp. 375–387.

Smoot, L.D., Pratt, D.T. (Eds.), 1979. Pulverized Coal Combustion and Gasification: Theory and Applications for Continuous Flow Processes. Springer, Berlin.

Smoot, L.D., Smith, P.J., 1985. Coal Combustion and Gasification. Plenum Press, New York.

Spijker, C. Kern, H. Held, K., Raupenstrauch, H., 2013. Modelling dust explosions. In: 2013 AIChE Annual Meeting, San Francisco, CA.

Sun, J.-H., Dobashi, R., Hirano, T., 1998. Structure of flames propagating through metal particle clouds and behavior of particles. Twenty-Seventh Symposium (International) on Combustion. The Combustion Institute, Elsevier, New York.

Sun, J.-H., Dobashi, R., Hirano, T., 2001. Temperature profile across the combustion zone propagating through an iron particle cloud. J. Loss Prev. Process Ind. 14, 463–467.

Sun, J.-H., Dobashi, R., Hirano, T., 2006a. Velocity and number density profiles of particles across upward and downward flame propagating through iron particle clouds. J. Loss Prev. Process Ind. 19, 135–141.

Takeno, T., Iijima, T., 1981. Theoretical study of nonsteady flame propagation in closed vessels. Prog. Astronaut. Aeronaut. 76, 578–595.

Thompson, P.A., 1972. Compressible-Fluid Dynamics. McGraw-Hill, New York.

van Wingerden, K., 1996. Simulation of dust explosions using a CFD-code. In: Proceedings of the International Symposium on Hazards, Prevention and Mitigation of Industrial Explosions and 7th ICDE.

Versteeg, H.K., Malalasekera, W., 2007. An Introduction to Computational Fluid Dynamics: The Finite Volume Method, second ed. Pearson, Upper Saddle River, NJ.

Vingerhoets, J., Farrell, T.M., Snoeys, J., 2013. Dust flame propagation in industrial scale piping. Part 2: CFD study of a conveying vessel-pipeline system. Proceedings of the 9th Global Congress on Process Safety. American Institute of Chemical Engineers, San Antonio, TX.

Westbrook, C.K., Mizobuchi, Y., Poinsot, T.J., Smith, P.J., Warnatz, J., 2005. Computational combustion. Proc. Combust. Inst. 30, 125–157.

Wilcox, D.C., 2006. Turbulence Modeling for CFD, third ed. DCW Industries, La Canada Flintridge, CA.

Zhong, S., Deng, X., 2000. Modeling of maize starch explosions in a 12-m^3 silo. J. Loss Prev. Process Ind. 13, 299–309.

Zydak, P., Klemens, R., 2007. Modelling of dust lifting process behind propagating shock wave. J. Loss Prev. Process Ind. 20, 417–426.

Appendix 1: Conversion Factors and Physical Constants

[Adapted with permission from Glassman et al., 2015, Appendix A; updated values as noted were obtained from Mohr, P.J., Taylor, B.N., Newell, D.B., 2012. CODATA recommended values of the fundamental physical constants: 2010. Rev. Mod. Physics 84, 1527−1605.]

$1\ \text{J} = 1\ \text{W-s} = 1\ \text{N-m} = 10^7\ \text{erg}$

1 cal (International Table) = 4.1868 J

1 cal (Thermochemical) = 4.184 J

$1\ \text{N} = 1\ \text{kg-m/s}^2 = 10^5\ \text{dyne}$

$1\ \text{Pa} = 1\ \text{N/m}^2$

$1\ \text{atm} = 1.0132\ 5 \times 10^5\ \text{N/m}^2$

$1\ \text{bar} = 10^5\ \text{N/m}^2 = 10^5\ \text{Pa}$

g = gravitational acceleration conversion factor = $9.80665\ \text{m/s}^2$

R_g = Universal gas constant = 8.314 J/mol-K = 8.314 kJ/kmol-K = 1.987 cal/mol-K

σ = Stefan−Boltzmann constant (updated) = $5.670 \times 10^{-8}\ \text{W/m}^2\text{-K}^4$

k = Boltzmann constant (updated) = $1.3806 \times 10^{-23}\ \text{J/K}$

N = Avogadro number (updated) = 6.022×10^{23} molecules/mol

h = Planck constant (updated) = $6.6261 \times 10^{-34}\ \text{J-s}$

c = speed of light = $2.997925 \times 10^8\ \text{m/s}$

Appendix 2: Table of Atomic Numbers and Masses

List of the Elements with Their Symbols, Atomic Numbers, and Atomic Masses[a]

Element	Symbol	Atomic Number	Atomic Mass
Actinium	Ac	89	(227)
Aluminum	Al	13	26.981538
Americium	Am	95	(243)
Antimony	Sb	51	121.760
Argon	Ar	18	39.948
Arsenic	As	33	74.92160
Astatine	At	85	(210)
Barium	Ba	56	137.327
Berkelium	Bk	97	(247)
Beryllium	Be	4	9.012182
Bismuth	Bi	83	208.98038
Bohrium	Bh	107	(262)
Boron	B	5	10.811
Bromine	Br	35	79.904
Cadmium	Cd	48	112.411
Calcium	Ca	20	40.078
Californium	Cf	98	(249)
Carbon	C	6	12.0107
Cerium	Ce	58	140.116
Cesium	Cs	55	132.90545
Chlorine	Cl	17	35.4527
Chromium	Cr	24	51.9961
Cobalt	Co	27	58.933200
Copper	Cu	29	63.546
Curium	Cm	96	(247)
Dubnium	Db	105	(260)
Dysprosium	Dy	66	162.50
Einsteinium	Es	99	(254)
Erbium	Er	68	167.26
Europium	Eu	63	151.964
Fermium	Fm	100	(253)
Fluorine	F	9	18.9984032
Francium	Fr	87	(223)
Gadolinium	Gd	64	157.25
Gallium	Ga	31	69.723

(Continued)

List of the Elements with Their Symbols, Atomic
Numbers, and Atomic Masses[a] *Continued*

Element	Symbol	Atomic Number	Atomic Mass
Germanium	Ge	32	72.61
Gold	Au	79	196.96655
Hafnium	Hf	72	178.49
Hassiuim	Hs	108	(265)
Helium	He	2	4.002602
Holmium	Ho	67	164.93032
Hydrogen	H	1	1.00794
Indium	In	49	114.818
Iodine	I	53	126.90447
Iridium	Ir	77	192.217
Iron	Fe	26	55.845
Krypton	Kr	36	83.80
Lanthanum	La	57	138.9055
Lawrencium	Lr	103	(257)
Lead	Pb	82	207.2
Lithium	Li	3	6.941
Lutetium	Lu	71	174.967
Magnesium	Mg	12	24.3050
Manganese	Mn	25	54.938049
Mendelevium	Md	101	(256)
Mercury	Hg	80	200.59
Molybdenum	Mo	42	95.94
Neodymium	Nd	60	144.24
Neon	Ne	10	20.1797
Neptunium	Np	93	(237)
Nickel	Ni	28	58.6934
Niobium	Nb	41	92.90638
Nitrogen	N	7	14.00674
Nobelium	No	102	(253)
Osmium	Os	76	190.23
Oxygen	O	8	15.9994
Palladium	Pd	46	106.42
Phosphorous	P	15	30.973762
Platinum	Pt	78	195.078
Plutonium	Pu	94	(242)
Polonium	Po	84	(210)
Potassium	K	19	39.0983
Praseodymium	Pr	59	140.90765
Promethium	Pm	61	(147)
Protactinium	Pa	91	231.03588

(Continued)

List of the Elements with Their Symbols, Atomic
Numbers, and Atomic Masses[a] *Continued*

Element	Symbol	Atomic Number	Atomic Mass
Radium	Ra	88	(226)
Radon	Rn	86	(222)
Rhenium	Re	75	186.207
Rhodium	Rh	45	102.90550
Rubidium	Rb	37	85.4678
Ruthenium	Ru	44	101.07
Rutherfordium	Rf	104	(257)
Samarium	Sm	62	150.36
Scandium	Sc	21	44.95591
Seaborgium	Sg	106	(263)
Selenium	Se	34	78.96
Silicon	Si	14	28.0855
Silver	Ag	47	107.8682
Sodium	Na	11	22.98977
Strontium	Sr	38	87.62
Sulfur	S	16	32.066
Tantalum	Ta	73	180.9479
Technetium	Tc	43	(99)
Tellurium	Te	52	127.60
Terbium	Tb	65	158.92534
Thallium	Tl	81	204.3833
Thorium	Th	90	232.0381
Thulium	Tm	69	168.93421
Tin	Sn	50	118.710
Titanium	Ti	22	47.867
Tungsten	W	74	183.84
Uranium	U	92	238.0289
Vanadium	V	23	50.9415
Xenon	Xe	54	131.29
Ytterbium	Yb	70	173.04
Yttrium	Y	39	88.90585
Zinc	Zn	30	65.39
Zirconium	Zr	40	91.224

[a]Approximate atomic masses for radioactive elements are given in
parentheses. [Based on 1993 IUPAC Table of Standard Atomic Weights of
the Elements]

Index